SUSTAINABLE FORESTRY: FROM MONITORING AND MODELLING TO KNOWLEDGE MANAGEMENT AND POLICY SCIENCE

Sustainable Forestry: from Monitoring and Modelling to Knowledge Management and Policy Science

Edited by

K.M. Reynolds, A.J. Thomson, M. Köhl, M.A. Shannon, D. Ray and K. Rennolls

www.cabi.org

CABI is a trading name of CAB International

CABI Head Office
Nosworthy Way
Wallingford
Oxfordshire OX10 8DE
UK

Tel: +44 (0)1491 832111
Fax: +44 (0)1491 833508
E-mail: cabi@cabi.org
Website: www.cabi.org

CABI North American Office
875 Massachusetts Avenue
7th Floor
Cambridge, MA 02139
USA

Tel: +1 617 395 4056
Fax: +1 617 354 6875
E-mail: cabi-nao@cabi.org

A catalogue record for this book is available from the British Library, London, UK.

A catalogue record for this book is available from the Library of Congress, Washington, DC.

ISBN-13: 978 1 84593 1742

Typeset by AMA DataSet Ltd, UK.
Printed and bound in the UK by Biddles, Kings Lynn.

Contents

Contributors

Albert Abee, *USDA Forest Service, EMC, 1400 Independence Ave., SW Mailstop: 3rd Floor Central, Washington, DC 20250, USA. Email: aabee@fs.fed.us*

James D. Absher, *USDA Forest Service, 4955 Canyon Crest Drive, Riverside, CA 92507, USA. Email: jabsher@fs.fed.us*

Frank Barrett, *College of Life Sciences, School of Biological and Environmental Science, Agriculture and Food Science Centre, University College Dublin, Belfield, Dublin 4, Ireland.*

Stephen J. Bathgate, *Forest Research, Northern Research Station, Roslin, Midlothian, Scotland EH25 9SY, UK.*

Livia Bizikova, *Adaptation and Impacts Research Division, Environment Canada, University of British Columbia, 2202 Main Mall, Room 441, Vancouver, BC V6T 1Z4, Canada.*

Kevin Boston, *Department of Forest Engineering, College of Forestry, Oregon State University, Corvallis, OR 97330, USA. Email: kevin.boston@oregonstate.edu*

Alice C. Broome, *Ecology Division, Forest Research, Roslin, Midlothian, Scotland EH25 9SY, UK.*

Gerard Buttoud, *French Institute of Forestry, Agricultural and Environmental Engineering, Nancy, F-54042, France.*

Angelo Carriero, *Servizio Foreste e Fauna, Provincia Autonoma di Trento, via Trener 3, 38100 Trento, Italy.*

Hugues Claessens, *Unit of Forest and Nature Management, Gembloux Agricultural University, 2 Passage des Déportés, 5030 Gembloux, Belgium.*

Piermaria Corona, *Dipartimento di Scienze dell'Ambiente Forestale e delle Sue Risorse, University of Tuscia, via San Camillo de Lellis, 01100 Viterbo, Italy. Email: piermaria.corona@unitus.it*

Flora De Natale, *Istituto Sperimentale per l'Assestamento Forestale e per l'Alpicoltura, ISAFA – CRA, piazza Nicolini 6, 38050 Trento, Italy. Email: denatale@isafa.it*

Hamid R. Ekbia, *Computer Science Dept., Appleton Hall, Room 228, University of Redlands, Redlands, CA 92373, USA. Email: hamid_ekbia@redlands.edu*

Patrizia Gasparini, *Istituto Sperimentale per l'Assestamento Forestale e per l'Alpicoltura, ISAFA – CRA, piazza Nicolini 6, 38050 Trento, Italy.*

George Gertner, *Department of Natural Resources and Environmental Sciences, Department of Statistics, W503 Turner Hall, University of Illinois, Urbana, Illinois, USA. Email: gertner@uiuc.edu*

Biing T. Guan, *School of Forestry and Resource Conservation, National Taiwan University 1, Sec 4, Roosevelt Rd., Taipei, Taiwan, 10617. Email: btguan@ntu.edu.tw*

Jeff D. Hamann, *Department of Forest Engineering, College of Forestry, Oregon State University, Corvallis, OR 97330, USA. Email: jeffery.hamann@oregonstate.edu*

Marc Hanewinkel, *Baden-Württemberg Forest Research Institute, Wonnhaldestr. 4, 79100 Freiburg, Germany.*

David W. Hann, *Oregon State University, College of Forestry, Department of Forest Resources, Peavy Hall, Corvallis, OR 97331, USA.*

Alan J. Harrison, *Forest Research, Northern Research Station, Roslin, Midlothian, Scotland EH25 9SY, UK.*

Debora L. Johnson, *Oregon State University, College of Forestry, College Forests, 8692 Peavy Arboretum Rd., Corvallis, Oregon 97330, USA. Email: debora.johnson@oregonstate.edu*

K. Norman Johnson, *Oregon State University, College of Forestry, Department of Forest Resources, Peavy Hall, Corvallis, OR 97331, USA. Email: norm.johnson@oregonstate.edu*

Michael Köhl, *University of Hamburg, World Forestry, Leuschnerstrasse 91, D-21031 Hamburg, Germany.*

Irina Kouplevatskaya, *ENGREF, Laboratory of Forest Policy, 14 rue Girardet – CS 4216 F-54042, Nancy, France. Email: kouplevatskaya@engref.fr*

Hugues Lecomte, *Walloon Forest Inventory, General Directory of Natural Resources and Environment, 15 avenue Prince de Liège, 5100 Jambes, Belgium. Email: h.lecomte@mrw.wallonie.be*

Valerie Le May, *Forest Biometrics/Measurements, Forest Resources Management Dept., 2045-2424 Main Mall, University of British Columbia, Vancouver, BC V6T 124, Canada. Email: valerie.lemay@ube.ca*

Manfred J. Lexer, *Department of Forest and Soil Sciences, Institute of Silviculture, University of Natural Resources and Applied Life Sciences, Peter-Jordanstr. 82, A-1190 Vienna, Austria.*

Nigel Lowthrop, *Hill Holt Wood, Norton Disney, Lincolnshire LN6 9JP, UK. Email: Nigel@hillholtwood.com*

Ronald E. McRoberts, *Northern Research Station, USDA Forest Service, 1992 Folwell Avenue, Saint Paul, Minnesota 55108, USA. Email: rmcroberts@fs.fed.us*

William H. McWilliams, *USDA Forest Service, 11 Campus Blvd., Suite 200, Newtown Square, PA 19073, USA.*

Alexander Mather, *Department of Geography and Environment, University of Aberdeen, Aberdeen, Scotland AB24 3UF, UK.*

Thorsten Mrosek, *Centre for Forest Ecosystems, University of Münster, Robert-Koch-Straße 26, 48149 Münster, Germany. Email: thorsten.mrosek@world-zentrum.de*

Giuseppe Scarascia Mugnozza, *Dipartimento di Scienze dell'Ambiente Forestale e delle Sue Risorse, University of Tuscia, via San Camillo de Lellis, 01100 Viterbo, Italy.*

M. Nieuwenhuis, *College of Life Sciences, School of Biological and Environmental Science, Agriculture and Food Science Centre, University College Dublin, Belfield, Dublin 4, Ireland.*

Maria Nijnik, *Socio-economic Research Programme, Macaulay Institute, Craigiebuckler, Aberdeen, Scotland AB15 8QH, UK. Email: m.nijnik@macaulay.ac.uk*

Risto Päivinen, *European Forest Institute, Joensuu, 80100, Finland.*

Michael P. Perks, *Forest Research, Northern Research Station, Roslin, Midlothian, Scotland EH25 9SY, UK. Email: mike.perks@forestry.gsi.gov.uk*

Enrico Pompei, *National Forest Service, Ministry of Agricultural and Forest Policies, via Giosuè Carducci 5, 00187 Rome, Italy.*

Michael Pregernig, *Institute of Forest, Environmental and Natural Resource Policy, Department of Economics and Social Sciences, University of Natural Resources and Applied Life Sciences Vienna, Feistmantelstrasse 4, A-1180 Vienna, Austria. Email: Michael.pregernig@boku.ac.at*

Thomas Puissant, *Unit of Forest and Nature Management, Gembloux Agricultural University, 2 Passage des Déportés, 5030 Gembloux, Belgium.*

H. Michael Rauscher, *USDA Forest Service, Southern Research Station, 200 WT Weaver Boulevard, Asheville, NC 28802, USA. Email: mrauscher@fs.fed.us*

Duncan Ray, *Ecology Division, Forest Research, Roslin, Midlothian, Scotland EH25 9SY, UK. Email: duncan.ray@forestry-gsi.gov.uk*

Andreas Reinbolz, *Baden-Württemberg Forest Research Institute, Wonnhaldestr. 4, 79100 Freiburg, Germany. Email: andreas.reinbolz@forst.bwl.de*

Keith Rennolls, *University of Greenwich, School of Computing and Mathematical Sciences, Park Way, Greenwich, London SE10 9LS, UK. Email: k.rennolls@gre.ac.uk*

Keith M. Reynolds, *USDA Forest Service, Pacific Northwest Research Station, Corvallis Forestry Sciences Laboratory, 3200 SW Jefferson Way, Corvallis, OR 97331, USA. Email: kreynolds@fs.fed.us*

Jacques Rondeux, *Unit of Forest and Nature Management, Gembloux Agricultural University, 2 Passage des Déportés, 5030 Gembloux, Belgium. Email: rondeux.j@fsagx.ac.be*

Christine Sanchez, *Unit of Forest and Nature Management, Gembloux Agricultural University, 2 Passage des Déportés, 5030 Gembloux, Belgium.*

Daniel L. Schmoldt, *USDA Cooperative State Research, Education and Extension Service, 800 9th Street SW, Washington, DC 20024, USA. Email: dschmoldt@csrees.usda.gov*

Andreas Schulte, *Centre for Forest Ecosystems, University of Münster, Robert-Koch-Straße 26, 48149 Münster, Germany.*

Charles T. Scott, *USDA Forest Service, 11 Campus Blvd., Suite 200, Newtown Square, PA 19073, USA. Email: ctscott@fs.fed.us*

Rupert Seidl, *Department of Forest and Soil Sciences, Institute of Silviculture, University of Natural Resources and Applied Life Sciences, Peter-Jordanstr. 82, A-1190 Vienna, Austria.*

Margaret A. Shannon, *State University of New York, Buffalo, NY, 14260, USA.*

M.J. Somers, *School of Biological and Environmental Sciences, University College Dublin, Ireland.*

Juan C. Suárez, *Forest Research, Northern Research Station, Roslin, Midlothian, Scotland EH25 9SY, UK.*

Alan J. Thomson, *Canadian Forest Service, Pacific Forestry Centre, 506 West Burnside Rd., Victoria, BC V8Z 1M5, Canada. Email: athomson@pfc.cfs.nrcan.gc.ca*

John H.M. Thornley, *6 Makins Road, Henley-on-Thames, Oxfordshire RG9 1PP, UK. Email: johnthornley@care4free.net*

Margarida Tomé, *Instituto Superior de Agronomia, Tapada da Ajuda, 1349–017 Lisboa, Portugal. Email: magatome@isa.utl.pt*

Harald Vacik, *Department of Forest and Soil Sciences, Institute of Silviculture, University of Natural Resources and Applied Life Sciences, Peter-Jordanstr. 82, A-1190 Vienna, Austria. Email: harald.vacik@boku.ac.at*

Jerry Vanclay, *Sustainable Forestry Programme, School of Environmental Science and Management, Southern Cross University, Po Box 157, Lismore, NSW 2480, Australia. Email: JVanclay@scu.edu.au*

Jerry J. Vaske, *Department of Natural Resource Recreation and Tourism, Colorado State University, Fort Collins, Colorado, USA.*

Edward D. Wallington, *Forestry Commission, Silvan House, 231 Corstorphine Road, Edinburgh, Scotland EH12 7AT, UK. Email: edward.wallington@forestry.gsi.gov.uk*

Bernhard Wolfslehner, *Department of Forest and Soil Sciences, Institute of Silviculture, University of Natural Resources and Applied Life Sciences, Peter-Jordanstr. 82, A-1190 Vienna, Austria.*

Preface

The importance to society of environmental services provided by forest ecosystems has significantly increased during the last few decades. Growing concern with deterioration of forests globally, from the 1980s, has led to increasing public awareness of the environmental, cultural, economic and social values that forests provide. Ideas about sustainable, close-to-nature and multifunctional forestry have progressively replaced the older perception of forests as only a source for timber. International impetus to protect and sustainably manage forests has come from global initiatives aimed at management, conservation and sustainable development related to all types of forests and forestry. A few of the more notable initiatives include:

- The 1992 Earth Summit in Rio de Janeiro, Brazil (United Nations Conference on Environment and Development, UNCED).
- Regional follow-ups to the Earth Summit such as the Montreal Process and Pan-European Process.
- The forest elements of the Convention on Biological Diversity (CBD).
- The Framework Convention on Climate Change (FCCC) and the Kyoto Protocol.

Since the original Earth Summit, forest management and forest research organizations around the world have been developing and testing new approaches to deliver sustainable forest management in three areas: timber production, environmental and ecosystem management and social forestry. The subsequent emergence of major international initiatives such as the Montreal Process and the Pan-European Process and the growing importance accorded to forest certification programmes in many countries around the world attest to international recognition of the importance of sustainable forestry within both the forestry sector and by the general public.

Many disciplines have been involved in the evolution of scientific knowledge to address the challenges posed by sustainable forestry, and have played important roles in advancing and applying new principles and practices in support of sustainable forest management. In 2003, the editors of this book, representing diverse sections of the International Union of Forest Research Organizations (IUFRO), began collaborating to organize a trans-disciplinary conference broadly focused on the theory and practice of sustainable forestry. The conference was held in April 2005 in Edinburgh, Scotland (UK), and this volume presents a compilation of what we believe are some of the more significant contributions. The conference addressed four broad themes relevant to sustainable forest management:

1. Science and policy.
2. Inventory and monitoring.
3. Statistics and modelling.
4. Information and knowledge management.

Major advances have been made within each of these themes in recent years. Although the work of each forest resource scientist may tend to fall predominantly within only one of these broad thematic areas, each theme has strong mutual interdependencies with all of the others. With this consideration in mind, the major objectives of the conference and the present book were to:

1. Review the current state of the art within each of the four major themes.
2. Foster dialogue across thematic areas concerning both strategic and operational approaches to integrate research on sustainable forestry.
3. Enhance and encourage international collaboration towards sustainable forestry practice worldwide.

Several organizations have contributed to this volume in terms of committing either financial or staff resources or logistical support. The editorial staff represents IUFRO sections 4.02 (Forest Resources Inventory and Monitoring), 4.03 (Informatics, Modelling and Statistics), and 6.12 (Forest Policy and Governance), as well as diverse organizations. Other organizations contributing support to the Edinburgh conference in particular include the European Forest Institute and the Environmental Systems Research Institute, UK.

Keith M. Reynolds,
senior editor, USDA Forest Service, USA
Alan Thomson,
editor, Canadian Forest Service, Natural Resources Canada, Canada
Margaret Shannon,
editor, University of New York, Buffalo, USA
Michael Köhl,
editor, University of Hamburg, Germany
Duncan Ray,
editor, Forestry Research, UK
Keith Rennolls,
editor, University of Greenwich, UK

Overview

K.M. Reynolds,[1] M.A. Shannon,[2] M. Köhl,[3]
K. Rennolls,[4] A.J. Thomson[5] and D. Ray[6]

[1]US Department of Agriculture, Forest Service, Pacific Northwest Research Station, Corvallis, USA; [2]State University of New York, Buffalo, USA; [3]University of Hamburg, World Forestry, Hamburg, Germany; [4]University of Greenwich, London, UK; [5]Canadian Forest Service, Pacific Forestry Centre, Victoria, BC, Canada; [6]Forest Research, Northern Research Station, Roslin, Midlothian, Scotland, UK.

The themes of the conference are reflected in the four major sections of this volume. Each section begins with an overview or synthesis chapter in which the authors have tried to pay particular attention to issues around strategic and operational approaches to integrating research efforts (objective 2 in Preface). An overview of the four major sections follows.

Science and Policy

Policy is not only an authoritative choice of values, it is also a representation of a particular way of seeing and understanding the world. Forest policy rests first on a definition of a 'forest' and is based upon a set of assumptions about the purpose of forests, the nature of forests and the place of forests in the natural and social world. But whose assumptions underpin forest policy? What role do scientists play in defining and selecting how we think about forests?

The values expressed in forest policy are generally understood to be ones of economic, political (read power), social and cultural origins. We understand forest policy to vary across countries and cultures and even within smaller regions or particular places due to differences in the valuation of forest resources. However, a critical role of science is to create representations of the world so that we can talk about what we see, hear, smell, taste and experience. These representations – theories, concepts, hypotheses, axioms, laws and so on – are often viewed by scientists to be 'reality'. However, reality is a bit more elusive – it is perceived through our ideas and science is only one source of these ideas.

Chapters in this section examine how the representations of forests, society, economy and political systems created by scientists are part of, and often wholly endogenous to, value choices. What we see is often what we want to keep or what we wish to be rid of. It is a fallacy to think that the involvement of scientists in forest policy is new. Forest policy not only emerged from scientific studies, it embodies science-based ideas of management. What is interesting, as Norm Johnson eloquently shows us, is that the disciplines represented by the scientists engaged in forestry and forest policy have changed dramatically over this past century. While science is often viewed as a source of neutral and objective information about the world we live in, Johnson effectively demonstrates that scientists have embedded their own values and preferences about how the world should be – their personal world view – into the 'scientific information' and so it is neither neutral nor objective. The fact that scientists have values and preferences is not surprising, nor is it unusual that their personal world view should affect their research and analysis; what is worrying is that scientists do not clearly acknowledge this fact when they provide scientific advice for policymakers. This chapter lays the ground for the rest of this section. Two linked questions animate the chapters in this section: What do scientists want from policymakers? What do policymakers want from scientists? While these questions are familiar, the answers are different when science-making is conceptualized within the policy and management systems rather than left outside as if it were an exogenous process that can be ignored at will.

The chapters in the science–policy theme are separated into two parts based on how policy processes have incorporated scientists and how scientists have engaged policymaking. In this way we address both questions – first, what policymakers get from science and, second, what scientists give to policymakers. The first set of chapters is introduced by Kouplevatskaya, who draws from her rich experience in Kyrgyzstan to develop a theoretical model of 'policy spirals' that integrates scientists, policymakers, managers and civil society. Of special interest in this chapter is the role of policy evaluation. Often policy evaluation is neglected in the policy process as world views take hold, as Norm Johnson demonstrated, or as policymakers grow complacent with the status quo. Thus, the critical role of policy evaluation is traced in this chapter in terms of transforming the context, re-engaging the role of the scientists, the recognition by policymakers and managers of the contributions of civil society to reframing problems and through it all creating integrated knowledge through collaboration.

The following three chapters provide examples of how research can contribute to the framing of policy questions and can set the stage for creating collaborative and participatory governance processes. Nijnik and Bizikova develop a critical analysis of sustainability as it is applied to countries in transition. The concept of sustainable forest management generally draws from the context of industrialized countries, in which forest protection and conservation tend to be key issues. However, when policies developed in this context are applied to countries in transition, as is happening now in Central and Eastern Europe as they seek membership in the European Union (EU), the privileging of environmental values over economic and social benefits leads to a diminishment of both. This chapter illustrates the kind of contribution scientific analysis can make to policymaking when scientists and policymakers are jointly engaged in evaluation of policy options, increasing the likelihood of developing policies and policy tools that work. This chapter also contributes to the discussion of integration by illustrating how the separated spheres of forestry, agriculture and climate change policymaking miss the opportunities to create improvements in each arena through a common policy approach to forests.

One key arena where scientists, policymakers and civil society have joined forces is in the development of sustainable forest management (SFM) policy at the global scale. Perhaps collaboration at the global scale seemed more obvious as all parties joined in debates prior to and after the United Nations Conference on Environment and Development in 1992. Or perhaps the scope and scale of the problem and desired solution simply overwhelmed the imagination and capacity of any one set of actors. Whatever the reason, the development of principles for sustainable forest management both in the governmental arena through various international processes and by the extra-governmental efforts of civil society presents a shining example of integration. Abee provides a historically situated analysis of the evolution of the Montreal Process Criteria and Indicators for Sustainable Forest Management, which is the primary governmental attempt to create a global policy framework. First, it is important to note that the use of 'standard setting' in the form of 'criteria and indicators' is a common approach to transnational problems (Meidinger, in press). Thus, the innovation was not in the policy approach itself, but rather in the creation of a collaborative and participatory process for integration of scientists who were invited to contribute to formulating policy and of policymakers forced to define problems amenable to scientific methods. The Montreal meeting of scientists and policymakers in September 1993 remains a distinctive moment in science–policy integration. The resulting set of criteria for policy guidance to countries for adopting SFM and the sets of indicators available for countries to utilize in measuring their progress towards sustainability remain a dynamic and vital instrument of policy integration through continuous review and improvement. Abee draws upon the documents produced to foster this process and so gives some insights from the policy side on how ideas and information were gradually incorporated into policy. Although this volume does not have a similar chapter tracing the evolution of extra-governmental attempts to define global policy standards for SFM that are implemented through voluntary adoption by landowners (including governments at times), analysis of the evolution of the Forest Stewardship Council (Meidinger *et al.*, 2003), the emergence of non-state market-driven governance (Cashore *et al.*, 2004) and the move of these processes toward a 'law making' framework reproducing governmental approaches to governance (Meidinger 2003) complements Abee's analysis.

However, as Rennolls and Reynolds demonstrate in their analysis of what a sustainable multifunctional forest might mean in the context of climate change, the evolution of policies for SFM is dynamic. They find that many of the underlying concepts are ambiguous, and that for the conservation of biodiversity in tropical forests, the main driver of the Earth Summit, the currently developed criteria and indicators are not adequate. They suggest that new forest models focused on biodiversity need to be developed. This critical assessment of the criteria and indicators framework parallels the arguments of Nijnik and Bisikova – global policy principles do not necessarily work everywhere and need to be adapted to place and circumstance – just the opportunity provided by integration.

The next set of chapters is introduced by Pregernig who analyses the changing nature of the forums wherein science–policy interaction occurs. Pregernig provides a conceptual introduction to the idea of science–policy integration through the mechanism of 'bioregional science assessments'. Bioregional assessments are by definition collaborative because they address transboundary problems and thereby engage policy actors from different venues and link them to scientists addressing as much of a complex and messy problem context as possible. Pregernig gives two brief examples from the USA and then turns to an analysis of the common features of these processes. Today, these types of processes are common around the world and are ensconced in an emerging policy framework of the Strategic Environmental Assessment in the Kiev Protocol.

How to understand the relationships among policy actors and scientists is important in order to access the appropriate people and groups. Mrosek and Schulte introduce a useful concept – the 'policy cluster' – in conceptualizing the relationships among policy actors within sectors. Cluster analysis is especially useful in forestry since there are clear clients (forest owners and industries) for forest policy and clear beneficiaries (land and forest owners, the public) for forest policies. Mrosek and Schulte, however, go beyond the simple usefulness of how the networks are organized into how they incorporate and utilize information – especially science-based information – and how they include (or exclude) scientists. Absher and Vaske complement this discussion of 'clusters' by illustrating how scientific information can help create relationships among a latent cluster when the issue is wildfire. Clusters are created by relationships, and relationships can be fostered through the integration of scientists and policymakers. Absher and Vaske demonstrate the need for this kind of integration if policies aimed at reducing fire losses by changing people's behaviour are to be successful. Fire is an interesting issue for science and policy since it is usually a surprise. Although fire risk assessment can inform us about the likelihood of a 'spark' leading to a fire, the actual occurrence, timing and place of wildfires are unpredictable – even when the fire is caused by human behaviour. Thus, how to create a 'fire-resistant' or 'fire-resilient' environment is of critical interest to policymakers, forest managers and people living in or near the forest.

However, as Nijnik and Mather show us, developing clusters and creating integration depend heavily on how important an issue is to the public and whether the 'policy cluster' – in this case the forest cluster – is viewed as competent to address the problem. In order to address the issue of how important a topic is to the public, they undertake two different kinds of analysis. First, they examine how well economic theories and tools fit with the kinds of problems faced by sustainable forest management. They conclude, 'Not very well.' The reason is simple: people's preferences are dynamic and continuously shaped by the context surrounding them as well as their personal values and situation. Thus, the authors turn to the analysis of a recent poll and show that there are a diversity of world views that encompass both preferences and normative expectations. In addition, they look at what kinds of priorities landowners are actually using when they make their applications for subsidies for forest and rural development. In both they find that the public has a mixed set of preferences and that it is this mix that must be accommodated. Thus, we return full circle – for scientific contributions to be useful to policymakers, they must attend to the whole of complex and messy problems and contexts and, for policymakers to effectively identify workable solutions, they must incorporate scientific analysis in the critical assessment of policy options and approaches.

The last chapter in this section takes us out of the lofty halls of policymaking and science and places us square in the middle of the forest. Lowthrop tells a wonderful and inspiring story of how the management of one small private forest is accomplishing the mix of objectives people desire from forests and has simultaneously contributed to rural development and sustainability. This story sums up the section very nicely by reminding us that it is individuals who must come together and, through respect for their complementary contributions to addressing the critical challenges facing us today, find innovative solutions. But the first step is always understanding what the problem is before setting off to solve it – herein lies the contribution of science to policy, and ensuring that scientists are pursuing problems of interest to society is why science must often be endogenous to the policy process.

Collectively, these chapters provide a framework for thinking about issues of knowledge management within different organizational and cultural contexts. They raise questions to pose when thinking about the implications of different kinds of analytical tools or approaches. They challenge the notion that policy is a linear process in which information, power and technical tools are simply inputs along the way. When science is within the policy system, the role of critical thinking and the necessity of interrogating different assumptions and concepts within and across disciplines becomes obvious. Thus, the notion of 'what is policy' is at the heart of this section, and it differs radically from conventional views that policy is merely the choice of means to serve desired ends.

Inventory and Monitoring

Forest inventory and monitoring programmes are a key element in providing objective information, and are thus an essential element of any strategy for the management, conservation and sustainable development of forests. Contemporary forest-resource assessments typically collect detailed information on a very wide range of forest ecosystem states and processes compared with traditional forest assessments, and much of this information depends on access to, and use of, spatially explicit data. However, new methods in spatial statistics and small-area estimation are facilitating the increased use of spatial data. Demands for increased information in general are requiring attention to the efficient use and integration of data sources and analysis techniques such as field assessments, remote sensing imagery or geographic information systems (GIS). Satisfying the information needs for sustainable management at different temporal and spatial resolutions has also led to new challenges for forest inventories.

Scott and McWilliams in the keynote presentation of the inventory and monitoring session of the conference presented recent developments of the US Department of Agriculture (USDA) Forest Service's Forest Inventory and Analysis (FIA) Program as an example of the impacts on survey designs created by the need for more current information on the multifunctional sustainability of forests. Besides the adaptation of the remeasurement cycle, the design was changed and now has three phases, which cover stratification, traditional forest survey and forest health indicators. The inventory now includes subsampling of all vascular plants, soils, lichens, down woody material, ozone damage and crown attributes. These changes enable FIA to provide a wide array of information on the sustainability of the multiple functions of forests. Another report from a national forest inventory is presented by Sanchez *et al.*, who demonstrate the integration of vegetation assessments in southern Belgium's permanent forest inventory.

Barrett *et al.* developed an inventory protocol with reference to criteria and indicators and local measures of sustainable forest management. With the introduction of SFM, forest owners and managers require an effective and efficient multi-resource inventory, data management and decision-support tools in order to produce management plans that are realistic, practical and sustainable. The PractiSFM research project addresses these needs. Barrett *et al.* describe the multi-resource component of the PractiSFM system, which has been designed in accordance with the criteria and indicators identified within the Irish National Forest Standard.

Johnson *et al.* deal with stand-level inventories, and show their importance for the sustainable management of multiple forest values in the presence of endangered species. The combination of a forest inventory and a growth-and-yield model allows for stand estimates that enable accurate aggregation and disaggregation of forest stands both for strategic planning and for detailed harvest scheduling. This approach facilitates the inevitable shifts in timings and locations of harvest as unforeseen events occur, such as the presence of endangered species.

Climate change, abandonment of land and migration from rural to urban areas result in the expansion of forested areas and subsequent landscape dynamics in the temperate and boreal region, while in the tropics and subtropics deforestation is progressing. It is widely accepted that many aspects of forest ecology need to be discussed in the landscape context, and require spatially explicit data and spatial reasoning. Corona *et al.* deal with change assessment on forest land by continuous inventories. They present a probabilistic estimation approach, based on land-use classifications on successive occasions and utilizing multi-temporal classification of ortho-corrected aerial photographs. They demonstrate the method in a test site in the Abruzzo region of Italy, and show their operational applicability as well as efficient statistical performance.

De Natale *et al.* present a method to describe the tree colonization of abandoned land. They use a two-stage sampling design. In the first stage, units are selected from ortho-photos with selection probabilities proportional to auxiliary variables describing the degree of fragmentation of forests. Second-stage units are randomly selected and surveyed in the field. The approach allows the sensitive detection of landscape dynamics caused by fragmented tree colonization. The authors apply their approach in a test area in the Italian province of Trento, located in the eastern Alps, and interpret their results with respect to factors characterizing tree succession and the effects of land abandonment on landscape patterns.

Hamann and Boston use spatial statistics to improve estimates of forest supply and its implications for the performance of the primary forest supply chain. They apply Kriging to estimate product volumes in unsampled areas of harvest blocks. For a case study, they show that the predicted variance of the sample estimated with Kriging was less than 1% of the variance estimated for the non-spatial sample. To demonstrate an advantage of spatially explicit methods, they optimize harvesting operations by an approach that minimizes the squared deviations between demand and predicted production from harvest blocks.

Finally, Wallington and Suárez describe the use of airborne LiDAR and X-band Interferometric Synthetic Aperture Radar systems to retrieve forest stand top height.

Statistics and Modelling

Most forest biometricians, whether statisticians or modellers, tend to work on specific problems with fairly well-defined objectives. They design surveys and experiments focusing on their precision and power, or they fit models to the growth, mortality and yield of a stand of trees of a particular species, or a given mixture of species, in a range of growth conditions. In general, forest biometricians have not become engaged in the processes of sustainable management for multifunctional forestry. Such issues seem to have been the domain primarily of those in the political and policy arenas, and technical support in these areas has been provided by inventory specialists, through the criterion and indicator (C & I) monitoring initiatives or by participatory research specialists, through techniques such as multi-criteria analysis (MCA).

Hence the invitation to become involved in a collaboration covering the main themes of this book, with the requirement to consider the interrelationships of the main themes in the context of sustainable forest management, presented a considerable challenge to forest biometricians. This challenge has required a considerable leap of the imagination on the part of forest biometricians, because it required consideration of the potential for application of statistical and modelling techniques in a policy area where biometricians have largely not previously been involved. Let us consider some of the main issues and examine the potential contributions of statistical and modelling disciplines.

Global warming and climate change have been with us for at least three decades and are increasingly serious issues. To understand the underlying processes and causes, we need to quantify and model the role forests play in the carbon cycle and determine their capacity for carbon sequestration. We need to be able to predict the effects of current and future forestry practices, in particular in the tropics, in order to find a path to sustainability. The generality and breadth of considerations and the global scale would seem to be outside the range of the rather simplistic (empirical) models often used by statisticians. Also, the kind of models required, which need to span vast scales of both time and space, are very different from conventional forest models, which normally have a restricted spatio-temporal range. Most of the broad-scale modelling work in climate change seems to have been conducted by earth and environmental scientists, while meteorological researchers continue to try to scale their models up to the global spatial scale and a timescale measured in centuries. Carbon accounting seems to be the most popular approach to quantifying and modelling the global carbon cycle. Statisticians and modellers might ask about the adequacy of data and information sources, the reliability of the model predictions, the adequacy of the concepts and models used and their scalability. However, they need to be included in the continuing policy debate if they are to ask these questions and possibly provide answers to some of them.

Forests are complex hierarchical systems, regardless of whether they are considered at the level of single stands or at the scale of the global forest resource. This complexity affects the way we should collect data and the ways in which we can model the forest ecosystem. Can we reasonably expect to be able to predict phenomena occurring at one level as emergent from lower-level system behaviour? How should we go about building hybrid models that integrate models at different scales? Forests provide some of the greatest repositories of biodiversity resources in the world. Such important habitats require protection in order to safeguard species and conserve genetic diversity. This challenge requires meaningful definitions and well-designed measurement methods so that we can analyse and model the changes that are taking place over time. Forest biometricians have started to consider such issues over the last decade or so, but need to work in closer collaboration with forest ecologists, landscape and climate experts and forest-policy scientists if their work is to have any impact on the issues of global forest management.

The processes of nature and forests are so complex that we might be justified in saying they contain a stochastic component and are intrinsically uncertain in nature. Even if we were to take a mechanistic view, we can never have complete data, only samples, so parameter estimates are bound to be uncertain. We may have more data from automatic digital data-collection instruments than we can handle, in which case we may need to sample our available data before we can analyse or model them. Hence our data, parametric estimates and the models we adopt are all subject to uncertainty and error, and such uncertainty will permeate through the decision-support systems that underpin management and policymaking. We therefore need to develop and use methods of analysis and modelling that will give us reliable measures of the uncertainty in our models and their predictions. The need for such uncertainty analysis would seem to be particularly strong for sustainable multifunctional forest management.

A consequence of the point made earlier, that biometricians have generally not been specifically involved in work on sustainable multifunctional forestry, has been that most contributions under the statistics and modelling theme were rather technical in nature and essentially did not address the core issue of sustainability. The result is that there are only a couple of chapters under this theme for this book.

Rennolls *et al.* offer an overview chapter on the contribution of statistics and modelling to sustainable forest management. It has been written by a team of authors that includes many of the world's leading forest biometricians and modellers. In this chapter they summarize the contributions made to date in the area of sustainable forest management, and some of the potential needs and challenges for the future.

The keynote chapter for the statistics and modelling theme, by John Thornley, provides a detailed account of the Edinburgh Forest Model (EFM). EFM is a compartment model, with the two central compartments being the tree/stand and the litter/soil. An environment compartment provides light input to photosynthesis in the model, and a water compartment tracks the exchanges between the two central compartments of the model and the atmosphere. Fluxes of C, N and CO_2 between the compartments of the model are modelled using differential equations. Besides being a tour de force in forest physiological modelling, this chapter by John Thornley presents, in a masterly manner, many of the good practices of the art and science of process modelling.

Information and Knowledge Management

There is considerable debate over the definition of knowledge management and how it differs from information management. In the call for submissions for the conference on which this book is based, we used the term in its broadest sense of the organization, creation and dissemination of knowledge, including related information management aspects. The call also focused on the specific areas of multiple-criteria decision models and decision-support systems (DSSs), including expert systems.

Chapters within this section of the book address many technical aspects of system development and deployment, and interactions with the policy process, while the section overview (Thomson *et al.*) focuses on the integration of knowledge management into institutional processes to achieve sustainable forest management.

Vacik *et al.* illustrate the manner in which the knowledge framework adopted by an agency can influence an analysis or assessment approach, based on an example combining the driving forces–pressure–state–impact–response approach of the European Environment Agency with simulation modelling and multiple-criteria decision making. Perks *et al.* then show how information systems must comply with sustainability standards that evolve in relation to changing legislation and policy. Sustainable forestry requires a degree of trust by policymakers and the general public in scientific findings: Reinbolz and Hanewinkel describe a virtual forester to guide people through complex information in a manner that creates trust and attempts to build an emotional connection with system users.

Rauscher *et al.*, in the keynote presentation of the knowledge management session of the conference, made the case that proficient problem solving depends on an adequate foundation of relevant and readily applicable knowledge and a coordinated approach. Thomson explores such coordination in depth within a 'knowledge ecosystem' context based on analogies with natural ecosystem processes. Ray and Broome then describe how one organization delivers up-to-date advice on sustainable forest management in relation to habitat and rare species protection under complex and constantly changing biological and legislative constraints. Finally, Ekbia and Reynolds discuss alternative approaches for use when the constraints of strictly rational DSSs are not sufficient for the decision context.

Getting the Parts to Work Together

We doubt that anyone reading this volume needs to be convinced of the complexity inherent in forest ecosystems and their management. Regardless of disciplinary background, if your own work is involved with any aspect of sustainable forestry, you have almost certainly had to personally confront that complexity at one time or another, and maybe even on a daily basis. Similarly, the scientific enterprise that drives sustainable forestry is, itself, a complex, self-organizing system, albeit one of human design. Unlike natural systems, which have been evolving over aeons, this human system has evolved in the evolutionary blink of an eye, as it were. As a self-organizing system of recent evolutionary origin, the contemporary scientific community collectively gets to decide what the components of this system are and the nature of the energy flows between those components that sustain it. This presents both challenges and opportunities, and this volume is about both.

As the organizers of the Edinburgh conference, the authors were obliged to make a first attempt at defining the major components of the sustainable forestry enterprise. However, conference delegates were given the charge of helping us define the flows needed to sustain the system. It is the discussion around these flows that lies at the heart of what we mean by integrated research to support sustainable forest management. As with any initial endeavour of this sort, it is probably fair to say that the work, taken as a whole, has succeeded in some respects but falls short in others. Certainly, not every chapter pertinent to a specific disciplinary area has attempted to address the dependencies between it and other disciplines. However, we think that the synthesis and overview chapters fill important gaps in this respect.

Finally, what can be said about the state of the art? The theory and practice of sustainable forestry has advanced rapidly in the past 15 to 20 years, but both continue to evolve rapidly even as this volume goes to press. The scientific community is still very much grappling with how to collaborate more effectively in this vast endeavour, but we hope that scientists, forest managers and policymakers will find some useful guideposts in these pages that will help to get all the parts working together to more effectively implement the goals of sustainable forest management.

Reference

Meidinger, E. (in press) The administrative law of global private–public regulation: the case of forestry. *European Journal of International Law*.

1 Science is Endogenous to Sustainable Forestry – Implications for Scientists and Policymakers

MARGARET A. SHANNON,[1] GERARD BUTTOUD[2] AND RISTO PÄIVINEN[3]

[1]State University of New York, Buffalo, USA, and University of Freiburg, Germany; [2]French Institute of Forestry, Agricultural and Environmental Engineering, (ENGREF), Nancy, France; [3]European Forest Institute, Joensuu, Finland.

Abstract

Policy is not only an authoritative choice of values, it is also a representation of a particular way of seeing and understanding the world. Science is a methodology for representing the world according to certain rules and procedures. The rules of politics and the rules of science could not be more different – policy opportunistically seeks 'windows' for making changes, while science systematically ponders data to make reliable generalizations. Making policy transparent brings a quality of science into policy, while making science responsive makes it adaptive and useful to policymakers. This chapter provides an overview of the science–policy challenge in sustainable forest management. The conclusion is simple – sustaining forests depends upon building collaborative partnerships among scientists and policymakers. The animating principle demanding this partnership is integration.

Introduction

As noted in the overview, forest policy rests first on a definition of a 'forest' and is based upon a set of assumptions about the purpose of forests, the nature of forests and the place of forests in the natural and social world. It is policymakers who choose what kind of forests society will have and it is scientists who create the images and representations of 'forests' among which policymakers choose. This fundamental fact about the scientific process – it develops over time an accepted representation of 'reality' and promotes it – is generally overlooked among scientists. Indeed, in policy processes scientific knowledge competes with other forms of knowledge – local indigenous knowledge, traditional cultural knowledge and experience – for power in making choices that will have authority in guiding future actions (Yankelovich, 1999).

Forests have always served as repositories for multiple representations and multiple values. Whether the 'dark' wood of literature, the place of danger in fairy tales, the home of forest-dwelling people and creatures or the 'breadbasket' of wood products, forests are repositories of culture and values. However, when we speak of 'forest policy', it is generally the economic values that come to mind – the trees we use for construction, the wood we use for energy – and the role of a primary production economy in creating capital for economic development. In the 20th century, the emergence of the scientific concept of a system led to a new representation – the forest 'ecosystem' – which expanded the set of values of forests to include the structure, composition and flows that constitute the 'ecosystem'. Yet 'forest ecosystem', while certainly more inclusive, still ignores the people living in, associated with and dependent upon forests. How is it possible to develop an integrated forest policy when we have no words to represent a forest with its people and its creatures, how its systems and flows sustain life or to remind us of its connection to the entire planet? This is the question that policymakers are asking.

Given the limits of language, it is not surprising therefore that forest policy traditionally included only the value of the timber for wood products. The non-timber forest products, essential to the survival of many people around the world, were simply ignored by policymakers – they were invisible in the representation of forests. Policymakers focused their concerns on the product flows from the forest that were valuable for capital creation, trade and human use. These values could be expressed in terms of the relative price a product could claim on the market. Indeed, when scientists began to coalesce around the concept of an ecosystem, policymakers resisted (Johnson *et al.*, 1999). A forest ecosystem had many potential values that did not enter the market, that indeed conflicted with the exploitation of forest resources and that greatly complicated the values to be considered in making policy.

The inclusion of the theme on science and policy in this book is indicative of the need to move towards integration in all its dimensions (Clark *et al.*, 1998). Science creates authoritative representations of the world through theories, concepts and methodology. Science claims are evaluated by other scientists based upon consistency with past ideas and proper application of method. Thus, in its ideal form, science is connected to society only at the beginning – when a problem in the world sparks curiosity – and at the end – when scientific information is used in policy – the scientist hands a report through the door to the policymakers and quickly returns to the laboratory so as to not be sullied by charges of political advocacy or guiding choices responsive to political power (Shannon *et al.*, 2000).

The overarching thesis of this book is that sustainable forest management cannot be adequately theorized in this isolationist mode of science nor can it inform practice until scientists are embedded in processes of policymaking and management practice (Meidinger *et al.*, 2003). This chapter and the chapters on this theme explore two linked questions: What do scientists want from policymakers? What do policymakers want from scientists? While these are familiar questions, the answers are different when science is conducted within the policy and management systems rather than developed in an external setting and merely given – through a small door – to the managers and policymakers as 'scientific information'.

Sustainable Forest Policy and Science – a Call for Integration

Sustainable forestry policy is technically, scientifically, socially and politically complex. Thus, the role of critical thinking and the necessity of questioning different assumptions and concepts within and across disciplines should be obvious. The argument of this chapter is that sustainable forestry practice requires a shift away from conventional views that policy is merely the choice of means to serve desired ends. Sustainable forestry necessitates the simultaneous formulation of ends and means, goals and policy instruments, objectives and criteria for evaluation. Since sustainable forestry aims at creating a new social practice across all of the existing institutions and structures of society related to forests, the policy process needs to create new knowledge about both the natural and social worlds. Thus, the role of science cannot be outside the policy process since the development and consideration of goals, objectives and criteria for evaluation depend upon the capacity to assemble what is understood and delve into what is unknown in order to develop an assemblage of information that creates accountable policy choices and legitimate bases for action. Thus, the question of how to integrate science and scientists within the policy process is a critical and unavoidable element of sustainable forestry.

In the case of sustainable forestry, policymakers must work on the whole problem in all its complexity. Simply representing and describing this complexity in economic, ecological, social, political, institutional and cultural terms engages scientists, stakeholders and knowledge holders in a 'knowledge creation' process. However, policymaking requires developing knowledge assemblages and templates (Watson-Verran and Turnbull, 1995) in order to represent the social and natural worlds clearly enough for action to occur. Thus, the challenges of science and policy in sustainable forestry are to create integrated and organized knowledge to animate the policy process within a policy community. From the point of view of scientists, integration is sometimes created by binding disciplinary pieces into a single document. But, from the point of view of policymakers, integration means not being surprised because all of the elements of the policy problem were not included in the analysis. While scientists will always claim that knowledge is limited, uncertainty is high and capacity to predict very low, policymakers need to act on immediate or imminent problems. So policymakers often complain that scientists oversimplify problems by ignoring context and complexity. Scientists complain that policymakers want a quick, certain and easy answer – preferably one that will magically solve the problem.

These mutual complaints provided the impetus for this chapter and for the other chapters on the science–policy theme. How can scientists contribute to problem structuring so that the whole problem and its context are considered? How can policymakers contribute to creating and organizing knowledge so that decisions are based on wisdom? What kinds of science–policy processes create the capacity for integration, transdisciplinarity, embedded analysis across scales of time and space, anticipatory analysis and action and open discursive deliberation among the community of enquiry and the policy communities?

Integration – the Common Demand

The call for integration in all forms – across disciplines, among policy priorities, between sectors of society, across countries with different circumstances and problems, around the world as problems affecting one region are recognized as part of the global policy context – is the single most important issue for scientists, policymakers, managers and civil society. However, what is missing is a rigorous definition of what it means to integrate different ideas, cultures and world views in the context of forest policy. In all too many instances, integration is created by 'paper clip' when analysis is done on pieces of a problem by disciplinary scientists devoid of communication and simply 'clipped' together. The chapters in this section share the common call for true integration, which requires communication, collaboration and participation as core elements of policymaking and policy evaluation, and where scientists are viewed as part of the process and not situated outside it in the ivory towers of the academy or safely sealed in research laboratories.

Using models to structure problems and anticipate science needs

Directly related to the core issue of integration is the necessity for modelling to be linked to problem structuring at the outset of a policy process. Most models respond to a received problem and thus are unable to influence the understanding of the problem and the ability to include the array of relevant characteristics of it so as to make models more robust and useful (Chertoff *et al.*, 1999). A participatory approach in policymaking and evaluation opens the opportunity for scientists to join with all others engaged in the policy process and contribute to structuring a problem that takes into consideration more of the 'parts' of a large, complex, ambiguous issue (Shannon, 1992). This call for integration at the problem structuring stage of policy is welcome news to policymakers, who must address the whole problem in all its messy complexity. When scientists, especially when working within a single discipline, provide thorough analysis of a small part of this messy problem, the policymakers ask 'What do I do with this?' as too much of the policy problem is neglected in the narrow – even if deep and accurate – analysis.

Collaborative processes for developing research agendas

One implication of meeting the needs of policymakers by addressing the complexity of real problems is that scientists cannot develop priorities for research in isolation (see Chornesky *et al.*, 2005). The need for a participatory process in developing research agendas and identifying priorities for funding is a clear call from these chapters. However, if scientists are to effectively research complex problems, they must do so in an integrated way. This means that the question, the research design, the methodology, the analytical tools, the frames of interpretation and the findings must all be produced within an integrated framework. To a large extent this means developing interdisciplinary approaches for complex problems, but, to a much more important extent, it means integrating the diverse world views of the participants – scientists and non-scientists alike. For scientists to engage in research on problems of direct concern to policymakers, policymakers and others who depend upon science-based information for decision making must be involved in the design of the research programme. Rather than feeling that their exclusive territory is being invaded, both scientists and policymakers need to recognize the contribution they each bring to the process (Daniels and Walker, 1996).

Integration does not mean a 'free lunch' or 'magic' solutions

It must be conceded, however, that many times policymakers seem to want 'magic' solutions from scientists, who seem to them lost in details and impervious to complex reality and hard choices. For this reason, new modes of science and policy collaboration are emerging around the world (Clark *et al.*, 1998). The chapters in this section describe some of these new forums for science–policy interaction and call attention to some of their strengths and pitfalls. The critical feature of the processes for science–policy interaction is that they embrace the questions that policymakers need answers to and harness the capacity of scientists to develop open and integrated frameworks that are capable of addressing complex questions.

All of these calls for new approaches to integration rest upon improved communication amongst all members of the policy community (Habermas, 1984, 1987). Thus, communication mediated by a science–policy process entails engagement by all in the processes of translation of scientific conceptual frameworks into policy-relevant and problem-relevant information. This means that policymakers and scientists are going to have to learn to talk to each other and create mutual understanding if the kinds of complex and messy problems faced by forestry are to be effectively addressed and perhaps resolved.

Anticipatory science and responsive policy

For scientists and policymakers there is another practical problem that can be solved: for science to be useful it needs to be ready when the problem erupts, but that means that science needs to anticipate problems, not simply wait to react to them. By working together to develop integrated research programmes to address complex, messy and diffuse issues before problems erupt and require immediate attention, policymakers are more likely to get the kinds of information they want when they need it and scientists are more likely to have funding based on future needs for information, thus facilitating the process of discovery and transformation in science (Dietz *et al.*, 1999).

For policy processes to integrate science and scientists they must be open, self-critical and reflexive and designed to interrogate facts and values through deliberation and dialogue (Forester, 1993). The culture of science easily incorporates the self-critical and open requirements of integration, but resists the reflexivity based upon questions and challenges from policymakers regarding relevance and usefulness. The culture of politics looks for the power, and easily incorporates reflexivity as a mode of seeking the centres of power, but resists the self-critical and open requirements in a desire to find a solution and move to the next problem. The differences in culture between science and policy are the greatest impediment to integration (Shannon and Antypas, 1996). While the chapters in this section illustrate different ways in which these impediments can be overcome and the potential for integration realized, most of them describe the lack of integration and the loss of policy options and opportunities.

Achieving anticipatory science cannot be achieved by scientists alone, since it is guided by prospective and future problems. Similarly, responsive policy cannot be created by policymakers alone, since information about what might be is necessary. Thus, without integration, sustainable forest management, which is a projection of a future into policy based upon scientific understanding of the forest, social values and the policy process, is impossible.

Common Themes among these Chapters

Scientists are needed immediately

Several aspects have changed in the position of science in relation to changes in society. First, the globalization of the language used in the international debate – sustainability, biodiversity, participation – requires translation into the context of policymaking (Eastwood, 2005). In addition, these new concepts do reflect a more integrated perspective, but they are also complex and ambiguous in that many values are included. As a result, policymakers have not yet completely understood or mastered the underlying concepts. The international dialogue on forests has generated an important call to scientists to bring precision and legitimization to the terminology in use when policies choices are explained. This issue of language and terminology is important partly because most of the policy issues related to the idea of sustainability were brought into the debate by environmental non-governmental organizations (NGOs) promoting a new style of expertise by arguing with science. Thus, the very concepts themselves are viewed by many – scientists, policymakers, managers, civil society – as normative expressions of values and not scientific concepts based upon theory and research.

Responding to this call is certainly not an easy task for scientists for several reasons: (i) basic scientific concepts used in defining sustainability and establishing policy solutions (such as ecosystem dynamics, social values, democracy, social change) are scientifically disputable, especially when considerations of environmental ethics are present; (ii) research very often leads to, and is encouraged by, peer review to provide complicated explanations. This tendency, of course, is the opposite of how policy seeks to create simple assertions and easily tested hypotheses; and (iii) cooperation between scientists and policymakers presupposes a new scientific behaviour, and this will require major changes in how research is organized, conducted and funded.

Secondly, the changes at the world level are so rapid (and not only in countries with economies in transition) that time pressures for policymaking and knowledge creation have become more intense than ever. Since the 1990s, the international community has accepted short time frames for major policy responses to critical, highly complex environmental problems. Correlatively, since policy decisions are made very rapidly, policymakers cannot wait for scientific results. Thus, scientific research as related to sustainability and sustainable forest management needs to be organized so as to perpetually adapt to emerging questions and problems and to provide results before a full analysis with scientifically acceptable levels of certainty is available. This social demand for science may have an important impact on what is considered by society as useful research in the coming years.

Which research topics need greater attention and funding?

Perhaps scientists need to reflect upon new paradigms and philosophies for forest sustainability. Surely these new paradigms should promote multifunctionality as a research concept. Defining sustainability as a complex of ecological, economic and social elements can easily lead to considering technical and policy means aiming at reaching it. However, sustainability is more like a guiding star illuminating the ways in which forests are needed to provide, in the same time and place, a variety of products, benefits, goods and services.

Multifunctionality is a major challenge in Europe and elsewhere, and clear concepts and theories are needed to guide public action (Nabuurs *et al.*, 2003). The concept of function, traditionally used in European forest policies as a norm, is a very disputable one, and a policy aiming at multifunctional management is advocated by some decision makers by pretending that these goals are objectively predefined by science (Krott *et al.*, 2000). The American 'multiple-use' notion seems specifically restricted to sociological issues, referring only to direct utilities of the forests to people, and yet it creates the same issue when decision makers assert that some kinds of values are more 'valuable' than others, especially those that are measured by prices and create political power though wealth creation (Shannon, 1987). The implicit and explicit values that underlie these concepts need to be open to disputation through interrogation of ideas, deliberation regarding both the science and the policy and collaboration in engaging researchers in providing greater credible definition for these terms, thus relieving them of their purely normative character (Healy, 1997).

'Integration' also refers to how, where and when various social values and utilities are provided by concrete management practices. Thus, the concept of 'multi-beneficiaries' may be more promising than multifunction since it can take into consideration both marketable and non-marketable goods and services provided by the forests, but it needs to be further specified and elaborated by scientists in order to be useful for decision makers (Buttoud, 2000). This example shows how more scientific discussion on basic, core concepts is needed.

A new idea is to develop a multi-theoretical framework for understanding change (mixed models) (Buttoud and Yunosova, 2002). In all countries in the world, forest policy is unlikely to achieve concrete progress based on specific disciplinary scientific results. Policies effectively aimed at achieving sustainable management of the forests will proceed from a multi-theoretical and multidisciplinary approach, which can more directly give the policymakers the scientific basis for public action. The need for a cross-analysis based on various theories and frameworks is derived from the large number of issues to be addressed at the same time by a forest policy, and also from the large number of people involved who share different – and sometimes opposite – views on priorities and objectives (Dube and Schmithusen, 2003). This is not a discursive topic: building up a multi-theoretical approach requires a rigorous deductive and participatory process engaging scientists from various disciplines.

Do we need new ways for conducting forest research?

A real challenge for scientists is not only in delivering the scientific results, but in presenting them in a useful way. This also means a new organization of the scientific work at an international level, which presupposes: (i) working with forest managers and practitioners in analysing their own needs and translating their technical questions into scientific ones (such a cooperative process between scientists and managers is usually difficult because, most of the time, the practitioners view the research as a way for reconsidering and negatively evaluating the activities they have carried out); (ii) working with social scientists, as one of the goals for scientific work is to question the demands of society (this cooperation remains weak in the forestry field because technical viewpoints still prevail); (iii) networking at both international and regional levels among scientists, in order to be able to act for decision-making support at the right levels at the time the policy processes are taking place (strong incentives in the EU have improved this linkage); and (iv) last but not least, discussions with politicians, who, after all, are among the major actors making policy. All these improvements in the science–policy linkage must begin by creating forums in which scientists and policymakers discuss critical issues of the day and for the future and pose questions of current and anticipated interest.

Hot questions for science

This role of scientists in the process of decision making brings new challenges to research institutions and to the conduct of science and the behaviour of scientists.

The instrumentalization of science

The support science may bring to decision making is often a questionable issue, especially in countries where forest policy is an old, institutionalized process. Many cases exist where scientists are considered simply to be alibis for politicians, stakeholders or bureaucrats willing to use them in order to justify their arguments by drawing upon the positive image of science. Finding ways in which research supporting decision making could bring more transparency to the policy process in line with the open culture of science could limit the use of scientific results as weapons.

What is good science?

In the academic world, good scientific research is basically constituted with a rigorous development of argumentation based on verifying facts through analysis. Formalization through mathematics is a must in the ideal. However, a policy decision is not only based on rational analysis, but also on collaborative action. In such a framework, it may be that 'good research' is research that is more or less directly usable for decision making in the present context, but which also enables the decision makers to adapt and modify the context of the public decision in a proactive manner, in order to take into consideration improved knowledge of the situation and ongoing processes. Science can inform a decision, but it can also alter the decision through the context of construction of knowledge that is endogenous to the society.

Is science neutral?

It should be clear that science and all underlying theory are not neutral – as Johnson discusses, it takes a position as to the nature of the world and its meaning. So a real challenge is, more than ever, ensuring the independence of the research structures. In all disciplines, including studies of the policy process itself, science will be best served by creating structures of accountability that recognize the role of scientists in policy and the implications of pressing social problems for the conduct of science.

A Practical Example of the Policy–Science Interface

Greater integration of scientists and policymakers requires forums where communication and dialogue, debate and deliberation can occur. Neither normal policy processes nor normal research processes work well as places for integration. Thus, a critical need for improving integration is the creation of new modes of governance wherein scientists and policymakers join together in structuring problems, creating new ideas and testing them in practice. The science–policy assessments, as discussed by Pregernig, are one such emerging forum in which scientists are engaged in assessments of important problems. Policy evaluation, as discussed by Kouplevatskaya, is another forum wherein a policy process is formulated on 'scientific rules' of critique, interrogation and disputation.

In Europe, the Ministerial Conference on the Protection of Forests in Europe (MCPFE) is creating a series of science–policy forums to improve science–policy integration for sustainable forest management. To date, conferences have been in Strasburg (1990), Helsinki (1993), Lisbon (1998) and Vienna (2003). The next conference is scheduled for 2007 in Warsaw, Poland.

In the 2003 Vienna Declaration, the Signatory States and the European Community committed themselves to 'make forest-related decisions based on science, and to take measures that support and strengthen research as well as increase interdisciplinary research'. This statement provides a basis upon which to design mechanisms to integrate research into the MCPFE process. Research can be employed to support the preparatory phases of ministerial conferences, to facilitate the implementation of commitments and recommendations and to monitor the progress towards sustainable forest management through evolving criteria and indicators, as well as to address specific issues and research needs raised by MCPFE.

The political commitments taken by the MCPFE are implemented through the resulting work programme. For example, since 1990 'research on tree physiology', 'research into forest ecosystems' and 'conservation of forest genetic resources' have been included in the work programme. Although a number of EU COST actions have also supported the issues taken up in the ministerial conferences, the implementation of Vienna resolutions is mainly carried out through organizing workshops and seminars. In the present work programme, the Vienna resolutions are supported by 17 workshops or seminars, four to five publications and about five ad hoc work groups engaged in various types of 'analysis'.

Before the Vienna Conference, new sustainability indicators were developed by an ad hoc working group. This type of modus operandi enables the compiling of existing information with little cost, but new information or even new data collection would need more time and resources to be successful. In addition, given the investment required, it should be expected that all countries cannot report some of the new indicators, and thus mechanisms to improve science resources are important for achieving sustainability. One illustrative example is the new criteria on 'cultural values of the forest'. In the Expert Level Meeting in 2004 it was concluded that 'we do not know what we do know' about this topic. Reporting of this indicator – and also of other indicators – would profit from an analysis of the state of the art in these areas of science now integral to sustainable forest management as well as on the costs and benefits of assessing and reporting

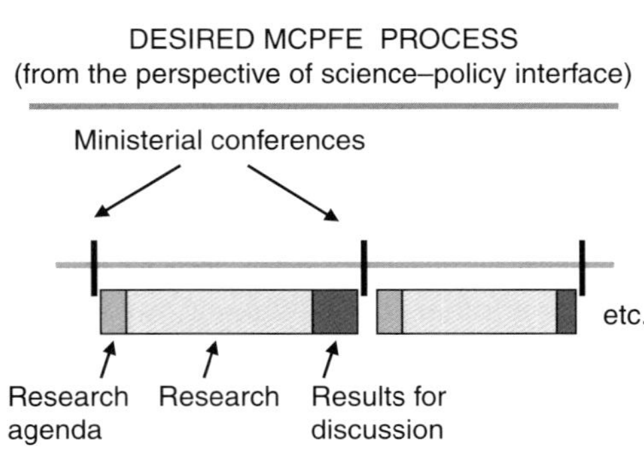

Fig. 1.1. An iterative science–policy process for Europe.

various indicators in the signatory countries in a harmonized way. Clearly, the question that could be posed – 'What do we not even know that we need to know?' – would benefit greatly from cultural and social research, and thus scientists from these disciplines need to be fully integrated into sustainable forest management.

As an example of how a policy process could become a forum for integration, the MCPFE process could facilitate improved collaboration and deliberation at the science–policy interface by, for example:

1. Organizing ministerial conferences every 3–5 years. This time span would allow research work between the conferences.
2. Based on the resolutions of the conferences, a research agenda could be set by the scientific community soon after the conference. The countries and the funding agencies should be included in the discussions in order to maximize the proper resources for the work.
3. The research itself can take place during the 2–3 years between the conferences.
4. The results would be brought into the discussion and preparation of the next conference in order to ensure that forest-related decisions are based on science and to continue to specify the guiding concepts like sustainability, biodiversity and participatory processes.

Unfortunately, to date, setting forth a research agenda has not been included in the work programme of the MCPFE; thus the opportunity to follow the 'desired' strategy has been lost for the time being. However, the scientific community has volunteered to discuss this agenda in the context of appropriate scientific meetings, often sponsored by IUFRO, another forum for integration of science and policy, in order to support a meeting on strengthening the science–policy interface in April 2006. While this gracious offer may suffice in the near term, it will not work for the long term. Meeting the needs of sustainable forest policy simply requires greater integration between scientists and policymakers, and governance processes need to recognize this need and create appropriate forums for dialogue and debate, discussion and deliberation among the full policy community when necessary and strategic actors always.

Conclusions

The inclusion of the science–policy theme in the midst of analytical and technical scientific discussions is a necessary but not sufficient step in fostering greater integration. Without the efforts of individuals to seek out new members of the sustainable forestry community, opportunities to create new forums for integration will be missed. The chapters in this theme section illustrate the many ways in which scientists are providing information of critical interest and use to policymakers.

However, perhaps even more importantly, when scientists are involved in policy discussions, they learn. They learn about new problems and complexities in old ones; they learn about when theories work and when they don't; they learn about how to communicate essential information in ways that those who need it can understand and act upon it; in essence, they learn. Science is designed to learn.

When the policy process embraces the learning culture of science, it too can learn. Policymakers learn when they are confronted by reasons why not to do something they want to do to meet constituencies' needs; they learn when actions expected to create changes fall flat in practice; and they learn when they can ask good questions to scientists.

The challenge of the integration of science and policy for sustainable forest management is to create new forums, and make better use of old ones, for engaging the demanding issues confronting policymakers, managers, forest landowners and civil society. In its essence, sustainable forest management is a conversation about the future.

References

Burgess, H. and Burgess, G. (1996) Constructive confrontation: a transformative approach to intractable conflicts. *Mediation Quarterly* 134, 305–322.

Buttoud, G. (2000) How can policy take into consideration the 'full value' of forests? *Land Use Policy* 17, 169–175.

Buttoud, G. and Yunusova, I. (2002) A 'mixed model' for the formulation of a multipurpose mountain forest policy: theory vs. practice on the example of Kyrgyzstan. *Forest Policy and Economics* 4, 149–160.

Chertov, O.G., Komarov, A.S. and Karev, G.P. (1999) *Modern Approaches in Forest Ecosystem Modelling*. Brill, Boston, Massachusetts.

Chornesky, E., Bartuska, A., Aplet, G., Britton, K., Cummings-Carlson, J., Davis, F., Eskow, J., Gordon, D., Gottschalk, K., Haack, R., Hansen, A., Mack, R., Rahel, F., Shannon, M., Wainger, L. and Wigley T.B. (2005) Science priorities for reducing the threat of invasive species to sustainable forestry. *BioScience* 55, 335–348.

Clark, R., Meidinger, E., Miller, G., Rayner, J., Monreal, S., Fernandes, J. and Shannon, M. (1998) *Integrating Science and Policy in Natural Resource Management: Lessons and Opportunities from North America*. General Technical Report PNW-GTR-441. US Department of Agriculture, Forest Service, Pacific Northwest Research Station, Portland, Oregon.

Daniels, S.E. and Walker, G.B. (1996) Collaborative learning: improving public deliberation in ecosystem-based management. *Environmental Impact Assessment Review* 16, 71–102.

Dietz, T., Chess, C. and Shannon, M.A. (1999) Who should deliberate when? *Human Ecology Review* 5, 45–47.

Dube, Y. and Schmithusen, F. (eds) (2003) *Cross-sectoral Policy Impacts Between Forestry and Other Sectors*. FAO Forestry Paper No. 142. FAO, Rome.

Eastwood, L. (2005) *The Social Organization of Policy: An Institutional Ethnography of UN Forest Deliberations*. Routledge, New York.

Forester, J. (1993) *Critical Theory, Public Policy and Planning Practice*. University of New York Press, Albany, New York.

Habermas, J. (1984) *Theory of Communicative Action*, Vol. 1: *Reason and Rationalization of Society*. Heinemann, London.

Habermas, J. (1987) *Theory of Communicative Action*, Vol. 2: *Lifeworld and System: a Critique of Functionalist Reason*. Beacon Press, Boston, Massachusetts.

Healey, P. (1997) *Collaborative Planning; Shaping Places in Fragmented Societies*. Macmillan, London.

Johnson, K.N., Shannon, M.A., Holthausen, J. and Sedell, J. (1999) The Forest Ecosystem Management Assessment Report: Lessons for Bioregional Science Assessments. In: Johnson, K.N. *et al.* (eds) *At the Crossroads of Science, Management and Policy: A Review of Bioregional Assessments*. Island Press, Covelo, California.

Krott, M., Tikkanen, I., Petrov, A., Tunytsya, Y., Zheliba, B., Sasse, V., Rykounina, I. and Tunytsya, T. (2000) *Policies for Sustainable Forestry in Belarus, Russia and Ukraine*. Brill, Boston, Massachusetts.

Meidinger, E.E., Elliott, C. and Oesten, G. (eds) (2003) *Social and Political Dimensions of Forest Certification*. www.forstbuch.de

Nabuurs, G.-J., Päivinen, R., Pussinen, A. and Schelhaas, M.-J. (2003) *Development of European Forests until 2050*. Brill, Boston, Massachusetts.

Shannon, M.A. (1987) Forest planning: learning with people. In: Miller, M.S., Gale, R.P. and Brown, P. (eds) *Social Science in Natural Resource Management Systems*. Westview Press, Colorado, pp. 233–252.

Shannon, M.A. (1992) Achieving a common approach to problems and directions. In: Adams, P., Atkinson, W. and Atkinson, W.A. (eds) *Watershed Resources: Balancing Environmental, Social, Political and Economic Factors in Large Basins*. Oregon State University, Corvallis, Oregon, pp. 38–44.

Shannon, M.A. and Antypas, A.R. (1996) Open institutions: uncertainty and ambiguity in 21st century forestry. In: Franklin, J.F. and Kohm, K.A. (eds) *Creating a Forestry for the 21st Century: the Science of Ecosystem Management*. Island Press, Covelo, California, pp. 437–445.

Shannon, M.A., Meidinger, E.E. and Clark, R.C. (2000) Science advocacy is inevitable: deal with it. *Reflections* 4, 8–9. Newsletter of the Program for Ethics, Science and the Environment, Department of Philosophy, Oregon State University, Oregon.

Watson-Verran, H. and Turnbull, D. (1995) Science and other indigenous knowledge systems. In: Jasanoff, S. *et al.* (eds) *Handbook of Science and Technology Studies*. Sage, Thousand Oaks, California, pp. 115–139.

Yankelovich, D. (1991) *Coming to Public Judgement: Making Democracy Work in a Complex World*. Syracuse University Press, Syracuse, New York.

2 Will Linking Science to Policy Lead to Sustainable Forestry? Lessons from the Federal Forests of the United States

K. NORMAN JOHNSON

Department of Forest Resources, 108 Peavy Hall, Oregon State University, Corvallis, USA.

> When men are most sure and arrogant they are commonly most mistaken, giving views to passion without that proper deliberation which alone can secure them from the grossest absurdities.
>
> (David Hume, Scottish philosopher)

Abstract

Science is an essential underpinning to sustainable forestry. When a particular scientific discipline is applied, though, the policymaker gets facts, concepts and values of the discipline all mixed up together, making it difficult to apply simultaneously the ideas from different disciplines. Nowhere is this more clearly seen than in the management of the national forests of the USA. Forest policy for these lands has been dominated, almost single-mindedly, by a sequence of major scientific disciplines and their associated world views. For much of the 20th century, the 'science' of classical forest management reigned supreme, with its focus on sustained yield of timber and the regulated forest. While proponents of preservation and wildness fought against this approach, their successes were largely restricted to areas of little commercial value. Starting in the 1980s, scientific disciplines underlying the three facets of sustainable forestry sequentially dominated the planning and management of the national forests. First, economics provided the intellectual framework for planning and management, with its focus on maximizing human wealth, benefit–cost analysis and efficiency. Next, ecology dominated planning and management, with its focus on conserving native biodiversity and ecological productivity. Recently, social science has become the ascendant discipline in national forest management, with its emphasis on the well-being of people and communities and democratic processes in decision making. Each new scientific wave has provided its own world view, often to the exclusion of those that came before it and unnecessarily restricting the choices and evaluations of decision makers. For policymakers to chart their own course and apply their own perspectives and values, we need to devise processes that utilize simultaneously the facts and concepts of the sciences supporting sustainable forestry without the dominance of a particular scientific world view.

 Sustainable Forestry: from Monitoring and Modelling to Knowledge Management and Policy Science (eds K.M. Reynolds, A.J. Thomson, M. Köhl, M.A. Shannon, D. Ray and K. Rennolls)

Introduction

World views, as related to the sciences, ethics, arts, politics and religions, are integral parts of all cultures. They have a strongly motivating and inspiring function. A socially shared view of the whole gives a culture a sense of direction, confidence and self-esteem.

As described by Heylighten (2000), in reviewing a book by Aerts *et al.* (1994), six key components of world views are: (i) theories and models for describing the phenomena we encounter; (ii) understanding of how the world functions and how it is structured; (iii) explanations of why the world is the way it is; (iv) descriptions of more or less probable future developments; (v) values addressing what is good and what is evil; and (vi) actions describing how we should act to solve practical problems.

Various forest scientists have noted how world views inevitably influence the different scientific disciplines their findings and their conclusions. Shannon (2000), as an example, points out that science advocacy is inherent in how scientists approach their work. Others, such as Mills *et al.* (2001), have stated that scientists can, and should, separate their values and perspectives from the advice they give policymakers.

In the history of the national forests of the USA, four different scientific disciplines have sequentially provided the dominant intellectual structures for managing these forests: (i) 'sustained yield' forestry; (ii) economics (merged with sustained yield); (iii) ecology and wildlife biology; and (iv) social science. In each case, these disciplines did not just provide facts and relationships. Rather, they provided complete world views, including value-laden concepts and decision guides that greatly coloured the way in which policymakers defined sustainability, the problems they saw and the solutions they sought.

My purpose in this chapter is to outline the sequential dominance of different scientific disciplines, the key attributes of their different world views and how each in turn has influenced federal forest management. After presenting this history, I summarize ways in which these disciplines have contributed to sustainable forestry and ways in which they have detracted from sustainable forestry. Finally, I offer the hope that the new approach to planning of the national forests, centred on social processes, can avoid the difficulties of the past by allowing perspectives from the different sciences to surface and be utilized in creating sustainable national forests for the future.

This chapter discusses the management of the national forests of the USA. It does, though, often use examples from the Pacific Northwest, the part of the national forest system with which I am most familiar.

Federal Forests: the Beginning

Originally, land policies in the USA were designed to shift all forest into private hands, and hundreds of millions of hectares passed into private ownership from 1780 to 1880. Goals for this transfer to private landowners, and subsequent settlement, were to solidify the nation's claim to these lands, provide income for a cash-poor country, pay for major projects like the transcontinental railway, contribute to the nation's economic development and develop a nation of responsible, landowning citizens (Dana and Fairfax, 1980).

Few restrictions were placed on private land use. In many places forests were converted to farms. From 1780 to 1880, over 30% of the forests in the east and south were converted to cropland. In other places, forests were rapidly liquidated to provide lumber for farmhouses and farms, with little thought for the future. Thus, the white pine forests of the upper Midwest were largely liquidated in the mid- to late 1800s.

In the mid- and late 1800s, forest policy became a national issue. In the east and Midwest, forests had been denuded, streams were often degraded and wildfires frequently burned unchecked. In the west, questions were raised about the role of forests in protecting water supplies and the adequacy of wood supplies for home builders and the nation (Dana and Fairfax, 1980). A series of analyses galvanized the concern of the public and Congress over adequacy of the management of forest resources by private owners. As an example, Hough (1878) began publishing his monumental volumes called an *Assessment on Forestry*, which gave the first status report on the nation's forests. Hough's reports did not present a reassuring picture: he raised the spectre of a coming timber famine if policies did not moderate the rapid consumption of the nation's forests. In these reports, he called for retention of forest lands by the federal government – perhaps the first government official to advocate such a position.

In response to these concerns, Congress passed the Forest Reserve Act in 1891 (16 USC 471), allowing the president to retain some forest in federal ownership but not specifying how these forests should be managed. A series of presidents used that authority in the late 1800s and the early 1900s to retain large areas of forest in the west.

Rise of 'Scientific' Forestry

The lack of management authority in the 1891 'Creative Act' caused consternation in many communities of the west. First, settlers could not (legally) access the timber within the new federal reserves. Secondly, it appeared that the lands would not make a contribution to economic growth as the act did not acknowledge any permitted uses.

After much debate, Congress passed the 'Organic Act' in 1897(16 USC 475) declaring that the purposes of the forest reserves were: (i) to improve and protect the forest within the reservation; (ii) for the purpose of securing favourable conditions of water flows; and (iii) to furnish a continuous supply of timber for the use and necessities of the people of the USA (Dana and Fairfax, 1980).

As with most laws, this one left much room for interpretation. Some hoped that the protection and preservation aspects of the language would be emphasized; others hoped that the use aspects would be highlighted. Gifford Pinchot, the first professionally trained forester in the USA, soon became Chief of the Forest Service and came down strongly on the side of use, albeit conservative use:

> In the administration of the forest reserves it must be clearly borne in mind that all land is to be devoted to its most productive use for the permanent good of the whole people, and not for the temporary benefit of individuals or companies. All resources of the reserves are for *use*, and this use must be brought about in a thoroughly prompt and businesslike manner, under such restrictions only as will insure the permanence of these resources. The vital importance of forest reserves to the great industries of the Western States will be largely increased in the near future by the continued steady increase in settlement and development. The permanence of the resources of the reserves is therefore indispensable to continued prosperity, and the policy of this department for their protection and use will invariably be guided by this fact, always bearing in mind that the *conservative use* of these resources in no way conflicts with their permanent value.
>
> You will see to it that the water, wood, and forage of the reserves are conserved and wisely used for the benefit of the home builder first of all, upon whom depends the best permanent use of lands and resources alike. The continued prosperity of the agricultural, lumbering, mining, and livestock interests is directly dependent upon a permanent and accessible supply of water, wood, and forage, as well as upon the present and future use of their resources under businesslike regulations, enforced with promptness, effectiveness, and common sense.
> (Letter from Secretary of Agriculture Wilson to Pinchot when he became Chief Forester of the USDA Forest Service, but apparently written by Pinchot – Dana and Fairfax, 1980, p. 82)

Pinchot's vision led to the sustained yield model that dominated forestry for most of the last century: his vision called for use of the resources of the national forests – especially wood, water and forage – to assist in the economic development of the west. Pinchot's model of conservative use – use at a rate that would not impair the permanent value of the forest – served as a guiding principle. It was expressed through the idea of a regulated forest, in which the same amount of timber would mature each year, and a sustained yield of commercial timber products. These concepts were utilized by generations of foresters (including the author) to calculate the allowable cuts on the national forests for much of the 20th century.

This vision was implemented through scientifically trained foresters with their 'science' coming from disciplines that supported the sustained-yield model of forest management – silviculture/mensuration (inventory, growth and yield)/ management (harvest scheduling). Forests would be grown as a crop with the goal of a regulated forest. Wildfire and pests would be controlled and suppressed. All lands that could produce commercial crops would produce them over time. Roads would be built in forests to provide access for timber harvest and to allow control of wildfire. Old growth would be a high priority for harvest so as to make way for fast-growing second-growth stands (Cleary, 1986; Hirt, 1994; Langston, 2005).

In the north-west, the national forests were inventoried to identify land productive enough to produce timber on a continuing basis. Then foresters used formulas to calculate the maximum harvest level over time from these forests consistent with concepts of sustained yield. Many of these calculations, though, were theoretical exercises, as there was relatively little demand for this timber in many parts of the west for many decades; private timber was cheap and available. After the Second World War, though, federal harvests ratcheted upward, reaching 20% of the softwood harvest in the USA, as economic prosperity combined with depletion of private supplies led mills to the federal woodpile.

Important facts to support this approach to forest management included an estimate of the amount of merchantable volume on each area and its growth. Also, because the sustained-yield model rested on long-term growth, the productivity of future stands under different intensities of management was important to estimate. Much effort was devoted to collecting this crucial information on growth and yield, such as Bulletin 201 (McArdle and Meyer, 1930).

As mentioned above, the facts and relationships about growth and yield were embedded in concepts such as the regulated forest and various procedures for calculating the allowable cuts. All these ideas had a strong foundation in European forest management, where early foresters like Pinchot studied. They reflected the goal of a balanced set of age classes, which would then produce a constant output of timber forever; this was presented as the appropriate way to manage forests (Davis and Johnson, 1987).

Underlying these facts and concepts were a set of world views – views about the way the world worked, about what the future held, about relationships that foresters accepted as truisms, about values that foresters held dear. First and foremost was the notion of an approaching timber famine that could be prevented only by adopting the 'scientific' approach to management that they advocated. In the late 1800s and early 1900s, when these ideas were gaining credence, trees were harvested without much thought of future forests (Dana and Fairfax, 1980). Studies like those of Hough (1878) raised alarm bells about the implications of continuing down this path. Thus, it is not surprising that the forestry profession anchored its work in the desire to thwart an approaching timber famine. Secondly, foresters believed that sustained-yield timber production was compatible with, indeed productive for, other forest values. Foresters were not totally blind to the multitude of values that people seek from forests, whether they are water, wildlife, recreation or spiritual renewal. They did believe, though, that these values could be protected through Pinchot's conservative use under the oversight of professionals. Thirdly, foresters believed that the sustained-yield model would provide for economic and social sustainability through the provision of regular, constant timber volume.

Key elements of this world view have been summarized by Duerr and Duerr (1975) as 'tenets of faith' – the strong long-term demand for wood products, the need for maintaining a sustained yield, the wisdom of the forest manager and, over everything, the primacy of timber production. As pointed out by Duerr and Duerr, this world view helped the foresters work through all the imponderables and uncertainties of their work.

In forestry schools there was no need to articulate these tenets of faith – they were implicit in the classes, the textbooks and the discussions (see, for example, Davis and Johnson, 1987). Thus, while often unstated, they dominated and directed forest management, and especially federal forest management, through the middle of the 20th century.

Passage of the Multiple-Use, Sustained Yield Act in 1960 (PL 86-517;16 USC 528) did little to change this world view. Timber primacy still drove the agency's priorities (Cleary, 1986), with foresters in charge, tremendous demand for the softwood timber from the national forests and the continuing concern over timber scarcity. The public's increasing desire for protection of wild places was satisfied largely by the Forest Service designating areas of low commercial value as backcountry and later by Congress designating these types of areas as wilderness.

The National Forest Management Act – a Challenge to Sustained-yield Forestry

The clear-cutting controversy over the national forests began in the 1960s, with public discontent expressed in lawsuits over the definition of multiple use. In response to these challenges, Congress held hearings and from them published the 'Church Guidelines', which specified the conditions under which clear-cutting would be acceptable (Dana and Fairfax, 1980). However, the Forest Service felt it still had the discretion in applying these conditions and continued to prescribe clear-cutting in the national forests. A local chapter of the West Virginia Izaak Walton League brought a suit in the district court to halt clear-cutting in the Monongahela National Forest. The suit alleged that the Forest Service harvesting practices violated the Organic Act of 1897. That law authorized the agency to sell dead, mature or large trees within national forests. It also required that such timber be marked and designated prior to sale and that it be cut and removed under federal supervision. The argument before the court centred on the meaning that Congress, originally and in subsequent legislation, intended with the words 'dead', 'mature', 'large growth', 'marked and designated' and 'cut and removed'. The district court held that by cutting immature and unmarked trees, as is necessary in clear-cutting, the Forest Service had exceeded its authority (Fairfax and Achterman, 1977).

Although Congressional leaders offered to amend the 1897 Organic Act based upon more recent scientific knowledge of silviculture, both the Forest Service and the timber industry refused, believing that the appeal court would agree with their interpretation based upon the deference usually given to an agency's interpretation of its statutory responsibilities. However, the appeal court agreed with the district court on the simple logic of 'the plain English' meaning of the terms. This led the Forest Service and the industry quickly back to Congress, but their new-found desire for legal changes was rejected. Instead, Congress began to examine the larger problems underlying the clear-cutting controversy. Thus, a 'quick fix' approach was rejected in favour of a comprehensive approach to resolving persistent problems and ambiguities in national forest management.

It is important to understand that simultaneously during the early 1970s Congress passed more environmental protection laws than ever before or since in the short span of 4 years. The new environmental protection extended to air, water and wildlife; for example, it prescribed the kinds of duties and responsibilities public and private actors had to ensure that the air and water were clean and, when polluted, restored to clean conditions, and that wildlife whose existence was threatened by human development was protected even if the economic cost was high. This willingness to assure the American people, and the world in terms of the numerous international conventions that underlay these laws, that environmental quality would be protected and restored manifested itself strongly after the unanimous passage of the National Environmental Policy Act in 1969 (NEPA) (PL 91-190; 42 USC 4321), which affirmed such a public duty.

In the case of federal forests, grasslands and other natural resources, however, the governing principle remained one of agency discretion to select 'that combination of uses that best meets the needs of the American people', as stated in the 1960 Multiple-use Sustained-yield Act – an act written by the Forest Service for itself, much like the Pinchot letter 55 years earlier. These two policy approaches now clashed as Congress set out to address the management of the federal lands. Numerous bills were introduced, but they all employed one of two different approaches: a prescriptive reform with specified conditions for decision making, or a more administrative reform that worked with the discretionary multiple-use framework.

The National Forest Management Act (NFMA) (PL 94-588; 16 USC 1600), which resulted from these deliberations, blended these two approaches. While it reaffirmed that the national forests should be managed for multiple-use and sustained-yield purposes consistent with the Multiple-use Sustained-yield Act, it also restated the Church Guidelines on Clear-cutting almost word for word in Section 6 and conditioned the use of clear-cutting on determining that it was the 'optimal' solution. The NFMA reinforced the analytical planning approach of the Renewable Forest and Rangeland Planning Act of 1974 (PL 93-378; 16 USC 1600), which formally NFMA amended by requiring that every National Forest prepare one integrated management plan for the entire forest, including all of its programmes. Many of the elements of the prescriptive approach bills were included as specific sections in NFMA, for example, the duty to provide for the diversity of plant and animal species, the incorporation of the Church Guidelines on Clear-cutting, and duties for soil and water protection, to name some of the most important ones.

The NFMA gave the Secretary of Agriculture considerable discretion in writing regulations on how to interpret and implement the act, although it did call for a 'Committee of Scientists' to help with the initial set. Every presidential administration since passage of the act has tried to rewrite these regulations (36 CFR Part 219), with four sets of regulations sequentially adopted over time (1979, 1982, 2000 and 2005). We shall discuss the last three sets (1982, 2000 and 2005) below as the guiding force in national forest planning and management since the passage of the NFMA. We shall not discuss the 1979 regulations as they were similar to the 1982 regulations and were quickly subsumed by them.

The Rise of Economic Forestry

Economic forestry generally analyses forest resources from the standpoint of maximizing their net benefit to humans. In doing so, economists place an emphasis on finding efficient solutions, that is, identifying management actions projected to achieve the highest net benefit given the values that the owner or manager places on different outcomes, the costs that must be incurred and the choices that are available. Where the outcomes cannot be valued, economic analysis focuses on finding the least-cost way to meet stated goals. Then, by varying the goal levels, trade-offs among the goals can be delineated. Techniques to maximize net benefits or minimize costs are standard items in an economist's toolkit (Lippke and Bishop, 1999; Davis *et al.*, 2001).

Most economic analysis in forestry considers the use of forests over long time periods. Thus, these analyses use methods to make comparable costs and benefits that occur at different times. A central axiom of economic forestry is that discounted net value measures the net benefit to humans from the use of forest resources. Net present value – the discounted value of the revenues and costs from the use of forest resources over time – is often used to estimate net benefit from an investment. Thus, maximization of net present value is often presented as a general guide to management (Davis *et al.*, 2001).

These principles became dominant in decision making in the federal government during the 1960s and 1970s. The publication of Duerr's *Fundamentals of Forest Economics* solidified the application of economic principles to forest management (Duerr, 1960). In addition, benefit–cost analysis became a standard part of federal decision making and was quickly extended to forestry (Teeguarden, 1987).

The 1982 NFMA Regulations: Economic Principles as the Driving Force

The 1982 regulations (US Dept. of Agriculture, 1982) embraced the principles of economics as the analytical approach to satisfying the NFMA. They state the overall goal as maximization of net public benefits, which broadened net present value to include non-market goods. Benefit–cost and trade-off analysis were the order of the day. To implement this economic perspective and the timber management provisions of NFMA, the Forest Service turned to a policy analysis instrument called the forest planning model (FORPLAN), a linear programming model that enabled quantitative analysis of the consequences of various planning assumptions (Davis and Johnson, 1987). FORPLAN was used to maximize or minimize an objective (usually net present value or timber volume) subject to constraints (usually an even flow of timber harvest and also restrictions on stand structures and harvest rates over time). The principal concern was still the sustainable timber harvest level, but an economic analysis of the implications of different levels could now be conducted; by varying the objective or the constraints, efficient trade-offs among those goods and services could be quantified and related to timber harvest. Prices could generally only be found for timber outputs, so most formulations valued only those outputs, and represented the remaining goods and services as constraints (Davis and Johnson, 1987).

Forest outputs were linked to national goals through the national Resources Planning Act programme. Each national forest was given a timber volume target to test in the forest planning process. Achieving that volume would enable the Forest Service to maintain and increase its overall contribution to the national economy. Insufficient budgets were considered the major impediment to achieving, simultaneously, high levels of timber harvest and high levels of recreation.

The transformation of sustained-yield forestry into economic forestry had less effect than might have been expected for two reasons. Key elements of sustained-yield forestry remained in place, such as achieving rotations at least as long as needed to maximize timber growth and maintaining timber harvest volume over time, with exception to these rules allowed only after extensive justification.

The Rise of Ecological Forestry

While the integration of economic perspectives into sustained-yield forestry provided the overall emphasis of the 1982 regulations implementing the NFMA, other provisions were included in the regulations that, in time, overwhelmed this approach. Most importantly, the regulations greatly strengthened the emphasis on fish and wildlife, with the requirement that:

> Fish and wildlife habitat shall be managed to maintain viable populations of existing native and desired non-native vertebrate species in the planning area. For planning purposes, a viable population shall be regarded as one which has the estimated numbers and distribution of reproductive individuals to insure its continued existence is well distributed in the planning area.
>
> (US Dept. of Agriculture, 1982, 36 CFR 219.20(b)(2))

When taken together with the Endangered Species Act of 1973 (ESA) (PL 93-205; 16 USC 4601), which required an ecosystem approach in order to preserve species threatened or endangered by a loss of habitat, this regulatory phrase embodied a strong, affirmative responsibility on the part of a land management agency to ensure the viability of species dependent upon habitats found in the national forests and grasslands or, if they were threatened or endangered, to ensure that they be recovered to the standard of 'viable populations'. In this way, forest planning and policy for the national forests began to slip away from the frame of economics and towards that of ecology and conservation biology.

Few at the time recognized the significance of the viability clause. Quantitative economic analysis with FORPLAN continued to be the order of the day and many plans were published using that approach, even though FORPLAN lacked the ability to do the spatial analysis critical to habitat assessment. In the late 1980s, environmental groups sued the Forest Service over the adequacy of protection of a wide-ranging species that favoured late-successional forests – the northern spotted owl. In the pivotal lawsuit, Seattle Audubon Society in 1989 sued the Forest Service for failing to adopt a credible conservation strategy for the northern spotted owl that would comply simultaneously with the NEPA, the ESA and the NFMA (Caldwell *et al.*, 1994). The courts agreed with the plaintiffs and issued a restraining order on the harvest of owl habitat until a credible strategy was developed to conserve the northern spotted owl. While that only directly affected national forests in Washington, Oregon and California, this judgement signalled a change in the role of the national forests in species conservation across the USA.

The injunction on the harvest of owl habitat led Congress to include ‘Section 318’ in the 1990 Interior Appropriations Bill, which set timber harvest levels for 2 years on the federal forests of the north-west. One moderately obscure provision of Section 318 called for the Forest Service to re-evaluate its protection strategy for the northern spotted owl. The Chief of the Forest Service had seen the plans developed by the national forests successfully challenged in court and wanted a new approach. He asked Jack Ward Thomas, a big game biologist with a Forest Service Research Station, to lead an effort to develop a conservation strategy for the spotted owl. Thomas's acceptance of this assignment began a fundamental shift in how we make decisions about federal forest resources and the role of ecological science and science teams in those decisions, both in the Pacific North-west and across the USA.

Thomas's assignment was to develop a scientifically credible conservation strategy for the northern spotted owl (Thomas *et al.*, 1990). With that assignment, the Chief and Thomas ushered in a major change in how we craft management plans. Ecological scientists had largely stayed on the sidelines during the development of forest plans under the NFMA during the 1980s. The call for scientifically credible strategies, however, thrust them into the spotlight as never before. While it is commonplace today to call for scientific credibility in forest management strategies, it was a new idea only 15 years ago.

Thomas gathered around him a cadre of scientific experts on the northern spotted owl, which came to be known as the Interagency Scientific Committee (ISC). After many false starts and attempts, they developed the first regional conservation strategy for federal lands in the Pacific North-west using the principles of conservation biology (Thomas *et al.*, 1990). The strategy created a network of relatively large reserves across the federal forests of the owl region, with additional restrictions on forest management on the land that fell in between (the ‘matrix’).

Release of their proposal came suddenly and with almost total surprise to the broader public, many of whom had believed that the Forest Service's proposed plans would protect species and ecosystems. No longer was wildlife protection seen as a constraint on timber harvest. Rather they turned wildlife protection into a primary goal for forest management into which timber harvest would have to fit. Now, species protection came first.

In addition, they had changed ecologists and wildlife biologists from bystanders in forest planning to the providers of regional strategies for the conservation of species and ecosystems. No longer could management agencies have significant freedom to select their approach to environmental protection. Rather, the ideas of an outside group (‘scientists’) would greatly constrain the forest management strategies that might be considered.

As the Thomas Report pointed out, the old-growth conservation issue was about more than owls and always had been. Spotted owls were only one of the thousands of creatures associated with old-growth forests. Wild fish stocks were also increasingly being viewed as at risk. North-west Congressmen and many others in Congress wanted a permanent solution to the problem. They hoped that it could be done in a way that would be acceptable to the timber industry.

They asked a group of scientists from a variety of disciplines to develop alternatives for conservation of old-growth ecosystems and the species within them. The scientists, nicknamed the 'Gang of Four', created over 30 choices for management of the federal forests of the owl region and analysed both their conservation and timber production implications, providing a marginal analysis of the trade-offs between timber harvest levels and risks to species and ecosystems (Johnson *et al.*, 1991). Their report confirmed that there would be no free lunch in solving the old-growth controversy in the Pacific Northwest. The North-west Congressmen, though, felt that the possible solutions had politically unacceptable costs and Congress did not act. Therefore, the issue was left to the newly incoming Clinton administration, which appointed yet another scientific committee – the Forest Ecosystem Management Assessment Team (FEMAT). While this scientific team included economists and social scientists, ecological considerations dominated the analysis, with the 'viability' requirement broadened to include all living things associated with late-successional forests (FEMAT, 1993).

The President chose an option that provided a high likelihood of survival for the targeted species, such as spotted owls, but did not provide a similar level of protection for numerous obscure species (mainly invertebrates) about which little was known. The President's plan allowed harvest at approximately 20% of historical levels – still mostly from old-growth forests. Government lawyers were reluctant to return to court until the plan was rewritten to also give a high likelihood of protecting all species. That could only be done by stopping all old-growth harvest. Thus, 'survey and manage' procedures were developed for species about which little was known, which made any action difficult and only possible after expensive surveys. Federal timber harvest declined to a trickle in the north-west.

While much attention and conflict centred on the Pacific North-west in the 1990s, similar lawsuits and processes unfolded on national forests throughout the USA. Thus, timber harvest from federal forests plummeted everywhere.

The 2000 NFMA Regulations: Ecological Principles as the Driving Force

Ecological forestry generally analyses forest resources from the standpoint of conserving native biodiversity and ecological productivity. Ecological forestry is distinguished by its emphasis on natural patterns and processes – understanding these natural patterns and processes, working in harmony with them and maintaining their integrity, even when it becomes financially difficult or inconvenient (Seymour and Hunter, 1999). As stated in Seymour and Hunter (1999), a central axiom of ecological forestry is that manipulation of a forest ecosystem should work within the limits established by natural disturbance patterns prior to extensive human alteration of the landscape. The key assumption here is that native species evolved under these circumstances, and thus that maintaining a full range of similar conditions under management offers the best assurance against losses of biodiversity.

The approach was codified with the second set of regulations adopted in 2000 (US Dept. of Agriculture, 2000). In them, the Clinton administration proposed an ecological approach to managing the national forests, which placed ecological sustainability as the first priority, used the concepts of maintaining natural structures, processes and functions as the foundation of the plan and called for both coarse and fine filter tests of ecological sustainability. For the fine filter (species) test, the administration adopted the following rule:

> Plan decisions affecting species diversity must provide for ecological conditions that the responsible official determines provide a high likelihood that those conditions are capable of supporting over time the viability of native and desired non-native species well distributed throughout their ranges within the plan area.
>
> (US Dept. of Agriculture, 2000, 36 CFR 219.20(b)(2))

Since we know very little about most species, that provision almost guaranteed that little activity would occur on federal forests and that planning resources would be largely consumed by viability analysis.

The 2000 NFMA regulations also introduced collaborative planning into national forest planning, following the guidance of a new Committee of Scientists (1999), but the overwhelming ecological emphasis in those regulations dominated any other perspective. At any rate, the regulations were immediately set aside once the Bush administration came into office, partly over concerns about their strong ecological focus.

The Shift in World View from Economics to Ecology

Between the 1982 regulations and the 2000 regulations, the major issue on the national forests changed from responsible, efficient timber production to the development of diverse, flexible and resilient policies to sustain the earth.

> The U.S. Forest Service, until the 1990s, epitomized the rational order of modernity. Economic efficiency was the principal criterion guiding the development of forest plans, producing a simplified landscape well ordered for the production of timber and other goods and services . . . Although the forests were generally well managed for timber production, the Forest Service came under devastating criticism for failing to protect the natural order of the forest.
>
> (Lee and Field, 2005, p. 300)

As the new ecological perspective took hold, in the early 1990s, foresters and economists lost their dominance of forest policy and forest policy analysis. Foresters and economists were often poorly equipped for the new approaches, most significantly because they did not identify with maintaining natural processes as the dominant goal in forest management, and also because they often lacked in-depth training in ecology, conservation biology and non-game fish and wildlife. The changing of the guard, from foresters and economists to ecologists and biologists, is apparent in the appearance and domination of new policy players (e.g. Jerry Franklin and Jack Ward Thomas), new policy reports (e.g. Thomas *et al.*, 1990; FEMAT, 1993) and new policy journals (e.g. *Conservation Biology*).

The change, though, was much more than a change in objective from commercial timber levels to protection of ecosystems and species. Otherwise, the Forest Service might have rearranged the objectives and constraints in FORPLAN and gone on its way. Rather, there was an almost complete replacement of one world view by another (Table 2.1).

While it might be argued that some of the differences were due to legal requirements, such as an elevation of threatened species to a pre-eminent position, there is more at work here. The differences described in the table do not just relate to differences in overall goals or analytical procedures, important as those differences may be. More fundamentally, they relate to differences in world view – in theories and models for describing the world, in our understanding of how and why the world functions and how it is structured, in descriptions of probable future developments, in values addressing what is good and what is evil and in actions describing how we should act to solve practical problems.

Key to these differences in world view is the instruction on how to act in the face of uncertainty about potential outcomes. Forestry is faced with an overwhelming amount of uncertainty, dealing as it does with long time periods, large wild or semi-wild landscapes and amazingly complex relationships. As pointed out by Duerr and Duerr (1975), the articles of faith you adopt to proceed in the face of these unknowns can have a profound influence on what you do: they provide the guidance to cope with the uncertainties that surround each decision.

Table 2.1. Comparison of sustained-yield/economic approach and an ecological approach. (Adapted from Davis *et al.*, 2001.)

Sustained-yield/economic approach	Ecological approach
Timber primacy	Primacy of species and ecosystems
Seeks plans that maximize achievement of goals	Seeks plans that give a mid-level achievement of goals (away from edges) to allow for uncertainty
Focuses on management goals to shape outcomes	Focuses on disturbance history to shape outcomes
Emphasizes what we know	Emphasizes what we do not know
Tends to ignore catastrophic events (fire, floods, insect outbreaks)	Tends to focus on catastrophic events
Has faith in continuous technological improvement	Suspicious of counting on technological improvement
Places burden of proof on those who wish to stop activities that might harm species	Places burden of proof on those who wish to undertake activities that might harm species (precautionary principle)

As an example, consider the 'precautionary principle', which came into prominence with the rise of the ecological perspective. The precautionary principle has been enshrined in many international agreements, especially in Europe. While there is no consensus definition of the precautionary principle, one oft-mentioned statement comes from the 'wingspread' conference in Racine, Wisconsin (1998): 'When an activity raises threats of harm to human health or the environment, precautionary measures should be taken even if some cause and effect relationships are not fully established scientifically' (Appell, 2001, p. 18).

As articulated by New Jersey governor Christine Todd Whitman (later Director of the Environmental Protection Agency):

> policymakers need to take a precautionary approach to environmental protection . . . We must acknowledge that uncertainty is inherent in managing natural resources, recognize it is usually easier to prevent environmental damage than to repair it later, and shift the burden of proof away from those advocating protection toward those proposing an action that may be harmful.
>
> (Appell, 2001, p. 19)

Applying these ideas to forestry, we face the question about who should have the burden of proof that proposed actions (cutting of trees, etc.) will not harm the environment – the person proposing the action or the person who wishes to stop it? With the shift from the sustained-yield/economic approach to the ecological approach, we saw a comparable shift in the burden of proof. This is most readily seen in the application of the 'Survey and Manage' provisions of the Northwest Forest Plan in terms of the prescriptions for management of species about which little is known – especially fungi, lichens and molluscs. These species were not known to be threatened or endangered; rather we lacked knowledge about their habitat needs. The Northwest Forest Plan assumed that timber harvest might harm the habitat for these species, and put the burden of proof that it would not on those who wanted to take action. This reversed previous approaches, which would have put the burden of proof on those who wanted to prevent actions.

The precautionary principle is but one example of how world view can profoundly influence the approach taken to forest management.

The Rise of Social Forestry

Social forestry generally analyses forest resources from the standpoint of sustaining the well-being of people, communities and society. A central axiom of social forestry is that the use of forest resources should directly benefit human and community well-being. Key elements in these benefits include the distribution of forest benefits, community capacity to accommodate change, the social acceptability of decisions and a decision-making process based on participatory democracy (Committee of Scientists, 1999; Maguire, 1999; Davis *et al.*, 2001).

More fundamentally, social scientists, such as Lee and Field (2005), editors of a new book on communities and forests, question the scientific/technical approach of economic and ecological forestry. They note that forestry's foundations were laid during an era in which rationality and the application of science were presumed to be sufficient for directing political action and that:

> the messiness of politics, passions, and human emotion were to be replaced by the rule of reason. The focus on rationality has persisted. Even today, politics are legitimated by chartering scientific studies and policy and management decisions by developing 'science-based' plans. Communities, especially territorial communities, are the recipients of 'rational' decisions made by experts.
>
> (Lee and Field, 2005, p. 291)

Further, they point out that massive public investments have been made in research and development activities to construct rational and orderly procedures for making decisions, including linear programming models like FORPLAN and conservation planning like FEMAT. Yet it may be that rational schemes are based on an illusion – that rationality can and will bring order to human conduct:

> The modernist assumptions of its dominant institutions are a poor fit for a pluralistic culture in which emotion takes its place beside reason, nature is re-enchanted with spiritual meanings, and broad consensus collapses in the face of localism and particularistic values and beliefs.
>
> (Lee and Field, 2005, p. 305)

Finally they urge encouragement of community-centred initiatives in technical innovation and conflict management, arguing that community building promises to embed creativity and innovation in social contexts governed by a social/ecological conscience and that sustainability, if expressed as a commitment to the future welfare of others and the environment, grows naturally out of well-functioning communities (Lee and Field, 2005).

Not surprisingly, social forestry generally sees an important role for 'community mobilization'. London *et al.* (2005) argue that supporting, not forcing, proper forest management can be seen as this new form of community mobilization that rose to prominence in the early 1990s, with community forestry or stewardship groups not merely seeking increased use of the forest but also tending to promote alternative forest management systems that meet multiple objectives in a sustainable framework.

One stellar example of this effort (described in London *et al.*, 2005) is the 'Quincy Library Group', in which disparate factions, formerly in conflict with each other, began to focus on where they agreed, as they tired of the struggle. Their goal was to promote forest health, ecological integrity, adequate timber supply and local economic stability. They did this by zoning the land into management areas, creating a new spatial order in which forest product harvest and forest protection were to coexist and suggesting a variety of innovative silvicultural regimes – regimes that might be less efficient in producing timber but which served a number of ecological and social functions. As pointed out by London *et al.*, the Quincy Library Group map can best be understood as a representation of the collaborative social relationships and arrangements that made the management plan possible. The Quincy Library Group plan, though, met a number of obstacles, including Forest Service indifference and resistance and the opposition of various sectors of the environmental movement based on territorial politics. Finally, the group persuaded Congress to pass a law mandating its plan, firmly interjecting them into the Forest Service's chain of command and, as might be expected, eliciting a variety of responses from the agency (London *et al.*, 2005).

As Langston (2005) notes, foresters had sufficient power during much of the 20th century for them not to need to acknowledge viewpoints other than their own. However, a set of escalating conflicts, brought on by environmental litigation on the national forests of the Pacific Northwest, have changed the power equation and eventually forced groups to embrace a political process in which stakeholders coming from different perspectives jostle against each other, argue with each other and listen to each other. Langston further argues that, with a lack of knowledge of how ecological systems work, this process moves us to better solutions than any one group could find on their own, and that we need a democratic process that creates a structure for useful conflict – a method for negotiating conflict.

These powerful ideas fly in the face of the process of planning for the management for federal forests under sustained-yield, economic or ecological forestry. In the previous approaches, there was a clear demarcation between facts and emotions, between the roles of the experts and the public, between the way the world works and people's values. In the postmodern world of social forestry, these clear demarcations become blurred and various forms of community mobilization and social negotiation take centre stage.

The 2005 NFMA Regulations: Social Principles as the Driving Force

While the social sciences may have been largely on the sidelines or window dressing in past planning efforts on the national forests, they will be central to forest management in the future. The Forest Service has made collaboration the keystone of their future planning in the newly adopted 2005 regulations, recognizing that little can get done without community support (US Dept. of Agriculture, 2005).

In these regulations, and the associated directives, the Forest Service calls for collaboration throughout the planning process, including a collaborative effort to describe the desired future condition. Since these new regulations position the desired future condition as the focus for management, the call for a collaborative process to delineate it represents a major commitment of the Forest Service to social processes in decision making. While the agency does not define what is meant by collaboration in these regulations and it also does not stipulate performance standards that would allow for external evaluation of success in this endeavour, expectations have been raised about its engagement of collaborative groups. Also, the agency has reduced its emphasis on species viability, which dominated previous planning. More than ever before, the Forest Service has embraced the 'wisdom of crowds'.

Has Linking Science to Policy Led to Sustainable Forestry?

With this history as context, I summarize below ways in which the different scientific disciplines that have sequentially dominated forest management in the national forests have both contributed to sustainable forestry and detracted from it. While this summary is necessarily qualitative and selective, I believe it captures the influences of these sciences.

The contribution of sustained-yield forestry

The application of science to forest policy in the Pacific North-west and throughout the USA grew out of the Progressive Movement around the turn of the century, in which technical–scientific approaches to management were trumpeted as a way to moderate resource exploitation. Sustained-yield forestry arose out of that movement and became the basis of forest policy for the management of the national forests in the Pacific North-west for almost three-quarters of a century. There is no question that sustained-yield forestry made important scientific contributions to forest policy. It caused policymakers to address the finite nature of the current stock and the importance of investment in reforestation, forest protection and future growth to maintain future timber supplies. In addition, it provided a mechanism to moderate the rate of timber harvest to help ameliorate the 'cut and get out' mentality then so prevalent in America. Finally, it provided a long-term vision of the forest of the future – the regulated forest – that would resonate with policymakers.

It must be acknowledged, however, that sustained-yield forestry created a mindset that would result in the orthodoxy of timber primacy and lead to future problems. The single-minded emphasis on suppression of all agents that threatened wood production resulted in the build-up in stand densities in our intermountain west, which, in turn, threaten the sustainability of the forests there. The single-minded emphasis on liquidation of old-growth forests to provide wood for mills and to free the land for the growth of thrifty young stands led directly to the ecological crisis that resulted in the collapse of national forest timber harvest levels.

The contribution of economic forestry

The merging of economic principles with sustained-yield forestry resulted from the emphasis on benefit–cost analysis in the federal government, the maturing and power of the forest economic profession and the emergence of quantitative analysis tools, like linear programming, which enabled the application of these ideas to forestry. This new approach emphasized the need to recognize the limitation of resources, the need to make reasoned choices through evaluating marginal benefits and costs and the need for efficiency in management. It harnessed the new power of computers to help in this analysis, especially to sort through the thousands of forest management choices to find the 'optimal solution'.

While this approach (to which the author devoted a decade of his life) provided a useful framework to think about some forest management problems, especially those related to timber management, the economic approach, and the process within which it was embedded, had two fatal flaws. First, there was no clear recognition that our environmental laws called for a protective approach to forest species and ecosystems, one that recognized all the uncertainties involved. Economic forestry tended to focus on what was known (as discussed above), to take a benign view of potential impacts of actions on what was unknown and to focus on the calculus of benefit–cost analysis without realizing that absolute requirements for protection might exist no matter what the cost. Secondly, the approach was largely impenetrable to most Forest Service personnel, let alone the public. People recoiled from the idea that results from a 'black box', i.e. the linear programming model FORPLAN, would determine the future management of the national forests.

The contribution of ecological forestry

Ecological forestry was propelled into the forefront of national forest planning by successful challenges to forest plans based on the sustained-yield/economic approach. It stayed in the limelight because it offered an approach to provide 'scientifically credible conservation strategies' – the new standard for legal sufficiency in forest planning. This new approach combined the principles of conservation biology with the mechanics of viability analysis from wildlife science. It overcame the previous neglect of the contribution to biodiversity of old forests and the important conservation contribution that the national forests make to the broader landscape. Ecological forestry caused a sea change in management of the national forests.

With the ecological world view that came with the approach, however, it proved much easier to stop activities that might damage species and ecosystems than it did to start activities needed to conserve them or to produce other benefits. Thus, the estimated timber volumes associated with conservation strategies did not materialize. Also, activities to reverse the build-up in stand densities that occurred under sustained-yield forestry were hard to justify, and relatively little activity occurred.

The contribution of social forestry

Ever since the passage of the NEPA in 1970, the public has had a role in federal forest management. Usually, the federal agencies would listen to the public about key issues, develop a proposed plan (and some alternatives to it) and then ask for comments. Thus, the public was asked to critique a plan they did not develop. As our society has become less willing to turn decisions over to experts, the 'notice and comment' approach of NEPA to public involvement became less and less satisfactory; people have shifted from wishing to be consulted to wishing to help make the decision. Thus, the demand for a power-sharing, collaborative approach to many decisions in modern life, including federal forest management, has grown over the past 40 years.

Social scientists have developed a set of principles that can help federal managers mobilize community groups and there have been successful applications of these principles in a number of instances (Wondolleck and Yafee, 2002). Thus, the Forest Service has built the latest regulations on the scientific foundation of collaboration and engagement with the public as mechanisms for increased democratic processes in public decision making.

Even more than with past changes in the dominance of particular scientific disciplines, this recent shift to the emphasis on democratic processes in planning raises questions about how federal forest management will proceed and what role will remain for the technical expertise and analyses. Social forestry fundamentally challenges the 'expert' approach to planning that has been utilized for over 100 years, and the future role for ecological and economic experts remains murky and somewhat unknown. The past major planning effort in the 1980s, which did try to negotiate with the public to some degree, resulted in plans that were unrealistic. While that negotiation occurred through the political process more than through community groups, the outcome might be instructive here. In retrospect, the plans did not recognize realistic conservation strategies for at-risk species and they did not recognize budgetary limits on what might be done. Thus, the Forest Service settled on plans that were infeasible both ecologically and economically. Feasible plans were not acceptable to the various public factions and interest groups because there would be both winners and losers. Only plans in which everyone was a winner, or lost only a bare minimum, were acceptable. While we can debate how much the agency itself was responsible for these outcomes, they still describe the possible pitfalls of a negotiation process if ecological and economic considerations are lost.

On the other hand, a major possible contribution of social forestry is the community mobilization discussed by London *et al.* (2005), in which communities come together to support forest management and suggest creative strategies, rather than obstructing management. The Forest Service desperately needs a broad spectrum of support regarding the lands they administer; this approach may help that happen.

Certainly the contribution of the social sciences to sustainable forestry on our federal lands will be closely watched in the next decade. Can they avoid the error of past scientific efforts that impose their singular scientific world view on the agency to the neglect of others that contribute to sustainable forestry? Can they help usher in planning and management that allow perspectives from the different sciences to surface and be given a fair hearing?

Ecological and economic forestry sees sustainability as a destination. Social forestry views sustainability more as a journey. Whether the journey has some sense of destination beyond the agreement of the participants will help determine the success of this new adventure in the management of our federal forests.

References

Aerts, D., Apostel, L., De Moor, B., Hellemans, S., Maex, E., Van Belle, H. and Van Der Veken, J. (1994). *Worldviews: From Fragmentation to Integration.* VUB Press, Brussels.

Appell, D. (2001) The new uncertainty principle. *Scientific American*, January, 18–19.

Caldwell, L., Wilkinson, C. and Shannon, M. (1994) Making ecosystem policy: three decades of change. *Journal of Forestry* 92(4), 7–10.

Cleary, D. (1986) *Timber and the Forest Service*. University Press of Kansas, Lawrence, Kansas.

Committee of Scientists (1999) *Sustaining the People's Lands: Recommendations for Stewardship of the National Forests and Grasslands into the Next Century*. US Department of Agriculture, Washington, DC.

Dana, S. and Fairfax, S. (1980) *Forest and Range Policy*, 2nd edn. McGraw Hill, New York.

Davis, L. and Johnson, K.N. (1987) *Forest Management,* 3rd edn. McGraw Hill, New York.

Davis, L., Johnson, K.N., Bettinger, P. and Howard, T. (2001) *Forest Management,* 4th edn. McGraw Hill, New York.

Duerr, W. (1960) *Fundamentals of Forest Economics*. McGraw Hill, New York.

Duerr, W.A. and Duerr, J.B. (1975) The role of faith in forest resource management. In: Rumsey, M. and Duerr, W.A. (eds) *Social Sciences in Forestry: a Book of Readings*. W.B. Saunders, Philadelphia, Pennsylvania, pp. 30–41.

Fairfax, S. and Achterman, G. (1977) The Monongahela controversy and the political process. *Journal of Forestry* 75, 485–487.

FEMAT (1993) *Forest Ecosystem Management: An Ecological, Economic, and Social Assessment.* Report of the Forest Management Assessment Team (FEMAT), 1993-793-071. GPO, Washington, DC.

Heylighten, F. (2000) What is a worldview? In: Heylighten, F., Joslyn, C. and Turchin, V. (eds) *Principia Cybernetica Web.* Principia Cybernetica, Brussels. http://pespmc1.vub.ac.be/worldview, accessed on 15 March 2005.

Hirt, P. (1994) *Conspiracy of Optimism: Management of the National Forests since World War II.* University of Nebraska Press, Lincoln, Nebraska.

Hough, F. (1878) *Assessment on Forestry.* Division of Forestry, US Department of Agriculture, Washington, DC.

Johnson, K.N., Franklin, J.F., Thomas, J.W. and Gordon, J. (1991) *Alternatives for Management of Late-successional Forests of the Pacific Northwest.* A report to the Agricultural Committee and the Merchant Marine Committee of the US House of Representatives. College of Forestry, Oregon State University, Corvallis, Oregon.

Langston, N. (2005) Resource management as a democratic process: adaptive management on federal lands. In: Lee, R. and Field, D. (eds) *Communities and Forests: Where People Meet the Land*. Oregon State University Press, Corvallis, Oregon, pp. 52–76.

Lee, R, and Field, D. (2005) Community complexity: postmodern challenges to forest and natural resource management. In: Lee, R. and Field, D. (eds) *Communities and Forests: Where People Meet the Land*. Oregon State University Press, Corvallis, Oregon, pp. 291–303.

Lippke, B. and Bishop, J. (1999) The economic perspective. In: Hunter, M. (ed.) *Managing Biodiversity in Forested Ecosystems*. Cambridge University Press, Cambridge, UK, pp. 597–638.

London, P., Starrs, F. and Fortmann, L. (2005) Power plants and forest plans: two decades of mobilization in a mountain forest community. In: Lee, R. and Field, D. (eds) *Communities and Forests: Where People Meet the Land.* Oregon State University Press, Corvallis, Oregon, pp. 78–96.

McArdle, R. and Meyer, W. (1930) *The Yield of Douglas-fir in the Pacific Northwest.* USDA Forest Service, Technical Bulletin No. 201, Washington, DC.
Mills, T., Quigley, T. and Everest, F. (2001) Science-based natural resource management decisions: What are they? *Renewable Natural Resources Journal,* Summer, 10–15.
Maguire, L. 1999. Social perspectives. In: Hunter, M. (ed.) *Managing Biodiversity in Forested Ecosystems*. Cambridge University Press, Cambridge, UK, pp. 639–666.
Seymour, R. and Hunter, M. (1999) Principles of ecological forestry. In: Hunter, M. (ed.) *Managing Biodiversity in Forested Ecosystems*. Cambridge University Press, Cambridge, UK, pp. 22–64.
Shannon, M. (2000) Science advocacy is inevitable: deal with it. *Reflections: Newsletter of the Program for Ethics, Science and the Environment*, Dept. of Philosophy, Oregon State University, Corvallis, Oregon. Special Issue 4 (April), 8–9.
Teeguarden, D. (1987) Benefit–cost analysis in national forest system planning: policy, uses, and limitations. *Environmental Law* 17(3), 393–427.
Thomas, J.W., Forsman, E.D., Lint, J.B., Meslow, E.C., Noon, B.R. and Verner, J.(1990) *A Conservation Strategy for the Northern Spotted Owl: a Report of the Interagency Scientific Committee to Address the Conservation of the Northern Spotted Owl*. USDA Forestry Service, USDI Bureau of Land Management, Fish and Wildlife Service, National Park Service, Portland, Oregon.
US Dept. of Agriculture (1982) National Forest System land and resource management planning: final rule (36 CFR 219). *Federal Register* 47(190), 43037–43052.
US Dept. of Agriculture (2000) National Forest System land and resource management planning: final rule (36 CFR 219). *Federal Register* 65(218), 67568–67581.
US Dept. of Agriculture (2005) National Forest System land and resource management planning: final rule (36 CFR 219). *Federal Register* 70(3); 1055–1061.
Wondolleck, J. and Yafee, S. (2002) *Making Collaboration Work: Lessons from Innovation in Natural Resource Management*. Island Press, Covelo, California.

3 Participation as a New Mode of Governance? Scientists and Policymakers Linked in a Double Spiral

IRINA KOUPLEVATSKAYA*

ENGREF, Laboratory of Forest Policy, Nancy, France.

Abstract

The international dialogue on forests has changed how scientists are involved in policymaking as well as the nature of their relationship to decision makers. Policy scientists and biologists brought new concepts and types of knowledge into the international lexicon, but soon found their statements utilized to justify reorientations of national and regional policies. This instrumentalization of scientists is especially clear in transitional countries, which are under strong pressure by the international community to introduce rapid and important changes in their policy systems in order to conform to international standards. The policy formulation process generally promotes policy changes. Seldom is it a linear process with a beginning and an end; rather, policymaking is iterative, allowing for a redefinition of the interests and positions of all actors involved in the process. Scientists are often viewed as a source of objective knowledge and judgement in the policy process, whose participation can help legitimize policy changes. However, just like all other policy actors, scientists are not neutral purveyors of objective knowledge, but value-laden people with their own world views. Thus, scientists like other policy actors adapt throughout the policy process as their own knowledge, their perspectives and even their world views change as a result of policy discussions. Thus, policymakers rely on scientists to represent 'reality' through scientific theories and concepts, but then, as new representations emerge through policy discussions, scientists must adapt their theories and viewpoints to the shared understanding created through policy dialogue. Often neglected in science–policy

* Ms. Kouplevatskaya is the former Deputy Programme Leader of the Kyrgyz–Swiss Forestry Support Programme, intercooperation Kyrgyzstan. This chapter was written in the framework of a doctoral thesis in forest science (forest policy) at the French Institute of Forestry, Agricultural and Environmental Engineering (ENGREF), Nancy, France, under the scientific direction of Prof. Dr Gerard Buttoud, who was also the consultative expert for the Kyrgyz–Swiss Forestry Support Programme in Kyrgyzstan.

research is a focus on how this mutual adaptation may promote power redistribution and affect the image of scientists.

This chapter is based on the 7-year participatory process of policy reform in Kyrgyzstan, leading to the adoption of a national forest programme (NFP) based on concepts derived from the international dialogue on forests. The policy process theory of the double spiral is used to explain the evolving link between scientists and policymakers, both of whom are compelled to redefine their mutual relations in a process of reinterpretation and renegotiation of goals and means. Indeed, the involvement of scientists resulted not only in an increase in rationalist approaches to policy analysis and policymaking, but surprisingly also in a communicative approach conducive to international principles of sustainable forest management. As a preliminary comparison, basic information collected from experience with participatory processes in various European countries tends to reveal the same iterative sequence formed with an outward spiral of expanding understanding followed by an inward spiral of focus and clarity evidenced in many situations.

Keywords: Forest policy, modes of governance, participatory process, national forest programmes, power redistribution, Kyrgyzstan.

What Might Policymakers and Scientists Expect from Each Other in an NFP Process?

The international debate on the concept of sustainable development focused global interest on the sustainable use of natural resources. Debates on the social values of forest products and services are often at the centre of major environmental controversies thus defining special tasks for forest policy scientists, distinguishing them from the tasks of the other scientists related to natural resources (biologists, ecologists, etc.).

Sustainable forest management focuses on the multiple functions of forests, the multiple beneficiaries of forest products and services and the necessity of integrating marketable and non-marketable goods and services. National forest programmes (NFPs) are expected to face the challenge of promoting a new mode of resource governance that will *inter alia* work with an expanded, although imprecise, definition of sustainable forest management (SFM); resolve conflicts of interest among beneficiaries; address various kinds of coordination problems of providing the multiple functions and benefits of forests; handle complexities of inter-sectoral and multi-level coordination among policy actors; and promote effective mechanisms for stakeholder participation. In this framework, forest policy research contributes to the policy process by generating theoretically informed propositions and hypotheses about the social, political and institutional preconditions for sustainable forest management.

Within the policy process, policy scientists are looked to for clarification of ambiguous concepts, like sustainable forest management and national forest programmes, and to generate methodologies for public decision-making procedures. In such a context, forest policy scientists form a link between the public, managers and politicians. The literature on theories of decision making schematically reveals three main groups of views related to the links between scientists and decision makers: the incrementalists, the rationalists and the cyberneticists.

1. The incremental decision-making school is associated with Charles Lindblom (1959) and his incremental 'muddling through' paradigm and stresses the role of scientific analysis as sufficient to solve the political problems faced by public administrators. Its premises are the following:

(a) Due to the fact that decision makers are cognitively constrained in time and other resources, including information, they can never achieve the ideal of comprehensive analysis of all ultimate goals and all available means, rather they typically 'muddle through' by focusing on proximate goals and known means. They usually come up with a solution – a means of achieving a desired end – that they believe 'suffices', even if it is not optimal, and then move on to the next goal. Thus, decision makers are cognitively constrained from pursuing complexes of long-term goals, and thus analysis is never sufficient to achieve the multiple goals encompassed by sustainability.

(b) Reality has a pluralist nature due to differing and conflicting social values, objectives and visions held by actors. It is difficult to ascertain the majority's preference or to find a preferable consensus; therefore public debate is rarely sufficient to solve the problem of cognitive limits of analysis and reasoning.

(c) Public policy is accomplished through decentralized bargaining in a democratic political economy. Incremental decision making holds a pluralistic view of a society as composed of competing interest groups who are lobbying the government for certain decisions. Decisions are constructed by a series of consultations largely based on people's actual experiences. Large decisions are distributed among a large number of independent actors, each pursuing their own interest (Lindblom, 1959; Friedman, 1987).

In such a framework, analysts are considered the only ones capable of making a comprehensive analysis with a general and objective view. However, for analysts to indeed be objective, their analysis must be founded upon scientific methodologies. The ultimate 'scientific analysis' is done by scientists themselves, and thus scientists are the source of policy analysis for decision making. However, the scientists are not viewed as policy actors, but rather as sources of information and analysis that are 'untainted' by politics.

2. The rationalist school of management, criticizing the 'muddling through' view of decision making, attributes great importance to the power and rationality of the decision maker and the predictability of human behaviour based on assumptions of a 'rational actor'. In this school, a rational decision maker bases a decision upon analysis and believes that solutions based upon what a 'rational actor' should do accurately predict policy outcomes. As commentators note, this assumption of a 'rational actor' ignores the actual relationships between ideology, values, events, goals and means (Gunton, 1984). The decision-making process is viewed as a logical rationalist chain: identification of a problem, development of goals, assessment of all possible solutions and the choice of a solution on the basis of the desired results in achieving the goal (Hudson, 1979). Scientific research in this case focuses on developing general theories of behaviour of natural, social and political systems. The assumption of scientific policy analysis is that the world, including human behaviour, is predictable and stable over time.

3. A third group of theoreticians, drawing from cybernetics, focuses on dynamics of social change affecting a decision-making process. They reject the absolutism of the two former traditional approaches: (i) that social changes are basically directed by the elite (voluntarism, as promoted by the rational school of management); and (ii) that social changes are brought about by society as a whole (pluralism, incremental school of decision-making process). Nevertheless, the proponents of cybernetics aim at combining the advantages of incremental and rationalistic planning in three levels of decision making (Etzioni, 1967):

(a) Fundamental political decisions are to be taken at the highest level in order to establish choices aimed at long-term goals (based on scanning of internal and external factors that relate to the problem and proposed solutions).

(b) Opportunistic, incremental decisions will be taken within the framework of the fundamental goals (short-term and middle-term policies).

(c) Periodic reviews of the incremental decisions and of the fundamental goals are necessary, based on the criterion of the achieved progress (review of strategy).

Thus the role of the scientists for the definition of long- and short-term goals (ends), analysis of internal and external factors (means) and evaluation of the progress (scientific judgement) is conceived as a part of the process of strategic political decision making.

The cybernetic framework is an implicit theoretical reference for many forest policy reform processes, and is the formal basis in the concept of a 'mixed model', developed for combining the involvement of the stakeholders and administration in a forest policy reform process in societies in transition (Buttoud and Samyn, 1999; Buttoud and Yunusova, 2002, 2003). The main feature of the mixed model is to involve all the stakeholders at each step of a rationalist sequence for defining and implementing new decisions, thus combining communicative and technocratic aspects and crossing top-down and bottom-up approaches to decision making. It creates a discourse between solutions and decisions, which may lead to a continuous negotiation. In this framework, the decision makers are confronted by the evolving reality and have to adapt to it.

Such permanent confrontation combined with mutual learning of all the actors involved in the process, followed by their adaptation, both to each other and to the changed reality, and a redefinition of initial positions and strategies, is an iterative process that progressively develops along spirals of learning and focus (Kouplevatskaya-Yunusova and Buttoud, 2005).

Following these theoretical implications, policy scientists have a variety of tasks: to develop theoretical and methodological advice for the decision makers; to bring to the process not only knowledge but also neutral and objective expertise; and to contribute their scientific judgement in addressing complexity, ambiguity and uncertainty. The role of policy scientists based upon the cybernetic theory and the 'mixed model' framework is elaborated by the example of the forest policy reform process in Kyrgyzstan, a former Soviet Republic.

The Case of Kyrgyzstan

Kyrgyzstan, a small mountainous country of Central Asia, is one of the few countries with a full process of formulation, implementation, evaluation and adaptation of a national forest policy. Since independence in 1991, the new economic and political situation in Kyrgyzstan has led to important changes in the general governance of the country. At first glance, the forest, which covers about 3–4% of the total area of the country's surface, did not present a huge challenge for the national policy decision makers. Nevertheless, due to the declared democratic transformation, the managing and decision-making practices for forests needed to be changed. In addition, at the same time, the transition to a market economy and collapse of the economic and political linkages with Russia resulted in a considerable reduction of the state financing of the forestry sector, which, together with the aggravated conditions of the people living in the direct proximity of the forests, posed societal pressure for changes in the forest management.

As a result of the reorganization of international development institutions at the world level in 1994, the Swiss Development Cooperation became one of the main donors to the country, with a support package including assistance to the whole forestry sector. The international dialogue on forests emphasized the need for analysing forest development in connection with ecological and socio-economic aspects; therefore these two directions were initially privileged by the donor. Scientists–ecologists were asked to evaluate the risk of overuse of the scarce national forests, drawing upon a global assessment of forests through new techniques in the application of satellite imagery. Very soon it became clear that the process of technical data collection would advance gradually and require a long time period. This was not adapted to the timing pressure on the decisions needed within the process of a rapid transition. At the same time, the results would provide the policymakers only with the data on the rate of deforestation, with no operational conclusions in terms of decision making in the new conditions. Since there was a strong requirement for an immediate change in the policy decisions, forest policy scientists were invited to assist in the introduction of a change in the process of decision making in the forestry sector, on the basis of a rigorous neutral analysis.

The Swiss Development Cooperation was strongly promoting the involvement of scientific knowledge for the social and economic aspects of the forest policy reform. From the very beginning of the project, there was a commonly shared understanding that the social factor might become prominent in the orientation of the public decisions in the sector. At the same time, most of the technical experts involved in the support/cooperation activities were not yet familiar with the conditions of post-Soviet Central Asia and thus were not equipped with an effective toolbox for forestry development in the transitional conditions. Expertise of policy scientists, in this framework, was fostered to provide ideas, knowledge and experience on how to initiate, implement, evaluate and adapt a forest policy process. In the course of the whole forest policy reform process, which began in Kyrgyzstan in 1997 and is still ongoing (Yunusova *et al.*, 2003), the position and role of the scientists has continuously changed. Through collaborative learning procedures both scientists and policymakers, and, at a lower level, other involved stakeholders, were brought to a permanent mutual adaptation.

Adaptation along spirals

As the state was still very strong in Kyrgyzstan and the social aspects, previously under-considered, were gaining more and more importance, the mixed model for the forest policy process was chosen as the most adapted framework (Buttoud, 1999b). The application of the mixed model permitted the combination of the rationalist sequence of decision making (including diagnosis of the present situation, listing of structured objectives and sets of possible means, selection of priorities and definition of strategic lines, procedures for regular follow-up and evaluation) with communicative iterativity (Buttoud and Yunusova, 2003), which focuses on building a conclusion through a negotiation at each step of the sequence, with feedback loops to the previous stages. All the involved parties brought to the process their own understanding of reality and their priorities and thus contributed to the definition of acceptable expected results. Thus, the forestry administration developed a comprehensive view of the situation to which it had to adapt. Participation in a decision-making process was a new experience for all of the actors and stakeholders involved in the process; learning – through gathering new information and new experiences – led to changes in thinking and doing (Weick, 1990, as cited by Shannon, 2002) and was a necessary step in the process, prior to any adaptation and decision taking. The emergence of mutual adaptation created a permanent dialectic between the phases of learning and the phases of appropriation of the results of the process (including the form of knowledge and information).

Due to the iterativity of the process, these phases traced a spiral because the feedback loops, according to the 'muddling through' paradigm, described 'a process of continuous adaptation through small steps instead of fundamental changes; from a solution of one problem to the definition of another one' (Lindblom, 1959). The spiral may have an outward direction and an inward one (Barstad, 2002). The outward spiral (the phase of learning) describes a situation where the various participants of the process redefine their own positions with no clear picture yet of the desired outcome. When all the actors of the process are open to any solution, the outward direction of the spiral is fed by continuous collaborative learning, knowledge generation and adaptation of solutions.

In contrast, an inward spiral (phase of appropriation) is a stepwise movement to a predefined solution, which could presumably be reached in a step-by-step progress. In an inward spiral, the decision-making process is taken under control by an actor who is faster to formulate his own interest in the process and push towards the newly defined goals, which usually leads to a solution conforming to the interest of a strong actor. The outward movement usually continues till the moment when one of the actors of the process forms a clear vision of a solution most profitable for him. This will be a critical point for the change to an inward spiral aimed at reaching this solution and controlled by the actor. Thus, a double spiral is formed, where the inward part may have a shorter circumvolution period (Kouplevatskaya-Yunusova, 2004).

The empirical experience shows that such a solution is practically never reached due to various subjective factors (poor appreciation of the reality, importance of stakes and bad governance) and objective factors (linked to the continuous changes in the society and permanent adaptation). Mutual learning in the process may also change both previously commonly defined goals and unspoken hidden agendas, thus leading to new solutions. The end of each double spiral necessarily gives the initial point for another double spiral. Many decision-making processes can be described with the application of this theoretical framework of a double spiral, in regard to the changes in the views and roles of the involved actors. This is the case in the links between decision makers and policy scientists jointly involved in the previously mentioned participatory process of the forest policy reform in Kyrgyzstan, where a double spiral may be illustrated by the changes in the roles and perceptions of the forest policy administration (policymakers) and scientists.

A pair of double spirals in forest policy formulation

The first stage in the process of the forest policy reform in Kyrgyzstan corresponds to the period from 1997 to 2001 and includes the preparation of the following basic documents: *Analysis of the Current Situation* (1998), *National Concept for Forestry Development* (1999), *Forest Code* (1999), *5-year Action Plan – National Programme LES* (2001) (Kouplevatskaya-Yunusova and Buttoud, 2005).

At the beginning of the process, the issue of forestry development in view of the transition to a market economy was ambiguous. The way in which economic and social components of sustainability could be integrated into forest management was a major issue, while the meaning of the concept itself was not clear for either the local actors, including the government, or the donor. Old top-down centralized decision-making procedures needed to be reconsidered in a new reality, when the state was administratively still strong but financially very weak; society was evolving rapidly; land-use issues and access to the natural resources were aggravated in the context of increasing poverty; and the data related to the forest resources were inconsistent.

In such conditions, the donor invited scientists to support the definition of a new framework for the forest policy. The phrase 'forest policy' initially was not accepted by the Kyrgyz decision makers, with the reasoning that it was much too close to 'politics' (in Russian it is the same word for both meanings), and foresters would prefer to stay away from politics. For the Kyrgyz government high officials, educated during the USSR time and traditions, science had great significance, but of a rather symbolic nature. In the traditional top-down system of decision making, 'the science' was used as a systematic reference for grounding important political decisions. However, science was considered to be based upon fundamentally ambiguous theoretical approaches with content understandable only by the scientists.

Urged by the importance of changes to be introduced and by the requirements from the donor, the Kyrgyz officials declared readiness for initiating a forestry sector reform process, but, following the habitual traditions, under the stipulation that there would be a scientific background. Thus, policy scientists joined the process of forest policy reform in Kyrgyzstan with clear terms of reference from the donor requiring results in terms of advising policy decisions for public governance, and a suspicious nod of politeness from the forestry administration, willing to have a neutral scientific justification. In the course of the process, the forestry administration (formerly a purely technical management) became more and more political in the broad sense of the word. Simultaneously, the role of the scientists and their attitude to the process were also changing. This development could be studied in parallel in the two double spirals presented in Figs 3.1 and 3.2.

Decision makers in forest policy formulation

The initial point of the outward spiral for the forestry administration (Fig. 3.1) was determined by the inherited system of centralized top-down planning in the forestry sector, oriented towards the achievement of quantitative results. During Soviet socialism, the role of state planning was fundamental; it determined the process, which was, accordingly, a deliberately political process (Davis and Scase, 1985). This practice was mechanically continued after the break with the Soviet Union, aggravating the situation by the fact that the state was still planning and controlling the implementation of activities that were justified neither by the financing from the state budget nor by objectively defined priorities and local potentials. The concept of sustainable forest management introduced by the donors through the advice of the scientists was incompatible with the situation in the Kyrgyz forestry sector in the mid-1990s. The requirements of sustainable management related to the conservation of biodiversity, productivity and regeneration capacity of the forests; to the relevant economic functions and development of all the values of the forests, including those with no direct market benefit; and to social sustainability linked with the role of the forests for poverty reduction

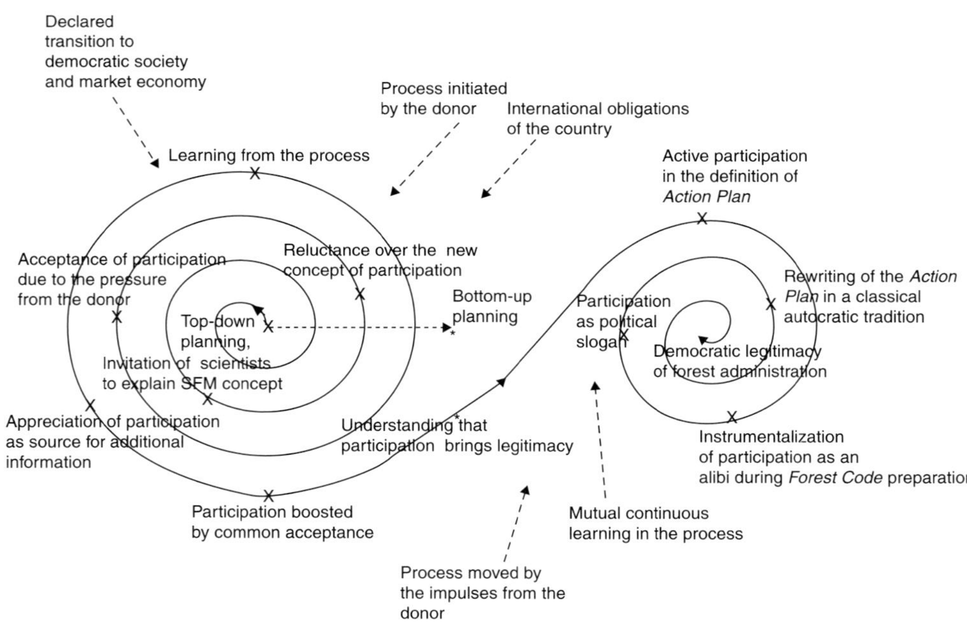

Fig. 3.1. The change in the perceptions and roles of decision makers during the forest policy formulation stage.

and integration of interest groups into the forest management and related decision-making processes: all indicated an urgency of reform of the forest policy in Kyrgyzstan.

A participatory approach for the forest policy reform, as one of the basic features of sustainable development and consistent with the democratic transformations of the society, was proposed by the forest policy scientists, but initially rejected by the Kyrgyz forestry administration. The main expressed reasons for an aversion to participation were linked with:

1. The high cost of the process: 'This money could be much better used for the creation of artificial forest plantations. This is the real indicator of forestry development in the country.'
2. A belief that other stakeholders and actors may not have sufficient special knowledge required for political decisions, while the forestry staff along with the hierarchy (foresters and forest rangers) would be interested in the decrease of the plans and would not give reliable information. 'If we let them plan, they will find many justified reasons for doing nothing.'
3. A fear that integration of the other ministries into the process (environment, finance, agriculture) would undermine the sectoral interests of the forest sector (risk of losing political power through sharing information). 'We do not interfere in their business: why open the gates for them to interfere in our place?'
4. The existing knowledge and experience at the level of the headquarters did not require any additional input from outside.

The donors insisted on the need for the broad participation of various actors and stakeholders and started the process with the facilitation of the policy scientists. The results received in the course of analysis of the present situation of the Kyrgyz forestry sector (during the year 1997) dissipated the fears of the forestry administration:

1. The donors confirmed that the money allocated for the participatory reform process would never be authorized to be used for plantations.
2. Other stakeholders, namely local authorities and village councils, gave no importance to the process; the local population and other forest users, including still a few private entrepreneurs, were occasionally represented by separate individuals and did not have a big say in the discussions; while foresters involved in the process did not bring too much criticism of the existing system (which was an unuttered fear), but, on the contrary, brought in some practical information.
3. Other ministries have gladly accepted the invitation to participate in the national working group on forest policy reform, with one symbolic meeting before launching the analysis, but did not attribute a big importance afterwards to the process itself.
4. The analysis of the situation was not considered to be a political decision and did not present a big challenge.

Consequently, the capacity of the participatory method to bring additional information has been appreciated by the forestry administration, which began to be more and more involved in the process and tried to broaden the number of participants, mainly from inside the forest service itself. In addition to that, there was a possibility for the forestry administration to have permanent contacts with all the hierarchical chains of the forest service, which permitted better control over the situation and organization of parallel meetings for discussing technical issues, after the workshops held for participatory forest policy reform. The presentation of the results of the *Analysis of the Current Situation* in the Kyrgyz forestry sector (1998), which was carefully prepared together with the forest policy scientists, brought significant public and political success for the Kyrgyz forestry administration. The President of the Republic participated in the presentation of the analysis and stressed the importance of both the forests – as 'the roof of the nation, the head for everything' – and the democratic initiatives of the forest service.

The introduction of participation in the forest policy decision-making process was conceived as a type of experiment, because in Kyrgyzstan, and specifically in forestry, it was a previously unknown phenomenon and nobody could precisely tell how it would evolve. The goal of any experiment is to learn something. In the case of the Kyrgyz process, the first lesson was learned by the forestry administration – the involvement of various actors could bring additional or missing information. Later on in the course of this incremental process, the scope of such information and, consequently, the knowledge was continuously increasing, and gradually transformed into adaptive management. It could be said that the Kyrgyz forestry administration was learning to manage by managing to learn (Bormann *et al.*, 1993).

Discussions with people brought optimization not only to the decisions but also of the management of the process. Appreciation from the top-level administrators gave legitimacy to the process and created an understanding that participatory policy formulation might help to improve the image and political status of the forest service. This knowledge led to the redefinition of objectives and the outward spiral acquired a tendency for changing inwardly. As the state forestry administration declared itself in the Kyrgyz political environment as a democratic reformer, for them, keeping this political status attained greater and more practical importance than sustainable forest management, which has remained an abstract notion, required by the donors for the continuation of financial support. That is why participation was still promoted as a process, while political decisions (*Forest Code, National Programme LES* (*5-year Action Plan*)) were approved with no regard to the results of the participatory process (Kouplevatskaya-Yunusova 2004, 2005; Kouplevatskaya-Yunusova and Buttoud, 2005). By the end of the described period of the forest policy formulation, participation became a political slogan and an alibi for the Kyrgyz state forestry administration – an instrument that guaranteed legitimacy of decisions facing the government, because they were democratically taken; obliged the executors for implementation, as they have been associated in the decisions; and gave a good image to the forestry service internationally vis-à-vis other potential donors, as a democratic administration, longing for sustainability.

Policy scientists during forest policy formulation

As argued by Hunt (1990), the major purpose of science is to develop laws and theories to explain, predict, understand and control phenomena. In the as yet unknown donor conditions of Kyrgyzstan, this was exactly what was needed. So the donor, together with the forestry administration, invited scientists to define the scope and needs of support for the Kyrgyz forestry sector. For the invited scientists the situation was also new, but they were equipped with theories and ready for an experiment.

A scientific fact-finding analysis showed that the growing poverty of the local population caused by the transition conditions in the country signalled the need to switch from ecological/conservation priorities, as previously propagated, towards sustainable forest management. Therefore, policy scientists were asked to define a conceptual framework and logic for sustainable forest management possibilities in Kyrgyzstan, to predict how it might develop and to provide monitoring over the process, with periodic insertion of guiding ideas when (or if) needed. The focus was on social and economic aspects of forest management as a requirement for sustainability.

The scarcity of the forest resource and its high protective importance due to mountainous conditions defined the multiplicity of (often opposed) interests in relation to forest management. The importance of the expression and representation of all the interests and views of the stakeholders was considered as the main demand of the new forest policy. Consequent to the democratic processes that were being developed in the country, the scientists have introduced a concept of participation as a basis for forest policy decision making. The top-down decision-making power of the state was still very strong in the country, while democratic processes, including capacity for public deliberation, are not rooted in the society yet. There was neither a tradition of nor a clearly expressed need for public deliberation, especially in a specific field like forest policy. In such circumstances pure bottom-up planning would be neither efficient nor sufficient; therefore the policy scientists proposed an adapted methodology, which combined familiar top-down decision-making processes with bottom-up participatory procedures.

The proposed framework of a mixed model (Buttoud and Samyn, 1999; Buttoud and Yunusova, 2002, 2003) allowed the forestry administration to retain its logical rationalist sequence of decision making, including the identification and classification of the principles and objectives, and, at the same time, combined this approach with the communication and negotiation of means with the other involved stakeholders.

The 'common interests approach' in a deliberative democracy framework combined with the constructive confrontation model (Buttoud, 1999a,b) formed the core of the forest policy reform process. For the policy scientists, the introduction of participation to a formerly centralized decision-making process was an interesting experiment. In conducting this experiment, they were seeking to design better policies but also to devise a better experiment. One of the 'hidden goals' of the experiment was to learn how decision-making theories might work in a country with an emerging democracy. For the policy scientists, the reactions of both decision makers and the other participants of the process were food for thought. The knowledge received from the process permitted the scientists to adapt themselves to the requirements from the decision makers and also to adapt the proposed methods to the situation.

After the success with the *Analysis of the Current Situation*, which was led by a mixed team of Kyrgyz and European forest policy and economists, the policy scientists earned the respect and confidence of the forestry administration and were inspired to develop participation and engage in the process. Such appreciation from the side of the forestry administration gave legitimization to the scientists and their proposals. The donor, mainly preoccupied with the technical expertise, left the forest policy process in the hands of the policy scientists, who continued to support the Kyrgyz forest policy formulation through the design and introduction of the procedures for the definition of the *National Concept for Forestry Development* with an adapted methodology. This exercise was followed and actively promoted by the forestry administration.

Forestry personnel and other stakeholders who joined the process adapted to the participatory procedures and grew to feel much more free to engage in the discussions. Indeed, they started to provide not only knowledge of the actual situation but also constructive and critical ideas. Even if the process was still under the very strong guidance of the policy scientists, there was a general feeling that the methodology proposed for participatory forest policy formulation was appropriate for the local conditions and quite efficient in application.

The first deviation came during the preparation of the draft of the new *Forest Code*. 'In the actual political conditions (in the view of forthcoming parliamentary elections) an urgent elaboration will simplify the approval of the new code. Besides a *Forest Code* is a purely technical legal exercise; therefore lengthy participatory procedures are not appropriate.' With these explanations, the forestry administration excluded the policy scientists from the process. Nevertheless, 'participation' was used as a 'password' during the lobbying for the draft *Forest Code* in the parliament, which gave clear notice of the deliberate instrumentalization of participation by the forestry administration. This change of appreciation for participation by the forestry administration and the use of participation as an alibi created uncertainty and puzzled both the donor and the scientists (Fig. 3.2).

In spite of this reaction, the participatory forest policy formulation process continued as initially planned and involved more and more participants. Contrary to the rather abstract nature of discussions for the *National Concept for Forestry Development*, the *5-year Action Plan* (*LES Programme*) included technical and practical aspects, which were much more familiar to the participants and which considerably facilitated their participation. After the 'hitch' with the *Forest Code*, the forestry administration regained their enthusiasm for promoting participation. 'The foresters and rangers will be responsible for the implementation of the plans; therefore it is up to them to define the plans.' Consistent with 'bottom-up' planning principles, the forest management units (*leshozes*) proposed a methodology for drafting their own 5-year plans based on local potentials and disadvantages. This exercise failed. There was neither sufficient experience in planning nor the professional knowledge for implementing it.

The forestry administration felt vindicated that top-down planning was more efficient, and the policy scientists learned that top-down willingness to engage in participatory processes must be complemented by bottom-up preparedness for it. The scientists were left alone to finalize the results of the participation and transform them into a draft *Programme LES* (*5-year Action Plan*). They needed

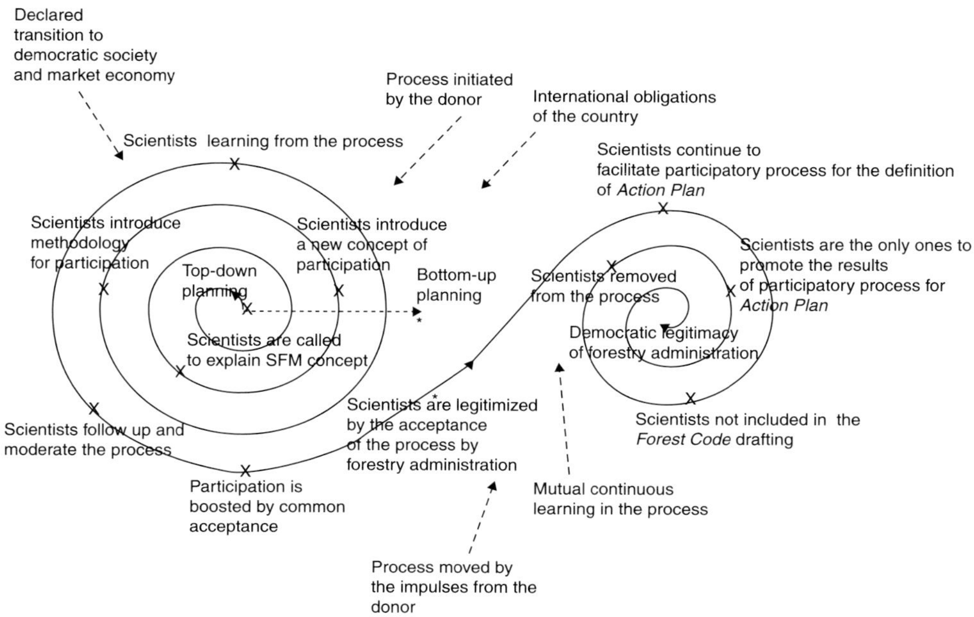

Fig. 3.2. The change in the perceptions and roles of policy scientists during the forest policy formulation stage.

to satisfy terms of reference and deadlines from the donor. The support for the forest policy process in Kyrgyzstan was suspended; the approved *Action Plan* did not take into consideration results of the participatory process (Kouplevatskaya-Yunusova, 2005; Kouplevatskaya-Yunusova and Buttoud, 2005) and the scientists were removed from the process by the donor.

Some outcomes

The outward movement of the spiral at the beginning of the process was determined mainly by the external factors: the process was initiated and 'pushed' by the donors; internal conditions of the country and international commitments of the state obliged the forestry administration to follow the requirements of the donors. The situation was open. The concept of participation was introduced and tested as a basis for forest policy definition, with no predefined agenda as to how the process should proceed, but with the establishment of democratic procedures for policymaking aimed at sustainable forest management, which was perceived as a potential expected result (ideal future).

During the course of the process, the forestry administration changed from aversion and negation of participation to its appreciation. Parallel to that, the attitudes regarding the role of the scientists and their proposals for the new way of forest policymaking were also changing. At the beginning of the process, the goal, defined in common by the donors, decision makers (forestry administration) and scientists (possibly with different degrees of awareness and consciousness), was to reach sustainable forest management through democratic planning.

But, in the course of the process, each of the three parties developed its own understating and priorities, which were not expressed directly but nevertheless significantly influenced the course of the process.

The head of the forestry service at the time was not a forestry specialist but an experienced politician, who understood that sticking to the old ways of management in the new conditions would weaken the administration. For a politician, the benefits from public involvement were easier to comprehend and appreciate. Hence the idea of forest policy decision making based on participation was promoted, incorporated into the process, declared as a principle for forest policy and, in the end, instrumentalized as an alibi.

The forestry administration, as a political actor, was the first to realize the possibility for instrumentalizing participation for its own benefit and changed the movement of the spiral inwards. The scientists did not immediately realize the change in the objectives and were still following the process of participatory policy formulation because of their obligations vis-à-vis the donor. Being bounded by the terms of reference agreed with the donor, at one moment the scientists were the sole actors pushing for the achievement of the commonly defined expected result. This was not effective because the other two actors (the donor and the forestry administration) unilaterally redefined their respective objectives. The process could not continue in its initially conceived state.

A pair of double spirals in forest policy adaptation

The period of forest policy adaptation covers the time between 2001 and 2004. After the approval (2001) of the *Programme 'LES'* (*5-year Action Plan*) in the form of a traditional top-down prescription of the number of hectares of forest plantations to be reached, no further steps were made at the national level towards forest policy reform. The status of the forestry administration was changed (it was transferred as a department to the Ministry of Environment and Emergency Situations), and the forest management units returned to reporting on the hectares of plantations. The donor continued support at the technical level only.

The policy scientists resisted the instrumentalization of participation and the break in the process. Science, above all else, is a critical and analytical activity and a scientist is pre-eminently a person who requires evidence before he or she delivers an opinion and, when it comes to evidence, is hard to please (Medawar, 1990). The situation in Kyrgyzstan provided evidence for two outcomes: (i) learning from the process by the forestry administration resulted in instrumentalization of participation and power redistribution; and (ii) the donor's decision to suspend the support interrupted the logic of the process and implementation of commonly elaborated decisions. Scientific recommendations were not implemented; therefore the policy scientists could not be responsible for the results. In addition, they did not like to be used as an alibi for failure. The policy scientists started to criticize both the forestry administration and the donor.

For the donor their international image was important (and it was at risk) not only because of the criticism from the scientists, but also because the forest policy formulation process in Kyrgyzstan started to attract the interest of the international community as the first experience of participatory policy formulation in an ex-Soviet republic. Clearly, the process needed to continue. All this resulted in changes in the perception of and roles in the participatory forest policy reform process by the decision makers and the scientists during the period of forest policy reformulation (Figs 3.3 and 3.4).

Decision makers in policy adaptation (Fig. 3.3)

An evaluation of the 5 years of forest policy implementation was required by the *National Concept for Forestry Development* (1999) and it provided a good opportunity for the forestry administration and the donor to recommence collaboration within the forest policy reform process. Following the same logic as at the policy definition stage; the forestry administration insisted on the methodological evaluation of the process from the same policy scientists, based on the fact that they had experienced success elsewhere with the proposed methodology. Engaging the same policy scientists strengthened the whole forest policy formulation process, partly because there was already a habit of working with these particular policy scientists. After all, why change the winning team?

The evaluation was to follow the same rules and approaches as the forest policy formulation process, meaning a reintroduction of participatory procedures. However, as a result of the learning during the policy formulation stage and the experience of implementation, the roles of the actors changed during the evaluation process. Both the donor and the policy scientists began with critiquing the appropriation of the results of the process by the forestry administration, since the latter was responsible for the results of the new policy. The forestry administration in turn wanted to regain its lost position and so supported the reanimation of participation in forest policy reform and assumed the leadership in the

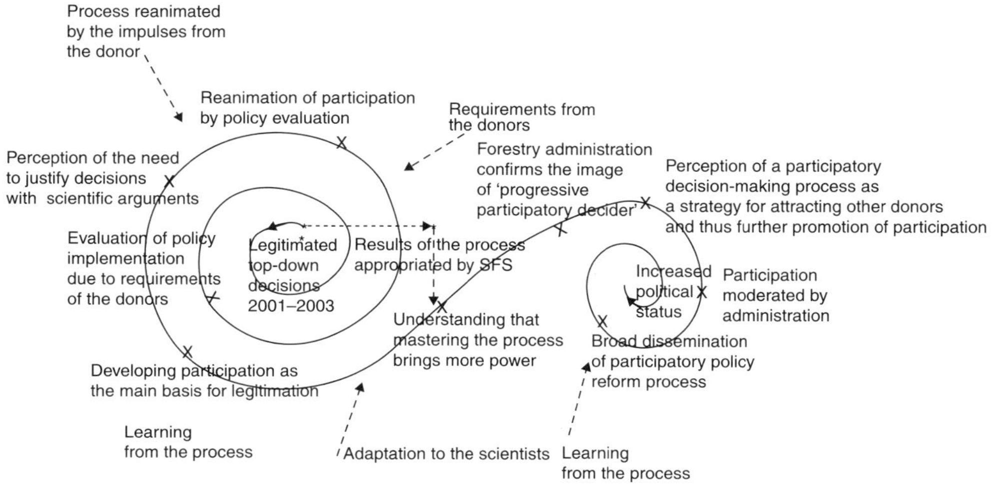

Fig. 3.3. The change in the perceptions and roles of decision makers during the forest policy evaluation stage.

process. A working group with representatives of the forestry administration, the donor and the policy scientists collected information through workshops and interviews, facilitated by the forestry administration, with no direct intervention from the policy scientists. Results from the workshops were complemented by a simple analysis of statistical and economic reports, prepared by the forestry administration. The same working group prepared a final report on the evaluation. In spite of the omnipresence of the forestry administration, the evaluation of the first 5 years' implementation of the *National Concept for Forestry Development* was a process open for broad participation of various stakeholders and actors (carefully selected by the forestry administration), including some new ones who were passive during the policy formulation stage.

The majority of the participants invited by the forestry administration understood the current situation. The local population lived in poor economic conditions, leading to increasing human pressure on the forests. Experiments with community forest management and leasing of the forest lands had not solved this problem. 'We need to involve those whose lives depend directly on the forest resources. If they are better informed about the problems of forestry and forest conservation, they may change their attitudes towards the resource and to the service.' Other ministries wanted to be part of the process: 'Good relations facilitate lobbying. Involvement in the discussion and decision will prevent opposition at the stage of approval.' In contrast to the passiveness of environmental NGOs during the forest policy formulation stage, 5 years later they provided a stronger voice and full engagement in environmental policy issues. 'Once they are convinced, with their help it is much easier to pass ideas through to the government.' Foresters and forest rangers, who were either excluded or only symbolically present during the policy formulation stage, were not fully engaged. 'If they are involved, they cannot criticize the decision afterwards.' Opponents and adversaries of the forestry administration were included. A fear of possible criticism was one of the main reasons for the initial aversion to the idea of participation. 'The floor given for an open critique decreases the risk of opposition.'

All these lessons learned during the policy formulation stage by the different players contributed to the appreciation of the importance of the broadest possible dissemination of the information about the process and propagation of the results. This well-organized instrumentalized participation brought the forestry administration the status of an independent service under the direct control of the President's office and an image of a pioneer of democratic transformations (including open evaluation and adaptation of a policy). All of these results created a new image of the forest service as a reliable business partner for international donor organizations, and power and control were consolidated at the top of the forest service hierarchy.

Scientists in policy adaptation (Fig. 3.4)

The break in the forest policy reform process in Kyrgyzstan (which exactly corresponded to the break in the direct involvement of policy scientists in the process) gave time for a scientific analysis of the process and its implications. Initially, the scientists came to the process with an assumption that the involvement of multiple actors and stakeholders would provide a basis for realistic decisions, leading

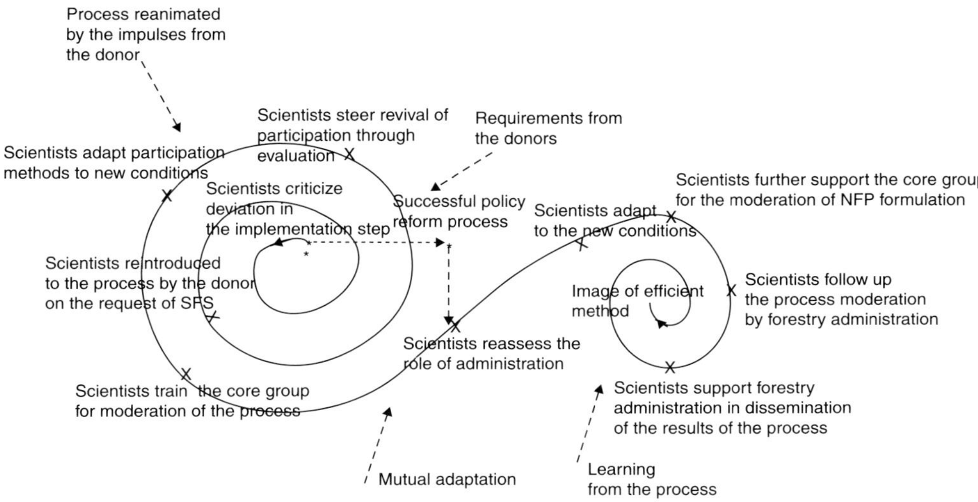

Fig. 3.4. The change in the perceptions and roles of policy scientists during the forest policy evaluation stage.

to the sustainable forest management, and would be in line with the democratic processes of a country in transition. The rhythm of the process depended from their perspective on the rhythm of involvement by the policy scientists. Obviously, the policy scientists brought in a methodology and techniques promoting participation and were very much involved in the facilitation of the processes, including adapting the techniques and methodology – although not the theory – to the changing situation. The choice of the theory, method, approach and concept adapted to a context was itself dependent upon political acceptance, and this is directly derived from the origin of the process itself (Buttoud, 2000).

The reality in Kyrgyzstan showed, on the one hand, that democratic behaviour (public deliberation) cannot be simply imported into a society, but that people need time for learning and becoming part of the practice. On the other hand, the rapid and organized learning from within the powerful structures (presumably those stakeholders who had bigger stakes and challenges in the course of the process) permits instrumentalization of participation for the benefit of those structures. A new policy concept and theory are needed to explain how the process was working, or how it has worked, and what should be adapted specifically for societies in transition. The short period of policy formulation gave some ideas for further theories, but it was not yet sufficient.

At this point there was a new call from the donor and forestry administration for methodological input for organizing policy evaluation. Even if the common goal was the appropriation of the process by the forestry administration, from the beginning it was generally agreed that public participation would be the major principle for the whole process. The policy scientists promoted the revival of participation through evaluation, but, contrary to the previous stage, were no longer directly involved in the process. The specialists from the forestry administration were trained in the methodology and techniques for policy evaluation and the organization of participation of actors and stakeholders, and all the responsibility for the process was given to a working group, comprised of those trained experts and a representative from the donor and headed by the deputy director of the forest service.

The role of the scientists at this step was limited to general observation and punctual advice and consultations when needed. It became clear during the preparation of the report on the results of the evaluation and the writing of a new edited version of the *National Concept for Forestry Development* based upon the results of participation that the input of the scientists was considerable, if not decisive, because of the need for a synthetic analysis and conceptualization. This work was always presented as the achievement of the working group. Furthermore, the scientists contributed to the dissemination of information about the experience of the forest policy reform in Kyrgyzstan at the international level and continue to support the forestry administration in disseminating all the achievements of the participatory process. As a result of the 8 years of the forest policy reform process, it can be concluded that a full policy cycle was implemented in Kyrgyzstan, including all the stages from the situation analysis to the national forest policy adaptation through evaluation. Indeed, now all of the basic documents for the forest policy are in place and the principles of participation are accepted as the basis for policymaking.

Some outcomes and conclusions

What the participants learned during the policy formulation stage made it likely that the actors would come back to the process in order to pursue their own specific interests. The forestry administration learned about the potentials of power that participation could bring. Ironically, the interests of the decision makers were to change as little as possible in their decision-making procedures in order to control the decisions themselves. Appropriation of the process and its results, which was an objective of the policy adaptation exercise proposed by the donor and promoted by the working group (mainly comprised of the representatives of the state forest service) as moderators of the process and by the scientists, gave the forestry administration a possibility to achieve both: maintain the power over decisions and develop a good image of a democratic authority, further strengthening its political status. Moreover, the involvement of the other stakeholders improved the links within the forest sector as well as with the other sectors, contributing again to the power and status of the forest administration. And, last but not least, democratic decision making attracted other donors to the forestry sector, with the forest service as a reliable partner.

Scientists are often expected to provide objective knowledge and unbiased judgement appropriate for legitimating both decisions and processes. For the process to become legitimate in the new conditions of a country in transition, it was important to demonstrate that the proposed theories were viable and the methodology appropriate. Clearly, this implies that scientists must adapt themselves and their methodology to current conditions. When the state administration is still very strong and public deliberation is not yet a habitual practice, scientists naturally orient themselves to the stronger actor. In this case, the adaptation of the scientists included the observation and analysis of the reaction of the policymakers to the proposed methodology, an improvement of techniques for participation based on this reaction and finally an adaptation of the participatory process itself to the needs of the administration.

There is also a third party to the story, the donor, which has been generally referred to as an 'external factor', but at the same time was a real decision maker and had a decisive role in the evolution of the process. The forestry administration was the local partner of the donor for the project. The reputation of the policy scientists as providers of a neutral and objective expertise served as a tool for moderating the reaction of the forestry administration. So the donor invited the policy scientists as a guarantor of objectivity, definers of the framework and moderators of the policy process. The unforeseen effect of such involvement (i.e. the strengthening of the position of the forestry administration and its getting out of control) did not fit with the agenda of the donor and led to suspension of support for the policy process.

The criticism from the scientists at the international level and mainly the need for the success of the project led to the revival of the donor's engagement in the forest policy reform, but with the obligation of achieving a result with the responsibility (especially in case of failure) shared with the policy scientists. The same scientists were invited as for the first stage, and the success of the participatory forest policy evaluation and adaptation inevitably made the donor the winner, who could now claim 'experience in forest policy reform in countries in transition'. Unexpectedly, the forest policy reform process in Kyrgyzstan based on a participatory approach brought a win–win–win outcome for everyone involved.

From the scientists' point of view, the participatory forest policy formulation in Kyrgyzstan was a practical test of theory and methodology. As in any experiment, the results can surprise the experimenter and it is up to a good scientist to recognize and pursue the implications. So, even if the conclusions challenge the classical image of a scientist, they need to be reported. Adaptation is one feature of such a scientific approach. In principle, scientists are ready to adapt themselves, and thus they can help promote a general adaptation by others. A policy formulation process, like planning at any level, is a procedure for promoting a change. To plan means to order actions you will carry out as needed and the change is intended to improve the present situation (Buttoud, 2000). Scientists, having a reputation of being 'neutral', may be called upon to provide an 'objective' analysis of the facts, which can then help the decision makers to decide.

The case study of Kyrgyzstan has shown that scientists are equal actors in the process and may have their own understanding of expected results and ways to reach them (means and ends). However, since scientists propose a theory (or methodology) for these processes, their values and perspectives can dominate the choice or the application of the proposed theory. Just like the other actors of the process, the scientists are not neutral; they too want to be winners and prove that their theories and methods are working. Can it be stated that the scientists were the principal motors for changing the direction of the movement of the spiral?

Does the analysis of the development of the double spiral lead to the conclusion that the double nature of the spirals was caused by the proposed methodology of the mixed model?

In politics, the use of science always entails a legitimization strategy, whether it succeeds or fails. In other words, politics cannot be avoided and a pragmatic rationalist approach is the only effective orientation (Antypas and Meidinger, 1996). The case study suggests that, when decision makers need validation for their actions, they seek out the scientists. Once the decision is made, the scientific expertise can become an alibi for the decision makers when it is challenged.

In Kyrgyzstan, the scientists have adapted theories to the reality and this adaptation has directly created a win–win–win situation. Another uncontrolled outcome was that the results of the policy evaluation indicated a need for a general reform of the forestry sector, which would include reorganization of the forestry administration. The donors have already engaged themselves in the support of this process. Are we in for a couple of new spirals? The context has changed with the 'tulip revolution' in Kyrgyzstan in March 2005. Was it fed, at least indirectly, by a process such as the participatory forest policy reform process? Or is there again, more globally speaking, a power redistribution between the decision makers winning from the public deliberation?

References

Antypas A. and Meidinger, R. (1996) *Science-intensive Policy Disputes: an Analytic Overview of the Literature*. www.law.buffalo.edu/homepage/eemeid/scholarship/scipol.html

Barstad, J. (2002) *Iterative Planning Processes: Supporting and Impeding Factors*. http://www.metla.fi/eu/cost/e19/barstad.pdf

Bormann, B.T., Cunningham, P.G., Brookes, M.H., Manning, V.W. and Collopy, M.W. (1993) *Adaptive Ecosystem Management in the Pacific Northwest*. USDA Forestry Service Gen. Tech. Report PNW-GTR-341, 22 pp.

Buttoud, G. (1999a) Principles of participatory processes in public decision making. In: Niskanen, A. and Vayrinen, J. (eds) *Regional Forest Programmes – a Participatory Approach to Support Forest Based Regional Development*. Proceedings 32, European Forest Institute, Joensuu, Finland, pp. 11–28.

Buttoud, G. (1999b) Negotiation methods to support participatory forestry planning. In: Niskanen, A. and Vayrinen, J. (eds) *Regional Forest Programmes – a Participatory Approach to Support Forest Based Regional Development*. Proceedings 32, European Forest Institute, Joensuu, Finland, pp. 29–47.

Buttoud, G. (2000) Linking theory with lessons learnt in forest policy planning: voiced thoughts on the role of science in N/RFP evaluation and formulation. In: *National Forest Programmes: Social and Political Context*. COST E19 meeting, SECF and Universidad Politecnica de Madrid, Madrid, pp. 9–14.

Buttoud, G. and Samyn, J.-M. (1999) *Politique et Planification Forestière: Guide pour la Formulation et L'évaluation*. Intercooperation, Berne, Switzerland.

Buttoud, G. and Yunusova, I. (2002) A 'mixed model' for the formulation of a multipurpose mountain forest policy: theory vs. practice on the example of Kyrgyzstan. *Forest Policy and Economics* 4, 149–160.

Buttoud, G. and Yunusova, I. (2003) The mixed model for decision making as a conceptual framework for IMP formulation. In: Brun, F. and Buttoud, G. (eds) *The Formulation of Integrated*

Management Plans (IMPs) for Mountain Forests. Proceedings of the International Research Course, Bardonecchia, Italy, 30 June–5 July 2002, Dipartimento di Economia e Ingeneria Agraria, Forestale e Embientale, University of Torino, Italy, pp. 19–31.

Davis, H. and Scase, R. (1985) *Western Capitalism and State Socialism*. Basil Blackwell, Oxford, UK.

Etzioni, A. (1967) Mixed scanning: a third approach to decision-making. *Public Administation Review* 27, 385–392.

Faludi, A. (1973) *A Reader in Planning Theory*. Pergamon Press, New York.

Friedman, J. (1987) *Planning in the Public Domain: from Knowledge to Action*. Princeton University Press, Princeton, New Jersey.

Gunton, T. (1984) The role of the professional planner. *Canadian Public Administration,* Fall 35–49.

Hudson, B. (1979) Comparison of current planning theories: counterparts and contradictions. *Journal of the American Planning Association*, October, 387–398.

Hunt, S. (1990) Truth in marketing theory and research. *Journal of Marketing* 54 (July), 1–15.

Kouplevatskaya-Yunusova, I. (2004) Spirals of forest policy development, or transformation of participation in an iterative process: the case of Kyrgyzstan. In: Buttoud, G., Solberg, Tikkanen and Pajan (eds) *The Evaluation of Forest Policies and Programmes.* Proceedings 52, EFI, Joensuu, Finland, pp. 103–114.

Kouplevatskaya-Yunusova, I. (2005) The evolution of stakeholders' participation in a process of forest policy reform in Kyrgyz Republic. *Swiss Forestry Journal* N10, 385–395.

Kouplevatskaya-Yunusova, I. and Buttoud, G. (2006) Assessment of an iterative process: the double spiral of re-designing participation. *Forest Policy and Economics* 8 (5), 529–541.

Lindblom, C. (1959) The science of 'muddling through'. *Public Administration Review* 19, 79–88.

Medawar, P. (1990) *The Threat and Glory: Reflections on Science and Scientists.* Oxford University Press, Oxford, UK.

Shannon, M. (2002) Understanding collaboration as deliberative communication, organisational form and emergent institution. In: Gislerud, O. and Neven, I. (eds) *National Forest Programmes in a European Context.* Proceedings 44, EFI, Joensuu, Finland, pp. 7–26.

Yunusova, I., Buttoud, G. and Grisa, E. (2003) Reforming forest policy in Kyrgyzstan: impediments and success of the process in extreme ecological and unstable socio-economic environment. *Austrian Journal of Forest Science* 20, 73–82.

4 The European Union Sustainable Forest Management and Climate Change Mitigation Policies from a Transition Country's Perspective

MARIA NIJNIK[1] AND LIVIA BIZIKOVA[2]

[1]*Socio-Economic Research Group, The Macaulay Institute, Craigiebuckler, Aberdeen, UK;* [2]*Adaptation & Impacts Research Division (AIRD), Environment Canada, University of British Columbia, Vancouver, BC, Canada.*

Abstract

In this chapter, the European Union (EU) sustainable forest management and climate change mitigation policies are examined in terms of the viability of their implementation in transition countries. Developing sustainable forest management practices and linking them to climate change mitigation offer multiple benefits for the transition countries, which possess relatively high potentials of carbon sequestration in terms of physical and economic characteristics. However, the still ongoing process of transition is lagging behind due to slow development of institutions, e.g. low capacities of local government and insufficient public involvement in decision making in forestry. Challenges and prospects for the implementation of sustainable forestry and related climate change mitigation policies in transition economies are analysed. The general conclusion is that for the successful performance of the policies in the countries in transition, the social and economic pillars of sustainable forest management are of particular concern.

Keywords: Climate change mitigation, carbon sequestration, policy scenarios, Ukraine.

Introduction

Sustainable forestry is considered to be a balanced management of forests that takes into account their role as a life supporting system as well as their role in meeting the needs of present and future generations for forests and their products without threatening their renewal capacity (IUCN/ENEP/WWF, 1991; MCPFE, 1993). For the sustainable forest management (SFM) concept to become operational, it was defined in terms of a set of principles and criteria. By the time of the UNCED (1992) conference, a number of European countries had developed principles for SFM and these principles were discussed in Rio. The Rio debates resulted in the *Statement of Principles of Forests*, a guiding document aiming to contribute to sustainable development in all of its dimensions – environmental, economic and social – in the management and conservation of forests. The economic dimension of SFM concerns the ability of forests to maintain production capacity, sufficient to meet current and future demands of people for forest products and services, through using the resources efficiently. The social dimension concerns both the institutions that are to be placed under the obligation to turn forestry toward sustainability. And, in addition, principles of socially acceptable equity in the distribution of employment opportunities, incomes and benefits from forestry, including cultural and ethical. The environmental dimension concerns maintenance of forest resources, their capacities, biodiversity, health and vitality, resilience of ecological systems, their integrity and ability to provide a continuous stream of environmental benefits (Nijnik, 2004).

Climate change is a complex issue, encompassing not only environmental questions, but also economic and social implications. Anthropogenic greenhouse gas (GHG) emissions that are mainly responsible for climate change have increased with intensification of economic activities, and are likely to lead to wide consequences (Spash, 2002). The Kyoto Protocol to the United Nations Framework Convention on Climate Change (UNFCCC) was adopted in 1997 to provide mechanisms to mitigate climate change at an international level (UNFCCC, 1998). The Kyoto Protocol became legally binding (for the 128 Signatories) on 16 February 2005 (UNFCCC, 2004) with the signature of Russia. The EU and member states ratified the Kyoto Protocol on 31 May 2002, and set an overall target of 8% (336 Mt CO_2 equiv.) that was distributed on a differentiated basis to individual member states under an EU burden-sharing mechanism (EC, 2002). This target can be achieved either by reducing emissions (reduction of sources) or by removing GHG from the atmosphere (enhancement of sinks).

Articles 3.3 and 3.4 of the Kyoto Protocol state that biological sources and sinks can be used for meeting the countries' commitments during the stipulated period (since 1990). Activities involving land-use changes can help to reduce GHG concentration in the atmosphere, by increasing biotic carbon storage on agricultural lands (such as carbon sequestration) and by decreasing GHG emissions. The potential activities include: reducing rates of deforestation; increasing land devoted to forest plantations; regenerating secondary forest; and indirectly the reduction can occur by producing wood, especially from short-rotation forest plantation (SRFP) for energy production. From a policy perspective, activities to achieve climate change mitigation must not compromise other policy goals, as in this case, the activities that involve forestry should fulfil the principles of sustainable forest management.

We begin with a critical analysis of the three dimensions of SFM relative to countries in transition. After suggesting that the economic pillar of SFM remains the key to sustainability in transition countries, we examine how rural development measures related to forest could support carbon sequestration and SFM under appropriate conditions. As examples of countries in transition, we compare Slovakia and Ukraine, to better understand the linkages between climate change mitigation potentials involving forestry and the policies and principles of SFM.[1] We conclude by connecting these policy options to the wider societal and institutional questions pertaining to carbon sequestration. The conclusions provide a synthesis of ideas and examples from the EU policies, which address SFM and climate change mitigation efforts from a transitional country's perspective.

Sustainable Forest Management Priorities

The EU adopted the *Forest Strategy* in 1998 to implement the principles and criteria of sustainable forest management. The Strategy articulates the priorities that should shape member state forest policies in Europe as follows: sustainable forest management and the multifunctional role of forests are overarching common principles; the importance of taking global and cross-sectoral issues into account is increased; and national forest programmes (NFPs) are expected to be the main instruments for implementing SFM principles (EC, 1998). As countries develop their NFPs, they specifically recognise that forests offer a wide range of benefits and contribute to the creation of opportunities for more people to enjoy forests as well as to help rural communities benefit from them.

While much has been done in the EU to manage forests sustainably, implementing SFM has often led to a shift in emphasis from the productive values of forests and their social and economic benefits to their environmental values and services. In the 2005 assessment of the implementation of the EU *Forestry Strategy*, the concern over the imbalance in addressing the different dimensions of SFM in terms of economic, social and environmental functions was explicitly noted. The report states that while there has been considerable progress in the environmental area, more could be done to better identify the major economic and social issues to secure sustainable management of forests in the long term (EC, 2005).

In transition countries, the regional socio-economic conditions mean that the economic aspects of forestry development and related social issues such as employment, poverty alleviation along with rural and community development are particularly important. In order for the countries in transition to accumulate financial assets, attract investment and economic development, the asset value of the forest is a critical resource for development and improved social conditions. However, these countries have inherited from the previous political system mechanisms of decision making that do not take into consideration the present and future conditions for wood production and consumption. Indeed, in many instances, the forests have not been considered an economic asset in market terms, but simply a state patrimony value. For instance in Ukraine,[2] commercial exploitation of nearly 50% of forest areas is limited yet the annual change in growing stock suggests that these forests could sustainably produce more wood products. To improve the efficiency of and capacity for timber production so that the country could export rather than import timber products, significant changes will need to be introduced into forest policy to promote economic sustainability.

Commonly, the economic dimension of sustainability in forestry requires cost efficient and long-term resource production, where the basic requirement is a constant or an increased flow of benefits from a forest whose structure and composition are maintained consistent with SFM criteria and indicators (Rice *et al.*, 1997). However, in the economies in transition, achieving economic sustainability is difficult because the forest is endangered through unregulated and illegal uses (Gensiruk, 1999). Despite official norms for forest protection, high interest rates, ineffective enforcement of laws, corruption and demand for wood in foreign markets provide incentives to entrepreneurs to cut and sell timber. Forest policy reform in transition countries needs to recognize that generally pro-environmental legislation may not result in sustainable forest management because forest enterprises are guided primarily by consideration of economic efficiency and profitability. Thus, a forest policy that officially focuses on environmental priorities and conflicts with the interests of forestry enterprises can easily lead to activities that will result in non-sustainable forest management (Komendar, 2001).

In order for sustainability to be consistent with economic efficiency the growth rate of the forest must be larger or equal to the interest rate (Nijnik, 2004). Maximizing the net present value (NPV) involves a comparison of the net benefits from postponing harvesting with the net benefits from harvesting timber and investing the profits. The objective of maximizing the NPV from the forests with moderate growth often promotes higher harvest levels than the net growth of forest stands. This focus on economic efficiency thus encourages the establishment of fast-growing commercial plantations instead of natural forests. This in turn endangers biological diversity and the health and vitality of forest ecosystems. In addition, often the establishment of forest plantations increases costs related to care and protection of monoculture forest stands that are less stable biologically, and these costs are not always included in the evaluation process (Nijnik, 2004). There is a threat therefore that in conditions of non-internalized externalities, consumption in forestry could be restricted to the consumption of economic goods and services, leaving the social benefits and environmental service components of SFM underestimated.

Generally, the economic criterion of SFM is less defined in the literature and in the indicator systems being employed. Useful indicators of SFM are, for example, rents. As a rule, rent is a measure of wealth and its proportion captured by the government indicates whether or not government revenue from forestry is sustainable. Recording of rent capture provides an indicator of the sector's sustainability because in a market economy (quasi) rent capture[3] accelerates technological development in forestry (Van Kooten, 1995). In forestry-in-transition, however, when forest resources are undervalued, rent-seekers tend to redistribute wealth according to their interests, which may do nothing to develop a more efficient use of the resources or technological innovation. The discrepancy between the prices of timber production and the prices for timber on the world market encourages rent-seekers to sell timber and capture the rents.[4] While with proper institutions exporting timber can be beneficial for the economy, since it decreases the budget deficit and provides resources to modernize the machinery, in economies-in-transition this may not be the case.

There is a need therefore in transition countries, where property rights are not always guaranteed, information is often asymmetric and natural resources are undervalued, to extend the frame of forestry toward a market economy. Forestry advances toward sustainability depend on the extent to which markets and social institutions function in delivering appropriate signals for more optimal allocation of various components of capital in space and time.

The weak institutions in countries in transition permit the *shadow economy* to reduce the tax base of the state and official foreign exchange holdings, leaving the overall performance of the economy in worse shape. Though the data on underground business in forestry in the countries in transition is unavailable, the situation is likely to be bad in the sectors of the economy that are coping with the extraction of natural resources. Losses to the national budget from the shadow economy could be substantial (Fonkych, 2000; Hryniv, 2001). Hence, for transition countries with weakened institutions and commodity prices that do not yet reflect the real value of resources and their scarcity, it is even more important, yet more difficult, to define and employ economic indicators of forest sustainability.

Meanwhile, the SFM addresses economic, social and environmental dimensions of sustainability as separate blocks which must be balanced. To ensure a sustainable multifunctional development of forestry, the economic policy reform focused on enlarging the efficiency in timber supply must be complemented by well-targeted measures to preserve forests and conserve their biodiversity and landscape values. The role of government is important in balancing economic objectives with social and environmental requirements (e.g. through the enforcement of legislation able to cope jointly with the needs of different forest owners). Governmental intervention in terms of public environmental and social policies is highly justified during the transition. Over time, and with institutional development, direct government intervention gives way to indirect guiding through institutional and policy design to further the advance of markets. However, the role of government in regulating the tenure, management, financing and production of public goods will remain even under conditions of a market economy.

Finally, an essential part of SFM policy reforms is the transition in governance. There are three basic mechanisms of governance[5] related to the economy: markets, hierarchy (authority) and collective action. Markets constitute governance by voluntary exchange between two parties. Hierarchy is governance by command-and-control instruments, when authority is assigned to one or many actors, on the basis of top-down approaches. Collective action is a coordination by common interest when people act together (Gerrard, 2000).

All over Europe, the importance of 'good governance' for the protection and sustainable management of forests is being increasingly emphasized, along with the need to develop and apply integrated approaches in forest policy formulation and implementation with wide public participation (EC, 2005). Prevailing institutions in the transition countries, however, constitute the hierarchy.[6] When rules of governance correspond primarily to government laws and organizations are mainly based within the governmental structure, the whole system of institutions does not maximize gains for the participants.

In such situations, policy reform necessitates replacing obsolete human capital and directing the motivations of policy actors to support of a market economy with active involvement of local communities and general public in decision making in forestry. It is important that forest policymakers, practitioners and environmentalists, all key stakeholders, and the public recognize the forest policy issues in their full importance and complexity. This should lead to a greater emphasis on wise use and protection of forests and to the realization of more joint and effective measures toward managing forests sustainably.

Rural Development Priorities

Forestry is naturally linked with various other land users. The economic, social and environmental dimensions of forest policy (including climate change mitigation initiatives) must therefore correlate with other land-use interests to fit with the general strategy of integrated rural development. Land used for forestry may compete with other land uses, and it is desirable therefore to plan and manage all land uses in an integrated manner. Integration takes place at two levels, considering on the one hand, all environmental, social and economic factors (including for example, impacts of various economic and social sectors on the environment, climate and natural resources) and, on the other, all environmental and resource components (i.e. air, water, natural resources etc). Integrated consideration facilitates appropriate choices and trade-offs, leading to sustainable (multiple) resource use.

Rural development policy has been the main impetus for the implementation of the EU forestry strategy at the Community level. It is also the basis for the linkages between multiple dimensions of SFM, including those relating to climate change mitigation. The overall principles of the forestry strategy at the theoretical level (e.g. multifunctionality and sustainability) are reflected in the rural development policy, which brings together economic, social and environmental objectives and transforms them into a coherent package of voluntary measures, giving an added value to the implementation of the forest programmes of the member states (EC, 1998) (Fig. 4.1).

Throughout the EU, an integrated approach to sustainable rural development offers a broad scope of measures supporting forestry. The forestry measures of the rural development programmes also contribute to the issues such as climate change mitigation and biodiversity conservation. In total, the EC financial support for forestry measures in the context of rural development amounts to € 4.8 billion for the period 2000–2006 (~10% of the Rural Development budget). In previous years, the aid for afforestation and reforestation provided by an EU initiative enabled the members to cover up to 75% of their tree-planting costs. The initiative was primarily focused on poor-quality land and, between 1994 and 1999, about 1 Mha was afforested (EC, 2002). This focus, however, has drawn the attention away from management of existing forest stands. The proposal for the Rural Development regulation for the period between 2007 to 2013 also emphasizes the importance of forestry (EC, 2005).

In transition countries, however, with newly re-established property rights, the lack of experience in balancing private and public forests, for instance, could decrease the potential of SFM management implementation (Schmithusen, 1996). For example, in Slovakia, the re-establishment of property rights was accompanied by a decentralization process. As a consequence, the restored forestry

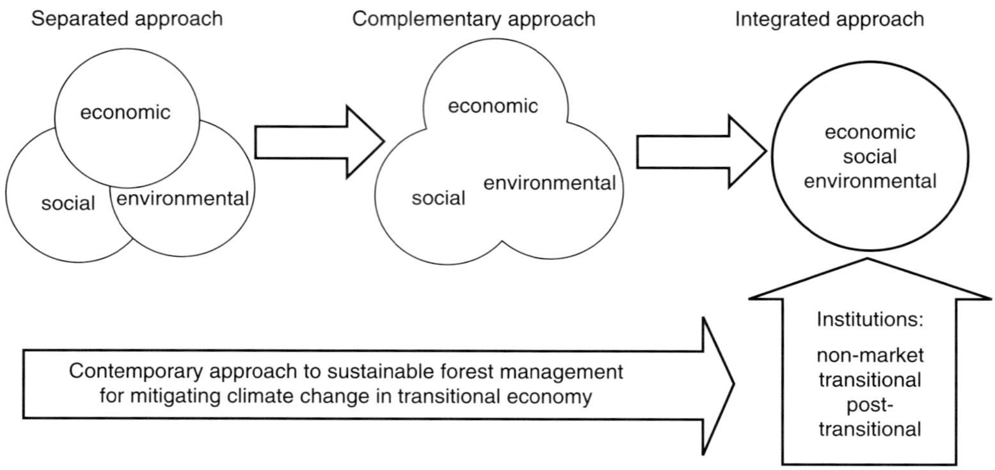

Fig. 4.1. Integrated approach to sustainable rural development (developed from Gray *et al.* 1995). Level of institutional development in transition countries and their ability to incorporate non-market values has significant impact on the ability to operationalize the integrated approach of SFM.

has come under the responsibility of relatively weak (in terms of available financial resources and decision-making power) regional governments. Moreover, major forest resources in Slovakia are located in areas where the economies are undergoing a huge recession, with the local gross domestic product (GDP) less than a half of the EU-15 average (GRA, 2002).

This once more necessitates proper sequencing and timing of the reforms: further insurance of property rights in transition economies; enhancement of good governance; promotion of collective action (public involvement); and the development of social capital through institution building. Only then will EU sustainable forest management principles be effective in achieving sustainability from a transition country's perspective.

Climate Change Mitigation Priorities

The EU forestry strategy and rural development policy have confirmed the role of the forest sector among the climate change mitigation measures that were agreed under the Kyoto Protocol. EU forests cover approximately 113 Mha and they are estimated to store 5 Gt C (18.3 Gt CO_2 equiv.) in their woody biomass. European forests act as an effective carbon sink. It is estimated that EU-15 forests sequester 63 Mt C (231 Mt CO_2 equiv.) annually (TBFRA, 2000), however, it should be noted that only a small fraction of this amount could be accounted for under the rules of the Kyoto Protocol because most of this carbon uptake is not 'additional' with reference to the 1990 baseline.

For the first commitment period of 2008 to 2012, the combined potentially accountable carbon (C) credits for the EU from agricultural and rural development measures (3.84 Mt C/year or 14 Mt CO_2 equiv/year) and forestry (capped at 5.17 Mt C/year or 19 Mt CO_2 equiv/year) would be approximately 9 Mt C/year or 33 Mt CO_2 equiv/year, which is roughly 10% of the corresponding EU emissions reduction target. Taking into account the EU 25 member states, a technical sequestration potential of 34 Mt C/year (125 Mt CO_2 equiv/year) may be reached in the long term (ECCP, 2004).

In terms of climate change mitigation priorities, the approach that includes the principles of SFM is broadly advocated. According to this approach, climate change mitigation measures must be socially and environmentally acceptable (EC, 1998), i.e. based on SFM principles. Though in the majority of forested countries in Europe, the carbon sink in trees counted under Article 3.3 of the Kyoto Protocol is minor, it may have relevance for the carbon budget of individual countries. The carbon sink can be significant in countries where present forests are relatively small, leaving areas available for further afforestation and reforestation (Chomitz, 2002).

The expansion of forests is also supported due to the added benefit of fuel and the availability of timber for use in wood products. The countries of Europe are committed to an increase in the use of renewable energy. The European Commission (EC) adopted a White Paper that stresses the necessity to raise the production of energy from renewable sources from 6% in 1998 to 12% in 2010 (EC, 1997). Taking into account the current share of renewable sources in the EU, as well as in transition countries, it is clear that this implies an increase in the use of biomass, in general, and of woody biomass, in particular, in energy production.

The potential of sequestered carbon, arising from energy crops and available to substitute for fossil fuel, is estimated in the range of 4.5–9 Mt C/year until 2012 (ECCP, 2004). To achieve these targets, there is a need for incentives to support renewable sources of energy and to establish a proper link between carbon sequestration and agricultural policies, mainly concerning land-use changes on setaside and marginal lands.

The transition countries have had relatively good experience with afforestation and forest management practices in the past. With time, however, afforestation[7] of marginal lands has become more difficult and expensive since the lands that were most suitable for tree planting have already been afforested. Partly, the area of new plantations has decreased due to the suspension of state programmes. However, even more importantly, the changing institutional and socio-economic environments in the transition period, including, for example, an absence of a well-defined and enforceable system of property rights on land, a shortage of investments, an absence of economic incentives for tree-planting activities and high interest rates, to name only a few of the most important, impeded afforestation rates.

The GHG emissions in transition countries have also fallen, with their average level at about 30% of the Kyoto commitments, which is valid even at the current level of economic development in these countries. The emissions reduction is primarily a result of the economic recession (often called 'hot air') in the early 1990s and to a lesser extent due to market reforms introduced in the last decade that have improved energy efficiency (Petkova and Faraday, 2001). The impact of 'hot air' and related potential to sell credits in the carbon market, as well as changes in the institutional and ownership structure related to agricultural and forest land, provide different conditions in transition economies for the implementation of climate change mitigation activities and policies involving agriculture and forestry.

Clearly, it is necessary to take into account the specific conditions in transition countries in order to improve the capacity of new member states to apply European Community policies, like SFM and climate change, and achieve the type of results expected in more developed European countries. In addition, the special context of countries in transition is a key element in fostering the integration of the accession countries.

Linking Sustainable Forest Management and Climate Change Mitigation Policies

Sustainable forest management policies designed to moderate carbon emissions are commonly considered as low-cost options for coping with climate change. Since the COP-7 Conference in Morocco (2001), afforestation, reforestation, forest management and soil carbon have become eligible climate change mitigation policies and afforestation has taken an important role as a carbon dioxide reduction policy. To analyse the SFM principles and the potential for implementation of strategies for coping with climate change in the economies in transition, we focus on Slovakia and Ukraine. The selected countries have different levels of socio-economic and institutional development, and different geographical locations with related physical conditions. Meanwhile these countries are in different relationships with the EU: Slovakia is a new member state, whilst Ukraine has started exploring its opportunities to join the EU.

The results of the analysis have shown that approximately 0.41 Mha of land is suitable for tree planting in Slovakia. This includes marginal agricultural land, abandoned agricultural land and land withdrawn from agricultural production in the last decade. In Ukraine, for various reasons, a total of 2.29 Mha of land is suitable for tree planting. These estimations indicate that with afforestation at this scale, timber production could be increased by 25–30%, and substantial environmental benefits could be achieved from the new forests.[8]

In Ukraine, where nearly 66% of the forest land (7.1 Mha) is publicly owned, tree planting activities are under the execution and control of the State Ministry (Committee) of Forests. The weaknesses of the command-and-control mechanism of afforestation include the lack of flexibility and economic incentives for encouraging tree planting. The control mechanisms, however, could be justified on efficiency grounds, if the savings in transaction costs exceeded the gains from using other coordination mechanisms or if there are economies of scale and scope that would not be realized otherwise (Van Kooten *et al.*, 2001).

Afforestation will enlarge social benefits, primarily to agriculture because of soil protection and improved hydrological forest functions and, to society in terms of climate change mitigation. Due to market failures, however, these social gains from afforestation (external benefits) will not be achieved, and welfare maximization conditions therefore will not be met without government regulation. The main reason is the discrepancy in the distribution of benefits and costs from forestry development. The establishment of forest plantations (also for carbon sequestration) is executed in the forest sector, while the soil protection (and climate change mitigation) benefits for instance, accrue to agriculture (society). The problem of 'who pays and who receives the benefits' cannot be solved through the market, since it is to be settled by government. Hence, despite the fact that afforestation costs, especially for wastelands, are relatively low in the transition countries, large-scale tree planting will not take place without government subsidies or foreign investment.

This argument has been proved in Slovakia, where with the cancellation of the state law and its framework, according to which, by the year 2000 the area of forest land would have been 50,000 ha, the actually afforested area appears to be just 877 ha. During the transition process, when, from previously state-owned forests, 42% were given back to former owners and more than 90% of claims were processed, the incentives aimed to support afforestation were ineffective. The afforestation process was negatively influenced by the unclear landownership and by the problems with the allocation of subsidies to landowners. Moreover, due to the historical development of economic, legal and social relationships, the average size of a private plot of land in Slovakia is 0.45 ha. This fragmentation of the land into very small ownership parcels also hampers afforestation programmes.

There is a pressing need, therefore, in Slovakia to develop and support a new approach to forest management planning in order to fit a wider range of forest sites together with the specific needs of various owners. As long as forest lands remain fragmented into small ownerships and there is an absence of long-term investments and appropriate incentives for tree planting activities, landowners are unlikely to undertake afforestation activities. In addition, government subsidies encouraging sustainable behaviour on the part of new landowners are continuously decreasing, thus the expansion of intensive economic activities will keep threatening forest sustainability instead of working as part of the solution (Kluvankova-Oravska, 2004).

It is reasonable, therefore, that to implement afforestation in Ukraine the authorities will use the production capacities of state forest enterprises. In addition, they may consider the option of giving certain lands suitable for tree planting to the farmers and cooperatives. Concurrently with the land, the farmers are to be given subsidies to enable them to carry out tree-planting activities. This will intensify the process of afforestation only under conditions of an appropriate level of subsidies that will provide real economic incentives to the farmers to plant trees on their land.

A very important task pertaining to SFM and climate change mitigation, which must be solved in transition countries, is to settle upon a proper structure of land (forest) ownership. The countries' legal documents have to redefine and enforce property rights in forest resources and wooded lands, for instance through the introduction of community–managed forests[9] and defragmentation of private plots of land, etc. There is also a need for simplified forest management guidelines for various owners within their lands (Bizikova, 2004). Furthermore, privatization of forests is not the only solution for transition economies to achieve SFM (Carlsson, 1999). The forest inventories may comprise public (state) forests and forests of municipalities, farmers, enterprises, organizations and institutions, as well as privately owned wooded land and land that is undergoing afforestation. An afforestation fund may include state, communal, and private land ready to be sold or leased to various stockholders for forest development.

Though there is a general consensus that the objective of the economic system transformation in transition countries is the creation of an effective market-oriented economy, consideration of marketable goods in forestry can never be separated from non-marketable goods (forest role in sustaining global carbon cycle and climate stability). The answer to the question as to whether private, common or public property in forests is needed, therefore, depends on the particular conditions and varies from situation to situation.[10] Moreover, in addition to sustainable management of marketable and non-marketable forest benefits, SFM includes sustainable labour management. Extensive forest privatization will cause job losses (Nijnik and Van Kooten, 2000) and for the social benefits it might be better to support a moderate pace in rationalization of work in forestry (Krott *et al.*, 2000).

The success of SFM and carbon sequestration policies involving forestry very much depends on the success of institutional transformation. However, the capacity to generate a simultaneous change in a whole range of political, economic and social institutions is limited. It is vital, therefore, to find a feasible time path for stepwise reforms that does not unhinge the macroeconomic balance (Eggertsson, 1994). The institutional framework 'able to handle such a dramatic change in property rights' as overall privatization of forests meanwhile might be absent in some transition economies (Carlsson, 1999), whilst an excessive forest privatization might also be unnecessary in some others. Thus, the creation of favourable settings for alternative forms of organization in forestry is important, as well as the creation of a selection process by which the winners between organizations are selected on an economic criterion, and those who minimize social costs win (Schrieder, 2000).

Some Considerations on Climate Change Mitigation Priorities in Transition Countries

Climate change mitigation opportunities that involve forestry are not viewed as priorities for the national climate change policy, forestry policy or rural development strategies in transition countries (GRA and GRF, 2002). However, as the above analysis demonstrates, carbon sequestration through afforestation represents an opportunity given declines in production and increases in abandoned land (Swinnen *et al.*, 1997).

Based on the analysis in the preceding section, afforestation of non-forested areas, increasing the level and efficiency of wood utilization, using biomass as a substitute for fossil fuels and the protection of existing carbon storage in forests could be relevant policy measures. However, without sufficient analysis (e.g. potential for carbon sequestration and renewable energy production on different types of land) and without implementing effective policy measures to encourage climate change mitigation through changes in land use and forestry nothing is likely to change (Ministry of Environment, 2001).

Ukraine provides an excellent example of this potential. Carbon trading includes the use of 'sinks' as a flexible policy mechanism to address goals of the Kyoto Protocol. One option would be to stabilize the collective emissions of Annex B countries at least cost. In order to analyse this option, the carbon uptake potential across these countries was assessed (Fig. 4.2).[11] The results indicate that, subject to the assumptions considered in the chapter,[12] especially high carbon uptake benefits would result from afforestation in the wooded steppe zone of Ukraine.

The analysis shows that the variation in carbon sequestration potential across different regions in both countries is related to the variety of conditions. The NPV of afforestation for carbon uptake in Ukraine is positive in the Polissja and wooded steppe, at 0% through 4% discount rate, and at 0%–2% discount rates in the Carpathian Mountains. In Slovakia, the highest potential for carbon sequestration is in the west, where higher-quality land is available and simulated cumulative carbon uptake is about 38.3 t/ha at a 4% discount rate.

The costs of carbon uptake[13] are 18.5–23 €/t (4% discount rate) and 8.5–14.2 €/t (0% discount rate) in Slovakia. They are the highest in the west where there is higher-quality land. In the central region, despite low opportunity costs of land, the low value of the carbon sequestered results in a negative NPV of carbon uptake for the assessed policy scenarios (Figs 4.2 and 4.3). In Ukraine, carbon uptake costs are 4.6–78.5 €/t, with 9.5 €/t on average (0% discount rate). When the benefits are discounted at 4%, the present value of carbon uptake costs is 7.2–173.3 €/t, with an average of 18 €/t of carbon.

The analysis of renewable energy scenario[14] has shown that, in Ukraine, the costs per tonne of carbon sequestered are €36.4 in the Polissja, €32.2 in the wooded steppe and €124.6 in the steppe, with a national average cost of 70 €/t (at 4% discount rate). In Slovakia, costs range from 37–48 €/t (4% discount

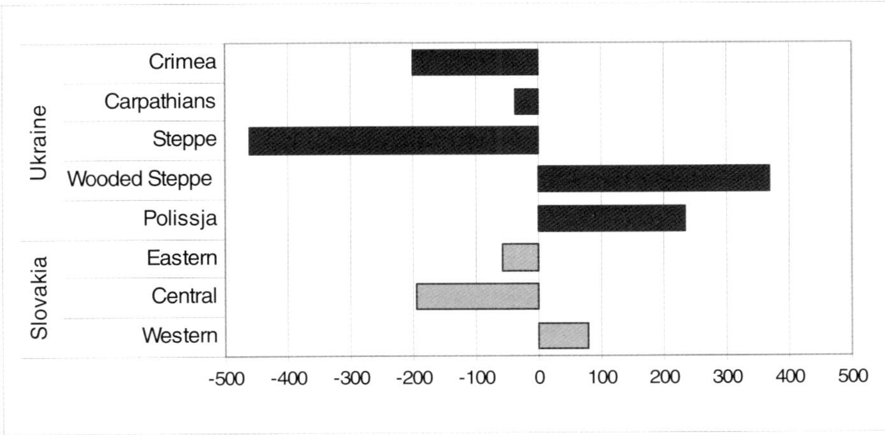

Fig. 4.2. NPV of benefits of carbon storage through afforestation in Slovakia and Ukraine (€/ha, 4% discount rate, carbon in permanent tonnes).

rate), but get much higher when the costs for energy production from the planted trees are included.[15]

Overall, the costs of carbon sequestration in both countries are relatively low in comparison with the corresponding estimates in some other countries of Europe. However, these costs are often higher than the value of the land. In areas that are strongly affected by the decline in agricultural production and land abandonment, the market prices of land are significantly lower than the prices set by the government (which are based on physical characteristics of the land), and this phenomenon is reflected in the estimates of costs presented in Fig. 4.4 (Bizikova, 2004).

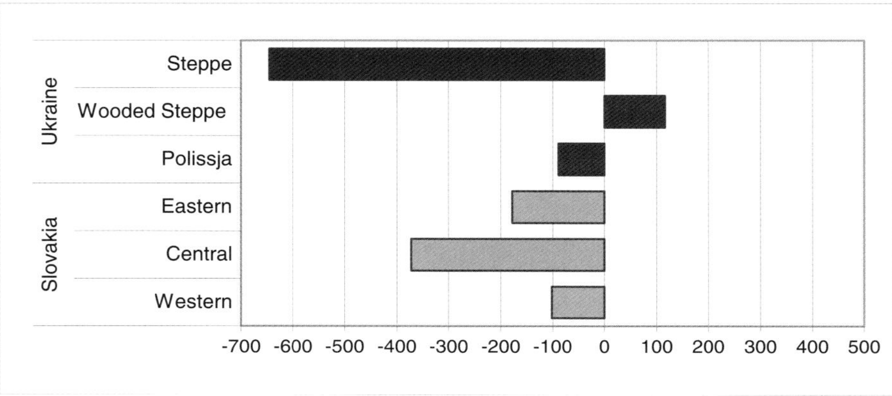

Fig. 4.3. NPV of benefits of the substitution of wood as renewable energy in Slovakia and Ukraine (€/ha, 4% discount rate, carbon in permanent tonnes).

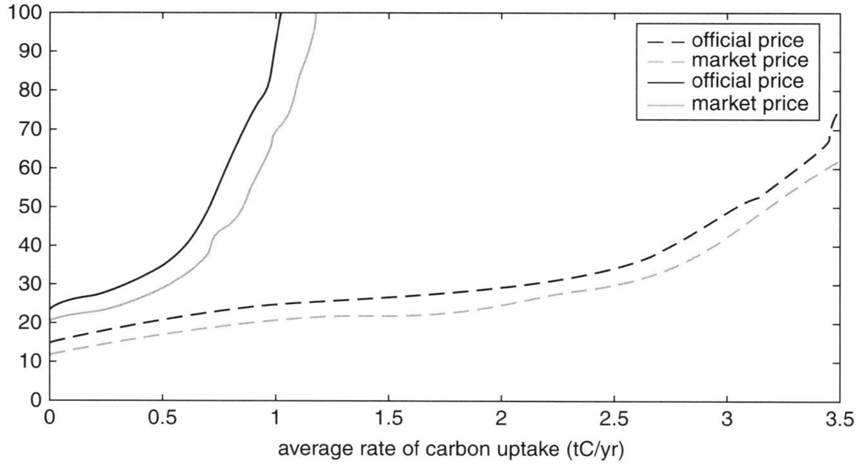

Fig. 4.4. Estimated costs per tonne of carbon uptake for afforestation (dashed line) and SRFP (costs of energy production are included, continuous line) in Slovakia, by using land prices set by government (official price) and market price (€/t, 4% discount).

The results suggest that the implementation of a renewable energy strategy based on wood produced by short rotation forest planting (SRFP) instead of the carbon storage policy would increase the costs of carbon uptake substantially. In the majority of cases, the costs would not be compensated by the returns.[16] Consequently, the alternative energy policy is even less viable for transition economies without external funding, except in the wooded steppe of Ukraine. The establishment of forest plantations for CO_2 emissions in transition economies, therefore, requires new sources of investment, and the question remains whether tree planting for carbon uptake would be a national or project-based agenda.[17]

Conclusions

The transition countries are moderate emitters of greenhouse gases (GHG), largely as a consequence of their economic problems in the course of transition. The countries, therefore, have reached their Kyoto Protocol targets and stand to profit from the sale of 'hot air'.[18] Many transition countries feel that they achieved emission reduction through significant economic hardship and that the reduction is real, so trade of these credits should not be restricted (Tichý and Billharz, 2000). Though the idea of 'hot air' selling sounds optimistic, the countries actually cannot count on it, as 'hot air' is a hot topic pertaining to the environmental effectiveness and economic efficiency of the Kyoto Protocol implementation. Since there is now a viable international emissions-trading market as well as the desire to maximize the seller's own financial revenues from trading, the policy of 'hot air' banking is utilized (Den Elzen and De Moor, 2002). 'Hot air' banking would significantly raise the permit price, and would increase the abatement efforts of Annex I countries, including Ukraine and Slovakia. In addition, the focus of transition countries on selling 'hot air' on the carbon market drives their priorities away from considering climate change mitigation through carbon sequestration.

Countries in transition to market economies are wide open to a range of opportunities for cleaner industrial and energy production. Due to substantial carbon emissions per unit of GDP, these countries also have a high potential for cheap joint implementation (Fankhauser and Lavric, 2003). However, even with this high joint implementation potential, the transition countries have insufficient institutional capacity (Nijnik and Oskam, 2004) for foreign investors to enter their business environment effectively. Hence over and above the emissions reduction, an enhancement of greenhouse gas 'sinks' and 'reservoirs' via climate change mitigation through afforestation efforts and SFM is important.

The essential potentials of carbon uptake through afforestation in Ukraine and Slovakia make the policy of planting trees to sequester carbon in these countries competitive with other policy measures of removing carbon from the atmosphere. The establishment of new forests, particularly in the wooded steppe zone of Ukraine, is a sound climate change mitigation policy, because of the potential of the newly planted forests to contribute to carbon uptake and of the availability of land suitable for tree planting with relatively low afforestation costs. None the less, the activities of enhancing terrestrial carbon sink offer a modest solution within emissions reduction measures for Ukraine and Slovakia.

The transformation process in the transition economy is faced with market failures and inadequate institutions. Over recent decades, Slovakia and Ukraine have both been faced with continuously shifting priorities in their rural development and SFM policies, driven by both internal and external forces. Today, the responsibilities for the environment and its components are divided between different institutions. Carbon sequestration policies require strong linkages between the sectors, and this fragmentation of responsibility, therefore, decreases both policy effectiveness and efficiency.[19]

The level of connection to rural development objectives is crucial for implementation of the strategies involving integrated approaches, such as climate change mitigation by forestry activities. The countries should capture the opportunity of bringing together the rural development and SFM priorities, those of climate change and generally, the issues of sustainable development. It is important to pay more attention to agricultural–environmental linkages and climate change-related measures and to integrate climate change mitigation forest policy to rural and regional developments. Also, there is a need for information campaigns, training facilities and pilot schemes to demonstrate SFM possibilities pertaining to climate change mitigation and to make them attractive for various stakeholders.

Land-use change and forestry activities, and more precisely those involving carbon sequestration, require a long-term perspective. The optimum offset policies must therefore link a long-term carbon sequestration in forestry with a long-term substitution of wood for fossil fuel in renewable energy projects. Overall, under the assumptions of this chapter and for the specific conditions of the examined countries, the climate change mitigation option through forests as 'storage' appeared to be more cost-efficient. The research results provide evidence that, under a renewable energy scenario, the costs of carbon uptake are not compensated by the returns in the majority of the cases. The main reasons are in cost-inefficiency of wood production, and often also in comparatively high land values, where the opportunity costs of maintaining forests on land for a long period appear to be too high. For more useful outcomes, the period under investigation must be substantially extended, so that a continuous process can be shown.[20] Besides, in the upcoming studies afforestation should be elaborated in view of economic, social and environmental benefits all at once, so that multiple gains from various forest values should be considered jointly.

Afforestation and utilization of biomass produced from the new plantations would provide substantial social benefits. To date, however, SFM policy measures and instruments for supporting biomass as a source of renewable energy are lacking in transition countries (GRA, 2002). There is a need, therefore, for further studies and for measures to be developed and implemented to enhance sustainable production of biomass and its use as a substitute for fossil fuels in the context of regional and rural developments in transition countries. Important research questions include: How can sustainability of biomass production (and trade) be defined in a broadly acceptable and controllable way? How can sustainability requirements be translated into guidelines and rules? How can we ensure the participation of various stakeholders in defining the sustainability requirements for SFM in transition countries? Most importantly, in these new research activities the aspects of renewable energy production, trade and use, as well as the aspects of SFM, sustainable land use and integrated rural development need to be fully interconnected in the analysis.

The research outcomes presented in this chapter have also shown that the establishment of forest plantations for controlling CO_2 emissions in the transition countries is not viable without new sources of investment. An elaboration of economic techniques for receiving credits from the world community for planting trees in these countries is, therefore, a new challenge for the future.

Acknowledgements

We are grateful for the support of the Scottish Executive Environment and Rural Affairs Department under the core project of the Macaulay Institute and of the EU Fifth Framework project of Integrated Development of Agricultural and Rural Institutions (IDARI) (QLRT-2001-02718).

Notes

1. The selection of transition countries Slovakia and Ukraine is based on the comparison of the countries' development path, which involves not only transition to market economy and restoration of private property rights, but also a need to set up a new institutional and organizational framework after achieving independence (Slovakia in 1993, Ukraine in 1991).
2. Annual wood increment constitutes 35 Mm^3, whereas harvested wood is only 12 Mm^3. In comparison e.g. in Austria, Hungary, Sweden or Switzerland 50–80% of increment is harvested.
3. Quasi rent is defined as returns that accrue to resources supplied out of human and human-created capital, and which are not attributable to natural capital. Only for human factors of production do they equal the difference between total revenue and total variable cost (van Kooten, 1995).
4. The lack of compensation for forest protection activities has already induced the 15% increase in timber production in private forests in Slovakia compared to state-owned (GRF, 2002).
5. See Nijnik (2004); Nijnik and Oskam (2004) for the analysis of institutions and governance in forestry-in-transition, with a special focus on Ukraine.
6. In Slovakia, after more than 15 years of transition, the forest laws and regulations remain a matter of government and big wood-processing entrepreneurs (SME, 2005).
7. Afforestation is an expansion of forest on land, which more than 50 years ago contained forests, but later, has been converted to other use. Reforestation is a restoration of degraded or recently (20–50 years ago) deforested lands (IBN-DLO, 1999). In this chapter, we do not make this distinction.
8. This is the maximum area suitable for tree planting. After the estimation of NPV of afforestation, the area available for tree planting was reduced at the account of lands, for which the opportunity costs appear to be comparatively high. Afforestation of up to

2 Mha of land would take roughly 25 years. It is hardly viable due to the shortage of investment and for other reasons (Nijnik, 2004).

9. For explanation of the differences between common property and community managed forests see Carlsson (1999). Using the Swedish forest commons, he argues that an introduction of community managed forests is an alternative to massive privatization as well as to undesirable continuation or strengthening of state forest management.

10. UK and Canada (BC) examples show accordingly success of afforestation on public lands and of SF.

11. For in-depth information on the potential for carbon sequestration in Ukraine, see Nijnik (2002, 2005), and in Slovakia, see Bizikova (2004).

12. Among the assumptions are as follows: afforestation and carbon storage scenario presumes one-time tree planting for 40 years, without considering the use of woodland after timber harvesting. This assumption comes with the idea that by harvesting the trees, using the revenues to cover future costs of establishing new forests and storing carbon, both the gains and losses in physical and monetary values are relatively balanced (Van Kooten *et al.*, 2000). The assumptions also include factors that would reduce the likelihood of achieving the expected carbon storing, such as a low risk that the trees will release their carbon too soon due to insect infestation, fungal disease or forest fires.

13. Costs are discounted at 4%.

14. The costs that are taken into account include tree-planting costs (including soil preparation), care and protection costs, opportunity costs of land, replanting costs and the costs of timber harvesting.

15. For Slovakia the cost of energy production from renewable sources is based on Ministry of Environment (2001). For the economics of renewable energy scenario in Ukraine see Nijnik (2005). The costs do not account for production costs of coal, for converting power plants to wood and changes in transportation costs.

16. The discounted returns from planting trees in the steppe and Polissja would be somewhat higher than the opportunity costs of the land, but they would be too low to cover the necessary silvicultural investments, care and protection costs and the costs of timber harvesting.

17. Being based on sustainability assumptions, transition countries should consider other options for fuel wood and wood products, such as export to the EU countries.

18. Recent scenarios for the emissions of CO_2 provide evidence that during the Kyoto Protocol 2008–2012 period, the surplus of Ukraine will not fall below 3 Mt of carbon per year (Victor *et al.*, 2001).

19. The situation is somewhat better in Slovakia, due to the accession process, during which the policies involving environmental *acquis* were adopted under supervision by the EU.

20. The time horizon is an important factor that influences the results. Another important factor is the discount rate employed in the models.

References

Bizikova, L. (2004) *Carbon Sequestration Potential on Agricultural Land-use Management in Slovakia.* Institute for Forecasting, Slovak Academy of Sciences, Bratislava.

Carlsson, L. (1999) *Towards a Sustainable Russian Forest Sector.* Panel 3-3, FR, 11 June, IIASA, Laxenburg, Austria.

Chomitz, K.M. (2002) Baseline, leakage and measurement issues: how do forestry and energy projects compare? *Climate Policy* 2, 35–49.

Den Elzen, M. and De Moor, P. (2002) Evaluating the Bonn–Marrakesh agreement. *Climate Policy* 2, 111–117.

EC (1997) *Energy for the Future: Renewable Sources of Energy*. White Paper for Community Strategy and Action Plan, COM (97)599 FINAL (26/11/1997), EC, Brussels.

EC (1998) *Communication from the Commission to the Council and European Parliament on a Forestry Strategy for European Union*. COM (1998) 649, 03/11/1998, EC, Brussels.

EC (2002) *EU Burden Sharing*. Council decision 2002/358/EC, EC, Brussels.

EC (2005) *Communication on the Implementation of the EU Forestry Strategy*, COM (2005) 84 final, SEC (2005) 333, EC, Brussels.

ECCP (2004) *Conclusions and Recommendations Regarding Forests Related Sinks and Climate Change Mitigation*. Final Report as seen at www.europe.eu.int/comm/environment/climate/

Eggertsson, T. (1994) The economics of institutions in transition economies. In: Schiavo-Campo, S. (ed.) *Institutional Change and the Public Sector in Transitional Economies*. Discussion paper No 241, World Bank, Washington, DC, pp. 19–50.

Fankhauser, S. and Lavric, L. (2003) The investment climate for climate investment: joint implementation in transition countries. *Climate Policy* 3, 417–434.

Fonkych, K. (2000) Rent-seeking and the interest groups in the period of transition in the Ukraine. *Economist* 3, 54–60.

Gensiruk, S. (1999) Ax is roaming in the Carpathians. *Rural News,* 11 January, Kiev (in Ukr).

Gerrard, C. (2000) The institutional perspectives on agriculture and rural development: a conceptual framework for policy makers, managers, and analysis. In: *XXIV IAAE Conference proceedings*, Berlin, pp. 1–11.

GRA and GRF (2002) *Green Report. Agriculture*. Ministry of Agriculture, Bratislava.

Gray, P.A., Demal, L., Hogg, D., Greer, D., Euler, D. and DeYoe, D. (1995) *An Ecosystem Approach to Living Sustainably: a Perspective for the Ministry of Natural Resources*. Discussion Paper, September, Queen's Printer, Peterborough, Ontario.

Hryniv, L. (2001) *Ecologically Sustainable Economy: Theoretical Problems*. University Publisher, Lviv, Ukraine (in Ukr).

IBN-DLO (1999) *Resolving Issues on Terrestrial Biospheric Sinks in the Kyoto Protocol.* Dutch National Research Programme on Global Air Pollution and Climate Change, Wageningen, The Netherlands.

IUCN/ENEP/WWF (1991) *Caring for the Earth: a Strategy for Sustainable Living*. IUCN/ENEP/WWF, Gland, Switzerland.

Kluvankova-Oravska, T. (2004) Cohesion policy and sustainable regional development. The case of the Slovak Republic. *Journal of Economics* 52, 851–864.

Komendar, V. (2001) The state committee of wiping out forests? *Rural News*: 22 May, Kiev, Ukraine (in Ukr).

Krott, M., Tikkanen, I., Petrov, A., Tunytsya, Y., Zheliba B., Sasse V., Rykounina, I. and Tunytsya, T. (2000) *Policies for Sustainable Forestry in Belarus, Russia and Ukraine*. Brill, Leiden, Boston, Koln.

Ministry of Environment (2001) *Third Report on Climate Change*. Ministry of Environment, Bratislava.

MCPFE, Second Ministerial Conference on the Protection of Forests in Europe (1993) Resolution HI: *General Guidelines for the Sustainable Management of Forests in Europe*. Helsinki, Finland. http://www.mcpfe.org/resolutions/helsinki

Nijnik, M. (2004) To an economist's perception on sustainability in forestry-in-transition. *Forest Policy and Economics* 6, 403–413.

Nijnik, M. (2005) Economics of climate change mitigation forest policy scenarios for Ukraine. *Climate Policy* 4, 319–336.

Nijnik, M. and Oskam, A. (2004) Governance in Ukrainian forestry: trends, impacts and remedies. *International Journal of Agricultural Resources, Governance and Ecology* 3, 116–133.

Nijnik, M. and Van Kooten, G.C. (2000) Forestry in the Ukraine: the road ahead? *Forestry Policy and Economics* 1, 139–153.

Petkova, E. and Faraday, G. (2001) *Good Practices in Policies and Measures for Climate Change Mitigation. A Central and Eastern European Perspective.* The Regional Environmental Centre for Central and Eastern Europe and World Resource Institute, Szentendre, Hungary.

Rice, R., Gullison, R. and Reid, J. (1997) Can sustainable management save tropical forests? *Scientific American* 276, 34–39.

Schmithusen, F. (1996) Tenure and joint resources management system on public forest lands: issues and trends. *Forstwissenschaftliche Beiträge der Professur Forstpolitik und Forstökonomie der ETH Zürich* 17, 35–55.

Schrieder, G. (2000) Theoretical concepts of institution sequencing and timing – how applicable are they to the transition process? In: *Institutional Sequencing and Timing in Transition Economies*, International Conference of Agricultural Economists, International Association of Agricultural Economists, Berlin, pp. 1–29.

SME (2005) *New Forest Code Initiative*. SME, Bratislava (in Slovak).

Spash, C.L. (2002) *Greenhouse Economics: Value and Ethics*. Routledge, London.

Swinnen, J.F.M., Buckwell, A. and Mathijs, E. (1997) *Agricultural Privatization, Land Reform and Farm Restructuring in Central and Eastern Europe*. Ashgate, Aldershot, UK.

TBFRA (2000) Temperate and Boreal Forest Resources Assessment. Forest Resources of Europe, CIS, North America, Australia, Japan and New Zealand. *UN-ECE/FAO Contribution to the Global Forest Resources Assessment 2000*, UN, Geneva.

Tichý, M. and Billharz, S. (2000) *Joint Implementation Projects in Central and Eastern Europe.* SEVEn – the energy efficiency centre, Centre for Clean Air Policy, Prague.

UNCED (1992) *Agenda 21,* UNCED, Rio, Brazil Chapter 37, part 1.

UNFCCC (1998) *The Kyoto Protocol to the Convention on Climate Change*. UNEP/IUC, Bonn, Germany.

UNFCCC (2004) *Japan Marks the Protocol's Entry into Force on 16 February*. UNFCCC, Bonn, Germany, as seen at http://unfccc.int/2860.php

Van Kooten, G. (1995) Can non market values be used as indicators of forest sustainability? *Forestry Chronicle* 71, 1–10.

Van Kooten, G.C., Shaikh, S. and Suchanek, P. (2001) Mitigating climate change by planting trees: the transaction costs trap. *Land Economics* 78, 559–572.

Victor, D., Nakicenovic, N. and Victor, N. (2001) The Kyoto Protocol emission allocations: windfall surpluses for Russia and Ukraine. *Climate Change* 49, 263–277.

5 Application of Criteria and Indicators of Sustainable Resource Management in the United States

ALBERT ABEE

United States Department of Agriculture, Forest Service, Ecosystem Management Coordination, Washington, DC, USA.

Abstract

Sustainable development, comprised of interrelated social, economic and ecological components, is a core value of the global community. We need jobs in rural and urban communities, commodities to support life processes, and a healthy environment inclusive of our heritage of plant and animal species. While sustainable development has gained worldwide prominence, difficulties remain in making progress towards that goal. Bridging information gaps in administratively fragmented landscapes, integrating environmental, economic and social issues, and the capacity to make consistent measures for assessing progress towards desired conditions that reflect sustainable development values are problematic. Within the USA and globally, we are seeing a convergence of how countries characterize and assess sustainable forest management. The Montreal Process (MP) framework of criteria and indicators (C & I) is helping to provide a unifying language of measures that clarify and better articulate understanding of key attributes that characterize the conservation and sustainable management of temperate and boreal forests. The USA is gaining experience in developing and applying C & I in both industrial and private forest certification processes as well as in state and federal forest assessment processes. The MP C & I are contributing to building both the intuitional capacity for and potential to: bridge administratively fragmented landscapes; foster dialogue and collaborative planning processes; and focus scarce resources on highest-priority areas. The MP C & I are also strengthening the linkages between countries and are contributing to the emerging mode of better science-based governance. This chapter discusses US progress in applying the MP C & I.

Keywords: Sustainable development, criteria and indicators, sustainable forest management, science, assessments, governance.

Introduction

Whether developed by federal, state, industrial or private sectors, ultimately the value of land and resource management plans will be based on, shaped by and assessed for their contribution to social, economic and environmental sustainability. Sustainability is an integral component of the evolutionary system. Our understanding of this continues to change over time. In one way or another, all countries, and, for that matter, all of life, are preoccupied with the notion of sustainability – and have been from their inception. The drive for 'survival', sustainability or sustainable development (however defined) comes from deep within the human spirit. As an attribute of the evolutionary system, sustainability can be characterized by a mere presence (enduring/continuous) or absence (extinction). Related to social constructs, sustainability in theological circles is embodied in notions such as 'nirvana' or 'everlasting life'. In the secular here and now, it is having an uninterrupted flow of resources to meet multiple definitions of 'basic needs'. What continues to change over time is not the drive to sustainably meet human needs to ensure respective definitions of well-being, but the way we characterize our needs, wants and desires, and the way we approach our work – sustainable development. This chapter discusses how a common language of criteria and indicators (C & I) is contributing to strengthening the USA's institutional capacity and potential for shared learning and decision making in seeking sustainable development.

Sustainable Development: a Common Driver for Life

Sustainable development was popularized by the World Commission on Environment and Development, chaired by Gro Harlem Brundtland (then Prime Minister of Norway) in the report, *Our Common Future*, which became known as the Brundtland Report. The report defines sustainable development as 'development that meets the needs of the present generation without compromising the ability of future generations to meet their own needs' (World Commission on Environment and Development, 1987). Thus sustainable development is an enduring ability to create jobs in rural and urban communities, the availability of commodities to support life functions and aspirations, and a healthy and productive environment providing a host of environmental services, and which is home to and inclusive of our heritage of plant and animal species.

In one way or another all countries are preoccupied with the notion of sustainable development and ways to characterize it. As understood in this chapter, the aspiring social goals of sustainable development – social equity, economic prosperity and environmental health – embody values about the kind of world in which we want to live and which we want to maintain for future generations (PCSD, 1996; Bosworth, 2001). Because human values are not fixed and depend on the social, economic and ecological context, there are multiple perspectives on what sustainable development means and how it should be achieved (Floyd, 2002). We see this in our conversations about such things as poverty eradication, global warming or how to address national security issues, for example. Thus, sustainable development is about awareness and choices regarding what to sustain, how, when, where and for whom.

The US sustainable development pathway meanders through complex, administratively fragmented landscapes (Abee, 2000). For example, Fig. 5.1 shows

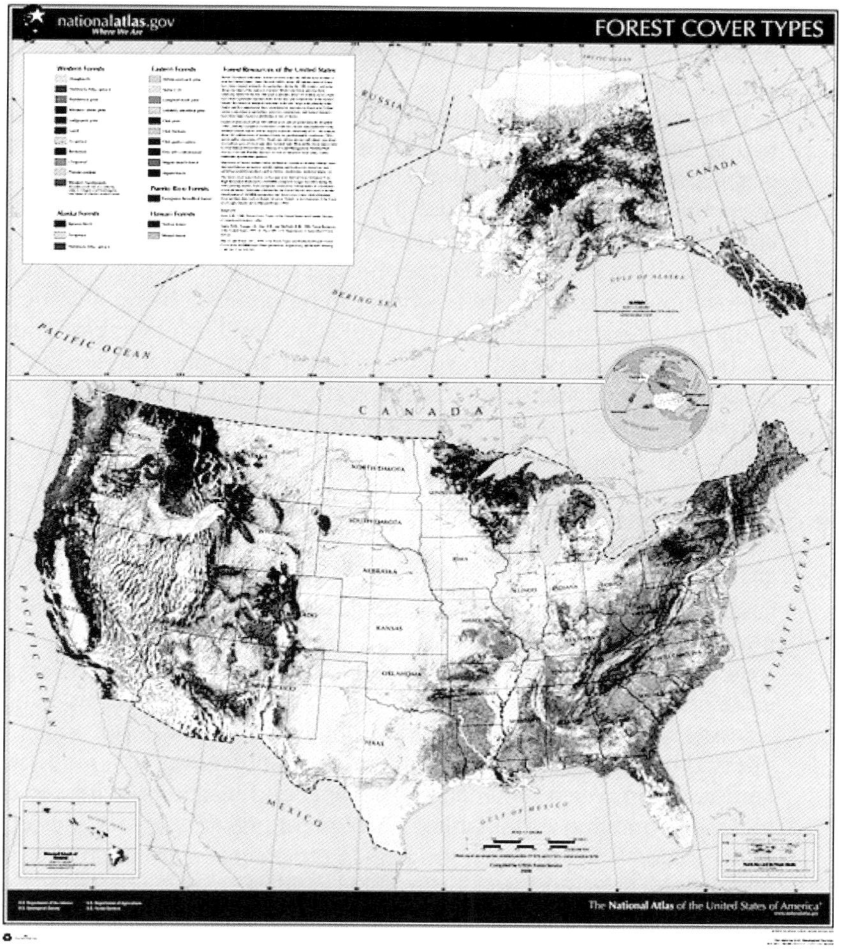

Fig. 5.1. US forest types.

that the 301,558 ha of US forest land cover the jurisdictions of the 50 states (several states have no US forest lands) (Smith *et al.*, 1997). Management is further complicated by decentralized management regimes. The Forest Service (FS) for example, has nine regional offices, 155 forest offices and about 550 district offices that manage lands in 42 states, Puerto Rico and the Virgin Islands. Each of the 50 states has a long tradition of establishing independent agencies, bureaux, departments and commissions through which public programmes are focused on forest resources (Ellefson and Moulton, 2000). Finally, landscapes are administratively fragmented by landownership patterns. For example, of the 197.9 million ha of commercial forest land, 58% is managed by over 10 million landowners, 13% by forest industries; 19% by FS and 10% by other public agencies (USDA Forest Service, 2003). These administratively fragmented forests provide a cauldron for experimentation and innovation in seeking to address shared needs, and emphasize the need for having a common language of measures across the landscape to facilitate collaborative processes designed to address shared concerns.

Sustainable development is not a destination: rather, it is viewed as a journey of incremental decisions modified through time along the path towards desired conditions. Desired conditions are the reference point for assessing progress towards the goal of sustainable development. Desired conditions may change over time. Assessing progress towards desired conditions requires judgement about the state of our communities, country and world, because it captures the relationship of ecosystem services to human well-being. Inherent in the sustainable development journey is a valuation of those tangibles and intangibles we believe should persist in space and over time, and the need to identify and agree upon key 'vital signs' (criteria and indicators) of sustainable development that serve as a barometer of the state of our values.

Use of Criteria and Indicators in the USA

The US legal and institutional framework of environmental law characterizing sustainable development is a long-term work in progress (Abee, 2000). An increasing number of legislative and administrative regulations developed by federal, state and local governments are embodying the notions and principles of sustainable development, and, while not exhaustive in their meaning, these collectively reflect early understanding that social health and public welfare are affected by, and dependent upon, natural resources and the management of the landscapes in which they occur. The legal and institutional framework mandating and promoting sustainable development precipitated decades of monitoring of associated C & I to gauge progress in protecting environmental values and regulating development through threshold guidelines and standards. For example, the USA General Accounting Office estimated that in excess of US$600 million is spent annually on monitoring conditions and trends of the nation's natural and environmental resources (GAO, 2004). Indeed, the scientific community and citizens in diverse locations and at all levels of society have developed a wide array of key indicator systems that provide economic, environmental, social and cultural information for local, state or regional jurisdiction [1]. However, despite decades of activity and billions of dollars of investment, no national system of C & I that enables the assembly of key information on environmental and social issues has been developed (National Research Council, 2000). Just as the scientific and medical community have established a 'dashboard' of key C & I that characterize and enable assessments of human health, resource partners need to do the same for natural resource management.

Criteria and Indicators as a Foundation

Social, economic and environmental sustainability issues, more often than not, cross multiple jurisdictions and are thus complex to solve or to remedy along administrative lines. International dialogues on energy, global climate change or biological diversity or domestic dialogues on employment, water quality and quantity, forest health or invasive species are examples of shared issues that require partnerships for resolution. The challenge for this century's generation of land managers is to bridge administratively fragmented landscapes to better enable collaborative efforts designed to address shared concerns. This requires new approaches, innovation and the creation of new alliances between parties that may not have worked together in the past. The ability to work together is often impeded because state, federal, private and industrial sectors have historically developed data standards and protocols independently of each other, which are not compatible or comparable with one another (GAO, 1994). Broadly accepted C & I are essential to collaborative assessment, planning and decision-making processes designed to evaluate alternative approaches to address shared concerns (GAO, 2005). Equally important is the process of embedding and integrating agreed C & I into broad-scale assessments and related land-use planning documents (GAO, 2000).

The importance of having a common language of measures in assessing the contribution of forests to human development was of particular interest at the 1992 United Nations Conference on Environment and Development (UNCED) in Rio de Janeiro. UNCED's Agenda 21, Chapter 11, suggested that countries 'formulate scientifically sound criteria and guidelines for the management, conservation, and sustainable development of all types of forests' (UN Conference on Environment and Development, 1992). Today there are nine C & I processes, including 150 countries that cover the diversity of the world's forests. The USA is a member of the Montreal Process.

Criteria and Indicators as a Common Language

The Montreal Process is widely seen as one of the success stories in strengthening the linkage between countries and in fostering the emerging mode of better science-based governance. The Montreal Process is the 'Working Group on Criteria and Indicators for the Conservation and Sustainable Management of Temperate and Boreal Forests' (Montreal Process Working Group, 1998). Twelve countries are members of the Montreal Process Working Group: Argentina, Australia, Canada, Chile, China, Japan, Mexico, New Zealand, the Republic of Korea, the Russian Federation, the USA and Uruguay. Membership in the Montreal Process is voluntary. Current members have a wide range of natural and social conditions. These countries have 90% of the world's temperate and boreal forests, 60% of all forests and 45% of world trade in forest products.

The MP began in 1994 and resulted in the collaborative development of C & I. Collaboration occurred both within countries and between countries and resulted in a framework of seven criteria and 67 indicators. The process for developing and revising indicators was an iterative process of formal and informal technical and political deliberations. From the onset, the US process for identifying C & I for inclusion into the MP C & I set included the substantial participation of scientists, stakeholders and decision makers. Involvement of the scientific community was especially important in providing the intellectual capital for integrating the social, economic and ecological components of sustainability embedded in the MP C & I framework. Involvement of the scientific community also resulted in countries recognizing the importance of assessing a country's legal and institutional framework for sustainable forest management. Contributions by individuals cannot be overstated but are not easily documented in iterative processes, as the originators of ideas quickly lose their identity. Interventions with 'power' move through technical and political processes, morphing into group consensus and adoption. For example, during the early phases of developing the MP C & I, the US working group on social and economic aspects listed suggestions for inclusion. These suggestions were subsequently discussed by a smaller group, to organize the ideas into 'indicators' for consideration by the full group. The subgroup participants noticed that most ideas did not fit into the traditional concept of 'social and economic' benefits and effects. It was clear that they represented another cluster of 'institutional and legal' factors. The social scientists in the subgroup proposed a new criterion and included within it indicators based upon the topics from the large group discussions (M.A. Shannon *et al.*, November 2005, personal communication). This suggestion moved forward in the iterative process. In the end, both domestically and internationally, stakeholders and managers agreed with those early recommendations of the scientific community, which resulted in the inclusion of Criterion 7: Legal, Institutional and Economic Framework for Forest Conservation and Sustainable Management. The MP resulted in the signing of the Santiago Declaration in 1995 (Montreal Process Working Group, 1999). The Santiago Declaration is a statement of political will that confirmed the commitment of participating countries to use the MP C & I in decision making to assess the sustainability of their nations' forests. This is not an international reporting requirement as such. Participating countries do not report to an international body. Rather, reports are for domestic application using the internationally agreed C & I.

The MP C & I framework provides a science-based framework composed of seven criteria and 67 indicators. The MP Working Group defines criteria as categories of conditions or processes by which sustainable forest management may be assessed. The criteria represent forest values that societies want to sustain or enhance. They fall into three general categories: vital functions and attributes (biodiversity, productivity, forest health, the carbon cycle and soil and water protection), socio-economic values and benefits (timber, recreation and cultural values) and the laws and regulations that comprise the forest policy framework. Indicators measure an aspect of a criterion. They identify scientific factors to assess the state of the forest and measure progress over time. Roughly half of the indicators measure economic, social or institutional concerns. While designed to be used nationally, the framework is multi-scale, including indicators with sub-national application.

At the XII World Forestry Congress in September 2003 in Quebec City, the 12 countries each released their first Country Forest Reports using the MP C & I (Montreal Process Working Group, 2003). The reports demonstrate that countries can assess respective forests using C & I in different forest ecosystems, landownership patterns and economic development. Additionally, they show that meaningful reports can be provided with less than perfect data, and can lead to a better understanding of the challenges that remain to strengthen the capacity to monitor and report using C & I. The Working Group approved a C & I revision process expected to be completed in 2005 and available for decision making by the Working Group in July 2006. For a complete listing of MP products see [3].

MP C & I Strengthen Capacity for Science-based Governance

Employing national C & I to assess country forest resource conditions is a good thing, but it is not sufficient to influence policies and decisions to achieve sustainable resource management. Just as the medical and economic sectors enjoy well-established and accepted frameworks to characterize the vital signs of human health and the economy, C & I for sustainable resource management need to be integrated into the everyday fabric of country governmental, industrial and private-sector programmes. C & I frameworks can be and are being used for a number of purposes including: monitoring forest management; accrediting or certifying forest industries with respect to how well they manage forests; guiding forest managers and policymakers to achieve better management of forests; and aiding forest institutions to prioritize resources by identifying areas that are most in need of remediation (Mendoza and Prabhu, 2000). While much work remains, many countries have made substantive progress in integrating C & I into country governmental, industrial and private-sector programmes and regulatory frameworks (Montreal Process Working Group, 2003). Together, this progress has contributed to the emerging mode of better science-based governance. Several examples of country involvement in the MP are described, with a broader amplification of US progress.

Australia

Australia's forests cover 21% of the continent and comprise about 4% of the world's forests (National Forest Inventory, 2003; United Nations Food and Agriculture Organization, 2005). Australia has three primary levels of government: commonwealth, state/territory and local. Politically, the responsibility for forest management and land allocation is with the six largely self-governing states and two mainland territories. Australian governments were quick to recognize the value of the Montreal Process framework for reporting and assessing sustainable forest management. However, Australia has realized that in order to report meaningfully within the C & I framework it had to be relevant to local social, economic and ecological environments (Howell *et al.*, 2005). Major milestones that promote sustainable forest management in Australia include:

- A major development was the implementation, for the first time, of Australia's sustainable forest management reporting framework in Australia's State of the Forest Report 2003.

- After extensive stakeholder and public consultation, in 1998, Australia's *Framework of Regional (Sub-National) Level Criteria and Indicators of Sustainable Forest Management* was published (Commonwealth of Australia, 1998). The framework, consisting of seven criteria and 74 indicators, is similar to the Montreal Process framework but reflects Australia's unique forest environment.
- Australian Forestry Standard. Australia identified economic, social and cultural parameters as key factors for assessing forests for certification. The Australian Forestry Standard utilizes a modified set of the Montreal Process criteria as the basis for the development of the standard. As a result forest managers and owners now acknowledge that, in managing forests for wood production, other aspects reflected in the C & I must be considered. These include environmental, economic and social values, with the goal of achieving environmentally responsible, socially acceptable and economically viable forest management.
- Several states have developed reports using the C & I. Australia's use of criteria and indicators has not only greatly improved the reporting of forest sustainability at the national level but also provided a basis for the monitoring and reporting of sustainable forest management at the regional and local levels (Wilson, 2005).

Australia, like many countries around the world, is using C & I to help achieve community benefit within the constraints imposed by ecological processes, ensuring that future options are not foreclosed.

Canada

Almost half of Canada's land is forest – representing 10% of the global forest area. Canada recognized at an early stage that successful development and implementation of C & I depend on the establishment and maintenance of strong linkages among those responsible for international, national, provincial and local activities. The Canadian Council of Forest Ministers (CCFM) provides leadership on C & I for SFM and sets the direction for the stewardship and sustainable management of Canada's forests. A core business activity of CCFM is to stimulate the development of policies and initiatives for the promotion of sustainable forest management in Canada. Major milestones of accomplishment for Canada include:

- Working with partners, Canada established a comprehensive national forestry database. This is used to provide forestry information for the general public and federal, provincial and territorial policy processes. The new operating framework permits and promotes enhanced coordination among the various national information and knowledge initiatives.
- The C & I are being used to help guide national-level research in Canada related to SFM.
- Canadians want a vibrant 21st-century economy and see forestry as an important part of that economy. The government is committed to enhancing

rural development by finding opportunities to add greater value to natural resources. Canada's national C & I framework features a number of indicators directly related to these goals, which offer tools to assist determining the progress towards the goals and provide guidance in making policy decisions.

- In Canada, forest management responsibilities rest with the provincial and territorial governments. Four provinces developed legislation or provincial strategies that require the use of C & I in assessing progress towards SFM. In some cases the legislation or strategies are explicitly linked to the C & I framework, using the criteria to identify important strategic directions and values. One province developed a resource evaluation policy to support its legislation, which outlines a provincial framework using C & I.
- At the forest management unit (FMU) level, several provinces and territories have developed forest management planning manuals that use C & I to assess progress towards goals and objectives.
- The industrial sector has embedded C & I as part of their annual operational plans and reports (within the International Organization for Standardization (ISO) system), and these are being incorporated into the development of their Forest Stewardship Council (FSC) certification systems. A number of indicators have a direct impact on a company's forestry operation practices and these are used to develop forest management plans.

Canada has seen the value of linkages among all levels of jurisdiction and seen improvements in the reporting of progress to SFM (Hall *et al.*, 2004).

USA

Forest covers about 23% of the USA and this represents about 6% of the world's forest cover. The USA has been a leader in broadening and deepening the application of C & I. Activities have both directly and indirectly contributed to the emerging mode of better science-based governance as well as the building of institutional capacity for potential shared learning and collaborative planning processes designed to address shared social, economic and/or environmental concerns. Evidence of this reality includes the following.

Political and constituent support for the MP C & I

Leaders of federal agencies, the states and environmental and industry NGOs recognized the value of the C & I as an important tool (American Forest and Paper Association, 1993; GAO, 1994; Keystone Center, 1996; Interagency Working Group, 1998). Early in 1997, the National Association of State Foresters (50 State Foresters) requested the national forestry agency, the USDA Forest Service (FS), to play a leadership role in using the MP C & I for integrating inventory, monitoring and assessment programmes, fostering sustainable forest management in all US forests (National Association of State Foresters, 1997). Early in 1998, a unique coalition of interest joined forces and requested the White House Council of Environmental Quality and the

Office of Management and Budget to place greater emphasis on the President's commitment to the SFM of US forests by the year 2000 (White House, 1993) through greater use of the MP C & I. Specifically, the letter requested the President to direct federal agencies with natural resource and monitoring responsibilities to collaborate with the FS to fully report on the MP C & I (National Association of State Foresters *et al.*, 1998). This unprecedented letter was signed by the National Association of State Foresters, World Wildlife Fund, American Forest and Paper Association, Society of American Forests, National Audubon Society and Global Forest Policy Project. This broad base of support has increased over time as shown by a letter to the Chief of the FS in September 2005 (NASF *et al.*, 2005). The September 2005 letter was signed by 14 national organizations from a broader coalition of interest, such as National Woodland Owners Association, the Wilderness Society and Defenders of Wildlife. The MP C & I are creating new alliances between parties that may not have worked together in the past. The FS is directly responsible for the management of 78 million ha of federal forests and grasslands of the national forest system. The agency's State and Private Forestry Program is responsible for working with the 50 state governments and private landowners and is thus indirectly responsible for promoting the sustainable management of another 206 million ha of publicly and privately owned forests and grasslands in the USA.

Forums for discussion, development and application of the C & I

In response to broad political and constituent support, and as part of the US effort to implement the MP C & I, in 1997, the FS organized a forum to discuss the development and application of the MP C & I (Dombeck, 1997). While the initial forum was specific to the MP C & I for boreal and temperate forests, managers noted that the C & I approach had broader application than to just forests (Bartlett and Maczko, 2002). Managers noted that the MP criteria were sufficiently broad to serve as an umbrella for multiple resources. Thus the initial forest forum led to the establishment of several round tables. These round tables have the same or very similar criteria but indicators have been embellished to reflect the different resources and stakeholder values and business requirements.

- Sustainable Forests Round Table. The Sustainable Forests Round Table was officially chartered in 1999 [4]. The round table includes federal agencies and non-federal organizations that meet regularly to discuss the C & I for forest management and conservation in the USA, how data for the indicators are collected and who is responsible for acquiring the data. For example, the round table sponsored workshops for technical experts to identify regional and national data sets and information gaps, to measure the C & I at the national level (Maille, 2000). Work group members found that nine of the 28 MP biological indicators have been part of FS sampling for 70 years (US Roundtable on Sustainable Forests, 2001).
- Sustainable Rangelands Round Table. The round table includes federal agencies and non-federal organizations working together to identify indicators of

sustainability based on social, economic and ecological factors, to provide a framework for national assessments of rangelands and rangeland use [5]. Round table partners have identified five criteria and 64 indicators that characterize key attributes of rangeland sustainability. Partners are working together to prepare an inter-agency 2010 national assessment using the agreed C & I.

- Sustainable Minerals and Energy Round Table. The round table includes federal agencies and non-federal organizations 'to support the nation's commitment to sustainable development' and to 'develop indicators of sustainability, based on social, economic, and environmental factors, to provide a means for assessing the status and trends of minerals/materials and energy systems' [6]. Partners have identified C & I that characterize the status of mineral and energy resources. A draft assessment using the C & I is currently under development.
- Sustainable Water Round Table. Round table participants are committed to interdisciplinary, inter-jurisdictional and cross-ownership collaboration that identifies and supports national, state and field-level activities to sustain water resources. Round table discussions and activities focus in part on criteria, indicators and methods for assessing the sustainability of water resources, as well as exploring, promoting and improving how this information is used to promote sustainable water resource management [7].

Although the round tables are diverse, they have some important similarities. They are comprised of: government and non-governmental organizations, including federal government agencies; tribal, state and local units of government; private landowners and citizens; industries and businesses; conservation and environmental groups; regional and community-based organizations; and researchers and academics. They enjoy tremendous political and constituent support. Although they are not decision-making bodies, they contribute to better decision making through the sharing of information and perspectives among individuals representing diverse interests and by promoting inter-agency and stakeholder application of the C & I. Their work has had a positive impact, such as enhancing collaboration to address public issues, and they have helped to inform decision making and improve research. The respective C & I frameworks include social, economic and environmental indicators with a multi-scale application. Additionally, the capacity of agencies represented at round tables to collect and report on indicators varies greatly. This is because either the data have not been collected in a traditional way or because there is a lack of scientific agreement on how to measure an indicator or collect data. The round tables have been instrumental in shaping how domestic forest, rangeland, water and mineral and energy resource sustainability assessments are characterized. For a complete listing of important accomplishments of the round tables, please visit their respective websites.

States

The National Association of State Foresters (NASF) is a guiding force for change in promoting sustainable forest management. About half of the states are now using MP C & I based protocols for assessment of their forests, as a tool to shape forest policy and practice, and to set priorities. Examples include:

- Oregon was the first state in the nation to embrace the MP C & I as a 'language for discussion and measurement'. In 2000, the Oregon Department of Forestry completed a 'First Approximation Report' [8] using the MP C & I. Oregon's First Approximation Report, modelled after the US First Approximation Report (Montreal Process Working Group, 1997), was the first step in an assessment process to determine data availability and data needs. Oregon selected a subset of indicators from among the 67 MP C & I to monitor and evaluate conditions and trends of Oregon's forests (Brown, 2004).
- Maryland's Department of Natural Resources used the MP C & I to develop their 2003 Strategic Forest Lands Assessment. Maryland incorporated the C & I into three computer-based models that can be used to assess the ecological and economic values of their forests as well as their vulnerability to loss to urban development (Horan and Wolf, 2004). The C & I contributed to prioritizing land for land purchase or easement agreements identified in Maryland's Green Infrastructure Matrix to retain values for citizens. The authors noted that the C & I provided data and clarity and enabled the public to see the larger picture: 'Trust was built as a result of our common language of measures'.
- The Northeast Area Association of State Foresters, an organization that represents the directors of 20 eastern state forestry agencies, adopted 18 indicators that span the seven criteria of the Montreal Process framework (USDA Forest Service, State and Private Forestry, 2002). Each of the seven criteria is represented by indicators [9].
- The development of the NASF Principles and Guides for a Well Managed Forest (National Association of State Foresters, 2003). Work on the Principles and Guides was strongly influenced by the framework provided by the MP C & I. The NASF developed these Principles and Guides as a means to assist in assessing the potential effectiveness of any system or programme's capacity to guide a forest owner or manager in efforts to achieve a well-managed forest while attaining their objectives [10].

Such application of shared criteria and indicators enables increased capacity for shared learning, collaborative planning and, moreover, improved performance on the ground. The NASF continues to encourage states to use C & I as demonstrated.

County government

At the time of writing, this is one of the most exciting developments for conservation of forest values. The Montreal Process Working Group [2] has always maintained that the C & I were developed for use at the national (or other large landscape) level. However, as noted, the C & I approach has provided a powerful conceptual framework for addressing sustainable forest management in many ways and scales of management. Over 3100 counties in the USA have vast tracts of forest land in private or county government ownership. Baltimore County is considered a national model for land use and forest protection, and is one of three counties in the nation to participate in a progressive, comprehensive new strategy for evaluating and promoting sustainable forests – from both an environmental and an economic perspective. The county has successfully completed a national pilot programme, Linking Communities to the Montreal Process Criteria and Indicators, and is moving into the full programme implementation stage, which includes sharing its experience with other local communities. Baltimore County Department of Environmental Protection and Resource Management has developed a Forest Sustainability Strategy collaboratively with citizens, advocates and business representatives. This plan includes guiding principles, goals, action steps and evaluation tools for 15 specific ecological and economic forest resource issues. Baltimore County's strategic plan is linked to the Montreal Process C & I on forest sustainability [11]. This is an excellent example of the multi-scale application of the MP C & I framework for improving conditions on the ground.

Industrial and private sectors

C & I assessment frameworks and related certification mechanisms bring together sustainability concepts as specific performance requirements and accountability processes at the management unit level. Some certification processes that use C & I are long-standing while others are parallel, co-evolved mechanisms to MP C & I. Certification systems enable managers to assess whether good management practices are being employed; they do not determine whether forests are managed sustainably. The MP C & I reflect more the sustainability of American forests, as trends captured in the C & I reflect the aggregated outcomes of both certified and uncertified forest activities. National-based systems incorporate both sustainability principles and C & I as templates to monitor long-term progress. Certification is a localized, performance-based system, relying on general standards that are independently set and use specific measures to monitor on-the-ground performance that conforms to sustainability principles. Examples of C & I application by the industrial and private sector in the USA include the following:

- The Sustainable Forest Initiative is among the strongest systems for guiding forest management towards sustainable outcomes, especially in North America. The Sustainable Forest Initiative (SFI) programme was adopted in 1994 as a condition of membership in the American Forest and Paper Association (AF&PA). The SFI programme is a comprehensive system of principles, objectives and performance measures, developed by professional foresters, conservationists and scientists, among others, that integrate the perpetual growing and harvesting of trees with the long-term protection of wildlife, plants, soil and water quality. This work is built on the MP C & I. SFI has helped bring sustainability to the forefront in the USA. There are currently over 55 million ha of forest land in North America enrolled in the SFI programme, making it among the world's largest sustainable forestry programmes. The SFI standard is overseen by the Sustainable Forestry Board, an independent non-profit organization responsible for maintaining and enhancing the SFI standard and verification/certification procedures [12].

- The Forest Stewardship Council (FSC) is among the strongest systems for guiding forest management towards sustainable outcomes. The FSC is an independent, non-profit-making, internationally oriented forest management certification system. Currently FSC standards for forest management have been applied in over 57 countries, including the FSC-US. The FSC was established in 1993 by a membership that is comprised of a balance of business, environmental and social interests. FSC accredits certifiers, who provide on-the-ground certification of sustainable forestry practices, as well as an associated 'chain of custody' certification protocol and label. The FSC system includes a set of principles and criteria for forest management that are applicable to all FSC-certified forests throughout the world. There are ten principles and 57 criteria that address legal issues, indigenous rights, labour rights, multiple benefits and environmental impacts surrounding forest management [13].
- Green Tree promotes responsible and sustainable woodland stewardship. The Green Tag programme was developed in 1998 by the National Forestry Association in cooperation with the Association of Consulting Foresters and the National Woodland Owners Association. Green Tag Forestry is a 'third-party' certification programme for non-industrial, private forest owners and is similar to the forest industry's Sustainable Forestry Initiative and to state/federal Forest Stewardship Incentive Programs. A Green Tag Forest is woodland whose stewardship has been certified as incorporating good forestry practices that ensure a balance of natural diversity and sustainable forest productivity, as defined in Green Tag's programme description [14].
- American Tree Farm is the nation's oldest programme committed to excellence in forest stewardship. The American Tree Farm programme started in 1941 and is committed to sustaining forests, watersheds and healthy habitats through the power of private stewardship. Certification of Tree Farms through the American Tree Farm System (ATFS), under the oversight of the American Forest Foundation, is the oldest and largest voluntary, third-party verification process in the USA. To be certified, property owners with more than 4 ha of forest land must have a management plan, actively manage the forest, protect it from fire and insects, protect water quality and provide for wildlife and recreation. A landowner's property is reinspected every 5 years to maintain Tree Farm certification status. Currently, ATFS has 13.4 million ha of privately owned forest land and 51,000 family forest owners who are committed to excellence in forest stewardship, in 46 states [15].

Certification mechanisms build trust and help society see the multiple dimensions of forests. They reinforce positive actions and cause society to look at the interrelationships of the economic, social and environment spheres. To date, forest certification has focused mostly on forest resources. It has been suggested that, in the future, market trading of ecosystem services could complement certification by focusing explicit attention on other goods and services, such as cleaner air and associated better human health, clean water, complements of forest-sequestered carbon, etc. – factors not automatically associated with sustainable forests, but certainly goods and services that forests create (Roussopoulos, 2005).

Enabling forest owners to actually capture what they pay for in the delivery of such environmental services, for example, may help provide market incentives to reduce the rate of forest loss due to conversion activity (Maloney, 2005).

Application of C & I to Forest Service planning processes

The USA is leading the MP countries in demonstrating the use of C & I as a means of linking together interrelated planning processes. The FS is the nation's leader on forestry issues and is using C & I to strengthen the relationships between: inventory and assessment; planning; budget formulation and execution; and monitoring and reporting (Abee, 2000). Figure 5.2 shows the role of C & I in linking FS planning and decision-making processes [16].

Column 1 (Fig. 5.2) reflects the fact that inventory and assessment information is collected and evaluated based upon selected C & I germane to informing approvals and interrelated decision-making processes. Information is derived from multiple sources and scales. The national assessment provides the context for landscape planning, and sub-national broad-scale assessments provide the context for project-level work. Criteria and indicators help shape the organization and information in national assessments.

Column 2 (Fig. 5.2) reflects the fact that the three-tiered planning process of the FS planning goals and objectives is designed to achieve desired conditions. The Chief of the FS is responsible for national planning, which is documented

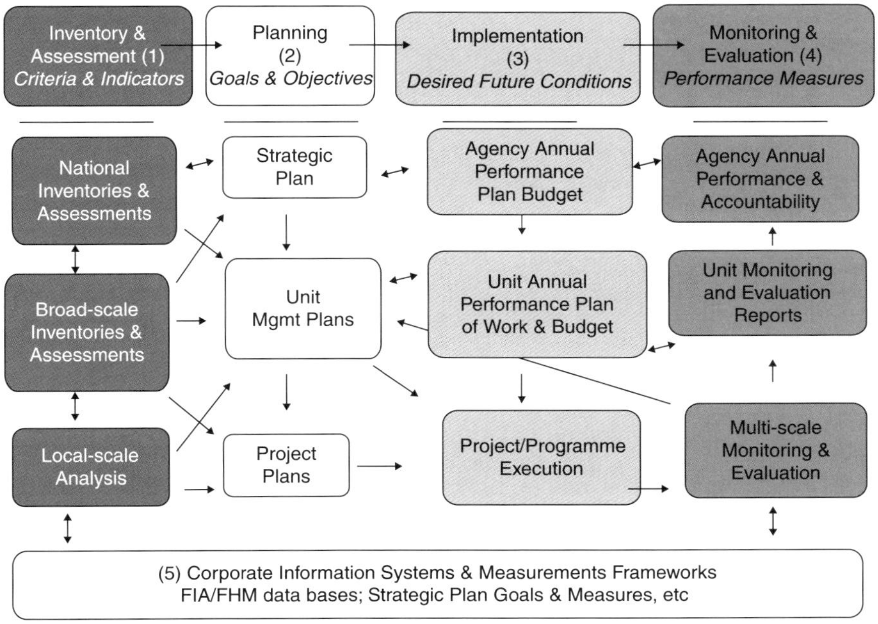

Fig. 5.2. Forest Service planning model.

in the FS Strategic Plan. After consideration of inventory and assessment information, the FS develops a Strategic Plan (SP) that discloses to the Congress and American public what the FS proposes to do with the appropriated funds received. Synergism occurs between local and national arenas in establishing strategic goals and objectives. The SP establishes sustainability goals, outcomes, performance measures and strategies that apply to National Forest System Lands as well as to the other FS mission areas (Planning Rule, 2004). The SP includes specific performance measures for gauging progress in achieving desired goals and objectives. The second phase of the planning process is the development of Unit Management Plans. Unit Management Plans (UMPs) translate the national vision into the context of the local planning area to establish management area goals, objectives, standards and guidelines. Thus the Strategic Plan establishes a national vision and context, but each UMP must have its own vision, including what is unique about the planning area, what the desired conditions are and how the UMP contributes to accomplishing the mission. In effect, the UMP is the tactical strategy for accomplishing the SP [17]. UMPs and project- and activity-level planning contribute to accomplishing the mission and are completed within the context of the Strategic Plan. The third phase of the planning process is project-level planning, which results in the implementation of site-specific activities designed to aid in achieving the goals, objectives, management direction and desired future conditions established in UMPs. Thus, it is through implementation of projects that UMPs are implemented. Projects are identified and evaluated using site-specific analysis guided by the National Forest Management Act and the National Environmental Policy Act and other laws and regulations. Project-level plans are designed to carry out specified work associated with accomplishment of annual goals.

Column 3 (Fig. 5.2) reflects budgets to move the planning area towards desired conditions. Priorities are established and Congress allocates the FS funding to implement actions designed to provide services and facilities to move the agency towards desired conditions. Priorities of the Strategic Plan are translated into annual performance plans, providing a basis for agency budget requests. The annual performance plan for a particular year establishes annual goals and objectives for what must be done in the near term in order to make progress towards the long-term desired outcomes, articulated in the Strategic Plan. The FS disaggregates congressional appropriation to field offices. For example, the National Office in Washington, DC, funds the regions, which fund the forests, which fund the districts, which execute project-level work in local communities. Unit goals, objectives and related project- and activity-level work are designed to move the agency towards the desired conditions covering local and national priorities. The aggregated outcome of project- and activity-level work reflects progress towards the desired conditions of both strategic and unit plans.

Column 4 (Fig. 5.2) reflects the fact that monitoring of performance measures is done to evaluate progress towards desired conditions. As an indicator of accountability, the agency tracks performance measures to gauge progress towards desired conditions: performance measures identified in the Annual Performance Plan are monitored and reported on annually; the measures in the Strategic Plan are monitored on a periodic basis, as described in the plan; and unit plans are also monitored to evaluate trends in resource conditions and to assess if project activities are having the desired effect. The agency's annual performance report includes information on what was accomplished relative to annual performance measures and the status of progress towards Strategic Plan objectives, based on multi-year trends. Monitoring provides data for informed decision making, particularly so that adaptive management is responsive to emerging needs and changing conditions.

The bottom row (5) (Fig. 5.2) reflects the fact that corporate information for analysis and reporting provides the foundation for consistent and effective information collection, analysis and evaluation at multiple scales for various reporting requirements and needs. Inventory and monitoring provide information necessary to evaluate the context and consequences of management options being addressed in the planning process and must be scientifically and legally defensible.

Aligning purpose to organizational structure requires some notion of internal consistency: that broad purpose leads to a specific action in a specific place. Figure 5.3 shows another perspective of how C & I enhance the performance orientation of FS planning. For example, the FS purpose is embodied in their mission statement: 'To sustain the health, diversity and productivity of the nation's forests and grasslands to meet the needs of present and future generations.' The mission statement draws upon the language and rationale of sustainable development. Criteria and indicators establish a planning framework and vocabulary to pursue the mission of sustainability (Wright *et al.*, 2002). In turn, the fulfilment of agency-wide objectives is carried out at field units, where unit plans align local needs and capabilities with national funding priorities. Project-level planning and related accomplishments are typically expressed as work outputs, which are strategically integrated to create desired outcomes and which contribute towards advancement of the FS mission.

FS planning processes have benefited tremendously from the Montreal Process. For example, in 2000, the FS published the first revision of the

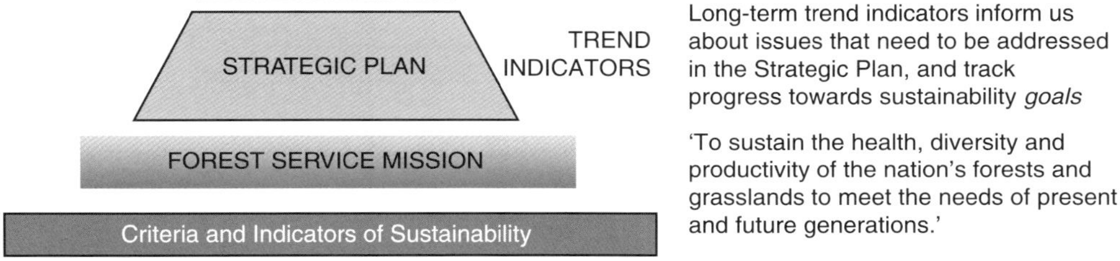

Fig. 5.3. Performance orientation model.

strategic plan with linkages to the MP C & I framework (USDA Forest Service, 2000). The connections between the Montreal Process and the 2000 Strategic Plan Revision were substantial with respect to indicators and the objectives. The 2003 SP update strengthened linkages between C & I derived from the Montreal Process and the agency's strategic goals and objectives (Grinspoon *et al.*, 2003). The objectives were largely based on contextual information from the National Report on Sustainable Forests (2003) and the 2000 RPA Assessment, both of which use C & I to assess ecological, social and economic status and trends in the nation's forest and rangelands.

Figure 5.4 shows how the FS initially blueprinted the relationship of C & I of sustainability (describing the Forest Service mission) to the annual programme of work. It begins with a warehouse of science-based knowledge – what is going on in the nation's forests and grasslands . . . what the probable condition trends are. Armed with this information, policy choices are made and priorities are set and expressed as long-term goals in the Strategic Plan. Each goal includes one or more specific and measurable objectives, which drive the annual programme direction, and in which funded work activities are aligned with these agency-wide objectives. While it looks as if there is a lot of information here, it is actually a very streamlined way of looking at the entire picture from purpose to programme, and demonstrating the alignment between the logic structure (the mission) and the business structure (the assets) (Brouha, 2004).

C & I strengthen the linkages between multi-scale planning and assessment processes (Abee and Hendricks, 1999). Local managers want to know how project-level work contributes to the agency mission and related desired conditions. Table 5.1 shows how national criteria can frame and guide the development of project-level activities and documentation. For example, MP Criterion no. 4 is

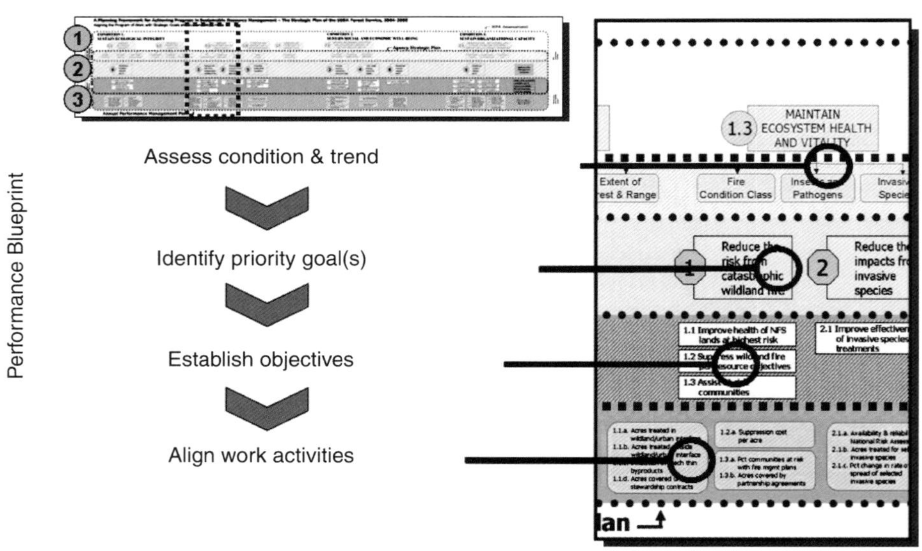

Fig. 5.4. C & I for performance-based management.

Table 5.1. Linking project activities to strategic objectives.

National criterion	Conservation of soil and water resources
Strategic plan desired outcome	Increase the area of forest and grassland watersheds in fully functional and productive condition
Strategic plan objective	Assess and restore high-priority watersheds and maintain riparian habitat in these watersheds
National performance indicator	Number of inventoried forest and grassland watersheds in fully functioning condition as a percentage of all watersheds FY 2002 baseline: 27% FY 2008 target: 41%
Unit management plan desired outcome	Increase the area of forest and grassland watersheds in fully functional and productive condition
Unit management plan objective	All class 3 streams and above in priority watersheds are at proper function condition (PFC) by 2020
Unit management plan performance indicator	Miles of stream in PFC FY 2005 baseline: 30% in PFC FY 2010 aspirations: 50% of watersheds at PFC or functional with upward trend
Project plan desired outcome	Improve Buck Creek watershed condition
Project plan objective	FY 2006: obliterate 3 miles of Mud Creek road FY 2007: implement new grazing management plan
Project plan performance indicators	Units of accomplishment, best management practices

Conservation and Maintenance of Soil and Water Resources. The Strategic Plan desired outcome is to increase the area of forest and grassland watersheds in a fully functional and productive condition. The national performance indicator is the number of inventoried forest and grassland watersheds in fully functioning condition as a percentage of all watersheds. In this example, subnational goals are expressed as achieving proper function condition on all class 3 streams and above. Local measures include miles of stream that meet the proper function condition. Existing conditions, benchmark goals and related activities are noted and serve managers as a reference point from which to gauge progress. It is important to design measures to assess progress towards mission-critical objectives to have the capacity to demonstrate programme effectiveness. The agreed-upon C & I from the respective round tables and other science-based processes mentioned here will continue to be considered in planning processes. As a result, policy objectives will be better linked to key social, economic and ecological conditions, as demonstrated in the 2004–2008 Strategic Plan update [18].

The FS planning process (as reflected in Figs 5.2, 5.3 and 5.4) is not a 'top-down' bureaucratic process without social engagement and related political processes. On the contrary, it is a very synergistic process that involves public participation throughout. Synergism occurs between national, regional and local levels. National priorities are largely summaries of needs, identified at local levels, that are broad in scope, and that are shared by multiple partners. For example, forest health and the need to reduce the risk of catastrophic fire are a national priority identified because many communities and land managers suffer catastrophic losses from wildfire. Local Unit Management Plans not only reflect such nationally derived priorities, but also localized priorities unique to respective planning areas.

C & I inform public policy and work at the project level

Figure 5.5 illustrates how C & I assessment information informs both policy and affected project-level work. Trend information confirmed that in recent years the US has experienced an increase in catastrophic fires (USDA Forest Service, 2003). Indeed, there have been significant loss of life and billions of dollars in damages. Trend information armed with fire condition class assessments of US forests enables informed decision making. President Bush and Congress met the growing forest health problem in the USA with the healthy forest initiative and the 2003 Healthy Forest Restoration Act [19]. Subsequently, congressional funding was provided to restore forest health and reduce the risk of catastrophic fire. Federal and state agencies responded accordingly. Figure 5.5 shows:

- how C & I encourage information linkages between national, regional and local scales;
- the national assessment of forest condition class identifying high risk states with significantly altered forest conditions (bottom right);

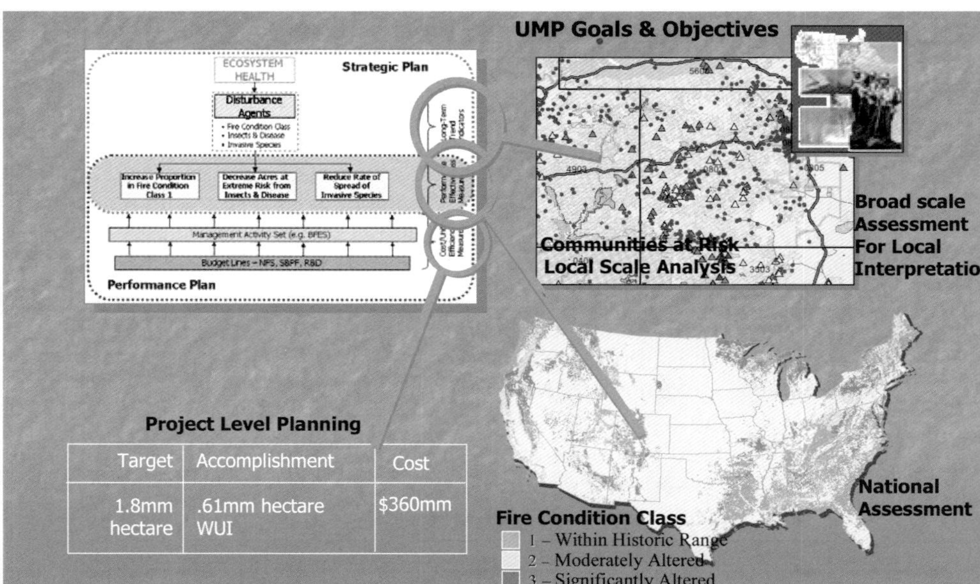

Fig. 5.5. C & I inform policy and related project work.

- how C & I were embedded in the FS Strategic Plan (top left). The desired condition is ecosystem health (MP Criterion no. 3), a long-term trend indicator is fire condition class (MP Criterion 3, indicator 3), a performance effectiveness measure is an increase in proportion of hectares in condition class 1;
- the results of a broad-scale assessment of communities at risk stepping down the results of the national assessment (top right). Broad-scale or more regionalized assessments are conducted to tease out specific watershed and related communities that have serious forest health issues. In this example, the SP identifies a reduced rate of hectares at risk from catastrophic fires as a performance measure of a desired condition;
- project-level opportunities identified within the Unit Management Planning Area (bottom left). These are funded to move the planning area towards the desired condition.

This is a practical example of how C & I can inform inventory, assessment, planning and budgeting processes. A shared set of C & I enables agencies to measure progress towards mission delivery; connects project-level plans and local activities and performance to strategic goals; helps to focus scarce resources to the highest-priority areas; and enables accountability.

Lessons Learned

The MP C & I is helping to bridge the administratively fragmented landscape. The C & I approach fosters collaboration to promote sustainable management as an important framework within many operational programmes in, for example: government, industry, universities and the private sector.

The changes noted, while not revolutionary, are more evolutionary in response to emerging needs that require new approaches to problem solving. Lessons learned in working to embed C & I in US operational programmes include the following:

- In seeking to incorporate C & I into agency operational programmes, build support within your own agency (Johnson *et al.*, 1999).
- Build support and develop partnerships with external customers. Broad public support results in management commitment.
- It takes energy and time to change the operational traditions of an organization. To provide momentum, establish an implementation team to represent the agency. Select highly motivated players who believe in, and want to be involved in, the effort. Maintain focus.
- Develop an action plan to identify specific tasks and individuals responsible for collaborative implementation within an achievable time frame for completion (Abee, 1999).
- Complexity and 'turf' or territorial issues can delay progress. Developing agreed protocols and data standards or developing mechanisms to provide compatibility between common but dissimilar data sets requires flexibility and a willingness to change.

- Begin with the end in mind. Keep agreed-to goals and objectives in focus, and build on consensus and common ground.
- Work with willing partners, respecting each other's unique roles, responsibilities and land-use objectives.
- Involve all interested parties early in the process. Be confident in enabling and encouraging shared responsibility for populating the national measurement framework with data. Do not impose constraints but define outcomes with quality assurance.
- Practitioners who use C & I information need to be involved in the development of the key questions the C & I are to inform. Managers need to know more about issues of scale, such as the relationship between national-level reporting and sub-national/forest management unit-level reporting.

Broadening and Deepening Application of C & I Information in the USA

Promoting sustainability through improved understanding

Social health and public welfare are affected by and dependent upon natural resources and the management of the landscapes in which they occur. C & I frameworks should continue to be refined to reflect ecosystem services and link to and establish the context for unit-level certification strategies. Use of C & I should promote the following desired outcomes.

C & I should enable better-informed decision making. Managers should promote an understanding of resource conditions, trends and relationships through the application of C & I. While broad support exists for the development and application of C & I, a level of concern exists within segments of the scientific community. The concern is often about the scientific basis for C & I processes, which are often value-based or otherwise influenced by political processes, rather than pure systems approaches, where linkages between specific environmental changes and the effect on human health and social and economic systems are better understood. The link between specific environmental changes, the effect on human health, social and economic systems and ecological condition is complex and often difficult to describe and remains a significant challenge. Determining causal relationships between specific management actions and changes in environmental conditions will remain problematic because such relationships cannot be fully understood. This does not diminish the need to strive to understand such relationships through systems approaches. The MP C & I framework of measures demonstrates great utility.

Education and training should focus on providing training in public involvement techniques to improve awareness of environmental and social benefits of sustainable development. An effort to develop extension activities translating science into everyday language and to bring criteria and indicator information to communities is key to success. There is a need to strengthen and clarify the relationships between national, regional and local-level monitoring and assessment using C & I.

A mantra developed by partners of the Sustainable Forest Round Table is that better data lead to better dialogue, which leads to better decisions. Round table partners should work towards: establishing shared monitoring strategies, identifying and agreeing on sources of information, establishing a collaborative national inventory platform to gather data and collecting and reporting on indicators specific to agency operations. The collaborative inventory platform envisioned would find 'homes' for each of the indicators – relative to the respective roles and responsibilities of the partners. Information and data processes should bring stakeholders together to collaboratively develop common data standards, formats and collection methods, and to develop public data-sharing and delivery systems.

Planning strategies designed to bridge a fragmented landscape

Managers should work towards reconnecting administratively fragmented landscapes by promoting planning strategies that are anchored in a framework of social, economic and environmental C & I. Agreed C & I should be embedded in federal environmental management systems to serve as a common monitoring method. Such efforts will foster opportunities that are more collaborative in design and which share responsibility among levels of government.

Promote sustainability with adaptive management

Plans, budgets and management needs should be responsive to new information and emerging needs revealed through C & I assessments. Managers should commit to a monitoring and evaluation system to assess progress from performance measures based on a common framework that is grounded in national C & I standards. Policy and guidance should be reviewed to assess statutes and regulations to remove barriers to collaboration and the development of partnerships.

Summary

Healthy ecosystems and sustainable economies are goals shared by national, county, state, private and industrial ownerships alike. Sustainable development should be viewed as having three equal and interdependent components: ecosystems must be healthy, economies must be sound and communities must be strong in order to fully meet the needs and expectations of people.

The US work environment is, from an administrative perspective, highly fragmented. Land managers face common issues that are beyond sole source solutions or remedy along administrative lines. The challenge for this century's generation of land managers is to bridge administratively fragmented landscapes to address shared concerns.

The legal and institutional framework mandating and promoting sustainability has precipitated decades of monitoring of associated C & I. Despite billions of dollars of investment, no national system of C & I has been developed to enable the assembly of key information on environmental and social issues.

There is convergence between how federal and state government, industrial-sector and private-sector managers characterize sustainable forest management in the USA. The MP C & I have promoted forums for discussion, development and implementation of C & I. This is an essential step for collaborative assessment, planning and decision-making processes to address shared concerns.

The US FS is using C & I to strengthen the relationships between inventory, assessment, planning, budget formulation and operations, and monitoring and reporting at multiple scales.

While a systems approach is important for understanding interrelationships, it should not be considered as the only approach for promoting dialogue and decision making. C & I measurement frameworks such as the MP C & I are invaluable tools to inform dialogue and decision-making processes.

There is a continued need to strengthen partnerships and integrated approaches to facilitate collaborative processes between federal and state governments, to enable the assembly of unified information on key social, economic and environmental issues.

C & I frameworks should continue to be refined to reflect ecosystem services and serve as the basis for unit-level monitoring strategies.

The MP countries have made progress in reporting on the state of their respective forests. Some countries have broadened and deepened the application of C & I to sub-national levels, as well as to regulatory frameworks.

The USA has made substantial progress in applying C & I for sustainable forest management by both federal, state and country government and industrial and private sectors. This has increased the US institutional capacity to bridge administratively fragmented landscapes, foster shared learning and collaborate to help provide key social values from US forests. This progress has contributed to the emerging notion of better science-based governance.

Acknowledgements

The FS Sustainable Development Issue Team contributed to the development and application of sustainability concepts identified herein. For more information please see http://www.fs.fed.us/sustained/msie4 Albert Abee, Chair.

Figures 5.3, 5.4 and 5.5 were developed by members of the FS Strategic Planning and Assessment Staff. http://fsweb.wo.fs.fed.us/rpa/ Paul Brouha, Director.

References

Abee, A. (1999) Reducing barriers to sustainability in a multi-jurisdictional environment. In: *Toward a Unified Framework for Inventorying and Monitoring Forest Ecosystem Resources: Mexico, Canada, US North America*. Proceedings RMRS-P-12, USDA, FS, Rocky Mountain Research Station, Fort Collins, Colorado, pp. 533–539.

Abee, A. (2000) Sustainable forest management: the application of criteria and indicator measurements in the United States Forest Service. In: Innes, J.L., Hickey, G.M. and Wilson, B. (eds), *International Perspectives on Streamlining Local-level Information for Sustainable Forest Management*. Information Report BC-X-400, Natural Resources Canada, Canadian Forest Service, Victoria, British Columbia.

Abee, A. and Hendricks, R. (1999) Possible application of Montreal Process national C & I at the sub-national level. Unpublished paper available from the Technical Advisory Committee, Montreal Process Working Group, Canadian Forest Service, Natural Resources Canada, Ottawa.

American Forest and Paper Association (1993) *Ecosystem Management: A New Approach to Federal Forest Management and Planning*. American Forest Products Association, Washington, DC.

Bartlett, H.I.R. and Maczko, K. (2002) The purpose and process of the Sustainable Rangelands Roundtable. In: *SRR Symposium Proceedings and Presentations for the Society for Range Management Conference in Kansas City, Missouri*, 17 February. http://sustainablerangelands.cnr.colostate.edu/SRM_Symposium.htm

Bosworth, D. (2001) *The Forest Service's Role in Fostering Sustainability*. Speech given by Forest Service Chief on 29 May. Society of American Foresters, National Capital Chapter, Washington, DC.

Brouha, P. (2004) Strategic planning for sustainable forests: the plan drives budgets which drive results. In: *Unifying Knowledge for Sustainability in the Western Hemisphere*. Symposium sponsored by the Consortium for Advancing Monitoring of Ecosystem Sustainability in the Americas (CAMES), Denver, Colorado, 20–24 September 2004.

Brown, J.E. (2004) Implementation of the Montreal Process: an Oregon case study. In: *Unifying Knowledge for Sustainability in the Western Hemisphere*. Symposium sponsored by the Consortium for Advancing Monitoring of Ecosystem Sustainability in the Americas (CAMES), Denver, Colorado, 20–24 September 2004.

Commonwealth of Australia (1998) *A Framework of Regional (Sub-National) Level Criteria and Indicators of Sustainable Forest Management in Australia*, Montreal Process Implementation Group, Department of Primary Industries and Energy, Canberra.

Dombeck, M.I. (1997) Letter to President, National Association of State Foresters. File code 3000, 23 July, USDA Forest Service, Washington, DC.

Ellefson, P.V. and Moulton, R.J. (2000) Fragmentation of forest resource agencies and programs: Challenges facing state and federal governments in the United States. In: *XXI IUFRO World Congress 2000 Proceedings, 7–12 August 2000*. IUFRO, Kuala Lumpur, Malaysia, pp. 589–597.

Floyd, D.W. (2002) *Forest Sustainability: the History, the Challenge, the Promise*. Forest History Society, Durham, North Carolina.

GAO (1994) *Ecosystem Management: Additional Actions Needed to Adequately Test A Promising Approach*. US GAO/RCED-94111. GAO, Washington, DC.

GAO (2000) *Forest Service Planning: Better Integration of Broad-scale Assessment Into Forest Plans is Needed*. GAO/RCED-00-56. GAO, Washington, DC.

GAO (2004) *Environmental INDICATORS: Better Coordination Is Needed to Develop Environmental Indicator Sets that Inform Decisions*. GAO-05-52, November. http://www.gao.gov/new.items/d0552.pdf

GAO (2005) *21st Century Challenges: Re-examining the Base of the Federal Government*. GAO-05-325SP. http://www.gao.gov/new.items/d05325sp.pdf

Grinspoon, E., Delfs, M. and Brouha, P. (2003) Strategic planning for sustainable forests: using criteria and indicators in the United States of America. Congress Proceedings, XII World Forestry Congress, Quebec City, Canada, September 2003, 21–28.

Hall, J.E., Bridge, S.R.J. and Haddon, B.D. (2004) Canada's experience in applying C & I to measure progress towards SFM – Perspectives from the national, regional and local levels. In: *Unifying Knowledge for Sustainability in the Western Hemisphere*, Symposium sponsored by the

Consortium for Advancing Monitoring of Ecosystem Sustainability in the Americas (CAMES), Denver, Colorado, 20–24 September 2004.

Horan, J.L. and Wolf, J.C. (2004) Maryland's strategic forest lands assessment – using indicators and models to help resource managers make decisions on the ground. In: *Unifying Knowledge for Sustainability in the Western Hemisphere*. Symposium sponsored by the Consortium for Advancing Monitoring of Ecosystem Sustainability in the Americas (CAMES), Denver, Colorado, 20–24 September 2004.

Howell, C.I., Wilson, A.D., Davey S.M. and Eddington M.M. (2005) *Sustainable Forest Management Reporting in Australia*. Bureau of Rural Sciences, Department of Agriculture, Fisheries and Forestry, Canberra, Australia.

Interagency Working Group on Sustainable Development Indicators (IWGSDI) (1998) Sustainable development in the United States: an experimental set of indicators. In: *SRR Symposium Proceedings and Presentations for the Society for Range Management Conference,* Kansas City, Missouri, 17 February 2002. http://sustainablerangelands.cnr.colostate.edu/SRM_Symposium.htm

Johnson, K., Abee, A., Alcock, G., Behler, D., Culhane, B., Holtje, K., Howlett, D., Martinez, G. and Picarelli, K. (1999) Management perspectives on regional cooperation. In: Sexton, W.T., Szaro, R.C., Johnson, N. and Malk, A. (eds) *The Interagency Stewardship Workshop: Common Reference for Ecosystem Management*. Elsevier Press, Oxford, UK, pp. 155–181.

Keystone Center (1996) *The Keystone National Policy Dialog on Ecosystem Management.* Keystone Center, Keystone, Colorado.

Maille, R. (2000) Sustainability roundtable builds partnerships. In: *USDA Forest Service International Programs Newsletter*, Issue No. 5, September 2000, Washington, DC. Available on the Internet: www.fs.fed.us/global/news/oldnewsletters/sep_00/welcome.html

Maloney, K.A. (2005) Partnering, cooperating, and collaborating for land and forest sustainability. Interview in April published in Sustainable Development e-News (SDe-News). http://www.fs.fed.us/sustained/special-feature-spring-2005.html

Mendoza, G.A. and Prabhu, R. (2000) Development of a methodology for selecting criteria and indicators of sustainable forest management: a case study on participatory assessment. *Environmental Management* 26, 659–673.

Montreal Process Working Group on Criteria and Indicators for the Conservation and Sustainable Management of Temperate and Boreal Forests (Montreal Process Working Group) (1997) *First Approximation Report of the Montreal Process*. http://www.mpci.org/

Montreal Process Working Group (1998) *Criteria and Indicators for the Conservation and Sustainable Management of Temperate and Boreal Forests*. Ottawa, Canada. Available on the Internet: www.mpci.org/rep-pub/1995/santiago_e.html#declaration.

Montreal Process Working Group (1999) *Forests for the Future: Montreal Process C & I*. Ottawa, Canada. Available on the Internet: www.mpci.org/rep-pub/1999/broch_e.html

Montreal Process Working Group (2003) *Quebec City Declaration. C & I for the Conservation and Sustainable Management of Temperate and Boreal Forests*. Montreal Process, Quebec City, Canada, 22 September.

National Association of State Foresters (1997) Letter to Chief of USDA Forest Service, Mike Dombeck, 18 June.

National Association of State Foresters (2003) *Position Statement, NASF WDc*. Website www.stateforesters.org

National Association of State Foresters, World Wildlife Fund, American Forest and Paper Association, Society of American Foresters, National Audubon Society and Global Forest Policy Project (1998) Joint letter to Council on Environmental Quality and Office of Management and Budget, 11 March.

National Association of State Foresters (NASF), Society of American Foresters, Pinchot Institute for Conservation, Wilderness Society, Nature Conservancy, American Forest and Paper Association, American Forest Foundation, National Woodland Owners Association, NatureServ,

Defenders of Wildlife, World Wildlife Fund, National Audubon Society, Sustainable Forest Partnership and American Forests (2005) Coalition letter to USDA FS Chief Dale Bosworth, 2 September.

National Forest Inventory (2003) *Australia's State of the Forests Report 2003*. Bureau of Rural Sciences, Canberra, Australia.

National Research Council (2000) *Ecological Indicators for the Nation*. National Academy Press, Washington, DC.

Planning Rule (2005) *Federal Register*, 36 CFR Part 219, 5 January 2005.

President's Council on Sustainable Development (PCSD)(1996) *Sustainable America: A New Consensus for Prosperity, Opportunity, and a Healthy Environment for the Future*. PCSD, Washington, DC.

Roussopoulos, P.J. (2005) Valuing ecosystems. Speech given on 2 February at Forest Leadership Conference, Toronto, Ontario.

Smith, B.W., Vissage, J.S., Darr, D.R. and Sheffield, R.M. (1997) *Forest Resources of the United States*. Forest Service General Technical Report GTR-NC-219, North Central Research Station, USDA, St Paul, Minnesota.

UN Conference on Environment and Development (1992) *Agenda 21: Rio Declaration. Statement of Forest Principles*.UNCED, Rio de Janeiro, Brazil.

United Nations Food and Agriculture Organization (2005) *State of the World's Forests*. FAO, Rome.

USDA Forest Service (2000) *USDA Forest Service Strategic Plan (2000 Revision)*. FS-682. WDC. http://www.srs.fs.fed.us/strategicplan

USDA Forest Service (2003) *National Report on Sustainable Forests – 2003*. FS-766. Washington, DC. http://www.fs.fed.us/research/sustain/

USDA Forest Service, State and Private Forestry, Northeastern Area. (2002) *Sourcebook on C & I of Forest Sustainability in the Northeastern Area*. NA-TP-03-02. Newtown Square, PA. http://www.na.fs.fed.us/sustainability/sourcebook.htm

US Roundtable on Sustainable Forests (2001) Roundtable on Sustainable Forests: C & I technical workshop results. http://www.sustainableforests.net/C&I_workshops/ci_index.html

White House (1993) *Presidential Decision Directive*. NSC-16, August. White House, Washington, DC.

Wilson, A.D. (2005) Criteria and indicators – a way forward for sustainable forest management. *Australian Annual Timber and Forest Review*, Australia Publishing Resource Service, Pty Ltd, Brisbane, Australia.

World Commission on Environment and Development (1987) *Our Common Future*. Oxford University Press, New York.

Wright, P.J., Colby, G.A., Hoekstra, T., Tegler, B. and Turner, M. (2002) *Monitoring for Forest Management Unit Scale Sustainability: The Local Unit C & I Development (LUCID) Test*. Report No. 5, Forest Service Inventory and Monitoring Institute, USDA, Fort Collins, Colorado.

Web References (All Accessed from October 2005 to January 2006)

[1] Government Accountability Office, Environmental Indicators, p. 55. www.gao.gov/new.items/d0552.pdf

[2] The Montreal Process. http://www.mpci.org

[3] The Montreal Process meetings. http://www.mpci.org/meetings_e.html#publications

[4] Roundtable of Sustainable Forests. www.sustainableforests.net

[5] Sustainable Rangelands Roundtable. http://sustainablerangelands.cnr.colostate.edu/

[6] Sustainable Minerals Roundtable. http://www.fs.fed.us/servicefirst/sustained/iemeindex.html
[7] Sustainable Water Resources Roundtable. http://water.usgs.gov/wicp/acwi/swrr/
[8] Oregon State First Approximation Report. www.odf.state.or.us/DIVISIONS/resource_policy/resource_planning/far/far/default.asp
[9] Northeastern Area Association of State Foresters, Sustainability Sourcebook. http://na.fs.fed.us/sustainability/sourcebook.html
[10] National Association of State Foresters – Position Statement. http://www.stateforesters.org/positions/P&G2003.htm
[11] Baltimore County Strategic Plan. http://www.fs.fed.us/global/baltimore
[12] Sustainable Forestry Initiative. www.aboutsfi.org
[13] Forest Stewardship Council. www.fsc.org/en/
[14] Green Tag Forestry. www.greentag.org/
[15] American Tree Farm System. www.treefarmsystem.org/
[16] Forest Service Manual Planning Code. http://www.fs.fed.us/emc/nfma/includes/directives/fsm1900_zero_code.pdf
[17] Forest Service Management Plan prototype. http://fsweb.wo.fs.fed.us/em/nfma/model.htm
[18] Forest Service Strategic Plan. http://fsweb.wo.fs.fed.us/rpa/
[19] Healthy Forest Restoration Act. http://fsweb.wo.fs.fed.us/hfra/

6 Indicators for Biodiversity of Tropical Forests: Problems and Solutions

KEITH RENNOLLS*[1] AND KEITH M. REYNOLDS[2]

[1]School of Computing and Mathematical Sciences, University of Greenwich, Park Row, London, UK; [2]USDA Forest Service, Pacific Northwest Research Station, Corvallis, Oregon, USA.

Abstract

Climate change has been with us for several decades. Some tropical forest tree species will be unable to disperse and will become extinct. Hence biodiversity loss is inevitable and strict sustainable forest management (SFM) is not possible in tropical forests. Some of the main international definitions of SFM consider joint optimization of a mix of competing criteria and indicators (C & I), and hence can only lead to compromise solutions in which further loss of biodiversity resources is inevitable. The complexity and diversity of tropical forest tree species means that any assessment of the state of a tropical forest is very uncertain, and almost impossible to achieve through routine cost-limited inventories. The adequacy of some current international C & I monitoring approaches are evaluated specifically in relation to the diversity of tropical forest tree species, and they are found to be inadequate for this purpose. The main weaknesses lie in the partial adoption of an inappropriate indicator methodology and insufficient consideration of uncertainty issues. It is suggested that the ecosystem approach of the Convention on Biological Diversity enacted within a modern modelling, statistical and scientific framework would enable understanding of the basic change process, and possibly aid in limiting losses of tropical biodiversity at a global scale.

Introduction

The main theme of this book is sustainable (multifunctional) forest management (SFM). The book brings together contributions from experts in four main disciplinary areas relating to this theme. The first of these disciplinary areas, Science and Policy, may be regarded as overarching because within this domain the most fundamental issues are issues of philosophy and politics at the national, international and global levels, and the general economic issue of how best to make use of the planet's limited resources. Many of the main concepts and policies relating to SFM have been formulated at the policy and political level,

* Corresponding author.

with suitable support from scientific and technological disciplines. However, in this process of policy formation for SFM, there needs to be greater linkage between the scientific and technological expertise on the one hand, and policy analysis and discussion on the other.

The other main disciplinary themes of this book may be regarded as part of the supporting scientific and technological disciplines in the SFM area. They are:

- Inventory and Monitoring – concerned with the technology and process of data collection.
- Statistics and Modelling – concerned with methodologies of data collection and data analysis, so that meaningful inferences and knowledge may be extracted from collected data.
- Information and Knowledge Management – concerned with managing information and knowledge in support of management, decision makers and policymakers.

In this chapter, we cross disciplinary boundaries to provide a critical review and analysis of the general concepts, policies and approaches of SFM, and do so with explicit reference to scientific and technological methodologies and techniques from the above-mentioned supporting disciplines. Inventories designed around criteria and indicators (C & I) have become the main operational means by which progress of the SFM process is monitored, and constitute a primary SFM methodology. It is important therefore to examine these programmes in terms of their efficacy. However, this task is too broad in scope to adequately address in a single chapter. Therefore, we shall focus our considerations on one particularly important issue that lies at the heart of the global SFM endeavour: SFM in tropical forests and the specific issue of conservation of tree species diversity. This specific theme has been chosen as indicative of the status, progress and problems of the larger SFM endeavour.

Some of the basic concepts of SFM, such as biodiversity and sustainability, have been confused at important policy levels. We have developed new, tighter and more explicit and useful definitions in the Appendix. In our view, the multifunctional framework within which SFM is pursued, while possibly justifiable from an economic or political perspective, causes considerable confusion of action and policy because of the differing scales to which each of the functionalities relate. Finally, we find that current C & I inventories are not a satisfactory means of pursuing SFM objectives, particularly with respect to the conservation of tree species diversity in tropical forests. We have no simple or easy alternatives to offer: the simple and the easy have already been tried, and are seen to be failing. Rather, we suggest that the only realistic way forward will be rather complex and difficult: the adoption of a model-based ecosystem approach.

The Threat to Global Biodiversity

Forests are one of the main repositories of global biodiversity, are a major carbon sink and provide a vast range of both timber and non-timber products for peoples of both developed and developing nations. The progressive clearance of tropical forests, the richest forests on the planet in terms of tree species diversity, means that global forest tree species biodiversity is under severe threat. Lang (2001) and [2] demonstrate a desperate situation in South-east Asian tropical forests, and the situation is becoming equally desperate in Africa and South America. Chape *et al.* (2005) have reviewed conservation efforts to conserve global biodiversity.

There have also been awareness and concern for a long time that climate change is driving changes to global ecosystems and that much of the Earth's biodiversity resource will be lost in the process:

> The current worldwide climatic deterioration may provide the tests (of a possible general theory of biodiversity that) we need; it may plunge us all into a vast, though undesired, ecological field experiment. If we observe the concomitant changes in the biosphere we may end up sadder and wiser in a very literal sense.
>
> (Pielou, 1975)

The threat demands the best efforts of humanity to attempt to avert or limit the impacts of climate change on global biodiversity, or at least to monitor those effects.

International Responses to the Threat to Global Biodiversity

The Rio Earth Summit

The Earth Summit in Rio de Janeiro in 1992 was widely heralded as a genuine ethical response from the world's leaders to the threat of global biodiversity loss, and the resulting Convention on Biological Diversity (CBD) was ratified by 175 countries. The CBD states that:

> biological diversity is about more than plants, animals and micro organisms and their ecosystems – it is about people and our need for food security, medicines, fresh air and water, shelter, and a clean and healthy environment in which to live. The Convention has three main goals: the conservation of biodiversity, sustainable use of the components of biodiversity, and sharing the benefits arising from the commercial and other utilization of genetic resources in a fair and equitable way. [5]

The CBD definition of biodiversity includes those aspects of global biodiversity that one would expect. One particular area, and the subject of this chapter, is the species diversity of trees in tropical forests. The CBD definition goes further than merely recognizing the importance of conservation of the conventional concept of biodiversity; it extends the definition of biodiversity to include the whole range of needs of the planet's human population.

The Ministerial Conference on the Protection of Forests in Europe (MCPFE)

The MCPFE [6] definition of SFM is wide in its coverage, aiming at long-term future maintenance at all spatial scales of the biodiversity, economic and social functions of forestry:

> The stewardship and use of forests and forest lands in a way, and at a rate, that maintains their biodiversity, productivity, regeneration capacity, vitality and their potential to fulfil, now and in the future, relevant ecological, economical and social functions, at local, national and global levels and that does not cause damage to other ecosystems.
>
> (Third Ministerial Conference on the Protection of Forests in Europe, Helsinki, 1993)

We see that the MCPFE also aims to maintain biodiversity in a context that satisfies the current and future economic and social needs of human populations.

What is Biodiversity, and What is it Not?

Biodiversity is generally understood to be the variety of biological life in a specified spatial domain at a given time. The biological entities included range through a large number of taxonomic levels from kingdoms to species, including *Homo sapiens*, but also ranging down to microbial and genetic levels. Conventional scientific definitions of biodiversity include the physical existence and abundance within the biological world, food and energy networks, and the dynamic and evolutionary processes that control how ecosystems adapt to changing environments.

Biodiversity measures

Quantification of biodiversity across its whole range of levels in a given region can become rather arbitrary, because it is necessary to weight the various taxonomic categories in any overall biodiversity measure. The weights might be chosen according to some attribute of the category: possibly size in terms of total number or total biomass, mean biomass or possibly even mean IQ. The appropriate choice of weights depends on the purpose of the exercise at hand.

If we focus on forest tree species that occur in a particular region, then the definition of a measure of biodiversity is simpler, but there are still many species diversity indices (Pielou, 1975; Magurran, 1988). Rényi's functional species-diversity measure (Rényi, 1961; Hill, 1973a) includes many of the popular ecological measures, including species abundance, the Shannon–Weaver index, Simpson's index and the Berger–Parker dominance index, and have been used by Rennolls and Laumonier (1999a, 2000, 2006), Kindt *et al.* (2001) and Kindt (2002) in the quantitative analysis of biodiversity in tropical forests. Species-diversity measures are essentially equivalent to the moments of the (transformed) species-abundance distribution, which is in itself probably the best characterization of species diversity. Observed distributions of species abundance have been fitted to the log-series distribution (Fisher *et al.*, 1943), the geometric distribution and the broken-stick distribution (MacArthur, 1960; Whittaker, 1972), and the log-normal distribution (Preston, 1962; May, 1975), and the fitted parameters of these distributional models have been used as biodiversity indices.

Major problems arise in the process of measuring biodiversity (in a population in a region), when the measurement is obtained by sampling. In general, sample values of diversity measures are biased, and the bias depends on the size of the sample (Fisher's α-diversity is an apparent exception). The problem arises because rare species are less likely to be observed or sampled than common species. Rennolls and Laumonier (2006) have suggested a simple way of alleviating this problem of sample bias.

Biodiversity is not . . .

The economic, social and cultural structures of human societies and their economic, social and cultural needs (now and in the future) are not normally regarded as part of the scientifically defined concept of biodiversity. The scientific definition of biodiversity becomes unclear if the needs of human beings are included within it, e.g. as in the CBD quotation in the last section. In the interests of clarity, for both science and policy, it is preferable to use the scientific definition of biodiversity, without including human needs. Human needs are crucially important, of course, but they should be regarded as conceptually distinct.

What is Sustainability?

We say that a given spatial domain and its current state are sustainable if, under a defined set of (sustainable) processes to take place in the future (in that domain, and elsewhere), they are maintained in essentially the same state in the future. Statements about sustainability necessarily involve predictions and projections about what is going to happen in the future. Such predictions are only possible by use of suitable models, but such models, and hence such predictions, are generally not available, at least not with any degree of certainty. It is for this reason that some prefer to use the term 'forest stewardship' rather than SFM (Palmer, 1996; [21]). Principles of good forest management are adopted by the Forest Stewardship Council ([21]), rather than any claim to sustainability, with its implications of steady-state stability.

Is sustainability possible?

Endemic forest tree species are adapted to their own particular, and very localized, environmental and climatic conditions. As global (and hence local) climate patterns change, there are, theoretically, only two possible survival strategies for an endemic species. First, an endemic species might be able to survive at its historical location by adapting to a new balance of competing species in the new local climate. However, because of the specificity of the adaptation/evolution of endemic species to local edaphic and climatic conditions, many endemic species will have insufficient plasticity to be able to do this. The only alternative survival strategy is that of dispersal to a new location in which the species is able to successfully compete and survive. However, endemic species have not, by definition, evolved an ability to disperse and colonize and it is unlikely that many would be able to adopt this survival strategy under climate change. Hence, there will be continuing and future extinctions of endemic forest tree species due to current climate change, just as there have been in the past. These effects will be most pronounced in tropical forests.

On a larger spatial scale, indigenous (or native) forest tree species have evolved or adapted to be able to compete and survive within specific environmental and climatic conditions. It is possible that climate change could pose the same dispersal challenge to widely distributed indigenous species, if the rate of current climate leaves no suitable dispersal corridor in space–time for such species.

The relative composition of forest tree species (i.e. the species distribution) will inevitably change at any location affected by climate change. Prasad and Iverson (1999 – ongoing) give detailed predictions of changes in spatial distributions of species that follow from predicted increases in atmospheric temperature. These predictions do not take into account the dynamics of species dispersal: the underlying model is 'steady-state' and based on the assumption that all species will be able to disperse sufficiently quickly to their new ideal ranges.

Because any strict definition of SFM is likely to include the conservation of species distributions, it should be admitted that SFM is unattainable in a strict sense (Moir and Mowrer, 1995). Attempts to conserve local forest biodiversity in the short to medium term, and within any SFM or forest stewardship programme, cannot negate the long-term effects of climate change either locally or globally. As Hulme (2003) points out, 'We all need to come to terms with climate change.'

Climate-adjusted SFM

Given that climate change is with us for the foreseeable future and that strict sustainability in terms of maintaining the status quo is not possible, it is reasonable to define a number of forms of 'climate-adjusted SFM'. Several such definitions are possible, from the viewpoint of evaluating models that might have potential for conservation and sustainability planning. See the Appendix for details. One such definition may be expressed in terms of a 'strong-rotational SFM' (SR-SFM) policy, which is defined in the context of no climate change. Hence, an SR-SFM strategy is one that will maintain a steady state (of all site and biodiversity attributes) over any periodically defined time points in repeated rotations, assuming no climate change. A 'climate-adjusted SR-SFM' (CA-SR-SFM) strategy is one that adopts an SR-SFM strategy at each point in time, as dictated by the current climatic conditions, without regard to likely future climatic change.

Although a CA-SR-SFM strategy is still suboptimal compared with a strategy that takes into account predicted changes in future climate, it has the attractive feature of being 'strictly rotational-sustainable' – that is, site and biodiversity are conserved as an explicit feature. It is rather ironic, however, in view of what has been said above about the ambiguities of multi-criteria scenarios, that an optimal climate-adjusted strategy would have to be couched in terms very similar to the multifunctional index as used in multi-criteria analysis (MCA) (Mendoza *et al.*, 2003) and as implied in the CBD statement given earlier (see Appendix for further details).

Measuring Biodiversity and Detecting Endangered Species in Tropical Forests

Continental floristic compositional differences in tropical forests depend primarily on the climatic history within which the species evolved, while at smaller scales floristic composition depends on relatively recent climate and edaphic conditions:

> no primary rain-forest species occurs in all the continental areas and very few in more than one . . . Endemic taxa confined to a small region or single locality may be evidence of past changes in climate . . . Many dipterocarps of the non-seasonal (but not the seasonal) parts of Southeast Asia are endemic to one island or an area within it . . . Within areas of a few hectares or less in extent, which can be assumed to be climatically uniform, the composition of primary mixed forest often varies in a complex and bewildering fashion.
>
> (Richards, 1998; Walsh, 1998; [4])

In tropical forests, there are many endemic species with localized distributional ranges, and relatively low dispersal rates. Such species are clearly at risk under climate change.

Most species are rare or very rare in tropical forests (Richards, 1998). For example, in a 3-ha research site at Batang Ule, Jambi, Sumatra, it was found that a total of 1897 trees of diameter greater than or equal to 10 cm were observed to fall into 497 identified species, (Trichon, 1996; Laumonier, 1997), and 216 of these species (43%) were represented by a single tree. Rennolls and Laumonier (1999b) used species–area relationships to estimate that the total number of species for an extended region like Batang Ule would be about 680 (diameter at breast height (dbh) $\geq$ 10 cm). If the minimum diameter is reduced, then the number of species in an extended Batang-Ule-type region was estimated to be in excess of 1000. These estimates indicate that 50% of species in an extended region are so rare that they are not likely to be observed in a reasonably intensive sample survey.

Established procedures for monitoring risk to biodiversity require the identification of those species for which the proportional representation of the range has radically decreased in the recent past, particularly if they are unique endemics. The IUCN (2001) defines a species as critically endangered, endangered or vulnerable in terms of observed massive reductions in stocking, normally of more than 70%. It might also be considered important for conservation purposes to monitor changes in the representation and range of species that are very rare, even though these measures might not have changed sufficiently in recent times to allow them to be classified as being endangered according to the IUCN criteria.

To fully characterize forest (tree species) biodiversity in a region at a particular time, it would be necessary to determine at least: (i) a complete species list; (ii) the proportions of tree species in the population of forest trees in the region; and (iii) the spatial range of each species; this might be localized in the region or cover the whole region and extend beyond it.

Precise estimation of the proportion of a rare attribute in a population requires very large samples: ‘any method of sampling that is adapted for general purposes is an expensive method of estimating the total number of a scarce type' (Cochran, 1977, p. 54). Hence, sample-based inventories with limited budgets will be unable to reliably estimate the stocking of most species, let alone estimate how much this stocking has changed in a given period, either in magnitude or in spatial range.

Finally, Newbery *et al.* (1999) found that tropical forests cannot be regarded as being in equilibrium, but rather that they are continually in a state of dynamic adjustment to previous events that have occurred from decades to centuries earlier. Such a finding questions the meaningfulness of attempts to monitor tropical forests in the conventional (sample-based inventory) manner for sustainability considerations.

Bubb *et al.* (2005), for UNEP-WCMC, in giving guidance on biodiversity indicators for use at a national level, recommended that an indicator is only worth considering if it addresses a question of interest to an identifiable stakeholder and if the data to generate the indicator can be obtained. In the context of tropical forests, this guidance probably amounts to a recommendation to do nothing, which is probably better than doing something that is not useful.

The Concept of Multifunctionality

Forests have always delivered many products and benefits, both timber and non-timber products, and have many uses and functions. That this has been so is now explicitly recognized by forest managers, so that current and ongoing management of forests has to be multifunctional and take into account these multiple functions simultaneously.

The extended definition of biodiversity by the CBD seems to be motivated by the desire to make it clear that there are multiple demands, needs, objectives and criteria, the satisfaction of which has to be considered simultaneously from a limited pool of resources. We have said above that such an approach clouds the definition of biodiversity, and that it is better to define each product and function separately and clearly. These may be combined into a suitable index (but not called biodiversity) in an appropriate way, ready for optimization in the management process (Mendoza *et al.*, 2003).

There are three important assumptions within the multifunctional and multi-criteria approach to SFM: first, that the system under consideration is closed (i.e. concerned with a particular closed domain); secondly, that all multiple objectives have to be satisfied as well as possible within this domain at this point in time; and, finally, that the resources and costs are fixed and related to the defined domain at this point of time. This is basically the approach adopted for SFM of a local forest management unit. Within such a restricted framework, it is normally not possible for all needs and objectives to be fully satisfied simultaneously, because they compete for an insufficient, finite and limited resource. Any proposed strategy will be a compromise, which will fail to fully satisfy each need and objective, so that each will be satisfied at least partially.

The assumptions of a closed and fixed domain, current time and fixed basis in resources and costs are all questionable in the context of SFM. Some functions of a local forest management unit will be almost certainly local in nature. That is, they normally need to be satisfied from within the defined domain. Examples are the timber that may be harvested, or the need for fuel wood of local forest-dependent communities. However, other functions, such as contribution to biodiversity conservation or contribution to regional landscape value, should not be considered at the scale of the local domain, because these services or products spread beyond the local domain to a wider spatial extent, and hence it is inappropriate to assume that they should be funded at the local level. Similarly, the various functions also differ in terms of appropriate time scales, and this means that decision making by traditional cost–benefit analysis over multiple criteria and functions is probably not the best way to proceed.

Tacconi (1995) makes a powerful case against the use of principles based on optimizing traditional economic utility in relation to biodiversity, and argues that 'conservation of biodiversity resources for future generations' is an ethical imperative, which should not be part of the utility trade-off. Of course, such conservation at the local and regional spatial scales has an economic cost. Interests in and benefits from biodiversity conservation in tropical forests are not restricted spatially to the local domains or regions in which the biodiversity conservation costs are incurred, but are shared globally. Hence, the responsibility for the costs associated with conserving biodiversity should be distributed globally.

Do Biodiversity Indicators for SFM Exist?

Biodiversity has to be maintained in any literal form of SFM. But, as pointed out in earlier discussion, under climate change this is not feasible. In tropical forests the measurement of biodiversity is impractical in a routine inventory context: detection of endangered species in tropical forests seems to be practically impossible.

A biodiversity indicator for SFM is defined (by common usage) to be a simple and easily measured forest attribute, which, when it changes, indicates that there is a change of the biodiversity status of the forest. Such an indicator, with such powerful and useful properties, would be highly desirable. We should ask, 'Do such biodiversity indicators for SFM actually exist, particularly for tropical forests?'

An analogous situation is common in mathematics. A problem is stated. The first question is always, 'Does a solution exist?' Often, it turns out that a reasonably stated problem or question has no solution, and this fact can be proved.

In the world of biodiversity monitoring, there can be no mathematical proof of the non-existence of a biodiversity indicator for SFM. However, from the considerations of the previous sections, it might be argued that there are no (simple and cheap) biodiversity indicators for SFM in tropical forests that satisfy all the requirements.

So why have simple biodiversity indicators for tropical SFM been suggested, when it seems they cannot satisfy the required properties of such indicators? Perhaps the answer lies in an inadequate methodology for developing indicators.

Indicator Methodology

Following the call of the CBD (Chapter 40 of Agenda 21, UNEP [12]) on countries and the international community to develop 'indicators of sustainable development', there has been much activity in this area. Inventories adapted for the MCPFE and the Montreal Process C & I are reasonably matched to the purposes for which they were designed, that is, for the monitoring of boreal and temperate forests. These inventories are generally extensions of well-established forest inventories in developed countries. There are none of the problems of species identification and very rare tree species that exist for tropical forests. In the temperate and boreal forests of the developed nations, relatively few tree species will be lost due to climate change. Though there are some problematic issues relating to efficient design and estimation, there are methods available by which they might be addressed (Rennolls *et al.*, Chapter 20, this volume). However, even in MCPFE and the Montreal Process, biodiversity indicators for SFM have only been validated to the extent that the chosen indicators relate to current biodiversity status. The indicators do not purport to be valid for the predictive role that is implicit in the sustainability assertion.

The situation is very different for inventories of biodiversity indicators in tropical forests, usually in developing countries, in which routine forest inventories are not an established feature. Though most sources stress the importance of a scientific and measurement-theoretic basis for any indicator and the need to evaluate uncertainty and to make accuracy statements, there is little evidence that adequate validation of indicators has taken place (Wade-Savage and Ehrlich, 1991; Hand, 2005). The FAO/Tarapoto Process Project TCP/RLA/3007, 'Validation of 15 priority indicators for the Amazon forest sustainability' [15] is an exception, though for its Indicator 13 (Criterion 12, Indicator d), 'contribution to the conservation of biological diversity', the 0–10 ordinal scale adopted is measured subjectively. Vanclay (2004) provides an example of good practice for bioindicator development in the wider context of tropical biodiversity.

There seems to be folklore associated with indicator methodology that is not entirely scientific in origin. There are analogies with the use of diagnostic indicators by clinicians and the use of performance indicators by management consultants: but in these areas there is a long historical record and clinical or economic theory to appeal to for justification (Bird *et al.*, 2005). Indicator species analysis (Clements, 1905) involves identifying the most characteristic species of a community of species. The approach has been extended by Hill (1973b, 1979) as part of a suite of techniques for community analysis. Apart from the arguments of Gleeson (1926) that species communities are chance occurrences, the techniques are not applicable when almost all species are very rare. Keystone species (Paine, 1966; Payton *et al.*, 2002) are those species regarded as playing a central and disproportionately large role in the stable structure of a community of species. The Commission for Sustainable Development [13] suggests that the indicator for biodiversity at the species level should be 'abundance of selected key species', to be used to 'represent changes in biodiversity, and the relative effectiveness of measures to maintain biodiversity'. However, techniques for identification of keystone species in tropical forests are problematic (Newbery *et al.*, 1999).

The techniques of rapid rural appraisal (RRA; McCracken *et al.*, 1988) have made much use of easily obtainable and cheap indicators as a basis for rapid assessment of status and condition, predominantly in rural communities where poverty is common and extreme and social factors predominate. Such indicators in RRA are often subjective and transient and are defined for convenience when there are no other alternatives due to constraints in financial resources. In participatory rural appraisal (PRA), the focus of the evaluation process is the human communities, and their social evaluation of the impact of policies aimed at alleviation of poverty in those communities. In such a social context, it is clear that socially based value judgements or indicators are the most appropriate measures for the purposes concerned. The use of multiple criteria analysis techniques in such a context (Mendoza and Prabhu, 2003) introduces a methodology to weight the multiple criteria and indicators involved in such social contexts, and so come to a suitable compromise or consensus decision. Biodiversity measures come into the balancing process of MCA in PRA, and it is appropriate that they do so in the local social context. However, Campbell (2001) posed some questions about the validity of the PRA methodology. Finally, participatory action research (PAR) is an approach for conducting research within society, based on postmodernist theories of sociology and anthropology. In such an approach, the scientific perspective is regarded as nothing more than a particular view of a part of (human) society. Scientifically objective sampling and measurement-based assessment are regarded as not being different in nature to measures and indicators based on subjective attitudes. PAR studies often have no baseline measurements, no aims or objectives stated prior to initiation, no replication and no measures of reliability or uncertainty. While the use of indicators in the RRA, PRA and PAR approaches (as described by the International Institute for Sustainability Development [14]) may be appropriate for exploratory or local community research, they do not seem to be a suitable basis for the objective monitoring of biodiversity, at either local, regional or global levels.

International C & I Monitoring Initiatives and Biodiversity Indicators

Inventory-based monitoring initiatives to support C & I are the main international actions that have been developed to address the issue of conservation of biodiversity and sustainable forest management since the Rio Summit. While such C & I initiatives have been largely developed and conducted in the context of the boreal and temperate forests of the developed nations (MCPFE and the Montreal Process), there are widespread arguments for the harmonization of standards of C & I inventories across the globe. Hence, what has been done in monitoring for biodiversity conservation in non-tropical forests has a direct bearing on what might be considered should be done in monitoring biodiversity conservation in tropical forests. We therefore consider how the indicators used in the MCPFE and the Montreal Process C & I inventory-based initiatives, as well as the International Tropical Timber Organization (ITTO) and Food and Agriculture Organization (FAO) C & I programmes (which have a specific tropical forest remit), might be regarded as adequate for addressing the task of monitoring tropical forest tree species diversity for conservation purposes.

MCPFE

Within Europe, the Ministerial Conference for the Protection of Forests in Europe (MCPFE [6]), first met in 1990 (pre-dating the Rio Summit), with the main function being to monitor European national biodiversity resources, and with the putative aim of achieving sustainable forest management. The monitoring methodology has involved traditional forest inventory combined with the C & I approach (see [8] for the UK report).

The MCPFE C & I methodology ('Improved Pan-European Indicators for Sustainable Forest Management' [7]) was adopted by the MCPFE Expert Level Meeting in Vienna in 2002. Its Criterion 4 for 'Maintenance, Conservation and Appropriate Enhancement of Biological Diversity in Forest Ecosystems' specifies:

> 4.1 Tree species composition: Area of forest and other wooded land, classified by number of tree species occurring and by forest type.
> 4.3 Naturalness: Area of forest and other wooded land, classified by 'undisturbed by man', by 'semi-natural' or by 'plantations', each by forest type.

These indicators, designed for European forests, are not appropriate for tropical forests, and were not meant to be.

The Montreal Process (MP)

Following the Rio Summit, a workshop in Montreal in 1993 focused specifically on criteria and indicators, and how they might help define and measure progress towards sustainable development of boreal and temperate forests. Countries with tropical forests (Brazil and the nations of Africa and of South and South-East Asia) were not initially included. The Montreal Process formally began in 1994, in Geneva, with the first meeting of the Working Group on Criteria and Indicators for the Conservation and Sustainable Management of Temperate and Boreal Forests [9]. The MP Indicators for forest tree species diversity are:

> 3.1 (a) The number of forest dependent species, (b) The status (threatened, rare, vulnerable, endangered, or extinct) of forest dependent species at risk of not maintaining viable breeding populations, as determined by legislation or scientific assessment.

These are realistic and measurable in the context of the boreal and temperate forests with which the MP is concerned, but not so for tropical forests. Abee (Chapter 5, this volume) describes how the MP C & I have been an integrative framework for multiple-resource and multiple-objective planning and decision making in the US. Seely *et al.* (2004) make use of a C & I approach, in conjunction with an ecosystem model-based approach, to demonstrate an effective multi-resource management strategy in Canada. However, the differing economic and social conditions in the developing countries, as well as the difficulty in monitoring biodiversity in tropical forests, mean that these North American models for the application of MP C & I are very unlikely to be relevant or applicable in developing countries.

Dudley and Stolton (2003) say:

> A global total of temperate and boreal forest 'undisturbed by man' of somewhere between 40 and 55% disguises the fact that the bulk of this is concentrated in a few, mainly northern boreal forest ecosystem types and that many temperate forests ecosystems have little or no natural forest remaining . . . For a substantial number of countries (and by implication for whole forest eco-regions) there are now no forest areas large enough or natural enough for country correspondents to consider them worth recording. Seventeen countries recorded no forest at all that is 'undisturbed by man' and a further 12 recorded less than 1 per cent.

While the legacy of forest diversity of the limited remaining forests undisturbed by man is clearly worth preserving in the temperate and boreal regions, it would seem that the Montreal Process or MCPFE C & I Inventories have little to offer the major challenges we face in relation to the monitoring or conservation of biodiversity of tropical forests.

International Tropical Timber Organization (ITTO)

The aim of ITTO is the promotion of the conservation and sustainable management, use and trade of tropical forest resources. ITTO stresses the importance of estimates of accuracy of reported indicators [10]. On species diversity conservation, its main criteria are:

> 5.3 Existence and implementation of procedures to identify and protect endangered, rare and threatened species of forest-dependent flora and fauna.
> 5.4 Number of endangered, rare and threatened forest-dependent species.

These are worthwhile goals, but it is clear from previous discussion that any practical implementation will be very problematic. Corresponding to Criterion 5.4, the first row of ITTO (2005) ([6], Table 23) requests, as indicators, the numbers of endangered, protected and endemic tree species, and the names of the five most important species from these threatened species. Details on rare species are not requested, even though the table header is 'Number of endangered, rare and threatened species'. IUCN (2001), in making a recommendation of what to do in circumstances of high uncertainty, says: 'A precautionary attitude will classify a taxon (species) as threatened unless it is certain that it is not threatened . . . Assessors should adopt a precautionary but realistic attitude to uncertainty when applying the criteria.' Following this principle, it seems that alarm bells should be ringing for forest management operations in virtually any natural tropical forest.

Food and Agriculture Organization of the United Nations (FAO)

FAO proposes the selection of indicator species to be simultaneously 'representative of the types of species found in the area' and to be 'those which are thought to be most susceptible to adverse effects'. This is a worthwhile aim, but, as we have seen, it is not really feasible. FAO seems to suggest verifiers as a cheaper monitoring measure than that provided by indicators, but verifiers are at least as problematic for tropical forests as the FAO indicators.

FAO guidance on the selection of biodiversity indicators is rather vague, and therefore unsatisfactory:

> Based on the information generated, and assessment of the probability that management actions will reduce risks to sustainability, it is recommended to apply a decision making system to determine if the targeted species can be deemed to be managed sustainably or, at least, cannot be deemed to be threatened under the prevailing or proposed forest management system. As the species selection process is intended to identify those species which are highly susceptible to loss of genetic variation, it can be concluded that, if these species are found to be sustainable in the management system applied, the other species (which by definition are less vulnerable and less susceptible), are also in the same condition.
>
> (FAO, 2002)

The FAO has sponsored and documented both the Tarapoto and ATO C & I proposals in South America and Africa, respectively. They both aim to monitor endangered species, in order to ensure protection. These activities of FAO are all very important. However, few details are given on proposed operational procedures for the measurement of these indicators, or the treatment of uncertainty.

Global Harmonization of the C & I Approach

In many areas of data collection and information sharing, there are good reasons why terminology and measures should be harmonized: interoperability and comparability of systems of measurement and assessment, as well as shared and common understanding, feature prominently. Having common standards would seem also to justify a harmonized policy and decision-making approach involving the allocation of resources between competing needs (Palmer, 1996).

However, different contexts will have differing measurement requirements. Harmonization or standardization of measurements is not appropriate if the differing contexts require different measures. In terms of species diversity, the boreal and temperate forests are fairly similar to each other and radically different from tropical forests, both quantitatively and qualitatively, and in terms of complexity and levels of uncertainty, system stability and likelihood of species extinctions. Hence, it seems unlikely that criteria and indicators for boreal and temperate forests can be adapted to tropical forests without drastic modifications, if at all.

SFM Certification

Kanowski *et al.* (1999) provide a good review of SFM certification issues. See also the Forest Stewardship Council [11]. However, it seems that some of the efforts to ensure certification of tropical forests for SFM make an assumption that the approaches used for C & I initiatives for boreal and temperate forests are appropriate for certification use in developing nations with tropical forests. Palmer (1996) states, 'It is, obviously, a major challenge to devise an assessment system for quality of forest management which can be applied impartially, objectively and equitably to all kinds of forests globally.' While such sentiments are admirable, it seems that, for biodiversity monitoring, a mixed strategy is likely to be better than harmonized certification efforts, which could be disastrous (Moir and Block, 2001).

The CBD Ecosystem Approach and the Need for a Modelling Framework

The fundamental conclusion of the previous sections is that biodiversity indicators for the tropics are currently inadequate. Another approach is needed. The ecosystem approach (EA), adopted by the CBD [16] as its primary framework for action, is 'based on the application of appropriate scientific methodologies focused on levels of biological organization which encompass the essential processes, functions and interactions among organisms and their environment'. This statement is a clear and reasonable adoption of the scientifically based ecosystem approach, with the intention that SFM should be based on an understanding of the whole system, and an understanding of its operation and the way in which it can be sustainably managed. Implementation of such a widely based scientific basis for SFM is rather more problematic than the intention, because fully developed scientific theories of complex dynamic ecosystems are not yet developed (Smith and Maltby, 2003; Schlaepfer *et al.*, 2004).

The development and use of such a holistic ecosystem approach and the development of a full and possibly predictive theory would demand an ecosystem modelling approach in order to capture quantitative and qualitative representations of the various processes over a range of biological levels of organization. Without mathematical and computational representations of processes, quantitative (and even qualitative) analysis and prediction are not possible. Modern science came with the birth of mathematical models of the real world. Plantation forest management is achieved entirely by the use of forest growth, yield and mortality models. Many modern biological and ecological theories are expressed in mathematical and statistical terms. Computers are the matrix of modern communication, of e-trade, of e-commerce and of e-society, and are also the means by which many modern scientific theories are implemented, tested and simulated. Hence, it is difficult to understand why (mathematical, statistical and computational) modelling approaches are never mentioned in the CBD documentation in relation to its proposed ecosystem approach. A unified ecosystem modelling approach is, however, probably the best option currently available for encapsulating an understanding of tropical forests, and the most appropriate context into which monitoring and research information can be placed and interpreted for the purposes of sustainable tropical forest management.

The history of the model-based approach to understanding ecosystems with the aim of maintaining them and conserving their biodiversity dates back to the International Biological Program (IBP) of 1964–1974 [17]. The aims of the IBP included 'biological studies focusing on the productivity of biological resources, human adaptability to environmental change, and environmental change itself', and these would seem to be shared with the CBD. Although there is no apparent model legacy from the IBP, much of the appropriate modelling methodology–infrastructure has already been well documented by the Scientific Committee on Problems of the Environment (SCOPE) [18]. The stable online contributions to SCOPE by Jeffers (1998) and Shugart *et al.* (1991) (in SCOPE 34 and 47, respectively) on modelling of long-term dynamics of ecosystems are substantial landmark contributions to the required modelling methodology. The most recent statements of the CBD suggest that C & I approaches are more mature than the ecosystem approach. The CBD seems to regard the way forward as being a convergence of the EA and the C & I approach. This may not be a suitable path in view of the above conclusions about indicators of tropical forest biodiversity for SFM.

Integrated Forest Ecosystem Models for SFM

The only way to move from data collected in the past to statements about what will happen in the future is with forecasting systems. Because sustainability is, by its very nature, a statement about the long-term future, predictive models are necessary in order to make meaningful statements about sustainability. Furthermore, the predicted long-term future steady state can only be reached through transitionary dispersal processes. Hence, systems and process models, applied to natural resource and management processes, should be considered essential components in any comprehensive approach to the evaluation of sustainability.

Politicians, policy- and decision makers, managers and administrators are understandably wary of scientific theories that are not transparent to them and that might potentially constrain their freedom to make judgements and policy decisions. Forest ecosystems and SFM are both highly complex, and a modelling approach that attempts to integrate the knowledge and understanding of such systems will not be the simple and transparent tool that policymakers and forest managers would prefer. It also has to be admitted that forest ecosystem models and SFM optimization techniques will often not be able to give clear and precise answers to the kind of questions that SFM might ask, because of the complexities and uncertainties of the situation, as well as the multiple criteria involved. This should not be regarded as a criticism of the ecosystem modelling approach to SFM. Rather, it is recognition of the complexity of the real world, and the continued need for judgements and decisions to be made by policymakers and managers, taking into account the best that science can offer in terms of advice and support.

SFM Ecosystem Modelling

In this section, we briefly review the state of the art in forest ecosystem modelling, and assess if it can provide a meaningful and useful framework for SFM policy and decision making.

Monserud (2003), in a review paper, considered the suitability of yield, gap, process and vegetation-distribution models for assessment of SFM in boreal and temperate forest stands. He used the biomass/net-primary-production SFM (BM/NPP-SFM) definition of sustainability (see the Appendix for definitions) and concluded that, while each performs adequately for the feature for which it was designed, none performs adequately for all, and therefore he recommended that hybrid models that combine the best features of yield and process-based models are needed. Rennolls and Blackwell (1988) used an integrated hybrid model involving a generic individual-tree process-based model together with stand-level constraints to evaluate the effects of fertilization treatments and insect defoliation. Liu and Ashton (1995), considering models for tropical forest management, commented, 'In the future, we expect to see more individual-based hybrid models which integrate gap models with growth-yield models and ecophysiological models.' Van Gardingen *et al.* (2003) used a model-based evaluation of the internal rate of return SFM (IRR-SFM) for tropical forests. Wallman *et al.* (2005) and Blanco *et al.* (2005) considered that SFM should use process-based and nutrient cycling models, respectively. The work of Hubble (1997, 2001) develops an individual-tree model that has the potential to faithfully characterize the species-diversity structure of a tropical forest. Thornley (Chapter 21, this volume) has provided an example of a state-of-the-art physiological process-based model that includes a soil component.

The use of forest models as a basis for multi-criteria decision analysis was considered by Huth *et al.* (2005) in the context of tropical forest. Seely *et al.* (2004) provided an impressive demonstration of the use of the forest ecosystem model of Kimmens *et al.* (1999) to develop a decision-support system for multi-objective forest management strategies.

The research referenced demonstrates that modelling methodology for forest ecosystems is well developed, and has been applied in many SFM contexts. The type of model that is most suited to SFM purposes in tropical forests is a hybrid model that is process-based and individual-tree-based and that includes a growth-yield component, a soil-nutrient component and a multilayer canopy model with associated species-diversity characteristics. However, such a model has not yet been developed, and, as far as we are aware, no person or institution has tried as yet. This is a challenge that should be put on the CBD agenda, if fully specified SFM forecasts (including biodiversity conservation) are to be developed. It is likely that extended international collaborative efforts will be needed to address this challenge. The Forest Model Archive (FMA [22]), proposed by Rennolls *et al.* (2002), is a possible infrastructure within which such developments could be managed. Such an FMA framework might also be an appropriate context within which to set up a range of SFM definitions that could form a set of benchmark criteria against which alternative models could be evaluated and compared.

Forest Ecosystem Modelling and C & I Inventories

A well-founded model-based scientific approach to ecosystem dynamics and SFM in conjunction with a well-designed inventory of appropriate measures is the ideal way in which the CBD SFM policies should be evaluated and monitored. The multiple contributions to Moldan and Billharz (1997) (SCOPE 58) on sustainability indicators provide a sound scientific basis for C & I methodology, but do not seem to have been noticed by ITTO and the FAO C & I inventory programme. Bossel (2001) has taken the systems modelling approach to indicator development to a high level of maturity.

If models were available that included species-abundance distributions (SADs) as an output, along with the more usual outputs of NPP, then it would be necessary for inventories that claim to monitor forest tree-species diversity to include measurements over time that relate to these SADs. Such measurements can only be sample-based measurements, but, together with unbiased estimation methods for species abundance, species diversity and SADs (Rennolls and Laumonier, 1999a, 1999b, 2000, 2006), a link can be made between inventory estimates and the state variables of an informative tropical forest ecosystem model. A similar comment also applies to monitoring indicators of soil nutrient status, so that possible nutrient depletion resulting from any particular forest management regime can be evaluated.

However, in a world of limited resources, monitoring resources are limited. It is clearly impossible to measure SADs and soils in detail on any large-scale continuing C & I inventory process. This does not mean that models that include such state variables should not be used. It is possible that auxiliary measurements of forest stands, obtained through remote sensing, topographic maps and local climate maps (precipitation, humidity and temperature) (Murai and Honda [4]; Chapman, 2005), could provide a sufficient form of data to meaningfully relate to measured sample SAD and soil-status variables, as well as other state variables of a stand. The remote-sensing and map data could then be used to spatially interpolate the forest ecosystems models. Some recent work along these lines has already been done in the spatial interpolation of stand assortment data from inventory sample-plot data (McRoberts *et al.*, 2002; Lemay and Temesgen, 2005).

Conclusions

International collaborative effort on SFM has taken place over the last 20 years on a number of fronts. However, it seems that these various international initiatives have not merged to provide a coherent and effective understanding and operational strategy. On the contrary, it almost seems that the various efforts have been going their own ways irrespective of each other. Extended scientific efforts, for example, those of SCOPE [3], the CBD and its ecosystem approach and the main international programmes, to define and monitor C & I for SFM hardly seem to correlate to each other, and apparently have inconsistent methodological bases.

The main conclusion is that the current indicator methodology of many high-profile C & I inventories is inadequate for monitoring biodiversity of tropical forests. We suggest that there needs to be a rapprochement of the scientific model-based approaches [19] with the operational C & I approaches. Any SFM statement is a prediction about the future. Models are essential if reasonable predictions relating to SFM and climate change are to be obtained. Dynamic forest ecosystem process models need to be developed and used to assist decision making if we as a species are to seriously and effectively address the issue of conserving global forest species.

Modern biotechnology might suggest that more can be done than merely monitoring and reacting. More proactive options such as conservation of genetic samples, possibly in seed banks, might be possible before extinction of species occur in the wild. Accurate monitoring of biodiversity resources and predictions of those species most at risk of extinction are prerequisites for such a genetic conservation programme. However, there is yet again a dichotomy in the magnitude of the challenges posed by such a programme and the likely costs, when considered for boreal and temperate forests, on the one hand, and tropical forests, on the other. It is likely that the only economically feasible option for tropical forests is the use of forest reserves (Chape *et al.*, 2005), though Rosenzweig (1995) has indicated that such a strategy could be ineffective in the long term.

Finally, it is all very well for the developed nations of the world to develop their own C & I monitoring programmes (i.e. MCPFE and MP) for temperate and boreal forests. It is also fine for the developed nations to develop their own forest ecosystem models for temperate and boreal forests (in which extreme biodiversity loss is not a threat), which they then use for multi-criteria SFM in the context of their own societies. However, significant future investments of resources in tropical forest conservation and tropical forest ecosystem modelling are needed if the aims of biodiversity conservation of the CBD are to be achieved at a global scale. The biodiversity resources involved are a global heritage, and the developing nations, in which such tropical forests exist, cannot conserve such forests or develop the required modelling and monitoring programmes alone and unaided. The responsibility must be shared internationally, with strong support needed from the developed nations for global conservation initiatives and in terms of the required research and development.

Acknowledgements

We appreciate useful comments on earlier drafts from Dr R.E. McRoberts and Dr D. Ray.

Appendix: Some Restricted Definitions of Sustainable Forest Management

Given that strict SFM is not possible under climate change, the following sequence of definitions is an attempt to move from the simplest but unsatisfactory definitions of SFM, through to definitions of SFM that are more meaningful under the climate change scenario in which we find ourselves.

1. BM/NPP-SFM

Biomass/net-primary-production SFM is demonstrated if it can be shown that the managed forest region can maintain its biomass production over ever-repeated rotations.

No explicit mention is made of soil nutrients or of biodiversity. The former is assumed to be stable and the latter is ignored.

2. IRR-SFM

A forest management practice is IRR-SFM if it can be demonstrated that a pre-specified internal rate of return can be obtained on the managed forest crop over one rotation of the management cycle.

This is an investment-oriented definition.

3. SS-SFM

Steady-state SFM is achieved if there is a steady state for all stand state variables.

State variables would have to include all biodiversity measures. The steady state has to be an average steady state, where variation is present, but the variation is essentially unchanging. The only ways that such an average steady-state is achievable is if: (i) the forest is in a naturally evolved steady state; or (ii) the forest is a managed 'normal forest', in the sense of Susuki (2005), and continuous-cover forestry (CCF).

Susuki (2005) says 'the normal wood in the wide sense is indispensable to the future of sustainable forest management'. A 'broad-band' steady-state forest sustainability definition would include continually applied management actions, such as selective removals or thinnings, and would involve the conservation of heritage, the continued access to traditional landscapes, the survival and maintenance of the species mix and biodiversity measures in the ecosystem, and the potential use of uniquely evolved biogenetic resources for a range of scientific and medical purposes. Note that the concept of a managed normal forest having a mosaic of compartments allows compartments to be clear-felled as long as long-term stability is maintained on average. These management systems seem to be equivalent to continuous-cover forestry (Von Gadow, 2002).

Rotational sustainability at the stand level

The unchanged status in all state variables is a stringent condition for forest managers, who will often aim to manage previously unmanaged forest resources for timber extraction. Most managed forest stands are managed on a rotation basis, and hence the minimal requirement for sustainability would be a definition in terms of one time-point within each rotation cycle at which to define the unchanged status of the forest.

4. SR-SFM

Strong-rotational SFM requires rotational sustainability of the forest biomass production, of the soil minerals, which will allow such long-term biomass production, and of biodiversity resources.

'Strong' is introduced into this definition, to contrast with BM/NPP-SFM, which might be regarded as 'weak-rotational' SFM. SR-SFM might be regarded as an 'ideal' sustainability target, since it includes all the concept of sustainability in the context of the rotational management treatments that normally have to be adopted in forestry. No explicit account is taken of climate change and its effects. Strict interpretation of SR-SFM would require local remediation of the climate change effects.

5. CA-SR-SFM

Climate-adjusted strong-rotational SFM is the adoption of an SR-SFM policy that would be appropriate under current conditions if there were no climate change.

Given that climate is changing for the foreseeable future, continuing to do nothing to a natural forest untouched by humans will result in change of the forest. It has to be recognized that on a global scale no other policy than no-intervention must be best for such natural forests. Remediation efforts would be impractical and ineffective in the long term. The only feasible remediation effort is to attempt to eliminate the driving climate changes, as was done in the Kyoto Protocol to the CBD.

Strict SR-SFM under climate change is not an acceptable option. CA-SR-SFM involves adopting the SR-SFM policy that would be appropriate under current conditions, under the assumption that there would be no further climate change or climate change effects. However, it may be the case that the best policy assuming no further climate change (or effects) will not be as good as the optimal forest management policy that takes into account predicted future climate change effects. Hence we come to our penultimate SFM definition:

6. OCA-SR-SFM

Optimal climate-adjusted SR-SFM is that policy which, taking into account future climate changes, maximizes an appropriately defined composite measure of biomass production (and other factors), soil nutrient status conservation and biodiversity conservation.

It is seen that the dynamic conditions enforced on us by climate change do not allow a strict and absolute requirement for the conservation of the nutrient status and the biodiversity of the forests concerned. While we have reached this definition as an optimum SFM policy, it can be seen to have fairly close similarities to the CBM and MCPFE definitions of SFM since it is couched in terms of simultaneous optimization of multiple criteria. Such a definition is problematic in its open-endedness. We therefore come to our final definition.

7. COCA-SR-SFM

Constrained climate-adjusted optimal SR-SFM requires that, at each time instant that an OCA-SR-SFM is applied, it does at least as well as a CA-SR-SFM policy in terms of the site condition and biodiversity conservation criteria.

Though COCA-SR-SFM is marginally suboptimal with respect to the composite measure mentioned in the definition of OCA-SR-SFM, it does provide the reassurance that economic, trade and poverty alleviation needs will not completely swamp considerations of site and biodiversity conservation factors.

References

Bird, S.M., Cox, D., Farewell, V.T., Goldstein, H., Holt, T. and Smith, P.C. (2005) Performance indicators: good, bad, and ugly. *J. R. Statist. Soc. A* 168(1), 1–27.

Blanco, J.A., Zavala, M.A., Imbert, J.B. and Castillo, F.J. (2005) Sustainability of forest management practices: evaluation through a simulation model of nutrient cycling. *Forest Ecology and Management* 213, 209–228.

Bossel, H. (2001) Assessing viability and sustainability: a systems-based approach for deriving comprehensive indicator sets. *Conservation Ecology* 5(2), 12. http://www.consecol.org/vol5/iss2/art12/

Bubb, P., Jenkins, M. and Kapos, V. (2005) *Biodiversity Indicators for National Use: Experience and Guidance*. UNEP-WCMC, Cambridge, UK.

Campbell, J.R. (2001) Participatory rural appraisal as qualitative research: Distinguishing methodological issues from participatory claims. *Human Organization* 60, 380–389.

Chape, S., Harrison, J., Spalding, M. and Lysenko, I. (2005) Measuring the extent and effectiveness of protected areas as an indicator for meeting global biodiversity targets. *Philosophical Transactions of the Royal Society B* 360, 443–455.

Chapman, A.D. (2005) *Uses of Primary Species-occurrence Data*, version 1.0. Report for the Global Biodiversity Information Facility, Copenhagen [19].

Clements, F.E. (1905) *Research Methods in Ecology*. University Publishing, Lincoln, Nebraska.

Cochran, W.G. (1977) *Sampling Techniques,* 3rd edn. John Wiley and Sons, New York.

Dudley, N. and Stolton, S. (2003) *Geneva Timber and Forest Discussion Paper 33: Biological Diversity, Tree Species Composition and Environmental Protection in Regional FRA-2000*. http://www.unece.org/trade/timber/docs/dp/dp-33.pdf

FAO (2002) *Criteria and Indicators for Assessing the Sustainability of Forest Management: Conservation of Biological Diversity and Genetic Variation*. http://www.fao.org/biodiversity/Forests_eco_en.asp

Fisher, R.A., Corbet, A.S. and Williams, C.B. (1943) The relation between the number of species and the number of individuals in a random sample of an animal population. *Journal of Animal Ecology* 12, 42–58.

Gleeson, H.A. (1926) The individualistic concept of the plant association. *Bulletin of the Torrey Botanical Club* 53, 7–26.

Hand, D.J. (2005) *Measurement Theory and Practice*. Arnold, London.

Hill, M.O. (1973a) Diversity and evenness: a unifying notation and its consequences. *Ecology* 54, 427–431.

Hill, M.O. (1973b) Reciprocal averaging: an eigenvector method of ordination. *Journal of Ecology* 61, 237–251.

Hill, M.O. (1979) *TWINSPAN. A Fortran Program for Arranging Multivariate Data in an Ordered Two-way Table by Classification of the Individuals and Attributes*. Cornell University, Ithaca, New York.

Hubbell, S.P. (1997) A unified theory of biogeography and relative species abundance and its application to tropical rain forests and coral reefs. *Coral Reefs* 16 (Suppl.), S9–S21.

Hubbell, S.P. (2001) *The Unified Neutral Theory of Biodiversity and Biogeography*. Princeton University Press, Princeton, New Jersey.

Hulme, M. (2003) Global climate change: climates of the future, choices for the present. In: Green, R.E., Harley, M., Miles, L., Scharlemann, J., Watkinson, A. and Watts, O. (eds) *Global Climate Change and Biodiversity*. University of East Anglia, Norwich, UK, pp. 2–4 http://www.jncc. gov.uk/page-2977

Huth, A., Drechsler, M. and Koehler, P. (2005) Using multicriteria decision analysis and a forest growth model to assess impacts of tree harvesting in dipterocarp lowland rain forests. *Forest Ecology and Management* 207, 215–232.

ITTO (2005) *Revised ITTO Criteria and Indicators for the Sustainable Management of Tropical Forests including Reporting Format*. ITTO Policy Development Series No. 15 [10].

IUCN (2001) *IUCN Red List Categories and Criteria: Version 3.1*. IUCN Species Survival Commission, IUCN, Gland, Switzerland, and Cambridge, UK. http://www.iucn.org/themes/ssc/redlists/redlistcatsenglish.pdf

Jeffers, J.N.R. (1998) *Practitioner's Handbook on the Modelling of Dynamic Change in Ecosystems*. John Wiley and Sons, New York. http://www.icsu-scope.org/downloadpubs/scope34/

Kanowski, P., Sinclair, D. and Freeman, B. (1999) *International Approaches to Forest Management Certification and Labelling of Forest Products: A Review* [20].

Kimmens, H., Mailly, D. and Seely, B. (1999) Modelling forest ecosystem net primary production: the hybrid simulation approach used in FORECAST. *Ecological Modeling* 122, 195–224.

Kindt, R. (2002) Methodology for tree species diversification planning for African agroecosystems. Thesis submitted in fulfilment of the requirement of the degree of doctor (PhD) in applied biological sciences. Faculty of Agricultural and Applied Biological Sciences, Ghent University, Ghent, Belgium.

Kindt, R., Degrande, A., Turyomurugyendo, L., Mbosso, C., Van Damme, P. and Simons, A.J. (2001) Comparing species richness and evenness contributions to on-farm tree diversity for data sets with varying sample sizes from Kenya, Uganda, Cameroon, and Nigeria with randomized diversity profiles. In: *IUFRO Conference on Forest Biometry, Modelling and Information Science, 26–29 June 2001*. University of Greenwich, London, UK. http://cms1.gre.ac.uk/conferences/iufro/proceedings/

Lang, C. (2001) Deforestation in Vietnam, Laos and Cambodia. In: Vajpeyi, D.K. (ed.) *Deforestation, Environment, and Sustainable Development: A Comparative Analysis*. Praeger, Westport, Connecticut, and London, pp. 111–137.

Laumonier, Y. (1997) *The Vegetation and Physiography of Sumatra*. Geobotany, Vol. 12, Kluwer Academic Publishers, Dordrecht, The Netherlands.

Lemay, V. and Temesgen, H. (2005) Connecting inventory information sources for landscape level analyses. *Forest Biometry, Modelling and Information Sciences (FBMIS)* 1, 37–49. http://www.fbmis.info/

Liu, J. and Ashton, P.S. (1995) Individual-based simulation models for forest succession and management. *Forest Ecology and Management* 73, 157–175.

MacArthur, R.H. (1960) On the relative abundance of species. *American Naturalist* 94, 25–36.

McCracken, A., Pretty, W. and Conway, G.R. (1988) *An Introduction to Rapid Rural Appraisal for Agricultural Development*. International Institute for Environment and Development, London.

McRoberts, R.E., Nelson, M.D. and Wendt, D.G. (2002) Stratified estimation of forest area using satellite imagery, inventory data, and the k-nearest neighbors technique. *Remote Sensing of Environment* 82, 457–468.

Magurran, A.E. (1988) *Ecological Diversity and its Measurement*. Chapman and Hall, London.

May, R.M. (1975) Patterns of species abundance and diversity. In: Cody, M.L. and Diamond, J.M. (eds) *Ecology and Evolution of Communities*. Harvard University Press, Cambridge, Massachusetts, pp. 81–120.

Mendoza, G.A. and Prabhu, R. (2003) Qualitative multi-criteria approaches to assessing indicators of sustainable forest resource management. *Forest Ecology and Management* 174, 329–343.

Mendoza, G.A., Prabhu, R., Nyirenda, R., Standa-Gunda, W. and Mutimukuru, T. (2003) A community-driven multi-criteria approach to developing indicators of sustainable resource management. *Journal of Forest Policy* 10(1), 1–21.

Moir, W.H. and Block, W.M. (2001) Adaptive management on public lands in the United States: commitment or rhetoric? *Environmental Management* 28, 141–148.

Moir, W.H. and Mowrer, H.T. (1995) Unsustainability. *Forest Ecology and Management* 73, 239–248.

Moldan, B. and Billharz, S. (eds) (1997) *Sustainability Indicators*. John Wiley and Sons, New York. http://www.icsu-scope.org/downloadpubs/scope58/

Monsuerud, R.A. (2003) Evaluating forest models in a sustainable forest management context. *Forest Biometry, Modelling and Information Sciences* 1, 35–47. www.fbmis.info

Newbery, D.M., Kennedy, D.N., Petal, G.H., Madani, L. and Ridsdale, C.E. (1999) Primary forest dynamics in lowland dipterocarp forest at Danum Valley, Sabah, Malaysia, and the role of the understory. In: Newbery, D.M., Clutton-Brock, T.H. and Prance, G.T. (eds) *Changes and Disturbances in Tropical Rainforest in South-East Asia*. Royal Society, Imperial College Press, London, pp. 39–58.

Paine, R.T. (1966) Food web complexity and species diversity. *American Naturalist* 100, 65–75.

Palmer, J. (1996) Monitoring forest practices. In: *Economic, Social and Political Issues in Certification of Forest Management*. University of British Columbia, Canada, and University of Pertanian, Malaysia.

Payton, I.J., Fenner, M. and Lee, W.G. (2002) Keystone species: the concept and its relevance for conservation management in New Zealand. *Science for Conservation* 203.

Pielou, E.C. (1975) *Ecological Diversity*. John Wiley and Sons, New York.

Prasad, A.M. and Iverson, L.R. (1999 – ongoing) *A Climate Change Atlas for 80 Forest Tree Species of the Eastern United States [database]. Northeastern Research Station, USDA Forest Service, Delaware, Ohio. http://www.fs.fed.us/ne/delaware/atlas/index.html*

Preston, F.W. (1948) The commonness and rarity of species. *Ecology* 29, 254–283.

Preston, F.W. (1962) The canonical distribution of commonness and rarity. *Ecology* 43, 185–215, 410–432.

Rennolls, K. and Blackwell, P. (1988) An integrated forest process model: its calibration and its predictive performance. *Forest Ecology and Management* 25, 31–58.

Rennolls, K. and Laumonier, Y. (1999a) Analysis of species hyper-diversity in the tropical rain forests of Indonesia: the problem of non-observance. In: Sassa, K. (ed.) *Environmental Forest Science*. Kluwer Academic Publishers, Dordrecht, Boston and London.

Rennolls, K. and Laumonier, Y. (1999b) Species–area and species–diameter curves for three forest sites in Sumatra. *Journal of Tropical Forest Science* 11, 784–800.

Rennolls, K. and Laumonier, Y. (2000) Species diversity structure analysis at two sites in the tropical rain forest of Sumatra. *Journal of Tropical Ecology* 16, 253–270.

Rennolls, K. and Laumonier, Y. (2006) A new local estimator of regional species diversity, in terms of 'shadow species', with a case study from Sumatra. *Journal of Tropical Ecology*.

Rennolls, K., Ibrahim, M.T. and Smith, P.A. (2002) A forest model archive? In: Rennolls, K. (ed.) *Proceedings of the IUFRO 4.11 Conference on Forest Biometry, Modelling and Information Science*. Greenwich, 2001. http://cms1.gre.ac.uk/conferences/iufro/proceedings/RennIbrSmithFMA.pdf

Rényi, A. (1961) On measures of entropy and information. In: Neyman, J. (ed.) *Proceedings of the 4th Berkeley Symposium on Mathematical Statistics and Probability*, Vol. 1. University of California Press, Berkeley, California, pp. 547–561.

Richards, P.W. (1998) *The Tropical Rain Forest: an Ecological Study*, 2nd edn. Cambridge University Press, Cambridge, UK.

Rosenzweig, M.L. (1995) *Species Diversity in Space and Time*. Cambridge University Press, Cambridge, UK.

Schlaepfer, R., Gorgerat, V. and Bütler, R. (2004) *A Comparative Analysis between Sustainable Forest Management (SFM) and the Ecosystem Approach (EA)*. Swiss Federal Institute of Technology, Lausanne (EPFL). http://www.iucn.org/themes/fcp/publications/files/esa_sfm/em_presentations/ rodolphe_mandat_sandra_eavs_sfm4_3_04.doc

Seely, B., Nelson, J., Wells, R., Peter, B., Meitner, M., Anderson, A., Harshaw, H., Sheppard, S., Bunnell, F.L., Kimmens, H. and Harrison, D. (2004) The application of a hierarchical, decision-support system to valuate multi-objective forest management strategies: a case study in north-eastern British Columbia, Canada. *Forest Ecology and Management* 199, 283–305.

Shugart, H.H., Bonan, G.B., Urban, D.L., Lauenroth, W.K., Parton, W.J. and Hornberger, G.M. (1991) Computer models and long-term ecological research. Chapter 12, in *Scope 47 – Long-term Ecological Research, An International Perspective*. http://www.icsu-scope.org/downloadpubs/scope47/chapter12.html

Smith, R.D. and Maltby, E. (2003) *Using the Ecosystem Approach to Implement the Convention on Biological Diversity: Key Issues and Case Studies*. Ecosystem Management Series No. 2. IUCN, Gland, Switzerland, and Cambridge, UK.

Suzuki, T. (2005) Gentan probability and the concept of the normal wood in the wide sense. *Forest Biometry, Modelling and Information Sciences* 1, 65–74. www.fbmis.info

Tacconi, L. (1995) Rethinking the economic analysis of forests: theory and practice. *Forest Ecology and Management* 73, 229–238.

Trichon, V. (1996) Hétérogénéité spatiale des structures en forêt naturelle de basse altitude á Sumatra, Indonesia. Doctoral thesis, Université Toulouse III.

Vanclay, J.K. (2004) Indicator groups and faunal richness. *Forest Biometry, Modelling and Information Sciences* 1, 105–113. http://www.fbmis.info/

Van Gardingen, P.R., McLeish, M.J., Phillips, P.D., Fadilah, D., Tyrie, G. and Yasman, I. (2003) Financial and ecological analysis of management options for logged-over dipterocarp forests in Indonesian Borneo. *Forest Ecology and Management* 183, 1–29.

Von Gadow, K. (2002) *Continuous Cover Forestry*. Kluwer Academic Publishers, Dordrecht, Boston and London.

Wade-Savage C. and Ehrlich, P. (eds) (1991) *Philosophical and Foundational Issues in Measurement Theory*. Lawrence Erlbaum, Hillsdale, New Jersey.

Wallman, P., Svensson, M.G.E., Sverdrup, H. and Belyazid, S. (2005) *ForeSAFE – an Integrated Process-oriented Forest Model for Long-term Sustainability Assessments*.

Walsh, R.P.D. (1998) Climate. In: Richards, P.W. (ed.) *The Tropical Rain Forest: an Ecological Study*, 2nd edn. Cambridge University Press, Cambridge, UK, pp. 159–205.

Whittaker, R.H. (1972) Evolution and measurement of species diversity. *Taxon* 21, 213–251.

Web References (All Accessed from October 2005 to January 2005)

[1] http://en.wikipedia.org/wiki/Kyoto_Protocol
[2] http://dte.gn.apc.org/srf1.htm
[3] http://www.icsu-scope.org/
[4] http://www-cger.nies.go.jp/grid-e/gridtxt/grid22.html World Vegetation Map (Murai and Honda; University of Tokyo)
[5] http://www.biodiv.org
[6] http://www.mcpfe.org/
[7] http://www.mcpfe.org/publications/pdf/improved_indicators.pdf
[8] http://www.forestry.gov.uk/website/pdf.nsf/pdf/fullsfi.pdf/$FILE/fullsfi.pdf
[9] http://www.mpci.org/members_e.html
[10] http://www.mpci.org/rep-pub/1995/santiago_e.html Montreal Process Criteria and Indicators
[11] http://www.itto.or.jp/live/Live_Server/963/ps15e.pdf
[12] http://www.fsc.org/en/about/about_fsc/certification
[13] http://earthwatch.unep.net/indicators/index.php
[14] http://www.un.org/esa/sustdev/
[15] http://www.iisd.org/measure/compendium/about.asp (IISD)
[16] http://www.rlc.fao.org/foro/valida/defaulte.htm
[17] http://www.biodiv.org/programmes/cross-cutting/ecosystem/
[18] http://www7.nationalacademies.org/archives/International_Biological_Program.html
[19] http://www.icsu-scope.org/
[20] http://www.gbif.org/GBIF
[21] http://www.affa.gov.au/corporate_docs/publications/pdf/forestry/certification/cert_label_review.pdf
[22] http://www.fsc.org/Forest Stewardship Council
[23] http://www.forestmodelarchive.info/Forest Model Archive (FMA)

7 Science–Policy Consultation as Boundary Spanning: the Interaction of Science and Politics in Two US Bioregional Assessments

M. PREGERNIG

Institute of Forest, Environmental and Natural Resource Policy, Department of Economics and Social Sciences, University of Natural Resources and Applied Life Sciences Vienna, Austria.

Abstract

Linking substantive knowledge and authoritative political decision making has proved to be a chronically difficult task. This chapter deals with the question of why that is so and what could possibly be done to overcome some of the barriers and close some of the gaps in the interaction between science and politics.

In theory, science–policy consultation is often framed in an indiscriminate way as the simple transfer of knowledge from science to policy. In the light of recent scholarship, however, the naive hopes of the 'scientification of the non-scientific world' turned out to be untenable, in both a theoretical and an empirical perspective. This chapter proposes a new conceptual model, the so-called 'boundary-spanning model', which frames the science–policy interface not as a sharp line of demarcation but rather as a fuzzy, dynamically shifting boundary.

Two extensive science–policy consultation processes are used to empirically substantiate this model. The science–policy advice processes investigated are two bioregional assessments that were carried out in the USA in the mid-1990s. The case studies indicate that science–policy interactions are best viewed as dynamic processes that evolve over time, occur sequentially and often iteratively and typically involve long-term interactions between scientists, policymakers, interest groups and citizens.

Introduction

Science has long been a political factor in society. In recent times, however, the interactions between science and politics have developed new qualities and unprecedented levels of intensity. Concurrently, expert scientific advice has been increasingly called upon to inform decision making, at all levels of policymaking (Glynn *et al.*, 2003). But linking substantive knowledge and authoritative political decision making has proved to be a chronically difficult task (Guston, 2001). Barriers and gaps in the interaction between science and politics result in the typical problems of decision makers not obtaining the information they need and scientists producing information that is not used (Cash *et al.*, 2002).

Also, in the field of forestry, the need for sound scientific information in the development of policies has grown significantly in recent years. Both forest policymakers and forest scientists have taken up the ensuing challenge. The International Union of Forest Research Organizations (IUFRO), for example, has set up a special 'Task Force on the Science/Policy Interface'. The mission of this task force is to develop a better understanding about the ways that forest research results influence the development and implementation of policies to protect, manage and utilize forest resources.

At the Fourth Session of the United Nations Forum on Forests (UNFF4) IUFRO had organized a side event on 'Supporting Informed Decisions: the Role of Forest Science in the Global and Regional Context'. At this event it was pointed out that forest-related scientific knowledge plays an important role in policy and decision making. At the same time, a strong potential to further strengthen interactions between the scientific community and policymakers as well as other users of forest-related scientific information was seen.

In this chapter, I want to critically discuss the chances and limits of effectively linking knowledge and action before the background of current theoretical reflections in the fields of political science and the social studies of science. I shall introduce and elaborate a new conceptual model, the 'boundary-spanning model', which strives to better describe and understand the role of science and scientists in policy processes. This boundary-spanning model will then be empirically substantiated with two extensive science–policy consultation processes that were carried out in the USA in the mid-1990s.

Theoretical Conceptualizations of Science–Policy Consultation

The interaction between science and politics or knowledge and action can be grasped theoretically in a number of ways. One of the classical conceptualizations of science–policy interaction is the 'knowledge transfer model'. Under this model scientists are brought into policy processes to impart their unique knowledge and wisdom to policymakers. Science and politics are linked in a way that could be best described with the phrase 'speaking truth to power' (Price, 1981).

The transfer model is associated with a picture of spatial separation between a place of knowledge production – science – and a place of knowledge use – politics (see Fig. 7.1). Thus, the main challenge is the way in which knowledge is 'transported' from one place to another (Nowotny, 1994). Here the 'transport routes of knowledge' are clearly traced out and are often conceptualized as 'one-way routes'. Decision makers and stakeholders are expected to have questions or demands, and scientists are expected to answer these questions or to meet these demands by providing policy-relevant solutions (Engels, 2002).

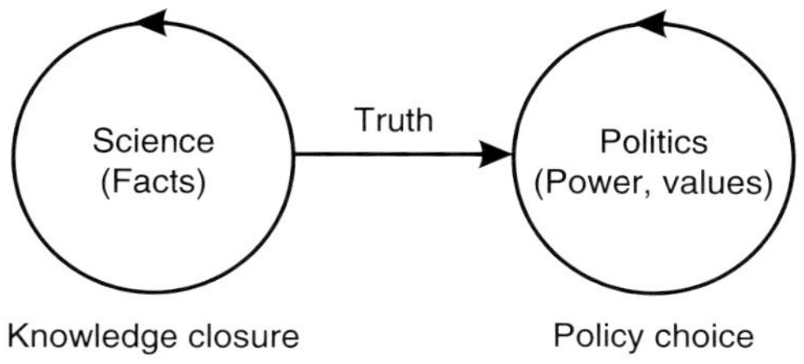

Fig. 7.1. Expert advice as 'speaking truth to power'. (Modified from Jasanoff and Wynne, 1998, p. 8.)

Under the transfer model, scientific advice is also viewed as the simple transmission of ready-made scientific results. Figure 7.1 indicates that, first, there is knowledge closure on the side of science, meaning that scientific questions are completely resolved and a finished product is handed over to policymakers. After that, policies are formulated.

In the light of recent scholarship, many of the assumptions on which the linear model of knowledge transfer is based seem questionable. Scientists can no longer – and probably never could – simply do the science and hope that someone else uses the information to make good laws that provide – in this case – for the sustainable management of forests (Cortner *et al.*, 1999).

A number of sociological studies show that scientific know-how on its way into practical fields is subject to various transformations (Beck and Bonß, 1989; Ronge, 1989). Processes of knowledge production and use are symbolic or communicative actions involving two or more parties who reciprocally affect the acceptance and rejection of knowledge claims through argument and persuasion (Dunn, 1993).

Under the transfer model, the question of where the boundary between science and politics is located, of where science ends and where politics begins, is rather unproblematic. As indicated graphically in Fig. 7.1, the social systems of science and politics are two completely separate, self-referential entities.

In contrast to that, the broader science studies literature conceptualizes the science–policy interface not as a sharp line of demarcation but rather as a fuzzy, dynamically shifting boundary (Gieryn, 1983, 1995; Jasanoff, 1987, 1990). Under this type of model, science–policy advice can best be described with the concept of 'boundary spanning'.

The boundary-spanning model implies that there are no invariant qualities that set science apart from other cultural practices and products. The separation of science from other knowledge-producing activities is rather a contextually contingent and interests-driven pragmatic accomplishment (Gieryn, 1995; Pregernig, 2005). Here, the boundary between science and politics is contested, negotiated and ultimately constructed by scientists and policymakers as they struggle to resolve the fundamental tensions of scientific advice in the policy arena, i.e.

> maintaining scientific credibility (by not politicizing the research) while assuring practical saliency (by producing information that is relevant and useful to decision makers) and doing so in a manner that secures political legitimacy (by being seen as fair and open to multiple participants).
>
> (Cash and Clark, 2001, p. 8)

While under the linear model of knowledge transfer the influence of science on policy is conceptualized in a quite straightforward way, the boundary-spanning model calls for a more differentiated look at some key concepts. Especially the notions of 'scientific expertise', 'scientific input' and 'policy impact' have to be modelled in a more refined way.

Under the boundary-spanning model, science–policy interactions are seen as dynamic processes that evolve over time, occur sequentially and often iteratively and typically involve long-term interactions between scientists, policymakers, interest groups and citizens (Miller *et al.*, 1997). As a consequence, the simple dualism between science, on the one hand, and politics, on the other, has to be replaced by a more refined gradation that distinguishes between different types of experts and expertise, from academic scientists, technical consultants and agency specialists to NGO experts and holders of local knowledge.

Analogously, scientific input into policy processes cannot be reduced to the production and transfer of a product – often in the form of a written report – but rather has to been seen as a social process. Much about what makes some exercises of science–policy advice more effective than others seems to be associated with the process by which they are developed, rather than just the product itself (Cash and Clark, 2001).

Finally, also in the evaluation of policy impacts a broader conceptual look is needed. Many policy actions are not 'decided' in a brisk and clear-cut style, but decisions rather take shape gradually; policies 'accrete' through small uncoordinated steps taken in many different places (Weiss, 1980). With that, the effectiveness of science advice can not only be evaluated through its ultimate impacts on the (ecological, social and economic) environment, but has to be assessed across a broad spectrum of policy-relevant factors: changing strategies and behaviour of key actors, putting an issue on the policy agenda or raising its visibility, mobilizing support, building actor networks or institutional capacity, identifying knowledge gaps and needs or building knowledge communities (Knott and Wildavsky, 1980; Rich, 1997; Cash and Clark, 2001).

Specific Research Questions and Empirical Design

Previous empirical research on science–policy consultation has frequently come to rather negative conclusions as to the impacts that science has in the application context. Often building upon the transfer model of knowledge use, those empirical studies were typically looking for an ideal situation where policymakers consider the findings of a particular study in the context of a specific pending decision and adopt the course of action recommended by (or derived directly from) the research (Knott and Wildavsky, 1980; Weiss, 1980). Knowledge utilization studies show that this kind of instrumental utilization seems in fact to be rare, particularly when the issues are complex, the consequences are uncertain and a multitude of actors are engaged in the decision-making process – as is the case, for example, in sustainable forest management (Webber, 1992).

It would be wrong to conclude from this that research is ignored in any case. It only does not come in the form of direct, instrumental use but rather in a more indirect way, which is indicated by the term 'boundary spanning'. Part of the deficiencies located in science–policy advice are attributable to an excessively narrow conceptual and empirical focus. Many of the case studies that critically analysed specific consultation processes have taken a predominantly project-oriented view, i.e. their unit of analysis was a specific research project, with clearly defined start and end dates. My guiding hypothesis is that many of the characteristics that make up a 'successful' science–policy consultation exercise cannot be determined empirically by just looking at 'the project' itself but, rather, it needs a broader conceptual framework.

In the following, I want to substantiate my methodological meta-hypothesis by reporting on two in-depth case studies from the field of forest and natural resource policy. The trans-disciplinary ventures I have investigated are two 'bio-regional assessments' that were carried out in the USA in the mid-1990s.

My investigations are mainly based upon 41 in-depth expert interviews with the key actors involved. With the interviewees belonging to different groups (university scientists, agency scientists, resource managers, policymakers, interest-group representatives) it was possible to point out the specific scope and context of action in which the different actor groups are operating. The interviews were built upon the principles of open interviewing and lasted between 45 min and 4 h. All interviews were tape-recorded, fully transcribed and finally analysed by means of qualitative content analysis. In addition to the expert interviews, publicly available records, media reports and transcripts of congressional hearings were analysed.

Description of Cases

Bioregional assessments, as a special type of science–policy assessments, are large-scale efforts to integrate a broad range of information about the social, economic and ecological conditions within a larger region in order to provide a basis for making decisions and taking policy action. They are bioregional, which means that they are ecosystem-based, delineated by natural processes and elements rather than by planning units and political jurisdiction (Herring, 1999). And they are assessments, i.e. social processes by which expert knowledge related to a policy problem is organized, evaluated, integrated and presented in documents and otherwise to inform decision making (Farrell *et al.*, 2001). Assessments are important forums for political negotiation and interaction between scientists, natural resource managers, policymakers and the public.

The following empirical analysis is based on two bioregional assessments, which have both dealt with large-scale mountain ranges: the Sierra Nevada Ecosystem Project and the Southern Appalachian Assessment.

The Sierra Nevada Ecosystem Project (SNEP) is an assessment of the Sierra Nevada ecoregion, which is located in the states of California and Nevada in the south-west of the USA. In 1992, there was growing public concern about the ecological health of that celebrated mountain range. A flashy newspaper story and strong lobbying by environmental interest groups led the US Congress to authorize a scientific review, which was intended to resolve the controversy about the condition of the range (Erman, 1999). The University of California was requested to coordinate this independent, scientific study. Pursuant to that request, a team of approximately 130 scientists, agency staff and consultants was formed. Three years and US$6.5 million later, a four-volume, 3200-page-long report was submitted to Congress (Erman, 1996).

The Southern Appalachian Assessment (SAA) dealt with the six-state southern Appalachian mountain area in the south-eastern part of the USA. Similarly to SNEP, the project grew from concerns about the state of the environment in the area. The final impetus for the assessment came from the upcoming revision of forest plans in several national forests of the US Forest Service. So, once again, it was not science that elicited the assessment but social – or, in this case, agency – needs. In 1994, the cooperating partners of the multi-agency Southern Appalachian Man and the Biosphere (SAMAB) Cooperative decided to collaborate on an assessment of the status and condition of the ecological resources in the southern Appalachian region. One and a half years and a relatively meagre US$1.8 million later, SAA published four technical documents and one summary report (SAMAB, 1996). The assessment was, for the most part, financed and carried out by the Forest Service and other state and federal agencies. Also most of the 150 scientists who participated in the assessment came from management agencies. With that, SAA is mainly seen as an agency effort and not so much as a scientific exercise (Meidinger and Shannon, 1996; van Sickle, 2001).

Results of Empirical Analysis

After having sketchily introduced my two bioregional assessments, I shall now try to undertake a twofold endeavour: first, I shall look into the empirical question as to if and to what extent science–policy assessments serve as forums for the creation of policy-relevant knowledge. Parallel to that, I shall try to develop and substantiate an expanded conceptual framework for the analysis of science–policy consultation processes.

Direct impact of assessments

Science–policy assessments are not designed to produce new, cutting-edge knowledge but they primarily serve the development of concrete solutions for practice. Assessments generate and/or collect individual research efforts to answer policy-relevant questions and otherwise provide technical advice for decision makers (Farrell *et al.*, 2001). So when evaluating a specific assessment process the central question is whether and to what extent the assessment has led to changes in the way public and private policies are formulated, i.e. what policy impact it had.

One of the main reasons why the bioregional assessments in the Sierra Nevada and the southern Appalachians had been commissioned was that the management plans in most of the national forests in the regions were awaiting revision. SNEP and SAA were, *inter alia*, intended to provide specific information for the US Forest Service on those upcoming forest planning activities. Now, what impact did the assessments actually have?

After the assessments were finished, the Forest Service drew heavily upon the assessment reports. In one of my interviews a Forest Service officer named SAA as 'a springboard for forest plans in the region'. And the Regional Forester stated in a news release that '[t]he new plans directly descend from the success of the Southern Appalachian Assessment'. Similarly, the Sierra Nevada Framework for Conservation and Collaboration, which in the Forest Service's own definition is an effort to better integrate the latest science into national forest management, strongly referenced the SNEP report.

Although the new forest plans have undoubtedly built on some of the data and analyses generated in the assessments, the overly positive picture presented above must also be seen as part of strategic rhetoric. One has to keep in mind that at that time, i.e. in the early 1990s, the Forest Service had come under heavy critique, especially from the environmental community. By making rhetorical references to comprehensive assessment processes that were perceived as highly credible and legitimate by most actor groups, the Forest Service hoped to get the new forest plans out of the firing line. At least in the first years, this strategy proved quite successful.

With the 2001 change in the presidential administration from Clinton to Bush, forest policy has gone in a more conservative direction. Likewise, the legitimizing function of the assessments has worn off and the Forest Service has come under critique again. This example shows that scientific closure, as mediated by the assessments, can under certain conditions actually contribute to political closure, as evidenced by forest planning activities in the Clinton years. But, in the end, momentous changes in the political environment can easily break this link again.

The legitimizing function of bioregional assessments seemingly depends, at least in part, on the constellation of interests in the larger political arena. The Clinton years were characterized by a hegemonic consensus on more conservation-oriented land-use policies, ensuring (too) easy reception of scientific claims that appeared to support these policies. Under the Bush administration, resource extraction and conservation interests have been sharply divided again and scientific claims have been subject to increased scrutiny by advocates from rival camps. This constellation very much resembles the prevailing destiny of science for policy: 'Far from promoting consensus, knowledge fed into such a process risks being fractured along existing lines of discord' (Jasanoff, 1990, p. 8).

How does the picture look when going beyond the immediate addressee of the assessments, i.e. the US Forest Service, and when asking about their impact in the broader political arena? An ad hoc appraisal also shows poor results here. For the Sierra Nevada an inside analyst has come to the conclusion that '[a]s yet, there has been no fundamental shift of national, state, or local policy or significant change in operations, because of the SNEP report' (Machida, 1999, p. 332) A critical appraisal of SAA would probably come to similar conclusions.

At first sight, these results would suggest that science–policy assessments remain without impact or that, at least, their influence is confined to situations where scientific findings resonate with the prevailing political climate and that in situations where political support is missing science remains without effect. Does this mean that science–policy assessments can easily be dismissed as nothing but *l'art pour l'art* exercises?

My in-depth, micro-level investigations produced some counter-evidence to such a flatly negative evaluation. One possibly comes to less sobering conclusions when taking a longer-term perspective and when not only looking at the macro level of interest-group politics and governmental power but also at the micro level of knowledge diffusion and policy learning. I found a great deal of evidence for SNEP and SAA results influencing policies, though often in very indirect and tortuous ways. In line with the guiding hypothesis formulated above, the oft-noted moderate effectiveness of science–policy assessments could, at least in part, be the result of a too restrictive, output-oriented view, in which scientific advice is conceptualized as the simple transmission of ready-made scientific results.

Assessments as communication processes

To come to a more realistic – and probably also more positive – picture of how bioregional assessments influence policy, a broader frame of evaluation is needed. Assessments have often been viewed as documents that convey information from scientists to policymakers; that means they have been reduced to the reports they produce. This kind of conceptualization overlooks the fact, however, that interactions between science and practice take place on different levels, make use of different media and occur in different phases of a research and/or policy process. With that, assessments are better viewed as broad communicative processes within and among particular scientific and political communities (Miller *et al.*, 1997).

Also the two bioregional assessments may be analysed most productively by taking a process-oriented view. During the preparation and formulation of the assessments but also during their utilization, information was communicated amongst a wide array of experts, policymakers and laypersons. The assessment processes themselves already took 1½ and 3 years, respectively. But, even before their formal start, there were lengthy discussions and intensive negotiations on how they should be designed and carried out (Erman, 1999; van Sickle, 2001).

Beyond that, SNEP and SAA were only two of multiple channels through which knowledge about the two bioregions was communicated among various scientific disciplines and policy actors. At the same time, land management agencies were involved in comprehensive land-use planning activities and parallel assessments were carried out, some completely independently from SNEP and SAA (either deliberately or incidentally), others in coordination with the two assessments (Ruth, 2000). If I had just looked at the two assessment processes themselves I would have missed a lot of contextual information and would have come to wrong conclusions as regards the degree of interconnectedness of the problems at hand.

Research on the policy impact of environmental assessments shows that short causal chains between assessments and policy or behaviour change are the exception (Mitchell *et al.*, 2004). It is not uncommon for 5 to 10 years to elapse between the release of an assessment report and the spin-offs that finally begin to be cited by decision makers as exerting a substantial influence on the policy process. Although both bioregional assessment projects were formally finished almost a decade ago, processes of 'disseminating' and 'using' the outcomes are still going on today. In the following, I shall give some examples of how assessments can alter policies in indirect and long-term ways.

Framing of policy issues

A helpful analytical perspective is to see assessments as exercises of framing. Following Rein and Schön (1991), framing is an activity of selection, organization and interpretation of a complex reality, 'so as to provide guideposts for knowing, analysing, persuading and acting. A frame is a perspective from which an amorphous, ill-defined situation can be made sense of and acted upon' (Rein and Schön, 1991, p. 263). Framing can play a crucial role in whether an assessment disrupts the existing equilibrium of goals, options and knowledge by convincing participants that current policies and behaviours no longer represent the best ways to achieve their goals (Clark *et al.*, 2002).

A first example of how the two assessments helped stabilize the framing of issues in public discourse is the reorientation of planning activities towards a more 'regionalized' approach. Before the assessments, each national forest – of which there are 11 in the Sierra Nevada and five in the southern Appalachians – independently prepared its forest plan. Years of intensive research, communication and negotiation as part of the assessment processes eventually led to greater visibility and strengthened perception of the Sierra Nevada and the southern Appalachians as distinct bioregions. With the introduction of a novel ontological entity, namely the entity of 'bioregions', land managers seemingly recognized that for many of the problems a range-wide, multi-forest planning approach was needed. In the end, bioregional issues found themselves embedded in emerging political and administrative frameworks. The new generation of forest plans is based on a more 'bioregional look'.

A second example of how assessments can frame – or reframe – natural resource issues for research and policy is the 'discovery' of water issues in and through SNEP. Before the assessment, forestry and timber-related topics very much dominated the discourse in the region. Forestry and wood products industries were seen as the major pillars of the local economy and job market. An economic analysis carried out under SNEP, however, showed that:

> [f]rom the perspective of the natural resources, water is the basis for most of the economic value. Timber, animal forage, other agricultural crops, and a range of recreational and residential services directly dependent of the ecosystem comprise the rest of the natural resource value. At the Sierra-wide level, a majority of the economic benefits from the use of the natural resource accrue to beneficiaries outside the region.
>
> (Stewart, 1996, p. 1054)

This economic analysis led to a definite paradigm shift in the way the Sierra Nevada was seen, both by its inhabitants and from the outside. Up to that time, the Forest Service as the main land user had been defining, from a resource point of view, what the value of the Sierra Nevada is. The comparatively low resource value of timber as calculated in the economic study clearly delegitimized many of the Forest Service's timber-related arguments.

The study also gave evidence for a striking gap between the resource value and the reinvestment value of water: only small portions of the economic benefits of water use have been flowing back into the region. This insight provided strong arguments for the predominantly rural Sierran counties vis-à-vis urban agglomerations, which are the main beneficiaries of water-based products and services.

Those two examples show how a rather simple economic calculus can introduce and stabilize new frames in public discourse, frames that can legitimize one line of argument while delegitimizing another and, in the end, leading to the empowerment of one set of actors and the concurrent disempowerment of another set of actors.

Linear models of knowledge transfer typically proceed from the assumption that what policymakers want is just what researchers are best qualified to supply, namely detailed data and findings. The sociological literature, however, shows that in many instances knowledge utilization is not deliberate, direct and targeted, but a result of long-term percolation of scientific concepts, theories and findings into the climate of informed opinion (Weiss, 1977).

The above-mentioned experiences of bioregional assessments clearly indicate that the less tangible outcomes associated with such assessments are at least as important as the more tangible outcomes, including written reports. The principal measurement of success is not whether a political counselling process has amassed an impressive collection of scientific reports, but rather whether it has contributed to improved mental models of the problem (Cortner *et al.*, 1999; Keating and Farrell, 1999). Frameworks, more than data, are the key to successful science–policy consultation (Johnson and Herring, 1999). The generative ideas of 'bioregions' and of water as an 'identity-giving resource' are just two cases in point.

Building of actor networks

What is true for 'ideas', namely a slow build-up of scattered impacts, also applies in an analogous way to the build-up of relations between actors. Miller *et al.* (1997) emphasize that ongoing assessment activities often help extend and connect actor communities, either by identifying the relevance of communities that had previously not widely participated or by enabling new communities to form. In both case studies, the initiation of a large-scale bioregional assessment stirred up the environmental NGO community and soon led to the formation of region-wide NGO alliances: the Sierra Nevada Alliance is a regional coalition of some 50 grass-roots and regional environmental groups that was set up in the run-up to SNEP. The Southern Appalachian Forest Coalition is an alliance of some 20 conservation groups, which was mainly established to provide grass-roots networking and other support systems and to develop or maintain involvement with the assessment initiatives.

The NGO community also maintained a high level of activity after the assessments were formally finished. They gained especially large momentum by providing a kind of 'translation service' for the assessment results. They published citizens' guides, i.e. small booklets summarizing the assessment results and providing perspectives on alternative views. They also created new GIS resources and easy-to-understand maps in order to facilitate work with grass-roots groups and local communities. With these citizens' guides, GIS maps and local workshops, NGOs made assessment results accessible to and understandable for a broader public.

NGOs not only helped to disseminate information but also used the assessments for their own activities and interests. Environmental groups used SNEP and SAA results in their formulation of bioregional strategies for the two mountain ranges, in their evaluation and critique of Forest Service planning documents and in their effort to increase public awareness of biodiversity issues on private lands. Private conservation funds used the assessments in identifying conservation and heritage resources and in setting conservation priorities, which eventually determined decisions of land purchases and easements.

These examples show that assessments can provide new policy actors with the resources and opportunity to interact and develop common ground with respect to policy choices. While policies had been dominated by resource-extraction interests for decades, the more comprehensive, bird's-eye view that comes with the idea of bioregionalism provided the environmental community with strong arguments for more resource protection. So, in the end, the assessments, to some extent, altered the prevailing power structures in the two regions.

Building of scientific communities

Assessments, as processes that unfold over prolonged periods of time, can play important roles in the emergence and growth of research and assessment communities. Ongoing assessment activities can be instrumental in bringing together diverse experts and enabling them to transcend geographical, political and disciplinary boundaries. The interaction of experts during assessments can stimulate the formation of entirely new 'knowledge communities', i.e. networks of scientists, assessors, policymakers, interest groups and citizens who interact around particular issues and often use assessments as part of their communicative processes (Miller *et al.*, 1997).

Both SNEP and SAA helped to build up extended networks of experts who learned to work with one another across disciplinary and institutional boundaries. Unlike academic peer networks, these assessment networks show a high degree of heterogeneity as regards the institutional backgrounds of their members. Academic scientists cooperate with scientists working in the research branches of federal agencies, with technical specialists whose job is to advise line managers in agencies on science-intensive questions, with non-university scientists in private research stations and with NGO science consultants. By their variety alone, such networks can mobilize different bodies of knowledge.

When SNEP and SAA were formally finished in 1996, expert communities didn't discontinue their policy-related activities. In the follow-up, a number of further assessments were initiated and carried through. I shall give here just two examples from the southern Appalachians. When SAA was in its final stages, the Southern Appalachian Mountains Initiative (SAMI) was started. SAMI provided a forum to develop regional air quality solutions and to resolve differences among institutional priorities of eight different states and dozens of different state and non-state institutions. The productive cooperation of so many actors would not have been possible if most of them hadn't already been brought together in SAA and experienced the benefits of collaborative interaction. The Southern Forest Resource Assessment (SFRA), a multi-agency effort led by the Forest Service, was commissioned in 1999 to document and analyse the factors impacting the forests of the south-eastern USA. SFRA built on the Southern Appalachian Assessment mainly in procedural respects. Especially public participation – with its wide array of opportunities for people to take part in the process – was very much modelled after SAA.

Finally, SNEP and SAA also left their marks in the institutional landscape as they helped to build scientific capacities. In the southern Appalachians, for example, a Cooperative Ecosystem Studies Unit (SA-CESU) was set up in 1999. SA-CESU is a network of federal agencies and universities engaging in cooperative research to provide research, technical assistance and education for resource and environmental managers. In the Sierra Nevada, there were lengthy discussions on the foundation of a similar unit that would coordinate research and application of scientific knowledge in the region. In the end, it took almost a decade before the Sierra Nevada Research Institute (SNRI) was installed at the newly opened university campus of University of California at Merced.

The last example impressively proves that real-world impacts often take a long time to materialize and that it is almost impossible to trace the exact roots of an event or achievement. But, while it is, of course, difficult to accurately describe the paths of interaction – let alone the causal links – between historical processes and contemporary circumstances, it is probably safe to say that the two bioregional assessments helped extend, enhance and connect various knowledge communities.

Conclusions

The chapter started out with the introduction of two alternative conceptual frameworks that describe the interaction between science and politics in general and the functioning and the effectiveness of science–policy consultation processes in particular.

The knowledge transfer model conceptualizes scientific advice as the simple transmission of ready-made scientific results from scientists to policymakers. This linear model is based on the notion of a direct, cascade-like 'scientification of the non-scientific world' (Beck and Bonß, 1984, p. 382). The value of the knowledge transfer model as a correct depiction of empirical reality has already been questioned at an early stage. Nevertheless, this unilinear approach of science in policymaking to some extent still dominates perceptions among policymakers and scientists alike (Weingart, 1999).

With the boundary-spanning model I propose an alternative conceptualization of the interaction between science and politics. Here, the science–policy interface is not seen as a sharp line of demarcation but rather as a fuzzy, dynamically shifting boundary. Science–policy consultation is best viewed as a dynamic process that evolves over time, occurs sequentially and often iteratively and typically involves long-term interactions between varied social actors.

A change from the linear transfer model to the boundary-spanning model also entails consequences on a methodological level. While the first implies a more product-oriented view, the second emphasizes that many of the characteristics that make up a 'successful' science–policy consultation process cannot be determined empirically by just looking at 'the project' itself but, rather, there is a need for a broader conceptual and methodological framework. With two in-depth case studies from the field of forest and natural resource policy, I could not 'prove' this hypothesis in a strictly statistical sense but my rather narrative descriptions of two bioregional assessments have provided some arguments that substantiate this hypothesis.

Previous empirical studies on science–policy consultation have usually examined ongoing or recently finished processes. In many cases, not surprisingly, these studies have come to rather negative conclusions as to the degree to which these efforts can live up to their normative expectations. In my own research, I have cast a more distanced eye on the role of science in policy, as I have revisited science–policy consultation processes almost a decade after they were formally finished.

The two case studies have shown that the impacts of science–policy assessments can be found in many places but that assessments may not produce direct and immediate results. To build up professional networks, to increase technical capacity and to ensure respect and credibility, all take repeated personal interactions over a longer period of time (Keating and Farrell, 1999). Assessments often alter the policy realm in ways that take time before their effects are evident in shifts in the policy debate or changes in the choices of policymakers and social actors (Mitchell *et al.*, 2004). Against that background, empirical studies that just focus on the 'projects' themselves must, more or less predictably, overlook many impacts and thus underestimate potential effects.

Over-simplistic theoretical models, like the transfer model, inevitably fall short of describing complex realities in an adequate way. The boundary-spanning model, on the contrary, seems more apt to grasp and make sense of a complex reality in which the boundaries between the realms of science and policy are relatively porous and flexible and, thus, practices of science–policy consultation do not simply come down to acts of 'speaking truth to power' but rather resemble intricate set-ups with multiple opportunities for formal and informal interactions and negotiations between scientists, natural resource managers, policymakers and the public.

References

Beck, U. and Bonß, W. (1984) Soziologie und Modernisierung: Zur Ortsbestimmung der Verwendungsforschung. *Soziale Welt* 35, 381–406.

Beck, U. and Bonß, W. (1989) Verwissenschaftlichung ohne Aufklärung? Zum Strukturwandel von Sozialwissenschaft und Praxis. In: Beck, U. and Bonß, W. (eds) *Weder Sozialtechnologie noch Aufklärung? Analysen zur Verwendung sozialwissenschaftlichen Wissens*. Suhrkamp, Frankfurt am Main, Germany, pp. 7–45.

Cash, D.W. and Clark, W.C. (2001) *From Science to Policy: Assessing the Assessment Process*. KSG Faculty Research Working Paper 01–045, Kennedy School of Government, Cambridge, Massachusetts.

Cash, D.W., Clark, W.C., Alcock, F., Dickson, N., Eckley, N. and Jäger, J. (2002) *Salience, Credibility, Legitimacy and Boundaries: Linking Research, Assessment and Decision Making*. KSG Faculty Research Working Paper 02-046, Kennedy School of Government, Cambridge, Massachusetts.

Clark, W.C., Mitchell, R.B., Cash, D.W. and Alcock, F. (2002) *Information as Influence: How Institutions Mediate the Impact of Scientific Assessments on Global Environmental Affairs*. KSG Faculty Research Working Paper 02-044, Kennedy School of Government, Cambridge, Massachusetts.

Cortner, H.J., Wallace, M.G. and Moote, M.A. (1999) A political context model for bioregional assessments. In: Johnson, K.N., Swanson, F., Herring, M. and Greene, S. (eds) *Bioregional Assessments: Science at the Crossroads of Management and Policy*. Island Press, Washington, DC, pp. 71–82.

Dunn, W.N. (1993) Policy reforms as arguments. In: Fischer, F. and Forester, J. (eds) *The Argumentative Turn in Policy Analysis and Planning*. Duke University Press, Durham, North Carolina, pp. 254–290.

Engels, A. (2002) European Forum of Integrated Environmental Assessment: evaluation of four science–policy interface workshops. Mimeo, Amsterdam.

Erman, D.C. (ed.) (1996) *Status of the Sierra Nevada: The Sierra Nevada Ecosystem Project*, 4 vols. Centre for Water and Wildland Resources, Davis, California.

Erman, D.C. (1999) Sierra Nevada Ecosystem Project: case study. In: Johnson, K.N., Swanson, F., Herring, M. and Greene, S. (eds) *Bioregional Assessments: Science at the Crossroads of Management and Policy*. Island Press, Washington, DC, pp. 304–320.

Farrell, A., VanDeveer, S.D. and Jäger, J. (2001) Environmental assessments: four under-appreciated elements of design. *Global Environmental Change* 11, 311–333.

Gieryn, T.F. (1983) Boundary-work and the demarcation of science from non-science: strains and interests in professional ideologies of scientists. *American Sociological Review* 48, 781–795.

Gieryn, T.F. (1995) Boundaries of science. In: Jasanoff, S., Markle, G.E., Peterson, J.C. and Pinch, T.J. (eds) *Handbook of Science and Technology Studies*. Sage, Thousand Oaks, California, pp. 393–443.

Glynn, S., Cunningham, P. and Flanagan, K. (2003) *Typifying Scientific Advisory Structures and Scientific Advice Production Methodologies: Final Report*. PREST, Manchester, UK.

Guston, D.H. (2001) Toward a 'best practice' of constructing 'serviceable truths'. In: Hisschemöller, M., Hoppe, R., Dunn, W.N. and Ravetz, J.R. (eds) *Knowledge, Power, and Participation in Environmental Policy Analysis*. Transaction Publishers, New Brunswick, pp. 97–118.

Herring, M. (1999) Introduction. In: Johnson, K.N., Swanson, F., Herring, M. and Greene, S. (eds) *Bioregional Assessments: Science at the Crossroads of Management and Policy*. Island Press, Washington, DC, pp. 1–8.

Jasanoff, S. (1987) Contested boundaries in policy-relevant science. *Social Studies of Science* 17, 195–230.

Jasanoff, S. (1990) *The Fifth Branch: Science Advisers as Policymakers*. Harvard University Press, Cambridge, Massachusetts.

Jasanoff, S. and Wynne, B. (1998) Science and decisionmaking. In: Rayner, S. and Malone, E.L. (eds) *Human Choice and Climate Change. Vol. 1: The Societal Framework*. Battelle Press, Columbus, Ohio, pp. 1–87.

Johnson, K.N. and Herring, M. (1999) Understanding bioregional assessments. In: Johnson, K.N., Swanson, F., Herring, M. and Greene, S. (eds) *Bioregional Assessments: Science at the Crossroads of Management and Policy*. Island Press, Washington, DC, pp. 341–376.

Keating, T.J. and Farrell, A. (1999) *Transboundary Environmental Assessment: Lessons from the Ozone Transport Assessment Group*. Technical Report NCEDR/99-02, NCEDR, Knoxville, Tennessee.

Knott, J.H. and Wildavsky, A. (1980) If dissemination is the solution, what is the problem? *Knowledge: Creation, Diffusion, Utilization* 1, 537–574.

Machida, D.T. (1999) Sierra Nevada Ecosystem Project: policy review. In: Johnson, K.N., Swanson, F., Herring, M. and Greene, S. (eds) *Bioregional Assessments: Science at the Crossroads of Management and Policy*. Island Press, Washington, DC, pp. 330–338.

Meidinger, E. and Shannon, M. (1996) The Southern Appalachian Ecosystem Assessment: Overview case study prepared for the Rennsalaerville Science and Policy Seminar, USDA Forest Service, 4–6 June 1996. Mimeo, Buffalo, New York.

Miller, C.A., Jasanoff, S., Long, M., Clark, W.C., Dickson, N., Iles, A. and Parris, T.M. (1997) Shaping knowledge, defining uncertainty: the dynamic role of assessments. In: Clark, W.C., McCarthy, J.J. and Shea, E. (eds) *A Critical Evaluation of Global Environmental Assessments*. Kennedy School of Government, Cambridge, Massachusetts, pp. 79–113.

Mitchell, R.B., Clark, W.C., Cash, D.W. and Alcock, F. (2004) Science, scientists, and the policy process: lessons from global environmental assessments for the northwest forest. In: Arabas, K. and Bowersox, J. (eds) *Forest Futures: Science, Politics and Policy for the Next Century*. Rowman and Littlefield, Lanham, Maryland. pp. 95–111.

Nowotny, H. (1994) 'Wissen entsteht im Kontext der Anwendung': Theoretische und praktische Anmerkungen zum Wissenschaftstransfer. In: Apeltauer, M. (ed.) *Wissen an der Börse: Bürgernahe Wissenschaft in Österreich*. Bundesministerium für Wissenschaft und Forschung, Vienna, Austria, pp. 31–37.

Pregernig, M. (2005) Wissenschaftliche Politikberatung als kulturgebundene Grenzarbeit: Vergleich der Interaktionsmuster in den USA und Österreich. In: Bogner, A. and Torgersen, H. (eds) *Wozu Experten? Form und Funktion wissenschaftlicher Politikberatung in gesellschaftstheoretischer und empirischer Perspektive*. Verlag für Sozialwissenschaften, Wiesbaden, Germany, pp. 255–278.

Price, D.K. (1981) The spectrum from truth to power. In: Kuehn, T.J. and Porter, A.L. (eds) *Science, Technology, and National Policy*. Cornell University Press, Ithaca, New York, pp. 95–131.

Rein, M. and Schön, D.A. (1991) Frame-reflective policy discourse. In: Wagner, P., Weiss, C.H., Wittrock, B., and Wollmann, H. (eds) *Social Sciences and Modern States: National Experiences and Theoretical Crossroads*. Cambridge University Press, Cambridge, UK, pp. 262–289.

Rich, R.F. (1997) Measuring knowledge utilization: processes and outcomes. *Knowledge and Policy* 10, 11–24.

Ronge, V. (1989) Verwendung sozialwissenschaftlicher Ergebnisse in institutionellen Kontexten. In: Beck, U. and Bonß, W. (eds) *Weder Sozialtechnologie noch Aufklärung? Analysen zur Verwendung sozialwissenschaftlichen Wissens*. Suhrkamp, Frankfurt am Main, Germany, pp. 332–354.

Ruth, L. (2000) Conservation on the cusp: the reformation of national forest policy in the Sierra Nevada. *UCLA Journal of Environmental Law and Policy* 18, 1–97.

SAMAB (1996) *The Southern Appalachian Assessment*, 5 vols. Southern Appalachian Man and the Biosphere Cooperative, Knoxville, Tennessee.

Stewart, W.C. (1996) Economic assessment of the ecosystem. In: Erman, D.C. (ed.) *Status of the Sierra Nevada: The Sierra Nevada Ecosystem Project – Volume III*. Centre for Water and Wildland Resources, Davis, California, pp. 973–1063.

van Sickle, C. (2001) Southern Appalachian case study. In: Jensen, M.E. and Bourgeron, P.S. (eds) *A Guidebook for Integrated Ecological Assessments*. Springer, New York, pp. 472–488.

Webber, D.J. (1992) The distribution and use of policy knowledge in the policy process. In: Dunn, W.N. and Kelly, R.M. (eds) *Advances in Policy Studies since 1950*. Transaction Books, New Brunswick, pp. 383–418.

Weingart, P. (1999) Scientific expertise and political accountability: paradoxes of science in politics. *Science and Public Policy* 26, 151–161.

Weiss, C.H. (1977) Research for policy's sake: the enlightenment function of social research. *Policy Analysis* 3, 531–545.

Weiss, C.H. (1980) Knowledge creep and decision accretion. *Knowledge: Creation, Diffusion, Utilization* 1, 381–404.

8 Cluster Organization in Forestry: Supporting Information and Knowledge Transfer in the Practice, Science and Policy of Sustainable Forest Management

T. MROSEK AND A. SCHULTE

Centre for Forest Ecosystems, University of Münster, Münster, Germany.

Abstract

Although one major function of forest policy is supporting sustainable forest management (SFM) research, implementation and evaluation, the capacity of traditional forest policy systems for integrating and optimizing all aspects of SFM seems to be limited. In addition to the different value systems and often competing interests of forestry stakeholders, limited or ineffective communication and cooperation between practitioners, scientists and politicians make it difficult to identify and solve key problems.

Cluster organization in forestry – the integrated view of all branches of forest and wood-processing industries – can support information and knowledge transfer across practice, science and policy. Cluster analysis can provide relevant information on important socio-economic and political aspects of SFM at different levels. Cluster management allows the effective and target-oriented transfer of information among all stakeholders in a participatory and reciprocal way, providing valuable input in the policy process. Through this transfer, scientists and practitioners can provide policymakers with the information they need to support SFM in a meaningful way.

Results from an extensive study of the forest and wood-processing industries cluster in the state of North-Rhine/Westphalia, Germany, are used to demonstrate the application of the cluster concept and consequent improvements in science and policy interaction in the field of forestry.

Introduction

One major function of forest policy is supporting sustainable forest management (SFM) research, implementation and evaluation (Krott, 2001; Köpf, 2002). In this context SFM can be defined by legislation and policy programmes at various levels (international policy agreements, national, sub-national and local legislation and policy programmes), forest management systems or SFM evaluation systems such as criteria and indicators (C & I) for SFM or forest certification.

 Sustainable Forestry: from Monitoring and Modelling to Knowledge Management and Policy Science (eds K.M. Reynolds, A.J. Thomson, M. Köhl, M.A. Shannon, D. Ray and K. Rennolls)

Suitable SFM standards include: the C & I for SFM of the Montreal and Helsinki Processes (Montreal Process, 1999; Ministerial Conference on the Protection of Forests in Europe, 2003); other C & I for SFM systems (Mrosek, 2005; Mrosek *et al.*, 2006); and different certification systems (Anon., 2004; PEFC Council, 2004).

Due to the complexity of the current SFM definition, the capacity of traditional forest policy systems for integrating and optimizing all aspects of SFM seems limited.

The different and often competing dimensions of SFM (ecological, economic and social objectives as well as objectives related to the policy and management framework) mean that stakeholders in the forest policy arena apply different value systems and often have competing interests (Krott, 2001; Köpf, 2002). The forest policy system can solve these land-use disputes to only a limited extent, particularly when faced with decreasing government resources. Although these general challenges are difficult to overcome, there are more specific problems that can directly be addressed by the forest stakeholders.

One research field that deserves more consideration is the transfer of relevant information and knowledge from science and industry to policymakers. The complex nature of international SFM standards and the dynamics of global markets for wood products require fast and efficient information and knowledge transfer. However, the policy priorities of more traditional forestry administrations may not necessarily reflect the actual needs of forest stakeholders for governmental support or the most recent scientific outcomes. This means that politicians are not always aware of the most urgent needs of forestry practitioners, and their decision making is not always based upon the best scientific knowledge. Such shortcomings are even more dramatic when coupled with decreasing governmental resources for policymaking in forestry and in general.

Under such policy and management conditions, limited or ineffective communication and cooperation between practitioners, scientists and politicians are major shortcomings. Such limitations in the information and knowledge transfer make identifying and solving key problems difficult. One example of an important problem insufficiently addressed by forest policy is the appropriate assessment and effective communication of the socio-economic importance of forestry.

In this chapter, cluster organization – the integrated view of all branches of forest and wood-processing industries – is introduced as a suitable approach for improving communication and cooperation among the various stakeholders in forestry. This form of organization is seen as suitable for supporting information and knowledge transfer across practice, science and policy in the field of SFM. The potential of the concept is illustrated with results from a case study of the forest and wood-processing industries cluster of the state of North-Rhine/Westphalia (NRW) in Germany.

Materials and Methods

The general scientific background for the concept of forest and wood-processing industry clusters was mostly formulated by Porter (1998) in the context of general economics. Whereas extensive literature is available for various industry branches (Armstrong and Taylor, 2000; Brenner, 2002; Maier and Tödtling, 2002; Scherer and Bieger, 2003; Schiele, 2003; Sölvell *et al.*, 2003), the general scientific background for cluster organization in forestry is very limited.

A relatively broad definition of the forest and wood-processing industry cluster was defined for the European Union (EU) by the European Commission (Kommission der Europäischen Gemeinschaften, 1999; Bundesrat, 2001). Examples of the early application of the cluster concept based on this EU definition can be found in Austria (TMG, 2005), Finland (Ministry of Agriculture and Forestry Finland, 2001) and Sweden (Anon., 2001).

The concept of forest and wood-processing industry clusters has seen further development in recent years. The modified and extended cluster definition integrates all industry branches relevant to the field of forestry. The different industry branches are presented in some detail in order to develop a conceptual framework for forest and wood-processing industry clusters (see Results below).

The method of cluster analysis allows cluster structures to be identified at different spatial scales (ranging from international, national, sub-national or regional to local levels). It includes data collection, analysis and assessment concerning all relevant aspects of forest resources, forest management as well as the utilization of timber and non-timber products. Cluster management can support the optimization of economic performance within a cluster and therefore contribute to the sustainable development of forestry-based regions (Schulte, 2003a, b; Mrosek and Schulte, 2004; Mrosek *et al.*, 2005; Schulte and Mrosek, 2006).

To improve information and knowledge transfer within the cluster, a transfer concept was developed and tested within both cluster analysis and management. This transfer concept was based on methods of general communication science (e.g. stakeholder participation, public relations, marketing), but also supported by methods specific to the general cluster concept. Among these methods were the approaches of corporate networking and cooperation (Howaldt *et al.*, 2001; Initiative für Beschäftigung OWL e. V., 2004; Stahl and Schreiber, 2004).

The NRW forest and wood-processing industries cluster case study used to illustrate the potential of the concept within this chapter is currently the only example of a large-scale cluster analysis in forestry within Germany (other studies at different spatial levels are in progress) (Mrosek *et al.*, 2005). This study, which took place from November 2001 to January 2003, applied the modified and extended EU definition of the forest and wood-processing industries cluster and methods of cluster analysis and management (Schulte, 2003a, b; Schulte and Mrosek, 2006).

Within the cluster analysis, data collection involved specific business surveys, expert interviews and general statistics from governmental institutions and industry associations. Although these statistics provided a suitable basic database, there was a lack of data in some areas (e.g. certain industry branches were not included in the existing NRW statistics) and the standards for data collection were not applicable to all variables (e.g. small companies with < 21 employees relevant to this study were not included in the existing NRW statistics). Additional surveys were conducted in order to identify companies in the forest and wood products industry cluster, with specific attention paid to small companies. Socio-economic data were collected for all identified companies, focusing on the total number of employees and the corporate annual revenue. Where it was not possible to collect this information on an individual company basis for a particular industry branch, estimates were derived from interview-based industry expert assessments (Schulte, 2003b).

In addition to socio-economic data analysis, a SWOT analysis (SWOT = strengths, weaknesses, opportunities and threats) (Fleisher and Bensoussan, 2003) was conducted to support the overall assessment of the cluster. Following a participatory research approach, this analysis also involved industry experts and other forest stakeholders.

Case study area

North-Rhine/Westphalia (Nordrhein-Westfalen (NRW)) is one of 16 states (*Länder*) in the Federal Republic of Germany. Located in the western central part of Germany, it borders the Netherlands and Belgium and covers an area of 34,082 km^2. Figure 8.1 shows the overall area of NRW with selected land cover features, as well as its geographical location in the larger contexts of Germany and Europe.

With about 18 million citizens, NRW is the state with the highest population in Germany. Accounting for 22% of the German gross domestic product

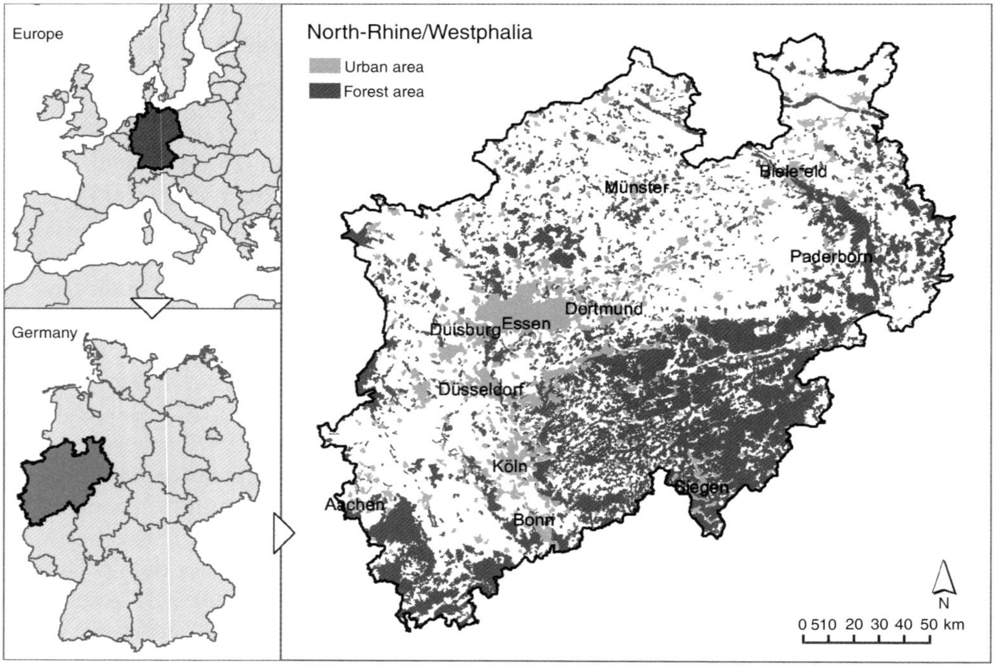

Fig. 8.1. Map of urban and forest land cover areas in NRW and physical location within contexts of Germany and Europe. (Data source: ESRI and Corine Landcover; GIS layout by Kies, 2005.)

(466.9 billion euros in 2004), NRW is also a region of major economic importance within Germany and the EU (Landesamt für Datenverarbeitung und Statistik NRW, 2003).

Sustainable forestry has a long tradition in NRW, as it does in Germany in general. Forest land (915,800 ha) covers 27% of the total land area in NRW. While 52.7% of the forest consists of deciduous stand types and corresponding tree species, the remaining 47.3% consists of coniferous stand types and tree species. The dominating tree species are spruce (*Picea abies* L.), 36%; beech (*Fagus sylvatica* L.), 16%; and oak species (*Quercus robur* L. and *Quercus petraea* Liebl.), 15%. The age class distribution of the forest is biased towards younger stands, especially in the coniferous stands, which is mostly a result of afforestation following the effects of the Second World War (degradation and deforestation). Forest landownership is dominated by private forest owners (64%), followed by municipal (20%), state (13%) and federal (3%) forest ownership. The privately owned forest area is characterized by a very large number of owners (> 150,000) and mostly small (≥ 200 ha) and very small parcels of forest land (< 200 ha). Concerning forest productivity, the merchantable timber volume (under bark) is 221 m^3/ha on average and 194.4 million m^3 in total. The current total annual timber harvest is 3.9 million m^3. Considering that the mean annual increment is 9.1 m^3/ha on average, the current timber harvest level is significantly smaller than the sustainable harvest level based on the annual allowable cut volume. Forest management is based on the principles of sustainability, multiple forest use and nature-oriented silviculture (e.g. Nature Oriented Forest Management Programme) (Ministerium für Umwelt und Naturschutz, Landwirtschaft und Verbraucherschutz des Landes NRW 2003a, b; Schulte, 2003a).

NRW is characterized by a high concentration of wood-processing industries, covering both the primary and secondary level and all types of wood processing. For example, the concentration of the wood furniture industry in NRW is of nationwide and international relevance.

Results

Conceptual framework for forest and wood-processing industry clusters

The concept of forest and wood-processing industry clusters includes the identification of all industry branches and their individual companies as well as related institutions. Within the cluster, these companies and institutions should be linked to each other by these characteristics:

- Close relationship to forest resources.
- Spatial clustering.
- High connectivity to each other (Mrosek and Schulte, 2004; Mrosek *et al.*, 2005).

Figure 8.2 shows the main branches of the forest and wood-processing industries cluster in NRW.

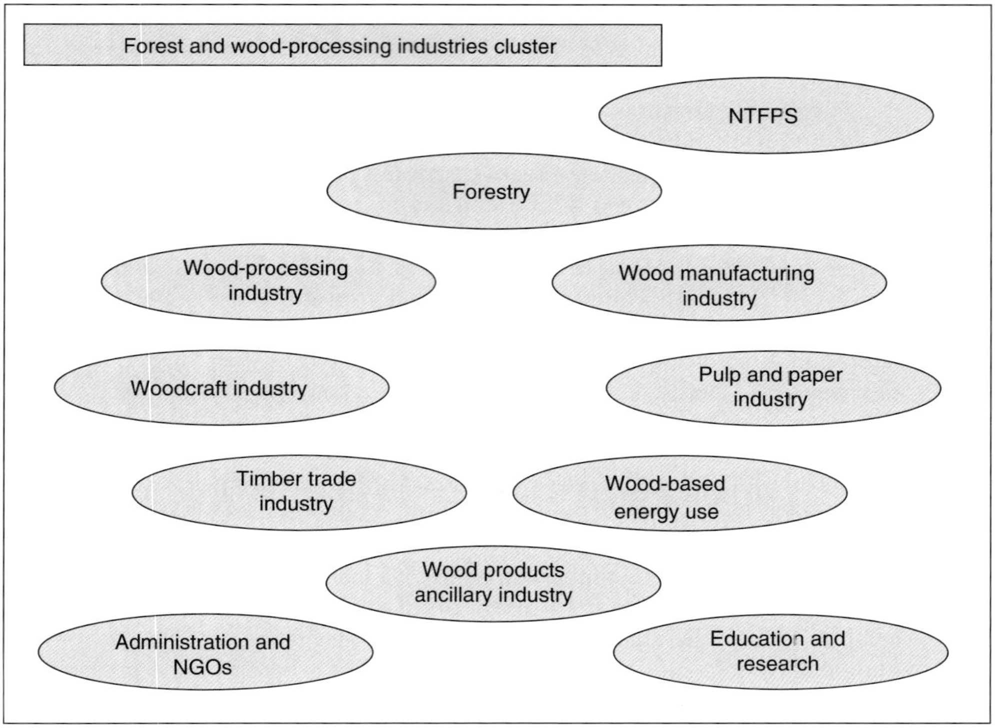

Fig. 8.2. Main branches of the forest and wood-processing industries cluster in NRW. NTFPS, non-timber forest products and services. (Modified and extended from Kommission der Europäischen Gemeinschaften, 1999; Schulte, 2003a, b; Mrosek and Schulte, 2004; Mrosek *et al.*, 2005; Schulte and Mrosek, 2006)

In addition to the traditional industry branches of forestry (e.g. the sawmill, wood-based panel and pulp and paper industries), the cluster concept also includes other businesses related to forests, such as non-timber forest products and services (e.g. hunting, tourism) and supporting institutions (e.g. the governmental administration, non-governmental organizations (NGOs), education, research). More information on these different industry branches is provided in the discussion of the definition of the forest and wood-processing industry cluster of NRW later in this section.

The overall goal of the cluster concept is to support generally all industry branches and their individual companies as well as related institutions. The main objective is the optimization of production and value-addition processes within and between different industry branches. For example, the concept can contribute to the competitiveness of forest industries by increasing productivity and innovation. A secondary objective is to support forestry stakeholders in developing an improved self-image for their industries. It can also help build more effective communications with policymakers, media and the general public.

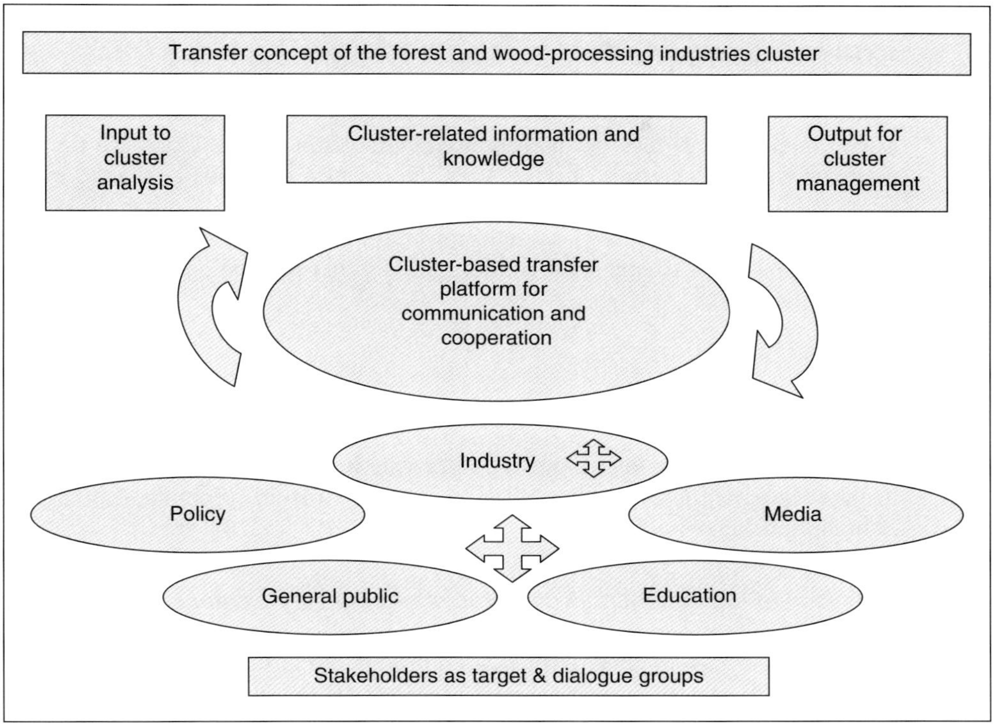

Fig. 8.3. The transfer concept of the forest and wood-processing industry cluster in NRW.

The two main components of the cluster concept are cluster analysis and cluster management. Cluster analysis can provide relevant information on important socio-economic and political aspects of SFM at different spatial scales, ranging from local to international levels. The case study presented in this chapter is based at the state (*Land*) level. Other studies (all still in progress) of cluster organization in forestry within Germany are available for the national (Fischer, 2004a, b), regional and local (Anon., 2005) levels. Although the strategies and tools for cluster management can be diverse (e.g. corporate networking to improve communication and cooperation, coordinating marketing and public relations campaigns, mobilizing political support, including the provision of funding), in this chapter the focus is on information and knowledge transfer.

Information transfer within the cluster should be participatory and reciprocal. In this way, cluster management can support an effective and target-oriented transfer of information between all stakeholders and provide valuable input in the policy process. More specifically, scientists and practitioners can provide the information that policymakers require to support SFM in a meaningful way.

Figure 8.3 shows the transfer concept of the forest and wood-processing industry cluster. The cluster-based transfer platform for communication and cooperation (e.g. in the form of networks or forums) is the basis for providing input for the cluster analysis (e.g. through stakeholder panels or expert interviews).

In this way, cluster-related information and knowledge (e.g. concerning the socio-economic importance of forestry) are created.

This information provides the basis for cluster management, which can take place in the form of recommendations to industry and policy decision makers, public relations activities or environmental education efforts. Once the transfer platform is established and the relevant information becomes available, increased communication and cooperation within and between all stakeholders can take place. In a reciprocal process of information and knowledge transfer, the stakeholders of the cluster serve a dual role as target and dialogue groups at the same time.

Case study of the NRW forest and wood-processing industry cluster

The suitability of the forest and wood-processing industry clusters concept is further demonstrated by the NRW case study results (Schulte, 2003a, b; Schulte and Mrosek, 2006).

Definition of the forest and wood-processing industry cluster of NRW

Sustainable forestry and the wood-processing industry have traditionally been of high importance in NRW. However, although the industry branches are well established in regional, national and international markets and suitable policy framework conditions exist, a holistic view of all industry branches and an appropriate representation to policymakers and the general public has been lacking.

Applying the EU cluster definition, the forest and wood-processing industry cluster of NRW is comprised of four main categories, which are summarized in Table 8.1.

Table 8.1. Primary categories in the forest and wood-processing cluster of NRW (timber trade and ancillary wood product industries not considered in this study). (Modified from Schulte 2003a, b; Schulte and Mrosek, 2006)

Cluster category	Industry branches
Forestry	Forestry companies and forestry consultants
Wood-processing industry (primary wood processing)	Sawmill, veneer, wood-based panel and other roundwood processing industries
Wood manufacturing (secondary wood processing)	Furniture, woodcraft, wood construction, wood-based packaging and other specialized wood-processing industries
Other wood-based industries	Pulp and paper, printing and publishing, fuel wood; also, timber trade and ancillary wood product industries

Socio-economic profile of the NRW forest and wood-processing industry cluster

Traditionally, the socio-economic contribution of forestry in NRW has been considered less important by politicians and the general public, particularly in comparison with 'high-technology' industries such as the mechanical, electronics or chemical industry. However, the results of this analysis and assessment of the forest and wood-processing industry cluster have started to change this perception.

With 260,000 employees (~4.5% of the entire employment in NRW) and an annual revenue of about 35 billion euros (Ministerium für Umwelt und Naturschutz, Landwirtschaft und Verbraucherschutz des Landes Nordrhein-Westfalen, 2003b; Schulte 2003a, b), the cluster is highly significant for the economy and the employment market. These numbers become even more significant when compared with important industry branches traditionally associated with the highly industrialized region of NRW. In this context, Fig. 8.4 shows the total employment provided by and Fig. 8.5 the annual revenue of selected industry branches in NRW in 2001.

The forest and wood-processing industry cluster leads in employment when compared with selected industry branches such as finances and insurance (225,000 employees), electronics (195,000 employees) and chemicals (132,000 employees).

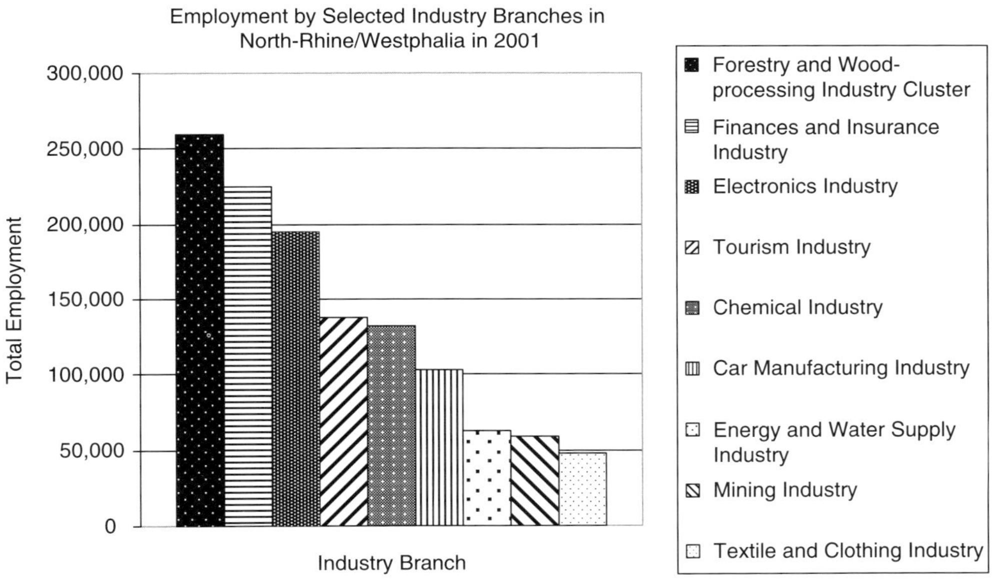

Fig. 8.4. Total employment (number of employees with social insurance registration) within selected industry branches in NRW in 2001, illustrating the socio-economic importance of the forestry and wood-processing industry cluster. (Modified from Schulte 2003a, b; Schulte and Mrosek, 2006)

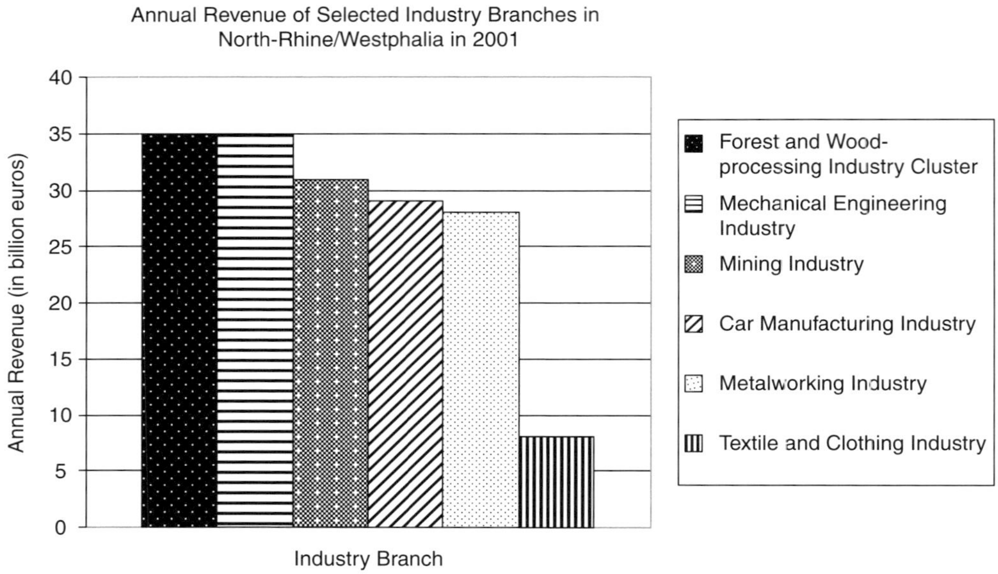

Fig. 8.5. Annual revenue of selected industry branches in NRW in 2001, illustrating the socio-economic importance of the forest and wood-processing industry cluster. (Modified from Schulte 2003a, b; Schulte and Mrosek, under review.)

In terms of annual revenue, the forest and wood-processing industry cluster creates revenue comparable to the traditionally important mechanical engineering industry (35 billion euros). It even has a dominant position when compared with other important high-technology branches, such as the car manufacturing industry (29 billion euros) and the metalworking industry (28 billion euros).

Improved information and knowledge transfer within the forest and wood-processing industry cluster of NRW

The case study provides various examples of how information and knowledge transfer can be supported through cluster organization in forestry.

In NRW, the general public traditionally did not recognize the significant role of the forest and wood-processing industry. In terms of marketing and interaction with policymakers, even the members and representatives of the forest and wood-processing industry cluster themselves usually underestimated the importance of their own profession and did not present their cluster effectively. Consequently, the forest industry and the wood-processing industry were not represented to their best advantage in the policy process and received only limited political support. Based on the new information available from the cluster analysis and cluster management efforts, the stakeholders reviewed their own position in the state economy and started to develop a more appropriate self-image. When this new information was presented within focused and multi-media-based public relations strategies, the forest and wood-processing industry cluster of NRW received high-level media and policy attention.

For example, the results of the research project were presented in the form of several hundred newspaper articles, several broadcasts and a few television shows, as well as through numerous personal presentations at conferences and special events. As an outcome of these research results and the recommendations made to decision makers in policy, industry and society, selected industry branches and selected regions with a high importance of forestry received political support. Follow-up initiatives were also initiated and additional research activities were launched. For example, based on the positive experiences at the state level, two cluster studies for municipalities within NRW (the municipality of Steinfurt and the city of Arnsberg) are currently under way (Anon., 2005) and regional initiatives are in preparation. Furthermore, a national study of Germany's forest and wood-processing industry cluster is also in progress (Fischer 2004a, b).

Discussion

Considering the significant problems in forestry (e.g. weak economic performance of small forest owners and companies, limited political support), the concept of forest and wood-processing industry clusters seems to be a suitable approach for the relevant industry branches to improve self-organization, communication and cooperation, develop a more appropriate self-image and more effectively present their needs to policymakers and the general public.

However, although a basic conceptual and methodological background exists in forestry science, the application of the cluster concept in forestry is still relatively new and further research and development are needed. Among the most important research needs are further development of the cluster definition and improved methods of cluster analysis and management. Also, further testing and implementation are necessary in order to gather more practical experience for refining the methods applied and for confirming initial conclusions. These efforts should focus on the main spatial scale of the concept, which is the regional level.

The significant resources required for a comprehensive cluster analysis and long-term cluster management pose one limitation to the application of the cluster concept in forestry. For example, this large-scale and innovative cluster study for NRW required funding of approximately 900,000 euros.

Regarding the transfer of information and knowledge, the cluster concept can improve communication and cooperation between stakeholders but the level of success often depends on specific circumstances. For example, cluster management is more likely to succeed where cluster structures already exist and when it is applied to regional industry branches and stakeholders that are already cooperating closely. This ability of forestry stakeholders to communicate and cooperate is valuable regional social capital.

Conclusions

The economics literature, experiences from selected international examples in forestry and the specific forestry case study results for NRW presented in this chapter suggest that the forest and wood-processing industry cluster concept is suitable for optimizing self-organization of different industry branches and improving communication and cooperation among various cluster stakeholders. In addition to developing a more appropriate self-image for the forest and wood-processing industry, it supports more effective and successful communication with policy, the media and the general public.

In both cluster analysis and management, the transfer platform for communication and cooperation is of key importance for successful information and knowledge transfer between practice, science and policy in the field of SFM. Therefore, in addition to traditional forest policy approaches, the concept of forest and wood-processing industry clusters can be used to support the implementation of SFM and sustainable development in rural areas.

Acknowledgements

The authors would like to thank their colleagues at the Research Group 'Forest and wood-processing industry clusters' at the Centre for Forest Ecosystems, University of Münster, namely Ms Dorothe Tesch, Mr Uwe Kies and Ms Anna Martinsohn.

The authors also thank Ms Denise Allen, University of British Columbia, Sustainable Forest Management Research Group, for reviewing and editing the manuscript.

This research was funded by the Ministry of Science and Research NRW (Ministerium für Wissenschaft und Forschung des Landes Nordrhein-Westfalen) and the Ministry of Environmental Protection and Nature Conservation, Agriculture and Consumer Protection NRW (Ministerium für Umwelt und Naturschutz, Landwirtschaft und Verbraucherschutz des Landes Nordrhein-Westfalen).

References

Anon. (2001) *Swedish Industry: Forestry and the Forest Products Industry.* Fact Sheet No. 129, Swedish Institute, Stockholm.

Anon. (2004) *FSC Principles and Criteria for Forest Stewardship.* Forest Stewardship Council, Bonn, Germany.

Anon. (2005) NRW: Zwei lokale Clusterstudien: Clusterkonzept erstmalig auf lokaler Ebene realisiert [Two local cluster studies initiated: first time application of the cluster concept at the local level in Germany]. *Holz-Zentralblatt* 19, 234.

Armstrong, H. and Taylor, J. (2000) *Regional Economics and Policy*, 3rd edn. Blackwell Publishing, Malden, Oxford, Carlton.

Brenner, T. (2002) Existence, emergence and evolution of local industrial clusters. Unpublished post-doctoral thesis. Friedrich-Schiller-Universität Jena, Jena.

Bundesrat (2001) *Unterrichtung durch das Europäische Parlament: Entschließung des Europäischen Parlaments zu der Mitteilung der Kommission über den Stand der Wettbewerbsfähigkeit der Holz verarbeitenden Industrie und verwandter Industriezweige in der EU (Drucksache 113/01)* [Report from the European Parliament: Decision of the European Parliament concerning the

Report of the European Commission with regard to the competitiveness of the wood-processing industries and related industries within the EU]. Bundesrat, Berlin.

Fischer, J. (2004a) Bundesweite Cluster-Studie Forst & Holz: Bedeutung der Forst- und Holzwirtschaft wird in der Politik und der eigenen Branche unterschätzt [Study on the German forestry and wood-processing industry cluster: importance of the forest and wood-processing industries is underestimated by politicians and industry representatives]. *Holz-Zentralblatt* 25, 343.

Fischer, J. (2004b) Bundesweite Cluster-Studie Forst & Holz gestartet: Interview mit Prof. Dr. Andreas Schulte, Wald-Zentrum, und HDH-Geschäftsführer Dirk-Uwe Klaas [Study on the German forest and wood-processing industry cluster initiated: interview with Prof. Dr Andreas Schulte, Centre for Forest Ecosystems, and Dirk-Uwe Klaas, Managing Director of the Main Association of the German Wood and Plastic Processing Industries and Related Industry Branches]. *Holz-Zentralblatt* 35, 440–441.

Fleisher, C.S. and Bensoussan, B.E. (2003) *Strategic and Competitive Analysis: Methods and Techniques for Analysing Business Competition*. Pearson Educational International, Upper Saddle River, New Jersey.

Howaldt, J., Kopp, R. and Flocken, P (eds) (2001) *Kooperationsverbünde und regionale Modernisierung: Theorie und Praxis der Netzwerkarbeit* [Networks and regional modernization: theory and practice of networking]. Gabler, Wiesbaden, Germany.

Initiative für Beschäftigung OWL e. V. (eds) (2004) *Unternehmensnetzwerke: Fragen der Forschung, Erfahrungen der Praxis* [Industry networks: research questions and practical experiences]. Kleine, Bielefeld, Germany.

Kommission der Europäischen Gemeinschaften (1999) *Mitteilung der Kommission an den Rat, das Europäische Parlament, den Wirtschafts- und Sozialausschuss und den Ausschuss der Regionen: Der Stand der Wettbewerbsfähigkeit der holzverarbeitenden Industrie und verwandter Industriezweige in der EU (KOM (1999) 457 endg.)* [Report of the Commission to the Council, the European Parliament, the Committee for Industry and Social Affairs, and the Committee for Regional Affairs: competitiveness of the European forestry industry and related industries]. Amt für amtliche Veröffentlichungen der Europäischen Gemeinschaft, Luxembourg.

Köpf, E.U. (2002) *Forstpolitik* [Forest policy]. Eugen Ulmer, Stuttgart, Germany.

Krott, M. (2001) *Politikfeldanalyse Forstwirtschaft: Eine Einführung für Studium und Praxis* [Forest policy analysis: An introduction for students and practitioners]. Parey, Berlin.

Landesamt für Datenverarbeitung und Statistik Nordrhein-Westfalen (eds) (2003) *Statistisches Jahrbuch Nordrhein-Westfalen* [Annual Statistical Report for Nordrhein-Westfalen]. Landesamt für Datenverarbeitung und Statistik Nordrhein-Westfalen, Düsseldorf, Germany.

Maier, G. and Tödtling, F. (2002) *Regional- und Stadtökonomik 2: Regionalentwicklung und Regionalpolitik* [Regional and urban economics: regional development and policy]. Springer, Vienna.

Ministerial Conference on the Protection of Forests in Europe Liaison Unit Vienna (ed.) (2003) *Improved Pan-European Indicators for Sustainable Forest Management: as Adopted by the MCPFE Expert Level Meeting 7–8 October 2002, Vienna, Austria*. Ministerial Conference on the Protection of Forests in Europe Liaison Unit Vienna, Vienna.

Ministerium für Umwelt und Naturschutz, Landwirtschaft und Verbraucherschutz des Landes Nordrhein-Westfalen (eds) (2003a) *Landeswaldbericht* [State forest report] [online]. http://www.munlv.nrw.de/sites/arbeitsbereiche/forsten/landeswaldbericht.htm March 2004.

Ministerium für Umwelt und Naturschutz, Landwirtschaft und Verbraucherschutz des Landes Nordrhein-Westfalen (eds) (2003b) *Clusterstudie Forst und Holz: Gesamtbericht* [Study on the forest and wood-processing industry cluster of North-Rhine/Westphalia: Overview report]. Schriftenreihe der Landesforstverwaltung NRW, Heft Nr. 17, Ministerium für Umwelt und Naturschutz, Landwirtschaft und Verbraucherschutz des Landes Nordrhein-Westfalen, Düsseldorf, Germany.

Ministry of Agriculture and Forestry Finland (2001) *Forest Industry in Finland* [online]. http://www.mmm.fi/english/forestry/industry/#Forest%20Cluster June 2005.

Montreal Process (1999) *Criteria and Indicators for the Conservation and Sustainable Management of Temperate and Boreal Forests*, 2nd edn. Montreal Process Liaison Office, Ottawa.

Mrosek, T. (2005) *Development and Testing of a Criteria and Indicators System for Sustainable Forest Management at the Local Level. Case Study at the Haliburton Forest & Wild Life Reserve Ltd., Canada*. Peter Lang Europäischer Verlag der Wissenschaften, Frankfurt am Main, Berlin, Berne, Brussels, New York, Oxford and Vienna.

Mrosek, T. and Schulte, A. (2004) Cluster Forst- und Holzwirtschaft: Problemstellung, Konzept, Erfahrungen und Ausblick [Forest and wood-processing industry clusters: situation, concept, experiences and outlook]. *Allgemeine Forstzeitschrift/DerWald* 23, 1261–1263.

Mrosek, T., Tesch, D., Kies, U. and Schulte, A. (2005) Cluster Wald, Forst- und Holzwirtschaft: Clusteranalyse und -management auf verschiedenen räumlichen Bezugsebenen [Forest and wood-processing industry clusters: cluster analysis and management at different spatial scales]. *Forst und Holz* 6, 239–243.

Mrosek, T., Balsillie, D. and Schleifenbaum, P. (2006) Field testing of a criteria and indicators system for sustainable forest management at the local level. Case study results concerning the sustainability of the private forest Haliburton Forest and Wild Life Reserve Ltd. in Ontario, Canada. *Forest Policy and Economics* 8, 593–609.

PEFC Council (2004) *PEFC Council Technical Document.* PEFC Council ASBL, Luxembourg.

Porter, M.E. (1998) *On Competition*. Harvard Business School Press, Boston, Massachusetts.

Scherer, R. and Bieger, T. (eds) (2003) *Clustering – das Zauberwort der Wirtschaftsförderung* [Cluster development – the key approach for supporting industries]. Paul Haupt, Berne, Stuttgart and Vienna.

Schiele, H. (2003) *Der Standort-Faktor: Wie Unternehmen durch regionale Cluster ihre Produktivität und Innovationskraft steigern* [Business location: how companies increase productivity and innovation potential within regional clusters]. Wiley-VCH, Weinheim, Germany.

Schulte, A. (2003a) Nordrhein-Westfalen zieht Bilanz für Forst und Holz: Cluster-Studie weist unerwartete volkswirtschaftliche Größe der Forst- und Holzwirtschaft aus. [Forestry in North-Rhine/Westphalia: cluster analysis shows unexpected results concerning the socioeconomic importance of forestry industry]. *Holz-Zentralblatt* 74, 1018–1019.

Schulte, A. (ed.) (2003b) *Wald in Nordrhein-Westfalen,* 2 vols [Forestry in North-Rhine/Westphalia]. Aschendorff, Münster, Germany.

Schulte, A. and Mrosek, T. (2006) Analysis and assessment of the forestry and wood-based industry cluster in the state of North-Rhine/Westphalia, Germany. *Forst Archiv* 4, 136–141.

Sölvell, Ö., Lindqvist, G. and Ketels, C. (2003) *The Cluster Initiative Greenbook*. Ivory Tower, Stockholm.

Stahl, T. and Schreiber, R. (2004) *Regionale Netzwerke als Innovationsquelle: Das Konzept "Lernende Region" in Europa*. [Regional networks as a source of innovation: the concept of the learning region in Europe]. Campus, Frankfurt, New York.

TMG (2005) *MHC-inside: Informationen aus dem Möbel- und Holzbau-Cluster* [MHC-inside: information concerning the furniture and wood construction cluster]. Oberösterreich Technologie- und Marketinggesellschaft mbH, Linz, Germany.

9 Modelling Public Support for Wildland Fire Policy

J.D. ABSHER[1] AND J.J. VASKE[2]

[1]USDA Forest Service, Pacific Southwest Research Station, Riverside, California, USA; [2]Department of Natural Resource Recreation and Tourism, Colorado State University, Fort Collins, Colorado, USA.

Abstract

Theoretically grounded explanations of wildland fire policy can be improved by empirically documenting the causal influences of support for (or opposition to) management alternatives. This chapter proposes a model based on the specificity principle (i.e. correspondence between measured variables) to empirically examine four common wildland fire policies in relation to three sets of causal influences. Two agency actions (prescribed fire and mechanical thinning) and two homeowner actions (defensible space and firewise construction) are analysed against socio-demographic, situational and psychological precursors. Data were obtained from a survey of Colorado residents (n = 532) living in the wildland–urban interface. The predictive validity of the independent variables was assessed using logistic regression. The acceptability of the two agency and two homeowner actions had significantly different patterns of social causes and linkages. Results supported the contention that socio-demographic, situational factors and psychological variables differentially influence support for agency or homeowner actions. Consistent with the specificity principle, the psychological measures were most useful in wildland fire policy analysis. Recognizing these causal influences can improve policy development, situated communications and local community involvement strategies. Overall, theoretical-based models of natural resource policies can facilitate understanding the causal mechanisms that drive support for (or opposition to) wildland fire actions.

Introduction

Recent catastrophic wildfires have reinforced the need for successful mitigation strategies that are coordinated across all levels of government (federal, state, county, local) and address the needs and concerns of affected homeowners living in the wildland–urban interface (WUI) (Blackwell and Tuttle, 2004). Despite the growing body of social science literature on wildland fires, knowledge gaps remain, especially with respect to wildland fire policy (Cortner and Field, 2004).

For example, to what extent are homeowners familiar with and do they approve of alternative agency-initiated wildfire mitigation strategies (e.g. mechanical thinning, prescribed burns)? Are WUI residents willing to adopt individual behaviours that can potentially minimize the consequences of a wildland fire (e.g. defensible space, firewise construction)? Do they consider these activities effective? Do aesthetics matter? Do homeowners feel that defensible space and firewise construction make their property safer? What influences individual behaviour and support for wildland fire mitigation strategies? To better manage wildland fire a deeper understanding of the underlying mechanisms is needed. Social science can facilitate this process by informing decisions and improving communication efforts.

The importance of merging science and policy can be traced back five decades to Lasswell (1951, p. 524), who noted that a healthy policy process 'brings to light factors that hitherto operated as a determining factor . . . but which had been operating unconsciously' (i.e. unknown to the policymaker). Work in policy arenas (e.g. housing or labour) has encouraged this approach by linking public beliefs and policy issues, particularly with empirical data (Hyman *et al.*, 2001). Natural resource managers have similarly recognized that the social sciences can inform the decision-making process (O'Laughlin, 2003). Cheng (2002), for example, emphasized that responding to wildfire risks is essentially a social process that blends scientific information with attitudes. Hoover and Langner (2003) noted 'the importance of understanding . . . attitudes, perceptions and beliefs about fire in developing feasible fire management strategies'. Despite this recognition of the potential contributions that the social sciences can make, scientifically based analyses of wildland fire policies are only starting to emerge in the literature (Cortner and Field, 2004).

This chapter examines the differential influence of three sets of predictors (i.e. socio-demographics, situational factors and psychological variables) on two agency policies (prescribed fire and mechanical thinning) and two homeowner actions (defensible space and firewise construction). The 'specificity' principle (i.e. correspondence between the measured concepts) provides the basis for explanations of differences in the strength of relationships between social and psychological predictor variables and the policies related to wildland fire mitigation management actions.

Predicting policy support

Combinations of underlying factors have been shown to influence support for wildland fire management alternatives (Taylor *et al.*, 1988; Hoover and Langner, 2003; Kneeshaw *et al.*, 2004a, b). In general, the wildland fire literature has addressed three general categories of predictor variables: socio-demographic, situational and psychological factors (Fig. 9.1).

Socio-demographic variables are commonly measured in social science surveys, and are frequently reported in wildland fire management studies. Variables such as age, sex, education and income have been shown to be related to residents' perceptions of wildland fires. Otani *et al.* (1992) showed that older

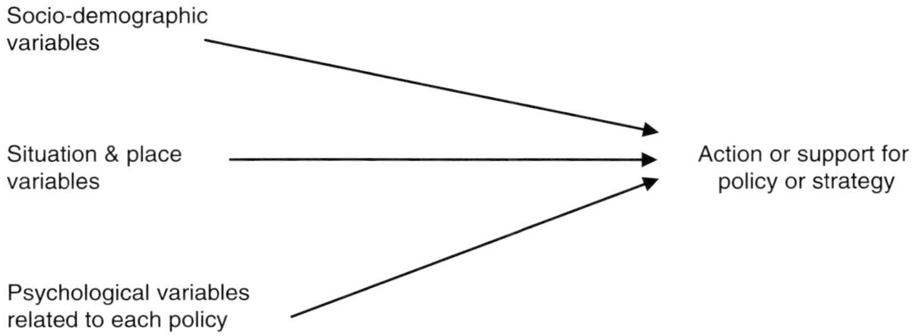

Fig. 9.1. Basic model of support for policy or action.

residents were more cautious in their interpretation of warning signs. A review of risk perception studies concluded that gender played a significant role and suggested that different meanings are socially constructed rather than genetically predetermined (Gustafson, 1998). Education and income have a role as well. Education level may be linked to knowledge about wildland fire and some homeowner mitigation strategies, such as firewise construction techniques, may be limited by income (Vogt, 2003; Vogt *et al.*, 2003).

Situational factors define a given context and influence what the public perceives as acceptable or feasible (Wittmann *et al.*, 1998; Zinn *et al.*, 1998). Large tracts of forested land often surround homes built in the wildland–urban interface (WUI). Protecting these private residences from fire is a primary consideration when managing wildland fires, and this protection influences homeowners' acceptance of fire management policy (Davis, 1990). Public support for fire management has been linked to whether the fire will affect private property (Manfredo *et al.*, 1990; Jacobson *et al.*, 2001). In areas where property damage from wildfires occurs frequently, residents supported the immediate suppression of fires that threatened personal property (Gardner *et al.*, 1987).

Studies of wildland fire beliefs and attitudes suggest that psychological variables are also important to understanding wildland fire policy support (Bright *et al.*, 2003; Vogt *et al.*, 2003, 2005; Winter, 2003; Brenkert *et al.*, 2005; Absher *et al.*, in press). The public often under- or overestimates wildfire risks (Beebe and Omi, 1993), and large attitudinal differences sometimes exist between experts and non-experts in risk situations (Zaksek and Arvai, 2004). Other research suggests that public expectations and understandings of wildland fire management in the WUI change over time and need to be affected by well-crafted public education programmes (Cortner *et al.*, 1990). More recently, however, McCaffrey (2004) concluded that such educational campaigns do not seem to be working, perhaps because of a lack of understanding or trust.

Not all of these classes of predictors (socio-demographic, situational and psychological), however, are likely to contribute equally to support for (or opposition to) agency wildland fire management polices or homeowner mitigation strategies (Fishbein and Ajzen, 1975; Eagly and Chaiken, 1993). Well-educated, wealthy homeowners (socio-demographic variables) living in the WUI (a situation variable), for example, may fail to perform defensible space activities because they do not consider the actions to be effective in protecting their home or worry about the aesthetic impacts on their property (psychological variables). Social–psychological theories offer an explanation for these disparities, suggesting that the 'specificity' principle (i.e. correspondence between the measured concepts) influences the strength of observed relationships between variables (Fishbein and Ajzen, 1975). This principle predicts that general socio-demographic variables (e.g. education, income) that are not issue-specific will explain less of the variability in support for agency (mechanical thinning, prescribed burning) or homeowner wildland fire mitigation strategies (defensible space, firewise construction) than more topic-specific psychological variables (beliefs about effectiveness and aesthetics of mitigation efforts). Situation variables such as proximity to a forest may raise awareness of the potential risks of wildland fires, but are less specific than the psychological variables. Consequently, the predictive power of situational variables should fall somewhere in between that of socio-demographic and psychological predictors. Previous research across a variety of natural resource management issues has supported this predicted pattern of results (Fishbein and Ajzen, 1975; Donnelly and Vaske, 1995; Whittaker *et al.*, 2006).

Methods

Data for this study were obtained from a mail survey of Colorado residents living in the WUI. The study population consisted of individuals over the age of 18 who reside in Boulder, Larimer, Gilpen, Grand, Jackson and Clear Creek counties. A random sample of resident names and addresses was purchased from a commercial sampling firm in the summer of 2004.

Mail survey administration

Four mailings were used to administer the survey beginning at the end of May 2004. Residents first received the 12-page questionnaire, a prepaid postage return envelope and a personalized cover letter explaining the study and requesting their participation. Ten days after the initial mailing a reminder postcard was sent to participants. A second complete mailing (questionnaire, prepaid postage return envelope and cover letter) was sent to non-respondents 10 days after the postcard reminder. To further increase response rate, a third complete mailing was sent 1 month following the second complete mailing. A total of 532 completed surveys were returned, with an overall response rate of 47% (532 returned/ 1200 sent – 56 non-deliverables).

As a check on potential non-response bias, a telephone survey was conducted of non-respondents (n = 100). Perceived effectiveness, approval and aesthetic impacts of prescribed burning and mechanical thinning variables were included in the telephone survey. For all the variables the Hedge's effect sizes were < 0.2, indicating a 'minimal' relationship (Vaske *et al.*, 2002). Non-response bias was thus not considered to be a problem and the data were not weighted.

Variables in model

The survey contained four separate dependent variables representing four different wildland fire mitigation strategies. Each was introduced with a short description and reinforced with captioned illustrations. Two strategies dealt with activities homeowners could adopt (i.e. defensible space and firewise construction) and two concerned agency activities (i.e. mechanical thinning and prescribed burns). Each of these dependent variables was analysed as a dichotomous variable. For the homeowner activities, respondents indicated whether or not they currently practised defensible space or firewise construction. For the agency action strategies, respondents rated three prescribed burn questions and four mechanical thinning items. These variables were originally coded on separate seven-point scales that gauged the respondent's perceptions of appropriateness, effectiveness and safety of the technique. After examining the reliability of the variables associated with each concept, an additive composite index of support was calculated for prescribed burning and mechanical thinning. Cronbach alphas were 0.83 and 0.87 for prescribed burning and mechanical thinning indices, respectively. For analysis consistency with the homeowner activity variables, these composite indices were collapsed into dichotomous variables, where 0 reflected opposition and 1 indicated support for each agency action (i.e. the 'neutral' point of the scale was used to differentiate the two groups, with neutral classified with the low support group).

Three sets of independent variables were examined. The socio-demographic predictors included: age (coded in years), sex (0 = male, 1 = female), total annual household income (1 = 'less than $10,000', 8 = '$150,000 or more'), and education (1 = 'less than secondary school diploma', 6 = 'advanced degree'). Four situational predictors were examined: Do you live at this address year round? (0 = No, 1 = Yes); How far from a forested area is this property? (coded in seven broad mileage categories); Do you own or rent this property? (0 = No, 1 = Yes); and How long have you lived in Colorado? (coded in years).

The psychological variables measured respondents' familiarity, perceived effectiveness and aesthetic impacts of prescribed burning and mechanical thinning. For defensible space and firewise construction, individuals were also asked to indicate whether or not the actions enhanced the safety of their property. Each of these items was coded on nine-point scales (e.g. 1 = not at all familiar, 9 = extremely familiar).

Results

Over three-quarters (79%) of the WUI residents practised at least one type of defensible space activity and nearly half (47%) engaged in some form of firewise construction (Table 9.1). Nine out of ten (90%) respondents approved of mechanical thinning and 82% agreed with prescribed burning activities.

The respondents were typically male (65%) and about 56 years old and had at least some college education and household incomes slightly over $70,000 per year (Table 9.2). These socio-demographic results differ somewhat from the

Table 9.1. Dependent variables: results of grouping action and support variables.

Dependent variables	Mean	SD	Cronbach's alpha
Defensible space[a]	0.791	0.410	n/a
Firewise construction[a]	0.471	0.499	n/a
Mechanical thinning[b]	0.902	0.388	0.83
Prescribed burning[b]	0.822	0.297	0.87

SD, standard deviation.
[a]Response scale is 0 = have done or currently practise or 1 = have not done or never practised.
[b]Response scale is originally from 1 = not at all/strongly disapprove to 7 = extremely/strongly approve, collapsed at 4 = neutral to become a dichotomous (0, 1) variable.

general population profile of the same region, but are typical of homeowners in WUI settings.

A large majority of respondents were year-round (84%) homeowners (93%) with many years of Colorado residency (average = 26.7 years). Because of the sampling design, most lived in, or very near, the forest. In terms of situation variables, the respondents were a stable, permanent group with a long-term association with forest lands.

Finally, the psychological variables showed a general trend for support/agreement, with averages ranging from 5.32 to 6.83 on a nine-point scale (i.e. all variable means were above the midpoint of the response scales). These means, however, were not strongly skewed to the upper end of the scale, suggesting that there were differences of opinion or self-evaluation on each policy aspect measured.

Separate logistic regression models were fitted for each of the three sets of predictor variables on each of the four criterion variables, or 12 logistic models in total (Table 9.3). Among the socio-demographic indicators, only age (mechanical thinning model) and income (firewise construction, mechanical thinning and prescribed burning models) were statistically significant. None of the socio-demographic variables statistically influenced defensible space activities. The socio-demographic variables explained at most only 2% (Nagelkerke R^2) in the personal mitigation strategies and 6% of the variance in the agency actions (Table 9.4).

Three of the four situational variables influenced participation in defensible space actions (Table 9.3); distance from the forest, home ownership and years living in Colorado were all significant in this model (Nagelkerke $R^2 = 7\%$). Only year-round residency was not significant in the defensible space model. In the firewise construction model, only distance from the forest had a statistical effect (Nagelkerke $R^2 = 3\%$). None of the four situational predictors influenced mechanical thinning. For the prescribed burning equation, only years living in Colorado was significant (Nagelkerke $R^2 = 2\%$).

The psychological variables explained more of the variability in both the personal and agency action models than either the socio-demographic or situational

Table 9.2. Summary of independent variables: socio-demographic, situational and psychological.

Independent variables	Variable type: scale points relevant to mean	Mean	SD
Socio-demographic			
Gender	Nominal: 0 = male, 1 = female	0.35	0.478
Age	Ratio: In years, 23–92 years	55.8	14.126
Education	Ordinal, 6 categories: 4 = some college, 5 = 4-year college degree	4.62	1.235
Income	Ordinal, 8 categories: dollar midpoints used	$70,338.35	$37,161.25
Situational			
Year-round resident	Nominal: 0 = no, 1 = yes	0.84	0.365
Forest proximity	Ordinal, 7 categories: 1 = within forested area, 2 = less than a mile	1.92	1.120
Home ownership	Nominal: 0 = own, 1 = rent	0.07	0.258
Years in Colorado	Ratio: 1–86 years	26.71	17.942
Psychological	All psychological variables are Likert scales, 9 points: 1 = not at all (attribute), 5 = moderately, 9 = extremely (not all points labelled).		
Familiar with policy			
Defensible space		6.72	1.575
Firewise construction		5.71	2.010
Mechanical thinning		6.02	2.030
Prescribed burning		6.61	1.700
Think it's effective			
Defensible space		6.27	2.111
Firewise construction		6.15	1.937
Mechanical thinning		6.83	1.580
Prescribed burning		6.74	1.456
Think it's safe			
Defensible space		6.00	1.985
Firewise construction		5.88	1.970
Mechanical thinning		n/a	–
Prescribed burning		n/a	–
Think it improves look			
Defensible space		5.32	2.041
Firewise construction		5.72	1.573
Mechanical thinning		6.30	1.986
Prescribed burning		5.65	2.047

Table 9.3. Socio-demographic, situational and psychological influences on policy variables: path ratio and significance from logistic regression of each set of variables on policy measure.

Independent variables	Personal actions				Agency actions			
	Defensible space		Firewise construction		Mechanical thinning		Prescribed burning	
	Exp(B)	Sig.	Exp(B)	Sig.	Exp(B)	Sig.	Exp(B)	Sig.
Socio-demographic								
Gender	1.01	0.971	0.93	0.709	1.20	0.612	0.87	0.584
Age	1.01	0.413	1.01	0.254	1.04	0.003	0.99	0.418
Education	1.09	0.409	0.92	0.390	0.78	0.110	1.06	0.543
Income	1.00	0.074	1.00	0.043	1.00	0.044	1.00	0.004
Situational								
Year-round resident	0.77	0.455	0.91	0.729	0.63	0.343	0.98	0.961
Forest proximity	0.73	0.001	0.82	0.024	1.11	0.483	0.94	0.567
Home ownership	0.39	0.013	0.58	0.140	0.81	0.711	1.46	0.446
Years in Colorado	1.01	0.039	1.01	0.091	1.00	0.748	0.97	0.018
Psychological								
Familiar with policy	1.63	< 0.001	1.64	< 0.001	0.823	0.027	0.84	0.045
Think it's effective	1.54	< 0.001	0.99	0.946	1.54	0.120	0.84	< 0.001
Think it's safe	1.25	0.034	1.15	0.078	n/a	n/a	n/a	n/a
Think it improves look	1.06	0.437	1.02	0.772	1.88	0.048	1.74	< 0.001

Table 9.4. Summary of sub-model tests: Nagelkerke R^2 and significance.

Independent variables	Personal actions: Defensible space	Personal actions: Firewise construction	Agency actions: Mechanical thinning	Agency actions: Prescribed burning
Socio-demographic	0.021 (NS)	0.017 (NS)	0.063 (0.042)	0.054 (0.004)
Situational	0.073 (< 0.001)	0.032 (0.021)	0.007 (NS)	0.020 (NS)
Psychological	0.441 (< 0.001)	0.270 (< 0.001)	0.390 (< 0.001)	0.393 (< 0.001)

NS, not significant. Significance of test in parentheses.

variables (Table 9.4). The Nagelkerke R^2 for the psychological models ranged from 27% (firewise construction) to 44% (defensible space). In the two agency action models, the Nagelkerke R^2 was 39%. Familiarity with the policy was statistically significant in all four logistic regressions. Perceived effectiveness influenced defensible space and prescribed burning. Aesthetics influenced approval of the two agency actions, but neither of the personal actions. For the two personal action equations, perceived safety was related to defensible space actions, but not firewise construction.

In summary, the results indicated clear differences in support patterns, both as individual variables and as categorical groupings for different types of wildland fire policies. Support for the specificity principle is mixed when considering these measures of support by predictor type because of a lack of clear differences in the socio-demographic and situational variables.

Discussion

Variables in each of the three classes of predictors can influence agency policy and individual homeowner behaviour. Consistent with the specificity principle, socio-demographic indicators had low predictive power. The strongest effects of socio-demographic variables were for agency action policies that affect multiple resource uses such as recreation, tourism, watershed and biomass outputs. Overall, these variables are probably indicative of other social processes such as previous agency–stakeholder communications and/or involvement in decision-making processes. However, such variables are not part of this data set and this supposition cannot be tested here. Future work should address the causal mechanisms for these broad classes of support indicators.

The situational variables also had limited explanatory power, and do not seem to be stronger or more frequent as significant predictors than the socio-demographic predictors. Notably, however, the pattern of significance was different. The influence of situational variables was most closely tied to policies that require homeowner actions to be successful, especially defensible space activities. Our findings suggest that many WUI residents have adopted some homeowner wildfire mitigation strategies and generally support agency-initiated efforts. These policy linkages emphasize the importance of localized communications and close working relationships with affected resident groups or communities (Nelson *et al.*, 2005).

The psychological variables displayed strong and consistent links to each policy. Even though the exact independent variable changes with the type of policy being predicted, the overall pattern supports a conclusion that an individualized approach is important for garnering support or behavioural compliance with a policy. Mechanical thinning and prescribed burns, for example, were closely tied to the psychological measures of support. For all four policies, familiarity with the policy was consistently a significant predictor of support. This suggests that better support for these policies, and perhaps compliance when action is required, may be possible if the communication strategy is tailored to the resident's individual lifestyle and crafted for separate market segments somewhat differently. The policy implementation phase needs to involve local people (volunteers or influentials) to generate positive social norms for a given community or residential area. For compliance with firewise construction and defensible space strategies, agencies should pay attention to the psychological drivers and to the situational variables such as proximity to the forest. Given the homeowners' costs associated with adopting firewise construction, and the barriers that these might pose to compliance, our results also suggest that residential land developers and the home construction industry should be an important target for communication efforts.

In conclusion, applying scientific data analysis methods to wildland fire policy is a relatively new endeavour. This work is an initial step that bridges traditional discursive policy analysis with a more empirical approach as espoused by Lasswell. The three-factor causal model of policy support implemented above provided a theoretical advancement and a practical analysis. Although encouraging, more work is needed to further develop a comprehensive model of policy support for wildland fire actions and to demonstrate its use in other geographical or resource settings. This general modelling approach should also be broadly applicable to other policy arenas, especially those focused on natural resource management or natural disaster issues. Recognizing these causal influences can improve policy development, situated communications and local community involvement strategies. In particular, these results especially point to the utility of including psychological determinants in the policy analysis model, and to the need to carefully assess the role of constituent influences for a specific policy. The use of social science data to inform wildland fire policy is rapidly evolving, and yet it is already bearing fruit for decision makers by specifying the different considerations that are important in affecting support, opposition or behavioural compliance to wildland fire policy.

References

Absher, J.D., Vaske, J.J., Bright, A.D. and Kneeshaw, K. (in press) Assessment and modelling of beliefs, value, and social norms related to wildland fire management in southern California. In: Narog, M.G. (technical coordinator) *Proceedings of the 2002 Fire Conference on Managing Fire and Fuels in the Remaining Wildlands and Open Spaces of the Southwestern United States, 2–5 December 2002, San Diego, CA.* General Technical Report PSW-189, Pacific Southwest Research Station, Forest Service, US Department of Agriculture, Albany, California.

Beebe, G.S. and Omi, P.N. (1993) Wildland burning: the perception of risk. *Journal of Forestry* 91, 19–24.

Blackwell, J.A. and Tuttle, A. (2004) *California Fire Siege 2003: The Story.* USDA Forest Service, Mare Island, California.

Brenkert, H., Champ, P. and Flores, N. (2005) *Mitigation of Wildfire Risk by Homeowners.* Research Note RMRS-RN-25WWW, USDA Forest Service, Rocky Mountain Research Station, Fort Collins, Colorado.

Bright, A.D., Vaske, J.J., Kneeshaw, K. and Absher, J.D. (2003) Scale development of wildfire management basic beliefs. In: Jakes, P.J. (comp.) *Homeowners, Communities, and Wildfire: Science Findings from the National Fire Plan.* General Technical Report NC-231, North Central Research Station, Forest Service, US Department of Agriculture, St Paul, Minnesota, pp. 18–25.

Cheng, A.S. (2002) Fire social science research: opening remarks. In: Omi, P.N. and Joyce, L.A. (comp.) *Fire, Fuel Treatments, and Ecological Restoration: Conference Proceedings.* USDA Forest Service Proceedings RMRS-P-29, USDA Forest Service, Rocky Mountain Research Station, Fort Collins, Colorado, pp. 209–212.

Cortner, H.J. and Field, D.R. (2004) Introduction: humans, fire, and forests – part III. *Society and Natural Resources* 17, 659–660.

Cortner, H.J., Gardner, P.D. and Taylor, J.G. (1990) Fire hazards at the urban–wildland interface: what the public expects. *Environmental Management* 13(1), 15–62.

Davis, J.B. (1990) The wildland–urban interface: paradise or battleground? *Journal of Forestry* 88, 26–31.

Donnelly, M.P. and Vaske, J.J. (1995) Predicting attitudes toward a proposed moose hunt. *Society and Natural Resources* 8, 307–319.

Eagly, A.H. and Chaiken, S. (1993) *The Psychology of Attitudes.* Harcourt Brace Jovanovich, New York.

Fishbein, M. and Ajzen, I. (1975) *Belief, Attitude, Intention, and Behavior: An Introduction to Theory and Research.* Addison-Wesley, Reading, Massachusetts.

Gardner, P.D., Cortner, H.J. and Widaman, K. (1987) The risk perceptions and policy response toward wildland fire hazards by urban home-owners. *Landscape and Urban Planning* 14, 163–172.

Gustafson, P.E. (1998) Gender differences in risk perception: theoretical and methodological perspectives. *Risk Analysis* 18, 805–811.

Hoover, A.P. and Langner, L.L. (2003) People and wildfire: an overview. In: Jakes, P.J. (comp.) *Homeowners, Communities, and Wildfire: Science Findings from the National Fire Plan.* General Technical Report NC-231, North Central Research Station, Forest Service, US Department of Agriculture, St Paul, Minnesota, pp. 1–3.

Hyman, D., Bridger, J., Shingler, J. and Van Loon, M. (2001) Paradigms, policies and people: exploring the linkages between normative beliefs, public policies and utility consumer payment problems. *Policy Studies Review* 18(2), 89–121.

Jacobson, S.K., Monroe, M.C. and Marynowski, S. (2001) Fire at the wildland urban interface: the influence of experience and mass media on public knowledge, attitudes, and behavioral intentions. *Wildlife Society Bulletin* 29, 929–937.

Kneeshaw, K., Vaske, J.J., Bright, A.D. and Absher, J.D. (2004a) Acceptability norms toward fire management in three national forests. *Environment and Behavior* 36(4), 592–612.

Kneeshaw, K., Vaske, J.J., Bright, A.D. and Absher, J.D. (2004b) Situational influences of acceptable wildland fire management actions. *Society and Natural Resources* 17, 477–489.

Lasswell, H.D. (1951) Democratic character. In: *The Political Writings.* Free Press, Glencoe, Illinois, pp. 465–525.

McCaffrey, S. (2004) Thinking of wildland fire as a natural hazard. *Society and Natural Resources* 17, 509–516.

Manfredo, M.J., Fishbein, M., Haas, G.E. and Watson, A.E. (1990) Attitudes toward prescribed fire policies. *Journal of Forestry* 87, 19–23.

Nelson, K.C., Monroe, M.C. and Johnson, J.F. (2005) The look of the land: homeowner landscape management and wildfire preparedness in Minnesota and Florida. *Society and Natural Resources* 18, 321–336.

O'Laughlin, J. (2003) *Policy Analysis and Natural Resources: Bringing Science to Policymakers.* University of Idaho, Moscow, Idaho.

Otani, H., Leonard, D., Ashford, V., Bushroe, M. and Reeder, G. (1992) Age differences in perception of risk. *Perceptual and Motor Skills* 74, 587–594.

Taylor, J.G., Carpenter, E.H., Cortner, H.J. and Cleaves, D.A. (1988) Risk perception and behavioral context: U.S. forest service fire management professionals. *Society and Natural Resources* 1, 253–268.

Vaske, J.J., Gliner, J.A. and Morgan, G.A. (2002) Communicating judgments about practical significance: effect size, confidence intervals and odds ratios. *Human Dimensions of Wildlife* 7, 287–300.

Vogt, C.A. (2003) Seasonal and permanent home owners' past experiences and approval of fuels reduction. In: Jakes, P.J. (comp.) *Homeowners, Communities, and Wildfire: Science Findings from the National Fire Plan*. General Technical Report NC-231, North Central Research Station, Forest Service, US Department of Agriculture, St Paul, Minnesota, pp. 63–73.

Vogt, C.A., Winter, G. and Fried, J.S. (2003) Antecedents to attitudes toward prescribed burning, mechanical thinning, and defensible space fuel reduction techniques. In: Jakes, P.J. (comp.) *Homeowners, Communities, and Wildfire: Science Findings from the National Fire Plan*. General Technical Report NC-231, North Central Research Station, Forest Service, US Department of Agriculture, St Paul, Minnesota, pp. 74–83.

Vogt, C.A., Winter, G. and Fried, J.S. (2005) Predicting homeowners' approval of fuel management at the wildland–urban interface using the theory of reasoned action. *Society and Natural Resources* 18(4), 337–354.

Whittaker, D., Vaske, J.J. and Manfredo, M.J. (2006) Specificity and the cognitive hierarchy: value orientations and the acceptability of urban wildlife management actions. *Society and Natural Resources* 19(6), 515–530.

Winter, P.L. (2003) Californians' opinions on wildland and wilderness fire management. In: Jakes, P.J. (comp.) *Homeowners, Communities, and Wildfire: Science Findings from the National Fire Plan*. General Technical Report NC-231, North Central Research Station, Forest Service, US Department of Agriculture, St Paul, Minnesota, pp. 84–92.

Wittmann, K., Vaske, J.J., Manfredo, M.J. and Zinn, H.C. (1998) Standards for lethal response to problem urban wildlife. *Human Dimensions of Wildlife* 3(4), 29–48.

Zaksek, M. and Arvai, J.L. (2004) Toward improved communication about wildland fire: mental models research to identify information needs for natural resource management. *Risk Analysis* 24(6), 1503–1515.

Zinn, H.C., Manfredo, M.J., Vaske, J.J. and Wittmann, K. (1998) Using normative beliefs to determine the acceptability of wildlife management actions. *Society and Natural Resources* 11(7), 649–662.

10 Analysing Institutions and Public Perspectives to Identify the Future of British Forests

M. NIJNIK[1] AND A. MATHER[2]

[1]Socio-Economic Research Group, The Macaulay Institute, Craigiebuckler, Aberdeen, UK; [2]Department of Geography and Environment, University of Aberdeen, Aberdeen, UK.

Abstract

The transition of forestry from material production to environmental objectives is analysed from the perspective of institutional economics. The research used a semi-qualitative approach wherein the quantitative method of Q-analysis was applied as relevant. New institutional frameworks and ultimately new prospects for British forestry development were identified through analysis of public opinion, forest policy and practice. The preference models developed here characterize current UK attitudes towards forest policy, providing some innovative perspectives on the areas of consensus and conflict between people regarding key directions for the future of forestry. The chapter concludes by demonstrating the importance of operationalizing social values into concrete forest management decisions by shifting from a timber management to a multiple resource management model and from a forest-focused to a people-focused concept.

Keywords: Q-methodology, climate change mitigation, public attitudes, the UK.

Introduction

British woodlands and forestry: historical overview

Britain was deforested early in its history and extensively (FC, 2001). Near the beginning of the 20th century, the forested area even in Scotland had fallen to under 5% (Smout *et al.*, 2005). Deforestation occurred as a result of the normal forces related to economic growth and of scant attention paid to environmental values.[1] Deforestation in the UK (as in some other countries such as Ireland and Denmark) occurred rapidly because land was accessible, interest rates were high, rural wages were low and there were opportunities for trade. The low stumpage fees contributed to the transfer of woodland into agriculture since even the

 Sustainable Forestry: from Monitoring and Modelling to Knowledge Management and Policy Science (eds K.M. Reynolds, A.J. Thomson, M. Köhl, M.A. Shannon, D. Ray and K. Rennolls)

modest rates of return that were expected from agriculture compared favourably to forestry at low stumpage prices. Industrial development contributed to deforestation as well, first with charcoal making and then through providing timbers for coal mining on which heavy industry depended. During, the First World War when importing timber from Norway or the Baltic states became difficult, any remaining woodlands were heavily depleted.

This timber crisis led to the establishment of a state forest service in Britain (the Forestry Commission (FC)), and a policy of creation of new forests by state planting. Before the new forests had matured, the Second World War led to similar pressures on the forests for war material. After the war, the afforestation programme was expanded. This programme has been remarkably successful. The forest area in Scotland trebled during the 20th century and in the UK it more than doubled. The Forestry Commission's annual round wood production increased by 700%. The policy of creating commercial plantations was justified by its protagonists based on several arguments: conifer monocultures are more productive and easily managed; straight-edged plantations minimize fencing costs, which often make up half the establishment costs; large areas bring economies of scale; and hardy pioneer species were a necessity for establishing forest ecosystems on degraded sites (Warren, 2000).

Following the Second World War, Britain pursued parallel, but partly conflicting, policies of agricultural and forest expansion. New forests were usually located on the least productive agricultural land, often in locations remote from the main centres of population. This was due to a combination of an emphasis on timber production, environmental constraints on species selection and most afforestation occurring in remote locations. These new forests were not connected to the urbanized population nor were they perceived to have strong values in relation to landscape, wildlife or recreation.

The forests we see today across the UK are a product of this past. Current public attitudes to forests were shaped by this history. So, although the forest area in the UK is comparatively small with a wooded cover of only 11.6% (FC, 2001) much of it was established under circumstances that no longer exist and for purposes that are no longer relevant, and in locations remote from the main centres of population. Due to the historically early and extensive deforestation in the UK, forest culture as manifested in folklore, literature and art is almost non-existent. Not only was the public not encouraged to visit the forests, the forests were viewed as the personal domain of the landowners. Once the dominion of kings and the landed gentry, class-based ownership of forests created a social as well as a geographical gulf between people and forests (Mather, 2003).

In the early 1990s non-material values of forests emerged as an important issue for the environment and people in the UK and elsewhere. The UNCED (1992) conference resulted in the non-binding Forestry Principles that served as a catalyst for change. The principle that 'forest resources and forest lands should be sustainably managed to meet social, economic, ecological, cultural and spiritual human needs of present and future generations' supports the concept of sustainable multifunctional forest management as an aspiration formally espoused by many countries. The Forestry Principles and the emergence of various international institutions and agreements have accompanied changes in dominant views on forests. When forests become multifunctional places of amenity consumption, recreation and wildlife observation rather than mono-functional places of wood production, the concept of sustainability in forestry[2] expands to cover sustainable multifunctional forest use (Table 10.1). Sustainable forestry then aims at delivering a socially acceptable distribution of incomes and benefits from forests while creating opportunities for more people to enjoy trees, woodlands and forests and helping rural communities benefit from them (SE, 2000).

A reduced emphasis on material production, combined with an increased importance of the provision of environmental and social services, characterized forestry in Britain[3] at the turn of the 21st century. Forestry today aims at meeting multiple and dynamically evolving social requirements, which reflect public preferences for multifunctional forestry and necessitate their satisfaction in a sustainable manner. Post-productivism is a descriptor of this transition in emphasis away from a policy concern with increasing material production and towards concern with the provision of environmental services.

A broad range of non-market forest benefits is currently recognized in Britain. They include: open-access non-priced recreation and leisure activities; landscape, amenity and countryside characterization; biodiversity; valuable habitats; measurable benefits in physical and mental health; carbon sequestration; absorption of air, water and noise pollutants; management of water resources; protection of archaeological features and historic sites; provision of opportunities for education, community activities and social inclusion; and regeneration and improvement of brown-field sites (FTA, 2004).

Table 10.1. Forest benefits. (Adapted from FAO, 1996.)

Production/goods (material)	Services (non-material)
Wood products	Environmental
Logs	Regulatory
Industrial round wood	Protection of soil
Pulpwood	Regulation of climate (global and local)
Paper and paperboard	Water regulation and purification
Chips and particles	Support
Wood-based panels	Of livelihoods and habitats
Fuel wood	Biological diversity
Non-wood products	Social
Berries, mushrooms, nuts, honey, game, birch juice	Recreational functions
	Leisure and tourism
Medicinal herbs	Game and fishing
Fodder for domestic animals	Landscape
Materials such as wool and skins	Information and reservoir
Decoration	Species and genes
	Sociocultural, spiritual
	Intrinsic natural values

This transition in forestry and forest policies is occurring around the world and raises important issues regarding the implications of a 'post-production forestry' within different countries. This chapter discusses some of the elements of post-production forestry in the UK by examining its emergence and current characteristics. This leads to a number of conclusions regarding the future of forestry in the UK and the kinds of institutions likely to be responsible for SFM.

Rules in Use: Do They Work for Post-Productivism?[4]

Methodological aspects

Neoclassical economic theory analyses the demand and supply sides of production, and suggests that a well-functioning market can account for the costs and benefits created in the economy by setting the right prices to guide resource use in a sustainable way. However, social gains from forestry are becoming increasingly multiple and they are often not included in economic models. Multiple forest values have a much broader spatial and temporal distribution than the distribution of the costs. It happens, therefore, that the recipients of forest benefits do not repay society in full for the costs of their activities and externalities negatively affect forestry. The reason for the existence of externalities can also be explained by the fact that the non-marketable public goods of forests possess the properties of non-rivalry and non-excludability that cause market failures (Slangen, 2000). For that reason, the role of good institutions (Shleifer and Vishny, 1998) to control the tenure, management, financing and production of public goods is incredibly important even in a well-functioning market economy.

In neoclassical economics, preferences are fixed and stable, and economic agents are rational. The value system and time preferences are exogenous and decided largely by the market. Institutions are exogenous as well, and their role in achieving economically optimal outcomes is neglected (Kant, 2003). With the inclusion of peoples' preferences for multiple forest functions in the decision-making process, attitudes and values held by the public become endogenous. There is a retreat from economist's neoclassical rationalism. When the traditional mono-functional concept of forestry fades and 'sustainable multifunctional forest management' is embraced, public involvement becomes increasingly influential.

The high level of participatory democracy is manifested in collaboration of key stakeholders, in the initiatives to involve the public in environmental decision making and in the extension of information and education (DETR, 2000). Consequently, governance that conceptualizes public perceptions, and first of all those of local communities, emerges. This governance is largely based on collective action[5] when people act together, driven by common interests. Sustainable forestry under these conditions focuses not only on sustainable timber management, but even more on sustainable management of non-marketable forest goods and the intrinsic values of the forest system.

Multifunctionality characterizes post-productivism, when social values of landscape amenities, wildlife and forest recreation to name a few are increasingly perceived as more important than commodity production. Post-productivism is a reflection of economic, social, cultural and environmental developments in modern Britain and elsewhere, and under these conditions, forestry cannot be considered through the foundational principles of neoclassical economics.[6,7] The boundaries of forest economics must therefore be extended towards the incorporation of multiple equilibrium points and new consumer choice theory in newly developed models (Kant, 2003). These multiple-use and multiple-criteria modes must be sensitive to institutions, the current state and the dynamics of which are becoming critical for addressing sustainability successfully in multifunctional forestry.

This chapter does not seek to push the boundaries of forestry economics. Instead, it employs methodological approaches of new institutional economics to examine some elements of forestry in transition. New institutional economists agree that there are two main approaches for examining institutional changes (Ostrom, 1990; North, 1993; Sabatier, 1998). The first approach considers 'action arenas' and focuses on actors, in our case in a forestry field. The actors have preferences and information-processing capabilities. According to this approach, substantial changes in the 'action arena' must be made to advance forestry development. The characteristics of the physical world and 'the rules in use' that constitute both the institutional environment and the attributes of the community should be modified (Ostrom *et al.*, 1994).

Among the ways to explore the mechanisms of institutional changes from this perspective is to consider changes as caused by the actors' responses to shifts in relative prices and preferences (Weimer, 1995). This approach argues that though inefficient institutions can exist because of path dependency, more progressive institutions are continually created in society (North, 1993). Another view within the actors' scheme explains the transition as a result of conflicting interests, when institutional arrangements change because of bargaining, and thus not all the changes are progressive (Knight, 1992).

The second approach to addressing institutional transformation, as in our case of forestry transition to post-productivism, is to focus on economic development as its primary cause. A system of institutions is then considered as endogenous and dependent on economic progress. It is argued that, at certain stages of societal development, economic progress exceeds institutional advance. When the gap appears to be broad enough, political and social preconditions for institutional transformation arise within society itself. Organizations and actors' perceptions are regarded as important, but they are not considered as the main cause of institutional changes, but rather as their consequences.

This research follows the 'actors' perspectives' approach of exploring institutional change.[8] The institutions are believed to represent themselves through a legal relationship between policy actors in the 'action arena' of forestry and the rules in use, both formal (policy and economic rules) and informal, are the driving forces that govern the patterns of interaction within the whole institutional system. The framework of forestry transition to post-productivism has been developed to provide a better understanding of the processes that are taking place in multifunctional forestry. The institutional environment and arrangements, and the interaction between policy actors in the 'action arena' of forestry, are shown in Fig. 10.1.

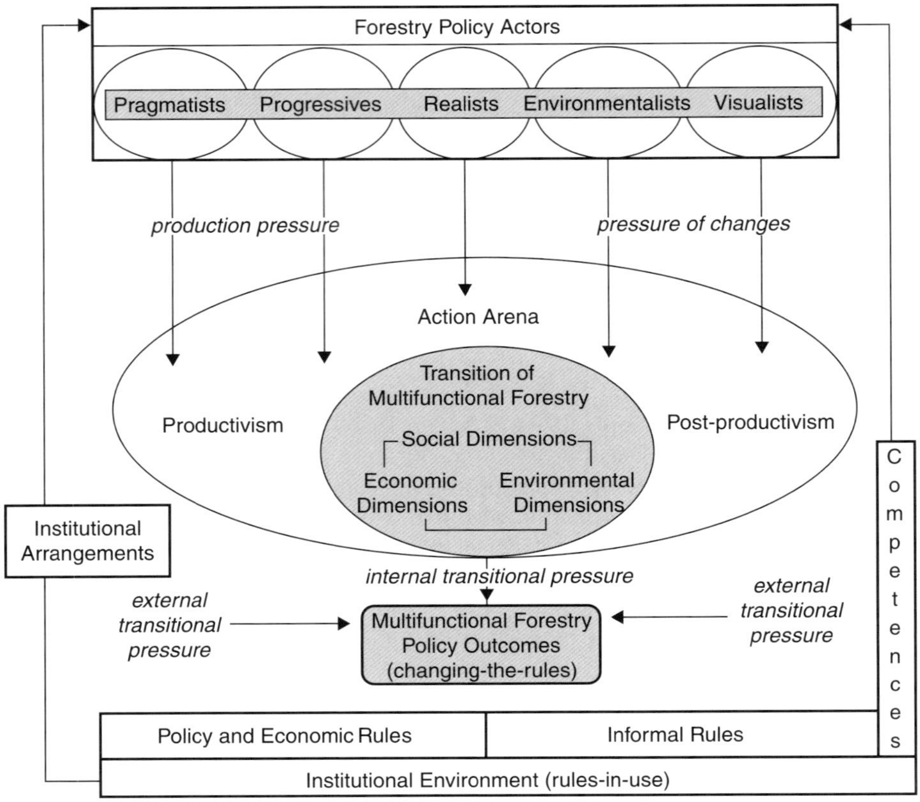

Fig. 10.1. The conceptual framework of forestry transition to post-productivism.

Institutional environment: 'the rules of the game'[9]

Traditionally, 'the rules in use', or policies relating to agriculture and forestry in Europe, have been oriented towards production. Yet starting from the 1980s, such schemes as environmentally sensitive areas (ESA)[10] and changes introduced in the reform of the Common Agricultural Policy (CAP) (EC, 1992) forced consideration of multiple objectives. Today, aid for woodlands development for multiple purposes is given by the European Union initiative, in which the Agricultural Guidance and Guarantee Fund finances up to 75% of the costs.

The expansion of forests was one of the main land-use changes in Britain during the 20th century. The change was driven primarily by a perceived need to increase timber production. However, over time the coniferization of some areas of native woodlands attracted opposition. By the mid to late 1980s there was strong resistance to the establishment of monoculture forest plantations on both environmental and social grounds. By 1992, the Forestry Principles (UNCED, 1992) had provided an external impetus to review forest policy and recast its objectives. These consisted of sustainable management of the existing woodlands estate and continued steady expansion of woodland (FC, 1991).

Today, the success of reforesting rural Britain depends on the integration of ecological, socio-economic and aesthetic aspects in landscape changes – on 'getting the right trees in the right places for the right reasons' (Warren, 2002). Until the late 1980s, the planting of broadleaved species was insignificant in terms of the proportion of the area. Since then, the proportion has risen steeply. Its trend serves as a rough proxy for post-productivism, since it is associated with a shift in emphasis towards landscape and wildlife objectives, and to the attraction of tourists to remote areas (Fig. 10.2).

In the early 2000s, the country-level forest strategies (FS) were introduced across the UK, with overarching themes such as sustainability, integration and partnerships and each with a set of 'directions' that could be categorized as multifunctional, and 'post-productive' in the sense that they re-directed policies focused primarily on timber production. The forestry strategies set out a government vision for the future of British forests, where sustainability has become an overarching principle, and forestry is acknowledged as offering a wide range of social benefits, including visual, cultural and ethical (Table 10.2).

Data from Table 10.2 complement the principal priorities of the measures within the Rural Development Programmes (RDP) (Table 10.3). The RDP priorities are split into the following three categories: re-structuring and competitiveness; environment and land management; and rural economy and rural communities.

The analysis of the priorities shows that forestry development and diversification of the activities are the first priority measures for Britain as a whole. Other measures with a common level of support amongst countries are environmental protection and landscape management. Also, for instance in Scotland, priority 1 measures comprise most of the significant woodland options, agricultural

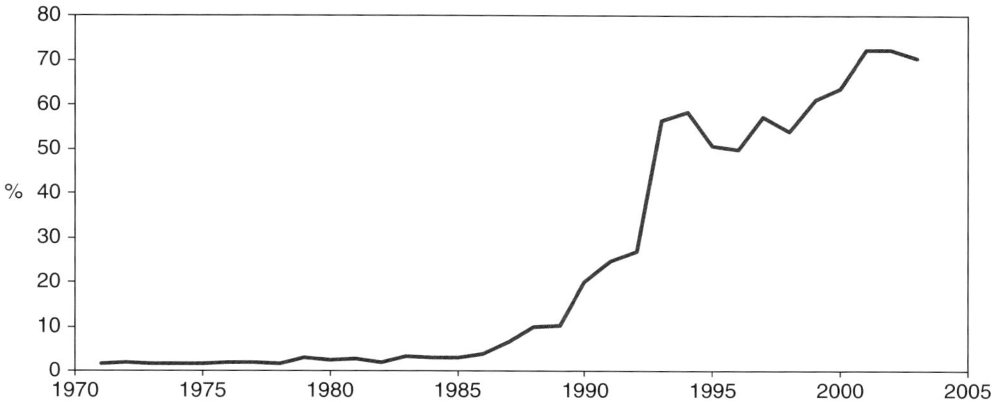

Fig. 10.2. Broadleaved new planting as percent of total new planting. Source: FC (2003a)*. *The estimates shown in Figs 10.2–10.5 were computed on the basis of data kindly made available by the FC.

Table 10.2. Forestry strategy objectives across countries. (From Forestry Commission (1998), Scotland Executive (2000) and Welsh Association of Governments (2001).)

England	Scotland	Wales
• Rural development	• Maximize value of the wood resource	• Woodlands for people
• Economic regeneration	• Create a diverse resource	• New emphasis on woodland management
• Recreation, access and tourism	• Positive contribution to the environment	• Location for world-class forest industry
• Environment and conservation	• Opportunities for more people to enjoy trees, woods and forests	• Diverse and healthy environment
		• Tourism, recreation and health

Table 10.3. Principal priorities of Rural Development Programme measures. (Based on EC, 2003.)

	England	Scotland	Wales
Restructuring/Competitiveness			
Agricultural water resource management	1	–	–
Development/improvement of infrastructures	–	–	1
Food quality – incentive scheme	–	–	2
Food quality – promotion	–	–	1/2
Investments in farm holdings	–	–	2
Investments in processing/marketing	–	–	1
Marketing of quality agricultural products	1	–	1/2
Training	1	–	2
Environmental/Land management			
Afforestation and forestry	1	1	1
Agri-environment/animal welfare	–	2	1/3
Environmental protection (agriculture, forestry and landscape management) and animal welfare	2	1/2	1
Areas less favoured and with environmental restrictions	2	1	3
Rural economy/rural communities			
Basic services for the rural economy and population	1	1	–
Diversification of activities to provide multiple activities or alternative sources of income	1	1	1
Encouragement for tourism and craft activities	1	1	–
Renovation/development of villages, protection of the rural heritage	2	2	–

diversification, plantations of short-rotation coppice and afforestation. Scotland has explicitly identified measures aimed at the protection of historic and archaeological sites and at the encouragement of tourism. There is quite a lot of evidence, therefore, that the development of the 'rules of the game' in British forestry as a whole, as well as across its countries, proceeds in the direction of multifunctionality.

Institutional arrangements: 'the play of the game'

The institutional arrangements in forestry in the UK are characterized by co-management arrangements among major policy actors such as central government agencies, local governments, community-based organizations, farms and enterprises. The devolved structure of government means that many of the principal regulatory frameworks and the policies most significant in terms of human–nature interactions are determined by the Scottish Executive and Parliament, the National Assembly for Wales and the Northern Ireland Executive and Assembly, or through government departments for England, but with inputs from regional planning groups and the public. Each part of the UK has its own approach to implementing delivery mechanisms of the schemes that support forestry as well as those relating to biodiversity conservation, landscape designations and protection of cultural heritage. Government agencies implement the policies according to existing rules and regulations, and the Forestry Commission summarizes standards and measures to manage forests sustainably.

The policy of woodland expansion is supported by direct financial instruments in four different programmes that subsidize 60% to 90% of the costs. Payments vary across land categories, tree species and distances of the land from settlements. An annual payment also encourages farmers to convert productive land and receive compensation to offset the forgone income. Subsidies are available for woodland expansion, restocking and stewardship, and are usually given for 10–15 years when broadleaved or local species are planted, and wood is not harvested for 30 years (SE, 2000). Guidelines are provided based upon the nature of the desired outcomes, such as a 'balance between forestry and other land uses', and the importance of multi-benefit forests for bringing diversification into the rural economy. The development of native woodlands is supported by the subsidies which provide 60% of costs when plans include removal of non-native vegetation, deadwood management, essential thinning and small scale felling to encourage regeneration (FC, 2004). Among the instruments employed in this new policy climate are woodland grant schemes (WGS).[11] In the procedures for applying for grants, applicants have to indicate relative priorities of the objectives related to their plans (Fig. 10.3).

Characterizing the profiles of objectives created by analysing these applications, and in particular of the ranking of 'producing wood and marketable timber' offers a means of quantifying the extent to which post-productivism has become established in landowner attitudes. Analysis of 'high' priority (HP) objectives (Fig. 10.3) leads to two main findings. First, the average number of high priority objectives per application is three, with a slight increase from 2.8 in

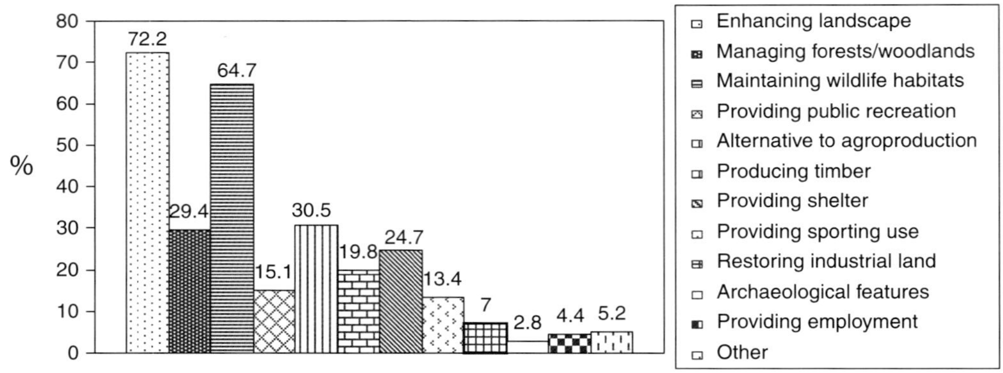

Fig. 10.3. High priority objectives identified in applications from 1995 to 2002 by percent of total high priority objectives (n = 18,755).

1995 to 3.2 in 2001. In other words, the new forests are more multifunctional. Second, objectives relating to material production are clearly overshadowed by those relating to the environment. Landscape and wildlife objectives are the most common of those classed as high priority. The classic 'productivist' objective of 'producing wood and marketable timber' is only sixth. The fact that it is no longer mandatory and is only in a modest position among landowner priorities suggests a significant shift to post-productivism. A similar pattern applies across the country ranging from regions that witnessed rapid 'productivist' afforestation in the 1970s to 1980s (Highlands and Strathclyde) to those that did not (south-west and south-east England).

Figure 10.4 shows that the cumulative number of applications which list non-material forest benefits as high priority is rising most rapidly. This fact also supports the trend toward post-productivism. Indeed, this relative tendency has been maintained over several years, leading to an increasing gap between productivist and post-productivism objectives (Fig. 10.5). Thus, these woodland grant applications reveal the behaviour of land managers as expressed in the creation of new forests and point to the transition of forestry towards post-productivism.

In addition to the evidence considered above, further signs of institutional transition exist in other forms, such as the 'felling to waste' of non-native species in the course of restructuring some state forests, the increasing attention to 'continuous-cover forestry', and the emphasis placed on 'community' forests to name only of a few of the kinds of changes occurring.

This institutional transition has led to significant changes in the 'action arena' of the countryside. Prior to the mid-1980s, farm forestry in Britain was largely unknown. The policy climate, with its sectoral structure in which agriculture and forestry were often in conflict, had inhibited coordination. While some farmers considered land sales for forestry, few would have contemplated direct

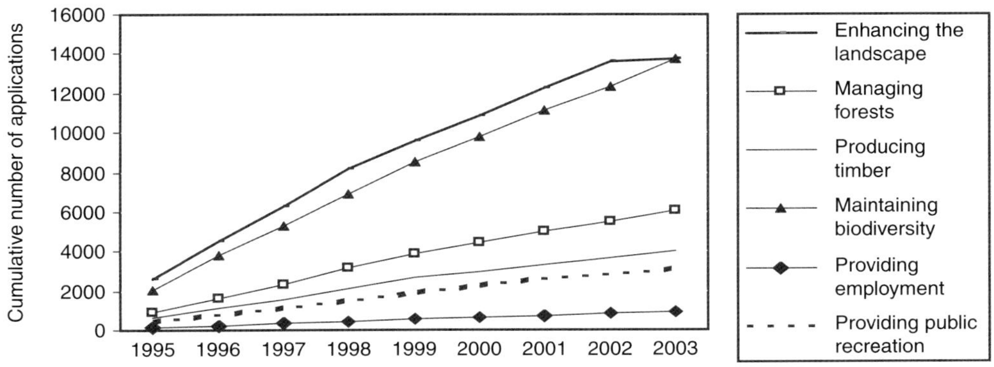

Fig. 10.4. Trends in high priority objectives listed in applications between 1995 and 2003.

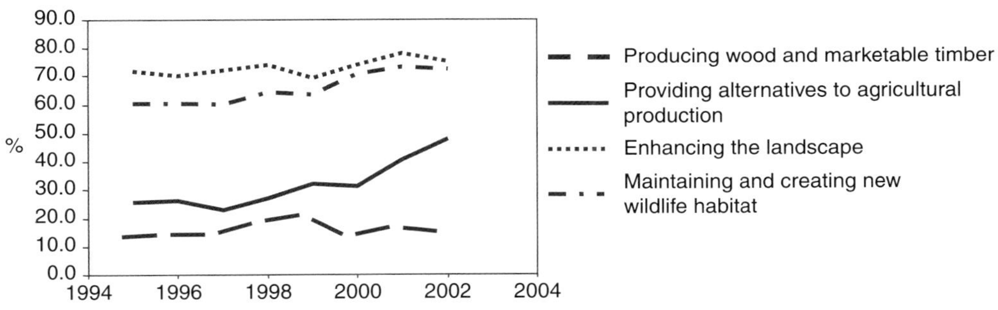

Fig. 10.5. Selected high priority objectives as percent of total high priority objectives.

involvement in afforestation. Since then, there has been significant shift towards land-use integration. This shift towards more integrated land use included a fundamental change of emphasis from production to the provision of environmental benefits. In retrospect, the suddenness of the transition to post-productivism is striking.

Public Involvement

Community involvement is becoming an essential mechanism for achieving sustainable forestry development. One of the critical issues is how forestry responds to the needs and opportunities of local communities, the people it is meant to serve. The human factor in the shift to post-productivism in forestry is important. The opinions of people interested in forestry transition and public support for the forest policy reforms rest largely on the competencies of main stakeholders and on their active involvement in implementing the changes.

A growing interest in multifunctional forestry has been observed in the UK. In the most recent Public Opinion Survey (FC, 2003b), 92% of respondents indicated at least one public benefit as a good reason for supporting forestry with public funding, with wildlife habitat and recreation being the most commonly mentioned benefits. The percentage of respondents believing there were good reasons to support forestry with public money was significantly higher in Scotland than in England or Wales. Most respondents across the UK wanted to have more woodland in their part of the country, and again the highest support of woodlands expansion (58%) was in Scotland. Two-thirds of respondents who wanted more woodland would like it to be mixed woodland. Support for multifunctional forestry is also provided in the recent consultation on the management of 'national' forests under the ownership of the Scotland Executive and managed by the Forestry Commission (FC, 2003a).

This section complements the mentioned Public Opinion Surveys. In the light of the Lisbon Resolution's awareness of the 'need for an increasing dialogue between the forest sector and the general public to define widely accepted objectives for forest policy' (MCPFE, 1998), the primary objective of our analysis is to understand public perceptions on the future of British woodlands.

Q-methodology (a sequential application of correlation and factor analysis of survey research data) was used to more specifically analyse the survey research data. This methodology is especially useful when input from local communities is important, as its results are driven by the public and are systematic and scientifically rigorous (Barry and Proops, 1999). Q-methodology (Brown, 1996) allowed us to examine empirically human subjectivity and to elucidate the range of attitudes that exists and the spectrum of sustainability definitions that emerge. The method made possible a nuanced analysis of attitudes and of the elements of which they are composed.

Six groups of attitudes towards the future of forestry development were distinguished in the course of this research (Table 10.4). All the groups have a very strong emphasis on the regeneration and preservation of forests, with the difference that the respondents who belong to group 3 highly prioritize native woodlands over conifer plantations, whilst the others don't really make any visible distinction between these types of forest.

How do people justify the necessity of woodlands development? The creation of new jobs in remote rural areas and the provision of employment opportunities related to the prospects of woodlands expansion through the development of tourism and recreation are central in people's reasoning (groups 1 and 6). People consider social security and balanced timber harvesting as the essential directions of a sustainable rural policy. They support the weak sustainability direction for forestry development rather than that of strong sustainability. By this we mean that they place a clear emphasis on woodlands expansion and simultaneously reject strict nature conservation, considering the social aspects of forestry development crucial. The difference between the most clearly defined group, the 'Pragmatists' (group 1) and the 'Realists' (group 6) is that the Pragmatists' vision is practical (socio-economic) and focuses on the enlargement of the wealth of rural communities and on the increase of income of forestry workers. In contrast, the Realists are oriented towards balancing socio-economic

Table 10.4. A synopsis of the outlook of survey respondents.

Group 1 Pragmatists	People have a strong focus on woodlands expansion and do not mind forest plantations. They support giving subsidies to farmers for planting trees. Creation of jobs related to woodlands possesses an important place in people's reasoning. Overall, these people reject strict nature conservation, considering the economic and social aspects of landscape changes crucial (supporters of the economic concept of weak sustainability and of productivism in forestry)
Group 2 Idealistic visualists	People are preoccupied with aesthetic values of landscapes and our rights to enjoy their beauty, and yet they would prefer not to pay for this with high taxes or entrance fees. The necessity of improving landscapes and of multifunctional forest development is highly recognized
Group 3 Radical environmentalists	This group most of all stands in support of the intrinsic values of nature and is primarily ecologically oriented. People prioritize native woodlands over plantations and support strict conservation of biodiversity and native woodlands (strong supporters of the concept of strong sustainability and of post-productivism in forestry)
Group 4 Progressives	People have no prejudice against non-native species and stand for stable timber production from plantations concurrently with the conservation of native woodlands. Their idea is that natural woodlands should be protected, and yet with the possibility to cut the trees down if the land is needed for a more important use
Group 5 Utilitarian visualists	Tree planting to improve landscapes is highly recognized. In addition to the beauty of landscapes, these people pay attention to the attraction of tourists to the remote areas (supporters of multifunctional forestry)
Group 6 Realists	Although close in attitudes to Group 1, these people are not as radical. Their concern is about social and environmental impacts of rural changes. They don't have a particular interest in a landscape's beauty, and yet are ready to pay for a more sustainable management of forests and for native woodland conservation.

objectives of forestry development with nature protection considerations and a special emphasis on biodiversity.

The second major reason for public support of the policy for expansion of woodlands which was distinguished in the course of this research was that people recognize the necessity of improving rural landscapes through multi-functional forestry development. The difference between the 'Idealistic Visualists' (group 2) and the 'Utilitarian Visualists' (group 5) is that people who are Idealistic

Visualists are preoccupied with aesthetic values of landscapes and with people's rights to enjoy their beauty, whilst the Utilitarian Visualists are also interested in economic and social aspects of landscape changes, for example, attracting tourists to the remote rural areas. The 'Radical Environmentalists' (group 3) recognize the intrinsic value of nature and are primarily ecologically oriented. Whilst all other groups do not have strong prejudice against the enlargement of wooded area at the expense of non-native species, Radical Environmentalists predominantly support biodiversity and native woodlands conservation and regeneration, and forestry transition toward post-productivism.

Although the scope of this chapter is limited, this research has defined the factors influencing current attitudes in the UK towards forest development and has explained these factors by analysing the respondents' socio-economic backgrounds. The results of the analysis did not distinguish any differences between subjective perceptions of female and male respondents. However, attitudinal differences do vary by age and living conditions of the respondents and still more by their competencies, such as work experience, occupation and other life history aspects.[12] The research outcomes have therefore provided insights into understanding why certain aspects of landscape development and forestry policies are unfavourably viewed by some people and favourably received by others. Through understanding the differences in importance accorded by the public to the integration of woodlands in rural landscapes, we become aware of people's priorities and of the factors that could hamper reforestation of rural areas in the UK.

The results of this analysis offer a useful contribution to the formulation and refinement of forest policy. It is clear that the public supports policies that promote tree planting and expansion of woodlands for multiple purposes. There is general public agreement that multifunction forest development is the right direction in which to go in order to provide multiple forest benefits to society, the economy and the environment.

This analysis of the public survey data demonstrates that the protection and improvement of national heritage, biodiversity and landscape amenity values are considered by people in the UK as very important directions of the forest policy. Although some of the people are in favour of native woodlands regeneration, whilst others have interests that come very close to the weak sustainability ideas,[13] there is a general consensus on the necessity of extensive integration of native woodlands in rural landscapes. Moreover the results of this analysis are proved by practice, as today public involvement in multifunctional forestry development in the UK includes thousands of people and various organizations and trusts.

Conclusions

This research has added empirical and theoretical evidence of the transition of forestry in the UK towards post-productivism. This transition is not interpreted as an end to production, rather it is considered as an outcome of social, economic and environmental changes leading to new concepts of sustainability and multifunctional forestry. Research of a different spatial focus or scale might provide different results. It is beyond the scope of this chapter to conceptualize post-productivism, and a definitive explanation of this phenomenon hardly exists. However, since essential institutional changes have occurred in UK forestry in order to shift from productivism to post-productivism, the challenge for theorizing about the primary causes and sectoral implications is obvious and huge.

It is expected that multifunction forestry ideas, which already prevail in the UK, will get even stronger over time, with a rising importance of native woodlands restoration and biodiversity conservation. More efforts should be given in the future to improving visual and ecological components of landscapes, and forestry will likely evolve towards commercialization of recreation and leisure activities as well as opportunities for public education.

To conclude, the poorly developed forest culture and low degree of forest consciousness on the part of most of the population in the UK (Mather, 2003) might at first sight seem an unpromising setting for the acceptance of woodlands expansion. On the other hand, however, it could be argued that it may be easier to inscribe the lineaments of 'new' forestry on the relatively clean slate represented by the setting than it would be in one in which forest culture is more deeply embedded. Perhaps this issue raises more general questions about the adoption of sustainable forestry development in relation to local historical and cultural settings.

Acknowledgements

The research was co-funded by the SEERAD under the core project of the Macaulay Institute and by the EC within the project 'Visualisation tools for public participation in the management of landscape change'. We are also grateful to the respondents for their help with the survey, and to Dr G. Hill for his comments on an earlier draft.

Notes

1. According to FAO and IGBP definition, deforestation is a 10% or more decline of forest crown cover and a conversion of the former wooded areas to another land use (IBN-DLO, 1999).
2. The concept of sustainability was formulated in German forestry literature in the 18th and 19th centuries. It started from a narrow idea of the yield of wood. In 1804, Hartig defined sustainability as follows: 'Every wise forest director has to have evaluated the forest stands without losing time to utilize them to the greatest possible extent, but still in a way that future generations will have at least as much benefits as the living generation' (Schmutzenhofer, 1992). Social and environmental dimensions of sustained forest use became recognized, especially in the USA and Germany, almost a century ago. The term 'human benefit sustainability' was introduced in forestry in the middle of the 20th century. Since the late 1970s, the concept of 'social forestry' has gained prominance in much of the developed world leading to a shift from a forest-centred to a people-centred paradigm (Wiersum, 1995).
3. It is also observed in (midland) Central Europe, North America and Australia (Mather, 2001).

4. For more discussion on multifunctionality in British forestry see Mather, 2001 and Nijnik *et al.*, 2005.
5. Basic mechanisms of governance are markets, hierarchy and collective action (Gerrard, 2000).
6. For more discussion see Kant (2003), as well as the ideas and principles developed by evolutionary, institutional and ecological economics.
7. Especially where multiple forest values are considered to be important, sometimes it is socially optimal to postpone harvesting depending upon non-timber contributions from forests. The results of simulation of timber rotation indicate that a profit maximizing objective does not always correspond to efficient behaviour, and therefore often neither Fisher nor Faustman rotation leads to optimal solutions and to highest social benefits (Nijnik, 2004). To take into account multifunctional use of forests the Faustman model was extended into the idea of maximization of the NPV of the revenue flows from both timber and non-timber forest outputs (Hartman, 1976). When a forest's ability to accumulate carbon is incorporated, the model is further extended (Van Kooten and Bulte, 2000). In most cases, the rotation ages of forest stands used for multiple purposes appear to be higher.
8. See Nijnik and Oskam (2004) for a discussion on governance in forestry. For explanation of attitudinal diversity of forest policy actors, see Nijnik (2004).
9. For more discussion, see e.g. North (1993).
10. In 1996/1997 to compare with 1987/1988, spending on ESA has increased 13.5 times in England and Scotland alone (adapted from Hanley *et al.*, 1999).
11. Following publication of Forest Strategy in 2000, the Woodland Grant Scheme (WGS) and the Farm Woodland Premium Scheme have been reviewed in Scotland, resulting in the new schemes – the Scottish Forestry Grants Schemes (SFGS) and Farmland Premium (FP). The WGS is largely closed now also in England, with the transition to the England Woodland Grant Scheme (EWGS), as seen on the website http://www.forestry.gov.uk/
12. 'Idealistic visualists', for instance, are mainly highly educated civil servants or middle-aged scientists, generally either women or men with children who have sufficient incomes but care about money.
13. See Van Kooten and Bulte (2000) on the definitions of sustainability accepted in this chapter.

References

Barry, J. and Proops, J. (1999) Seeking sustainability discourses with Q methodology. *Ecological Economics* 28, 337–345.

Brown, S. (1996) Q methodology and qualitative research. *Qualitative Health Research* 6, 561–567.

Department for Environment, Transport and Regions (DETR) (2000). *The UK Programme.* SE 2000/209, DETR, London.

European Commission (EC) (1992) *The European Community Aid Scheme for Forestry Measures in Agriculture*. Regulation No. 2080/92, EC, Brussels.

European Commission (EC) (2003) *Planting Seeds for Rural Futures*. Salzburg, November 2003, as seen at: http://www.europa.eu.int/comm/agriculture/events/salzburg/index_fr.htm

Food and Agriculture Organization (FAO) (1996) *European Timber Trends and Prospects into the 21st century*, Timber and Forest Supply Papers, FAO, New York.

Forestry Commission (FC) (1991) *Statement of Forestry Policy for Great Britain.* FC, Edinburgh, UK.

Forestry Commission (FC) (1998) *England Forestry Strategy*. FC, Cambridge, UK.
Forestry Commission (FC) (2001) *Forestry Statistics*. FC, Edinburgh, UK.
Forestry Commission (FC) (2003a) *Forestry Statistics*. FC, Edinburgh, UK.
Forestry Commission (FC) (2003b) *Public Opinion Survey*. FC, Edinburgh, UK.
Forestry Commission (FC) (2004) *The UK's Forestry Standard, the Government Approach to Sustainable Forestry*. FC, Edinburgh, UK.
Forestry and Timber Association (FTA) (2004) *Fast Forwarding Forestry: Manifesto Action Group on Climate Change and Non-Market Benefits*. FTA, Edinburgh, UK.
Gerrard, C. (2000) Ten institutional perspectives on agriculture and rural development: a conceptual framework for policy makers, managers, and analysis. In: IAAE Conference, Berlin.
Hanley, N., Whitby, M. and Simpson, I. (1999) Assessing the success of agri-environmental policy in the UK. *Land Use Policy* 16, 67–80.
Hartman, R. (1976) The harvesting decision when a standing forest has value. *Economic Inquiry* 16, 5–12.
IBN-DLO (1999) *Resolving Issues on Terrestrial Biospheric Sinks in the Kyoto Protocol, Dutch Nation*. Research Program on Global Air Pollution and Climate Change, IBN-DLO Wageningen, The Netherlands.
Kant, S. (2003) Extending the boundaries of forest economics. *Forest Policy and Economics* 5, 39–56.
Knight, J. (1992) *Institutions and Social Conflicts*. Cambridge University Press, Cambridge, UK.
Mather, A. (2001) Forests of consumption: postproductivism, postmaterialism and the post-industrial forest. *Environment and Planning C* 19, 249–268.
Mather, A. (2003) The future. In: Smout, T.C (ed.) *People and Woods in Scotland: a History*. Edinburgh University Press, Edinburgh, UK, pp. 214–224.
Ministerial Conferences on the Protection of Forests in Europe (MCPFE) (1998) *State of European Forests*, as seen at http://www.mcpfe.org/secure/k-tools/
Nijnik, M. (2004) To an economist's perception on sustainability in forestry-in-transition. *Forest Policy and Economics* 6(3-4), 403–413.
Nijnik, M. and Oskam, A. (2004) Governance in Ukrainian forestry: trends, impacts and remedies. *International Journal of Agricultural Resources, Governance and Ecology* 3(1-2), 116–133.
Nijnik, M., Mather, A. and Hill, G. (2005) From multifunctional forestry to post-productivism? In: Gatto, P. *et al.* (eds) *The Multifunctional Role of Forests: Policies, Methods and Case Studies*, University of Padova, Padova, Italy, pp. 1–15.
North, D. (1993) Institutes and credible commitment. *Journal of Institutional Economics* 149(1), 11–23.
Ostrom, E. (1990) *Governing the Commons*. Cambridge University Press, New York.
Ostrom, E., Gardner, R. and Walker, J. (1994) *Games and Rules and Common Pool Resources*. University of Michigan Press, Michigan.
Sabatier, P. (1998) The advocacy coalition framework: revisions and relevance for Europe. *Journal of European Public Policy* 5(1), 98–130.
Schmutzenhofer, H. (1992) IUFRO birthday. *IUFRO News* 21(1-2), 3.
Scottish Executive (SE) (2000) *Forests in Scotland: the Scottish Forestry Strategy*. Stationery Office, Edinburgh, UK.
Shleifer, A. and Vishny, R. (1998) *The Grabbing Hand: Government Pathologies and their Cures*. Harvard University Press, Cambridge, Massachusetts.
Slangen, L. (2000) *Framework for Analysing Institutions of Sustainability*. Wageningen University, Wageningen, The Netherlands.
Smout, T., MacDonald, A., Watson, F. (2005) *A History of the Native Woodlands of Scotland 1500-1920*. Edinburgh University Press, Edinburgh, UK.
Warren, C. (2000) *Managing Scotland's Environment*. Edinburgh University Press, Edinburgh, UK.
Weimer, D.L. (ed.) (1995) *Institutional Design*. Kluwer Academic Publishers, Dordrecht and London.

Welsh Assembly Government (WAG) (2001) *Woodlands for Wales*. WAG, Cardiff, UK.
Wiersum, K. (1995) 200 years of sustainability in forestry: lessons and history. *Environmental Management* 19(3), 321–329.
Van Kooten, G.C. and Bulte, E. (2000) *The Economics of Nature: Managing Biological Assets*. Blackwell Publishers, Oxford, UK.
UNCED (1992) *Agenda 21*, Chapter 37, para. 1. UNCED, Rio de Janeiro, Brazil.

11 Economic Conservation – Hill Holt Wood: the Three Legs of Sustainability in Practice

N. LOWTHROP

Hill Holt Wood, Norton Disney, Lincolnshire UK.

Abstract

The concept proposed in this chapter is a holistic approach that is appropriate to the location, the community and the market availability. The sustainability of the forest has three elements and the proposal is that these cannot be taken in isolation. To achieve environmental sustainable management of forests there has to be integration of the local community and an economic return. The example of Hill Holt Wood, a successful and now nationally recognized environmental social enterprise, will be used to show how it is possible to involve the community in forest management and policy. The evolution of the project and its origins will be described and in particular the economic returns achieved on a small ancient woodland will be discussed. The potential for replication and the lessons learnt that could have relevance to forestry on a worldwide scale will be covered.

Introduction

In the UK, conservation land management has been, and continues to be, a fringe activity compared with intensive production. Recent changes in farm subsidy payments are moving in the right direction but their impact will still be peripheral. For many years what many would regard as sustainable land management has been supported by grant funding and/or goodwill. Perhaps some of the best examples of sustainable land management have been the traditional estates,[1] often planning management policy in terms of tens or even hundreds of years, rather than the more typical, at best, 5-year plan. Estates are often also socially aware but in a paternalistic, controlling way, not in any sense under community control. This should not, however, detract from the underlying model as a basis for future policy.

The proposition that forms the basis of the Hill Holt Wood concept is that for long-term sustainable management of the countryside a balance needs to be

 Sustainable Forestry: from Monitoring and Modelling to Knowledge Management and Policy Science (eds K.M. Reynolds, A.J. Thomson, M. Köhl, M.A. Shannon, D. Ray and K. Rennolls)

achieved between the social, environmental and economic legs of sustainability. Too often a business viewpoint is totally economic, the farmer/forester does not understand the social leg and the environmentalist assumes that the work they do is so worthy that it warrants funding. Why not a balanced perspective that in turn provides opportunity? The provision of benefits for the community encourages support to 'buy into' the project, providing the resources to environmentally manage and in turn enhance the experience and quality for the community.

What is proposed is not a formulaic approach but a principle that could potentially be applied to any site anywhere. The key is to identify the appropriate site-specific management, engage the community and seek out the market opportunities.

Economic Conservation

Following attendance at an inspirational conference (Second World Game Ranching Symposium, held at Edmonton, Canada) featuring successful projects in Africa, Central and South America, a business was established to apply the principle, as seen by the author, demonstrated by these projects to the wider management of the British countryside.

ECONS = Economic conservation

E = Economic
C = Community (Social)
O = Opportunity
N = Natural (Environmental)
S = Sustainability.

In principle, this concept could be applied to any land management but for many reasons, not least that it provides a wonderful environment in which to live and work, the author chose woodland, in particular ancient deciduous woodland. Woodland also appeared to provide the widest range of opportunities for adding value, greater diversification and the ability to 'hide' development. Interestingly, when the concept was proposed to forestry advisers and other 'experts', the general view was that broadleaf ancient woodland was of no real cash value other than for rough shooting and amounted to a rich man's hobby.

After a 4-year search, an existing business was sold and a 14-ha ancient woodland purchased, namely Hill Holt Wood. The woodland had been heavily felled by the previous owners, a large timber company, who removed over 200 mature oak trees, in the process damaging the drainage system. In addition the woodland had difficult access on to a major trunk road, invasive non-native rhododendron (*Rhododendron ponticum*) and bramble (*Rubus fructiosa*) taking advantage of the opened canopy. Oak (*Quercus robur/petraea*) was the dominant tree species, with ash (*Fraxinus excelsior*), birch (*Betula pendula*) and large, old and unmanaged hazel (*Coryllanus avellana*) coppice stools.

The initial target was to purchase the wood, install the necessary infrastructure and build a house for the equivalent of a four-bedroomed detached house in one of the surrounding villages. At the time (1995) this would have equated to £100,000.

In addition, a balance between reduced living costs and income generated would provide for a quality of life on a very modest turnover. All this, of course, assumes overcoming the major hurdle of planning permission,[2] including persuading the planners that a viable income could be achieved from 'forestry'.

Now jump forward to the present position (2005), with the start of construction of a log cabin home overlooking a lake, an award-winning profitable social enterprise employing 17 staff, enhanced environmental management and free public access. A marker has been put down for a different approach to countryside management.

A holistic approach raises a multitude of issues to be faced as accepted paradigms are rejected: building in what is now recognized as ancient woodland, accepting a range of activities as forestry or ancillary to forestry, proving that an efficient business could operate in such an environment, developing community support into community control and not to mention groundbreaking building designs being accepted by building control. Government departments and agencies tend to operate in 'bunkers' and are only interested in their area of activity. Unless the operation falls largely into their sector, departmental staff tend to express little interest or suggest 'more appropriate' departments. To break down this reluctance to recognize innovation and a holistic approach, a project has to excel in each individual sector to attract interest and support and to promote an alternative approach.

Forestry

In the case of Hill Holt Wood (HHW) a key obvious starting point is the government's forestry agency, the Forestry Commission (FC). Throughout Britain there are many woodland projects that share elements of the HHW operation. Many receive significant core funding, including considerable funds from the Forestry Commission. These projects are often set up as exemplar environmental and community-based developments. Conferences and network events are regularly organized and a key element is often the justification for core funding from government, arguing that public benefits require public funding. Unfortunately, the public purse is restricted and, worthy though these projects may be, the likelihood is that the available resources will decline rather than increase. This results in the exemplar projects fighting for survival rather than rolling out to encompass more woodland in positive management. In addition there are a number of 'alternative' woodland ventures, often involving basic living and traditional skills. These alternative schemes depend on a minority of dedicated idealists prepared to live in almost medieval conditions. Apart from being likely to remain on the fringe of society, these projects often suffer from a difficult relationship with local populations and, in particular, planners.

HHW had to find a way to be seen, when not one of the 'closed shop' sponsored projects, but also to avoid being viewed as yet another fringe activity. It took a number of letters to the government minister over a period of months to trigger a senior-level visit from the Forestry Commission to the wood. This visit finally recognized that HHW may have a contribution to make towards future forestry policy, in particular the social agenda. A fruitful and mutually beneficial relationship has now developed between FC and HHW. Forest Research has produced a comprehensive research report on HHW (O'Brien, 2004), a series of research papers and the involvement of HHW in a series of conferences. The author would categorize the style of management as eco-forestry, close to the model propagated in western Canada.

The Business Model

HHW started as a small private partnership business with enthusiasm to ensure a strong social leg to the business. A community group, the Hill Holt Wood Management Committee, was quickly established to provide a link between the surrounding villages and the owners of the wood. Woodland has the capability to take hold of people as a very special environment. This enthusiasm led to a very strong link, which developed a momentum over the following 5 years. At the same time the agenda for FC was changing, with a drive to encourage managers to reach out to their local communities. The term 'social forestry' came into use with the aim of involving communities in the planning and management of their local woodlands. The Arnstein scale (Arnstein, 1969) was put forward as a method of assessing the level of community involvement. At this stage Forest Enterprise, the operational arm of FC, scored low on the scale, 3 or below, and there was an obvious reluctance on the part of many managers to promote the development of greater community consultation, citing financial and time restraints. The general view could be summed up by 'This is just a passing fad, keep your heads down until it passes and then we can get back to some real forestry.' It was the view of the author that this was not a passing phase but the only future that might lead to truly sustainable forestry.

The accepted management models were government, private or voluntary/charity. The owners and local community decided to seek advice to establish what other options might be available for social control. The new and burgeoning sector of social enterprise that was registering on the government's horizon seemed to fit the bill perfectly. HHW became an environmental social enterprise. The full title is actually a community-controlled, not-for-profit, membership organization limited by guarantee.

The community board of HHW would regard the business as more than profit – in fact, 'triple bottom line' or non-distributive profit and definitely not 'not for profit', which is a misleading term. The first 3 years of operation as a social enterprise has seen a total turnover of £830,000, a net surplus of £100,000 and a staff of 17, all within a 14-ha woodland. The anticipated turnover for 2005/06 is £460,000, with the first replica in place and two more planned. Total funding to date, including the previous partnership, is £75,000 and this is largely capital. This high turnover is achieved by using the woodland environment as a 'life changer', working with young people with extreme disadvantages: basically solving urban issues in the rural environment. This does not exclude also adding value to woodland products and every effort is made to have a package of income streams that enable a viable business.

The concept applied at HHW is not and cannot be regarded as formulaic or a franchise, but should be site-specific dependent on the views of the owners, the wishes of the local community, the environmental requirements or restrictions and the local market opportunities. Two similar projects are under consideration in close proximity to HHW, each with very different income streams. Business profitability will enable the comprehensive environmental management of the sites and additional public benefits such as quality public access.

Additional information on the structure and governance of the business, the income streams and the policy implications may be found in the Forest Research report.

Planning and Construction

The mission statement of Hill Holt Wood is 'Proving the value of ancient woodland in the 21st century' and sustainable management and development are at its core. In Britain the density of population places a high value on green space, with planning controls that are particularly strict. Any form of development in ancient woodland would normally be regarded as outside any permissible development. Agricultural and forestry tenure for a dwelling would have to be proved by significant profitability based on the primary product, excluding any added value. HHW campaigned with strong community support for the activities carried out at HHW to be accepted as forestry due to the national changes in policy emphasis. No longer would timber production be seen as the *raison d'être* for forest management. Environment, landscape, education, employment, access, recreation and social control are now higher on the agenda than pure silviculture.

A groundbreaking decision by the local district council granted permission for a house within the woodland for 'forestry, social and educational reasons' and in the process accepted that the activities at HHW were forestry or ancillary to forestry. Who is more appropriate to live in woodland than the managers of that woodland? This also addresses another key issue facing rural Britain: that low-paid rural workers can no longer afford housing in the countryside and often 'reverse' commute from urban areas, with the rural housing passing to the retired, urban commuters and second homes. Social enterprise can be the vehicle to develop and own such affordable housing, ensuring that profit from planning gain cannot be realized by individuals and locking the development into appropriate community use and ownership.

The underpinning philosophy of the business of sustainable development has led the company to build using low-impact techniques. Buildings of straw, timber, earth and recycled materials have been constructed so far. All the buildings are 'off grid', with local heating and power supply, avoiding the disruption of bringing services to the wood. A particular feature of the site has been the use of twin-vault earth composting toilets, which produce a useful by-product for the site tree nursery. This in turn has led to an additional income stream designing and constructing similar toilets for other isolated sites. To upgrade the quality of such buildings a government-sponsored Knowledge Transfer partnership will bring a graduate architect to the business under the supervision of the local University of Lincoln. A target for this programme will be to develop designs that utilize locally grown timber, a much underemployed construction resource in Britain.

Site buildings are visually in keeping with the environment and are now designed to be removable if no longer required. All recent buildings have been constructed on stilts (timber piles) to avoid the use of concrete. The small area of HHW restricts the number of buildings and is beginning to restrict the growth of the business, as the economic leg has to be balanced against environmental impact. Expansion will be by land acquisition as enhanced land management takes over from monoculture farming. Findings, supported by Forest Research, indicate that the 'lived in' woodland at HHW is found to be welcoming to a wider range of visitors than the majority of similar woodlands. The visitors tend to fall into the categories of families, the retired and women, rather than the male dog walkers that are identified as making up 70% of the visits to small woodlands (Burgess, 1995). This is almost certainly due to the perceived and actual safety of a lived-in and working woodland as opposed to the neglected unmanaged woodland often traditionally favoured by environmentalists. The latter often suffers from a range of antisocial activity, particularly when in close proximity to centres of population.

Environment

HHW when purchased was typical of many neglected British woodlands, with invasive species, a lack of regeneration, difficult access and unmanaged coppice. The most environmentally acceptable intervention is very low-tech manual clearance and management. This is often prohibitively expensive but in the case of HHW the main income streams, training and re-engaging disaffected youth, also provide the labour to carry out the required work. All the woodland management, improvement of public facilities and construction of buildings, forms part of the development and training of the young clients. As the youthful client base is developed, motivated and re-engaged with society, so is the woodland environment healed and returned to a vibrant habitat. In spite of, or possibly because of, the high level of intervention and activity, at least 11 listed rare and protected species are resident at HHW, a number likely to increase. The range of bird species resident and the actual numbers of birds have dramatically increased as the variety of habitat, territory definition and tree age structure increase. A management plan based on the economic life of the oak (in the region of 200 years) is being prepared. Within this very long cycle a series of shorter cycles involving ash, birch and hazel coppice will operate. The social control of the woodland encourages and supports this long-term view of management. Such a visionary and strategic approach is rarely within the scope of private ownership and does not fit the cycle of government elections leading to short-term thinking.

Enhanced environmental management supports the social benefits leading to increased public and customer support, generating the income streams that allow the environmental management – the three-legged stool of sustainability in practice – a virtuous, mutually supporting triangle. Key to the success of this approach is the development of understanding and trust between the managers and the community through openness and transparency.

The Future

Already HHW has been recognized by an assortment of government departments as contributing to potential solutions to a range of issues: the Department for Environment, Food, and Rural Affairs (DEFRA), including the Forestry Commission, with regard to rural policy, the Department of Trade and Industry (DTI) for social enterprise, the Department for Education and Skills (DfES) for an alternative education programme and Department of Communities and Local Government (DCLG) for groundbreaking planning and sustainable development. Perhaps the most significant contribution of the project may yet be the 'big issue' facing the present government and that is youth crime, disaffection and antisocial behaviour. The police force local to HHW recognizes the change in a significant proportion of the young clients participating in the project who are no longer involved in crime but are now contributing members of society. This is another 'profit' of HHW, which receives no financial support at present from government or society. If this proves to be a way to significantly reduce the antisocial behaviour problem that faces all urban areas and if this can be financially supported, further income streams will flow to rural areas and in particular for sustainable environmental management. HHW has identified the potential to work more closely with youth crime officers, with the probation service and even with prisons to take advantage of the benefits of the woodland environment.

The very basic eco-structures at HHW have, despite their simplistic design, attracted considerable interest and a significant demand for their replication. Linking to a university through a Knowledge Transfer Partnership (KTP) will enable HHW to develop the concept of the buildings to a higher level. The KTP 2-year programme brings a high-quality graduate architect to the project under the guidance and support of the university (Lincoln) and the university gains practical training and research opportunities. HHW in the near future intends to develop a construction business to offer straw, earth, timber and other unusual structures to a wide range of clients. This in turn will enhance and extend the skill opportunities available to the trainees. Already two earth composting toilet blocks have been constructed, with three more ordered. Ideally this business will utilize timber from HHW and other sites managed by the project.

Another key objective is replication, in terms of both similar projects in other areas and also projects that replicate the concept, but not the actual activities, in close proximity to HHW to show that this could be an approach that will work on a large scale, not just occasional well-separated localized projects. Two miles to the west of HHW a much larger scheme in a commercial conifer forest is planned, with another based on eco-burials 2 miles to the east.

The continuing growth of the business has led to a review of the structure of the organization and much development to ensure that the organization is not dependent on the founders and is capable of itself being long-term sustainable. In a rural area such as the North Kesteven District of Lincolnshire, a business employing 19 full-time staff is a significant employer. In terms of the rural economy HHW has achieved its objective of a significant income and, more importantly, a surplus by managing an area of countryside in an environmental way. The business needs to continue to develop initiatives, think laterally and take an entrepreneurial approach to continue as a successful enterprise.

Lessons to be Learned and Policy Implications

(adapted directly from Liz O'Brien, 2004)

HHW has taken a holistic approach to its work, which focuses on people and communities, management of the environment and the creation of a sustainable business.

HHW provides a model of a woodland social enterprise. Elements of the project could be encouraged and enabled to take place in other areas as part of the drive for rural redevelopment and diversification.

HHW has become increasingly well known at both a regional and national level due to effective networking and partnership working, as well as word of mouth. There is a growing awareness amongst environmental and educational professionals that this type of project crosses a range of government policy areas. While this is beneficial in terms of the public benefits provided, it is difficult for a small enterprise such as HHW to deal with a wide range of organizations and government departments that cover these issues (health, education, environment, youth crime, etc.). One-stop shops that provide advice, guidance and information for social enterprises could assist in the situation.

The role of social enterprises in rural development and rural diversification is becoming an issue of increasing interest. The government sees social enterprises as a way of providing public goods. What is not clear is how these organizations can or should be helped and supported in this provision.

Because HHW is a lived-in and worked woodland it attracts people to the site who might be otherwise concerned with accessing woodland alone, such as women. Woodland organizations and woodland social enterprises can play an important role in rural development. Forestry's contribution to the rural economy has been shown to be significant if, for example, tourism, recreation and house prices are taken into consideration.

Current planning policy acts as a barrier to organizations such as HHW that want to create a lived-in and working woodland. This is due to strict restrictions on the consideration of the construction of dwellings and a lack of definition over which activities are considered ancillary to forestry, and which therefore require no planning permission. Through persistence, determination and building relationships with local planners HHW has been able to change attitudes and construct a variety of buildings on site.

Benefits can be gained for a social enterprise such as HHW by involving the community in the decision-making processes. A commitment by the community and staff of the business to the objectives of the enterprise can provide the momentum for success and are key assets in delivering benefits.

Many respondents felt that HHW contributed to wider issues such as social inclusion, helping disadvantaged young people, building community capacity and undertaking sustainable management of the environment. While the type of education and training being carried out at HHW may work in other settings, the woodland habitat has a number of advantages: (i) the ability to absorb activity without seeming crowded; (ii) calming and therapeutic effects of trees and woodlands on the young people being trained, some of whom have emotional and behavioural difficulties or special needs; and (iii) providing opportunities for a variety of different training activities, such as coppicing, charcoal manufacture and making wood products, as well as recreational activities.

In addition, HHW links urban with rural issues, seeing the urban majority as customers. Too often country people object to bending to the wishes of urbanites, but for the rural economy to succeed it has to find 'products' and a market.

Planning can be used as a positive tool to support community business. Lack of investment capital may to some extent be compensated by planning gain.

Low-impact development utilizing appropriate local materials, without connection to mains services and low embodied energy are possible and acceptable to local communities.

The importance of auditing the social and environmental 'profits', as well as the financial audit, will become increasingly apparent. This could be the key to gaining funding (investment) in return for assessed and valued public benefits.

The importance of leadership in providing vision and inspiration is necessary for similar projects to emerge. Each site should be seen as unique and individual to instil a sense of ownership for the staff, participants and local community. Similar projects should not be imposed, but allowed to grow organically, developing support and scale as they progress.

Conclusions

Hill Holt Wood is a small but successful social enterprise that has achieved successful outcomes in a series of policy areas. It is the view of many with an understanding of the project that the approach can be replicated. Nothing at HHW is new but the mix and approach are the key to success. HHW is very definitely not a franchise and is rarely likely to work if the model is taken as a standard format. The addition of the economic strand and a strengthening of the social element should be seen as the way forward for environmental improvement. Communities when given control are prepared to be innovative, involved and take a long-term view of their environment. The rural minority have to see the urban majority as customers and cater for the needs of that marketplace to secure an income for the long-term sustainable future of the countryside and, in particular, our forests.

Hill Holt Wood: proving the value of ancient woodland in the 21st century.

Notes

1. In the UK context a traditional estate would describe an area of land in the ownership of a single family for generations, often associated with a significant house or stately home. Such estates usually provide houses for estate workers and often include a village or villages. These family estates are able to take a very long-term view of land management and also have a social element, although they are non-democratic and somewhat paternalistic.
2. In the UK pressures on available land have led to the development of a strict system of planning for all development. Although there are still some concessions for agriculture and forestry, development in woodland would under normal circumstances be considered impossible.

References

Burgess, J. (1995) *Growing in Confidence: Understanding People's Perceptions of Urban Fringe Woodlands*. Countryside Commission Technical Report CCP457, Cheltenham, UK.

O'Brien, E. (2004) *Hill Holt Wood Social Enterprise and Community Woodland*. Forest Research report. Forestry Commission, Edinburgh, UK.

Bibliography

Budiansky, S. (1995) *Nature's Keepers*. Weidenfeld and Nicolson, London.

O'Brien, E. (2005) Bringing together ideas of social enterprise, education and community woodland: the Hill Holt approach. *Scottish Forestry* 59, 7–14.

Tabbush, P. and O'Brien, E. (2003) *Health and Well-being: Trees, Woodland and Natural Spaces*. Forestry Commission, Edinburgh, UK.

12 Measuring Sustainability Using the US Forest Inventory and Analysis Program

C.T. Scott and W.H. McWilliams

Forest Inventory and Analysis, USDA Forest Service, Newtown Square, Pennsylvania, USA.

Abstract

In 1998, citing the need for more current information, the US Congress directed the USDA Forest Service's Forest Inventory and Analysis (FIA) Program to shift from a periodic to an annual inventory. The remeasurement cycle is now 5, 7 or 10 years, depending on the region and a state's contributions to funding. The new national design has three phases, which cover stratification, traditional forest inventory and forest health indicators. Field protocols, data storage and retrieval systems, estimation procedures and reporting requirements are being standardized nationally and documented. The inventory is also more comprehensive as it now includes the subsampling of all vascular plants, soils, lichens, down woody material, ozone damage and crown attributes. These changes enable FIA to address a host of sustainability factors such as those enumerated in the Montreal Process Indicators of forest sustainability. Consistent national sampling and estimation protocols allow issues related to sustainability to be addressed at the national, regional and state levels.

Introduction

At the direction of the US Congress, the USDA Forest Service's Forest Inventory and Analysis (FIA) Program shifted from a periodic to an annual inventory of the nation's forest in 1998. All plots have been made permanent, and the remeasurement cycle is now 5, 7 or 10 years, depending on the region and a state's contributions to funding. The shift to a survey with annual panels with shorter remeasurement cycles is generating more current information, and has increased the accuracy of data collected on the status of, and changes in, the nation's forest resources. A more comprehensive inventory, which includes forest health indicators, provides data that address a number of Montreal Process Indicators (MPI) of forest sustainability.

FIA addresses sustainability at the national, regional and state levels. Nationally it provides data on 14 of the 67 MPI for sustainability and is working on nine others (USDA Forest Service, 2004a). For example, FIA monitors forest areas by forest type, standing volume and removals, erosion on forest soils, ozone (O_3) damage, forest biomass and carbon. Regionally, in collaboration with federal and state partners, FIA monitors forest health, e.g. crown indicators and mortality, and their relationship with areas of pest outbreaks. In the north-eastern USA, FIA data are used by states to monitor forest sustainability, because this information can be compared across state boundaries. Common regional and state-level indicators of sustainability have been identified for use in managing forests and developing policy. At the state level, FIA produces comprehensive reports, which include data on volume by species and diameter class, ownership productivity, disturbance, growth-to-removals ratios, stand structure, species composition and regeneration.

Changes in FIA

Citing the need for more timely and comprehensive information by land managers, the forest industry and policymakers, the US Congress directed FIA as part of the 1998 Farm Bill to make fundamental changes and develop a strategic plan with respect to implementing these changes. As a result, FIA has shifted from a periodic survey (in which states were measured one at a time) to an annual survey, in which plots are divided systematically into panels so that subsets of plots in every state are measured every year. FIA has also stabilized an erratic remeasurement cycle that varied by state and region to a 7-year cycle in the east and a 10-year cycle in the west. States now have the flexibility to reduce the prescribed cycle to 5 years if they contribute additional resources. Twenty-two states currently provide resources to achieve 5-year cycles.

In addition, FIA now:

1. Samples all forest lands. Previously, many wilderness areas, 'unproductive' forests and national forest had been excluded.
2. Is changing from a regional to a national emphasis. The sampling frame, plot design and field protocols are now consistent nationally. At the same time, sustainability indicator processes are demanding additional regional data, which FIA will provide more consistently.
3. Has developed and is implementing standardized sampling and estimation procedures. Examples include a field manual, data processing and estimation methods, a database and Internet access to the data and estimates.
4. Has expanded from a focus on timber to one that includes comprehensive monitoring of forest ecosystems.
5. Works in partnership with the National Forest System, State and Private Forestry and state agencies. States contribute roughly US$10 million annually.
6. Has shifted from producing occasional reports based on periodic inventories to publishing comprehensive, analytical reports for each state every 5 years.

Three-phase Design

Historically, different inventory methods were used by FIA in different regions. The new design establishes a consistent sampling and estimation framework to be implemented nationally. The selected design includes three phases: stratification with satellite imagery, a ground-plot network for traditional forest inventory and a subset of the plot network augmented by forest health indicators.

Phase 1: Stratification

Sampling with stratification improves the precision of estimates by grouping similar plots into strata. Until recently this was accomplished using aerial photography. Thousands of photo points were classified manually. Today, most FIA regions use satellite imagery, and classification schemes are becoming more automated. Although the general approach to stratification is standardized nationally, each region retains the flexibility to choose the stratification attributes that are most appropriate for its conditions, e.g. elevation in the mountainous western USA.

Phase 2: Traditional forest inventory plot network

Phase 2 consists of FIA's basic set of ground plots on which traditional forest and tree attributes are measured by field crews. During the conversion to an annual inventory, a hexagonal grid was superimposed over the USA. Each cell covers 2400 ha, and grid centres are spaced about 5 km apart (Fig. 12.1). Existing plots from previous periodic designs were assigned to each hexagon to produce a systematic sample across the country. If more that one existing plot fell within a cell, one was chosen randomly and the rest were dropped. If there were none, a new plot was placed randomly within the cell. These plots are permanent and are scheduled to be remeasured on the prescribed 5-, 7- or 10-year cycle. All regions are currently filling in the new national sampling grid.

To accommodate the requirement for an annual inventory, the hexagonal grid was divided systematically into panels, each with complete spatial coverage. The number of panels in a given state is based on the desired length of a measurement cycle. Because each panel covers the entire population of interest, population estimates can be obtained from a single panel or any number of consecutive panels within a measurement cycle. With 15,000 to 20,000 forest plots measured annually, regional and national trends will be available on an annual basis.

The ground plots that make up FIA's Phase 2 plot network are clusters of four points arranged as in Fig. 12.2. Each point in the cluster is surrounded by a 7.3 m fixed-radius subplot, in which trees 12.7 cm or larger in diameter at breast height (dbh) are measured. All four subplots total about 1/15 ha. Regions can use a larger diameter limit for the 18.0 m radius macro-plot. Each subplot contains a 2.1 m fixed-radius micro-plot, in which seedlings and saplings less than

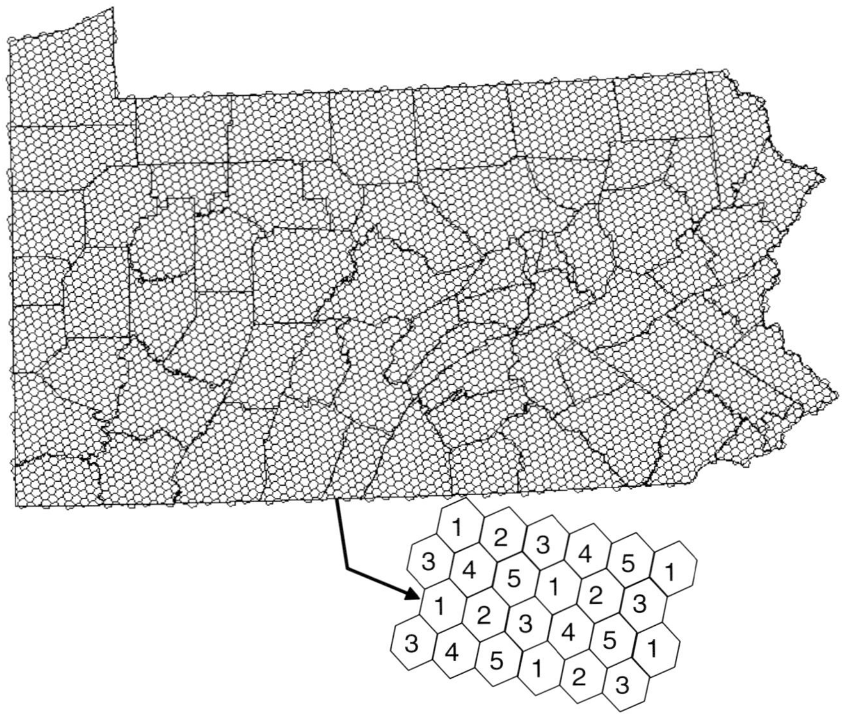

Fig. 12.1. Example of national FIA hexagonal grid applied to Pennsylvania, USA; plots are divided into five annual panels.

12.7 cm dbh are tallied. Micro-plots are offset from subplot centres to minimize trampling. FIA plot design is based on that implemented in 1990 by the national Forest Health Monitoring (FHM) Program. FIA regions began switching to this design in 1995, 3 years before the passage of the 1998 Farm Bill and prior to the establishment of the national sampling grid.

In addition to tree measurements on these plots, data are collected on the area or setting in which the trees are located. To facilitate dividing the forest into various domains of interest for analytical purposes, plots are mapped by condition class. Condition classes are defined by a series of predetermined discrete variables – land use, forest type, stand size, regeneration status, tree density, stand origin, ownership group and disturbance history.

Phase 3: Forest health indicators

The 1998 Farm Bill mandated the integration of the FIA and FHM plot networks. FHM plots were incorporated systematically into the newly formed hexagonal grid as FIA's Phase 3 subset of plots. This subset, divided into five panels, constitutes 1/16 of the total number of Phase 2 ground plots. The Phase 3 grid is

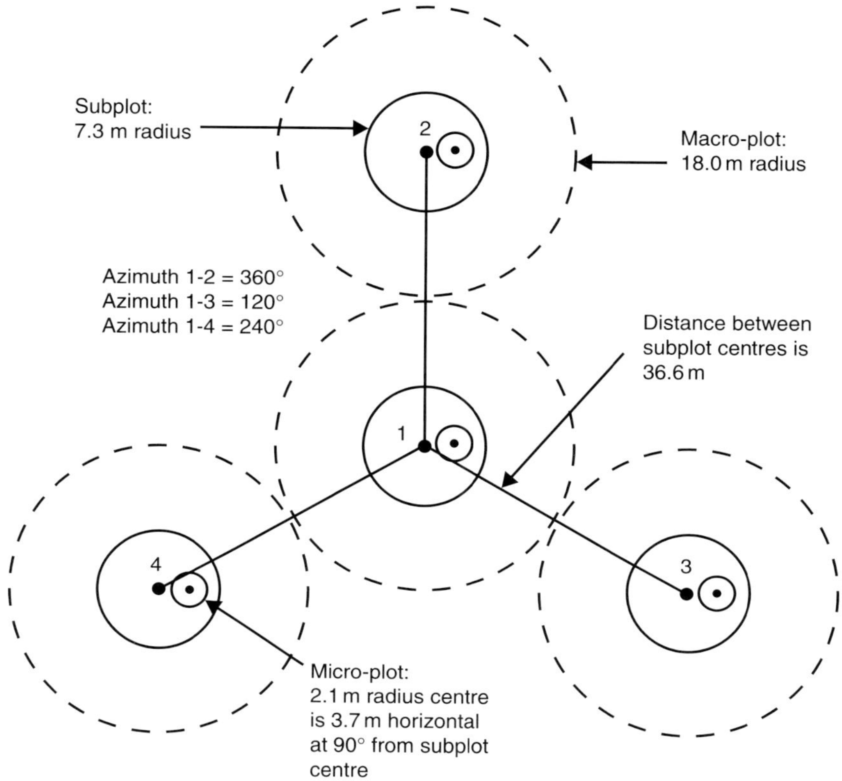

Fig. 12.2. FIA Phase 2 plot configuration.

relatively sparse, because these plots are relatively expensive, and many of the attributes associated with a forest health indicator must be measured during the growing season. As a result, panel measurement cannot be spread over the entire year.

Phase 3 plots include all of the features and attributes of Phase 2 plots, but also additional protocols associated with the sampling of six 'indicators' of forest health (Fig. 12.3): crowns, downed woody material, soils, lichens, ozone and understorey vegetation. Crown density, dieback and transparency are measured on all sample trees. Coarse and fine downed woody material are sampled along three 7.3 m transects radiating from each subplot centre. Litter and soil samples are taken in the zone between plot centre and the three other subplots. Lichen species and abundance on trees and other surfaces are recorded within a 36.6 m radius of the plot centre. Ozone exposure is estimated by evaluating damage to sensitive species within open field conditions. The vegetation indicator, which has been pilot tested, will be fully implemented nationally in 2006. Individual species are recorded on three 1 m^2 quadrats within each subplot. Cover by height class is also recorded, by species, within each subplot.

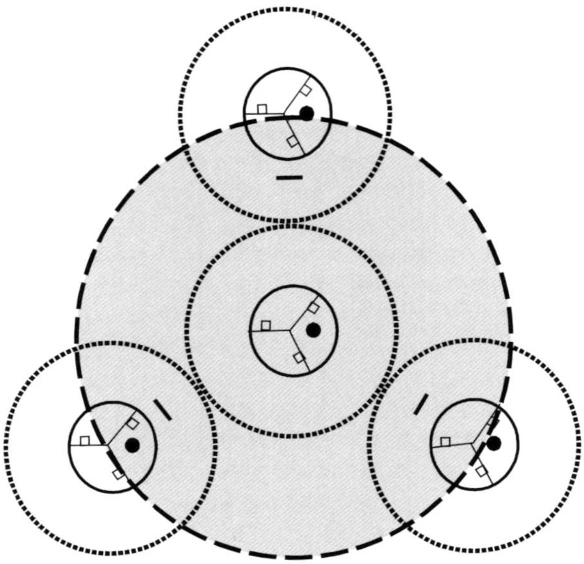

○	Subplot	24.0 ft (7.3 m) radius
●	Micro-plot	6.8 ft (2.1 m) radius
	Annular plot	58.9 ft (18.0 m) radius
	Lichens plot	120.0 ft (36.6 m) radius
□	Vegetation plot	1.0 m^2 area
▬	Soil Sampling	(point sample)
—	Down Woody Debris	24 ft (7.3 m) transects

Fig. 12.3. FIA Phase 2/Phase 3 plot design.

Implementation Status

A strategic plan was developed in conjunction with the 1998 Farm Bill to guide the implementation of the new requirement. States were scheduled for conversion to the new system over a 5-year period. States that provided financial support or field assistance to reduce to 5 years the federally funded 7- or 10-year cycle were among the first to be converted. Currently, all but five states and interior Alaska have implemented the new design.

Attaining national consistency has required considerable effort. National technical teams have been established to ensure that FIA collects, stores, compiles and reports forest-resource information consistently across the USA. To date, these teams have:

1. Published national sampling design and estimation methods for Phase 2 (Bechtold and Patterson, 2005).

2. Developed national field manuals for Phases 2 and 3 data collection protocols (USDA Forest Service, 2004b, c).
3. Developed data collection, storage, retrieval, processing and reporting systems (http://fia.fs.fed.us/tools-data/).

Several publications that describe standardized estimation and analysis methods for the Phase 3 forest health indicators are being prepared.

The new design includes a rigorous quality assurance/quality control (QA/QC) programme, which incorporates three checks of field data. Hot checks consist of on-site interaction between field crews and their trainers during data collection. These checks verify that crews understand all data-collection protocols. The other checks are not conducted in the presence of field crews. Cold checks are remeasurements of recently completed plots by QA/QC staff. Data from the original crews are available for review by the QA/QC staff in the field. Crews receive scores and additional feedback based on the quality of their data. Blind checks are remeasurements taken when data from the original crew are not available to the QA/QC staff. Blind checks made on roughly 3% of the plots provide estimates of repeatability. Data from blind checks are used to produce FIA's national QA reports.

Measuring Sustainability

Data from each of the three phases can be used to address a variety of sustainability issues at various scales.

Phase 1

Geo-statistical methods are being applied in different ways to develop a variety of spatial products. One by-product of Phase 1 stratification is forest/non-forest maps, which are useful in assessing forest fragmentation. In combination with imagery, FIA uses road networks to compute road density. This aids in identifying and delineating areas fragmented by disbursed networks of houses under tree cover. FIA data are also being used to develop more detailed maps of land use and land cover. Over time, these maps can be used to assess changes in land use/cover. While FIA has produced tabular estimates of area by forest type for decades, 250 m MODIS (moderate resolution imaging spectroradiometer) imagery is now being used to develop national forest type maps.

Phase 2

FIA ground plots have been the primary source of forest statistics in the USA for 75 years. With the remeasurement of permanent plots during the 1960s, detailed change data, which allowed calculations of components of change, added a new dimension to the inventory. The emphasis on monitoring versus current inventory has continued to evolve. In addition to volume, current status and change estimation can be expressed as biomass and carbon. Changes affecting carbon pools are of particular interest.

FIA also assesses stand structure and species composition and changes in them. For example, US forests are generally maturing, i.e. the area in pole-timber stands is decreasing, but it is increasing in saw-timber stands. In regions dominated by hardwoods, species composition is shifting from oak species towards red maple (*Acer rubrum*). By measuring tree damage and monitoring trees over time, we can identify relationships between damage and mortality, e.g. beech bark disease and mortality in the north-eastern USA.

Phase 3

FIA collects a variety of forest health attributes that are used to assess forest sustainability. Tree-crown attributes are used as early indicators of health problems. Lichen diversity is used to monitor air quality over time. Some lichen species serve as indicators of old-growth conditions. Field and lab chemical analyses of soils are used to assess factors such as erosion, acidity, calcium, organic matter and carbon. Measures of litter and duff are combined with those of downed woody material to estimate above-ground carbon pools, fuel-bed loading and wildlife habitat. Data on ozone damage help relate ozone exposure data and drought conditions to potential damage to vegetation (ozone damage is reduced during times of drought). Research is needed to determine the effects of ozone on forested conditions. Finally, sampling all vascular plants aids in monitoring plant biodiversity and assessing non-native invasive species.

Much of the analysis of forest health has been focused on one attribute at a time, perhaps arranged in a table by one or two other attributes (row and columns). However, there is increasing interest in combining the indicators in ways that enhance our understanding of ecosystem dynamics. For example, data on downed woody material, soils, vegetation and tree biomass can be combined to assess carbon pools, while crown rating and tree damage can be used to predict tree mortality. Geographic information systems can be used to add contextual information, which, in turn, can be used in geo-spatial analyses to identify relationships between attributes and across the landscape and ultimately over time. For example, the forest/non-forest map (Phase 1) could be used to map mortality (Phase 2) and beech bark disease (Phase 3), as well as their relationship.

Since the late 1940s, FIA has monitored primary wood-using mills throughout the nation to monitor timber removals. These studies have been augmented by periodic studies of active logging operations to estimate the source of wood in the forest used to deliver primary forest products such as saw logs, pulpwood, veneer logs and other material to the mills.

Since the mid-1970s FIA has conducted studies of private forest landowners to determine their goals and objectives. These studies are also being annualized to provide valuable trend data on the 10 million private forest owners from whose lands more than half of the nation's forest products are derived annually.

Future Direction

FIA continues to work in several areas to enhance its ability to assess sustainability, and make the data useful to forest managers, scientists, policymakers and other citizens. It is developing statistical methods to extrapolate the Phase 3 information to Phase 2 plots (and Phase 1) to better relate and localize the valuable Phase 3 information. Similarly, FIA is developing techniques for scaling up research results associated with air pollution and climate change. This effort also assists in efforts to create nationwide maps for a variety of attributes. Techniques are being developed to localize estimates (small area estimation). As the US forest inventory evolves, there is considerable interest in integrating other metrics and other populations of interest, for example, expanding to other treed lands in urban areas, farm fencerows and riparian zones. FIA's strategic plan for the next 5 years will address these and other issues.

References

Bechtold, W.A. and Patterson, P.L. (eds.) (2005) *The Enhanced Forest Inventory and Analysis Program – National Sampling Design and Estimation Procedures.* General Technical Report 80, US Department of Agriculture, Forest Service, Southern Research Station, Asheville, North Carolina.

US Department of Agriculture (USDA) Forest Service (2004a) *National Report on Sustainable Forests – 2003.* FS-766, US Department of Agriculture, Forest Service, Washington, DC, 139 pp.

US Department of Agriculture (USDA) Forest Service (2004b) *Forest Inventory and Analysis National Core Field Guide*, Vol. 1: *Field Data Collection Procedures for Phase 2 Plots, Version 2.0.* Internal report file at USDA Forest Service, Forest Inventory and Analysis, Washington, DC, or http://fia.fs.fed.us/library/field-guides-methods-proc/

US Department of Agriculture (USDA) Forest Service (2004c) *Forest Inventory and Analysis National Core Field Guide*, Vol. 2: *Field Data Collection Procedures for Phase 3 Plots, Version 2.0.* Internal report file at USDA Forest Service, Forest Inventory and Analysis, Washington, DC, or http://fia.fs.fed.us/library/field-guides-methods-proc/

13 Vegetation Diversity Assessment in Southern Belgium's Permanent Forest Inventory

C. Sanchez,[1] H. Claessens,[1] T. Puissant,[1]
H. Lecomte[2] and J. Rondeux[1]

[1]*Unit of Forest and Nature Management, Gembloux Agricultural University, Gembloux, Belgium;* [2]*Walloon Forest Inventory, General Directory of Natural Resources and Environment, Jambes, Belgium.*

Abstract

Southern Belgium's permanent forest inventory (IPRFW) has been running since February 1994. Its main objective is to identify and control the state and evolution of Wallonia's forests. Because of the growing awareness of the importance of biodiversity and the potential ability of a national forest inventory to assess and monitor it (Köhl, 1996; Tomppo, 1996), since 1997 the inventory has been compiling exhaustive listings (*relevés*) of the vegetation present in each woodland-based field plot. The inventory was designed to use this floristic information to determine the forest type of each sample unit, characterizing its vegetation diversity and monitoring its evolution across the repeated measures of the inventory, normally taken every 10 years. The inventory's current methodology for vegetation description is based on the well-known phytosociological *relevé* method of Braun-Blanquet. This chapter presents only a few examples of data processing results. Furthermore, several problems related to field operations and data processing are also analysed in order to enhance the next inventory cycle scheduled for 2008. The proposed methodological adjustments are designed in such a way that the data and results obtained are useful not only from the point of view of biodiversity, but also for everyday forest management practice. This chapter summarizes the current status of the regional forest inventory in Wallonia with respect to vegetation diversity issues and highlights the need for further methodological progress.

Introduction

The follow-up procedures of the resolutions of international ministerial conferences on policies for nature conservation and forest protection suggest that sustainability, with special focus on biodiversity, has to be implemented in forest management. The concept of forest biodiversity is not easy to define, especially in the framework of large-scale forest inventories for which operationalization of the concept itself depends on international consultations.

The European National Forest Inventory Network (ENFIN), launched in Vienna in 2003, provides an excellent starting point for meeting such objectives, as does the COST E43 programme 'Harmonisation of National Forest Inventories in Europe: Techniques for Common Reporting', supported by the EU since 2004.

The Walloon Region woodland covers 545,000 ha, of which 487,000 ha is considered as productive forest land. A permanent forest inventory (IPRFW) was launched in 1994 to provide large-scale forest information for use in forest policy and planning of forest industries in Wallonia. This inventory is based on the sampling and measurement of permanent sample plots over a 10-year cycle. The sample units (three concentric circular plots with a radius of 18 m, 9 m and 4.5 m according to the type of data collected – see Fig. 13.1) are centred on the intersections of a 1000 m × 500 m grid (Rondeux and Lecomte, 2001). Approximately 10,800 of these sample units are located in woodland. In 1997, sustainability with respect to biological diversity was taken into account, and various observations were integrated to characterize forest biodiversity, including dead wood, stand structure, forest edges, status of tree health, soil properties and ground vegetation. By processing these ecological data, a second objective is to characterize biodiversity and estimate its evolution across the repeated IPRFW measures. However, these ecological observations were simply added to the existing methods of data collection, instead of conducting a thorough review of the inventory's methodology and internal organization (Rondeux, 1999).

The present chapter only concerns investigations on vegetation. Its aim is to examine the methodological aspects of vegetation assessment, briefly present current and potential data processing and results, and discuss several problems related to phenological aspects in plant communities. The final objective of this analysis is to propose several methodological adjustments to improve future observations dealing with vegetation and data processing for the next inventory cycle, which starts in 2008.

Vegetation Assessment in the Regional Forest Inventory

Objectives of vegetation assessment

Vegetation surveys were already carried out to a certain extent at the beginning of the first inventory cycle in 1994. All ligneous species were identified, and a restricted number of dominant and indicator herbaceous species were registered in a listing. This limited survey was considered an adequate response to its assigned objective at that time, which was to determine the vegetation type of the sample unit site. In 1997, new objectives were assigned to the inventory concerning the ecological aspect of sustainability:

- To characterize forest biodiversity and the diversity of vegetation types.
- To contribute to the assessment of long-term evolution in biodiversity in a context of global change and human activities.

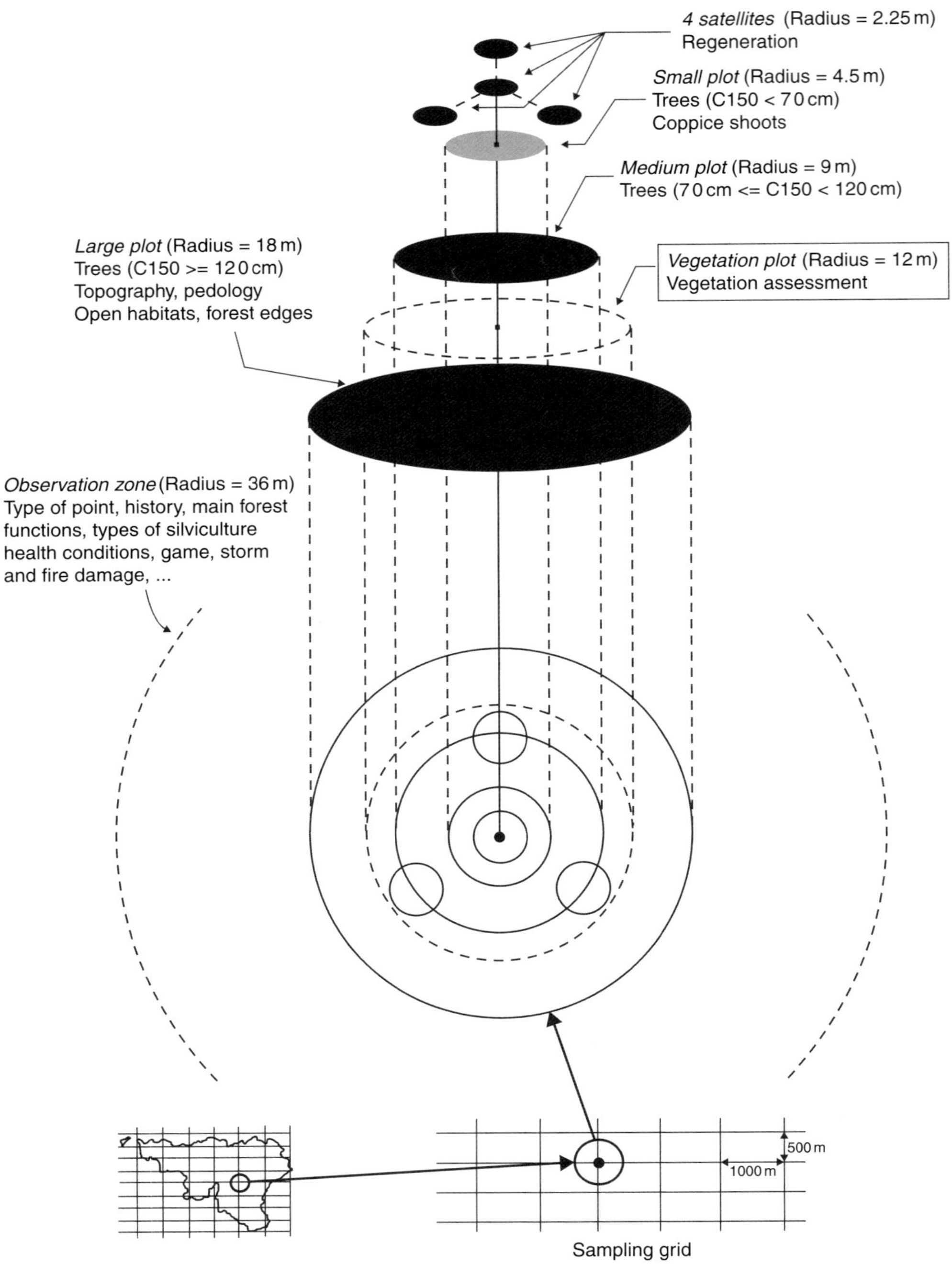

Fig. 13.1. Structure of a sample unit and type of data collected.

Methodological aspects of vegetation assessment

The vegetation sample area was chosen after testing the suitability of the two largest dendrometric plot surfaces (Fig. 13.1): the 9 m radius circular plot (255 m^2) and the 18 m radius circular plot (1018 m^2). Despite the practical interest in using one of the existing plots, both of these surfaces appeared to be inappropriate to the floristic study. The 9 m radius plot was too small to give a good description of the tree strata of all forest habitats, especially for dense land cover, and the 18 m radius plot was unnecessarily big, largely exceeding the minimal area required to fully represent forest species from all strata. It was therefore decided to set up a new 12 m radius circular plot (452 m^2) for the phytosociological *relevé*. Moreover, this plot size matches the reference values suggested by Mueller-Dombois and Ellenberg (1974) and Noirfalise (1984) for temperate forest *relevés* including tree layers.

As for the species studied, all phanerogams and vascular cryptogams are taken into account in the *relevé*. Some terricolous bryophytes, especially indicator species, are also recorded. Plants encountered in special places (tree stumps, paths, tracks, soil disturbance, dead wood, rocks, etc.) are also noted and given an indicator of their special location. Unidentified species are noted as such and are sampled and stored for subsequent identification. The nomenclature system used for vascular plants follows the national plant guide (Lambinon *et al.*, 1992). As indicated in Braun-Blanquet's method, the species list is enhanced with information on the quantity and dispersion of each species. Each record is allocated a two-digit coefficient. The first digit indicates the plant's abundance, whilst the second is an evaluation of its spatial distribution. The values of this coefficient are detailed in Braun-Blanquet (1965). The vertical structure of the vegetation is also recorded. Separate records are made for the species that occur in several vertical strata. Ligneous and semi-ligneous species are divided into three height categories: above 10 m, between 3 and 10 m and below 3 m. Herbaceous species are listed in the herb layer. A visual estimate of the extent of land cover for each layer is also recorded as a percentage of the plot area.

In order to determine the updated state of the Walloon forest, the inventory covers the whole region on an annual basis. Every year, one-tenth of the sample units are visited, according to a distribution per forest range. The dates of observations in the sample units are determined, mainly according to staff availability and work schedule. Furthermore, in order to facilitate tree and stand measurements, broadleaved forests are preferentially visited in the winter, as the absence of leaves allows more accurate height measurements.

All inventory data are recorded on site on a laptop computer. The encoding application program consists of 16 different forms. The vegetation form (Fig. 13.2) is divided in three parts, one for each layer: ligneous, herbaceous and bryophytic. Each of these sub-forms is comprised of a four-column table:

- 'Species' column = species name
- 'Ab' column = species abundance
- 'Di' column = species spatial distribution
- 'Pertur' column = note on any special location of the species

Therefore, one line corresponds to one species and one layer.

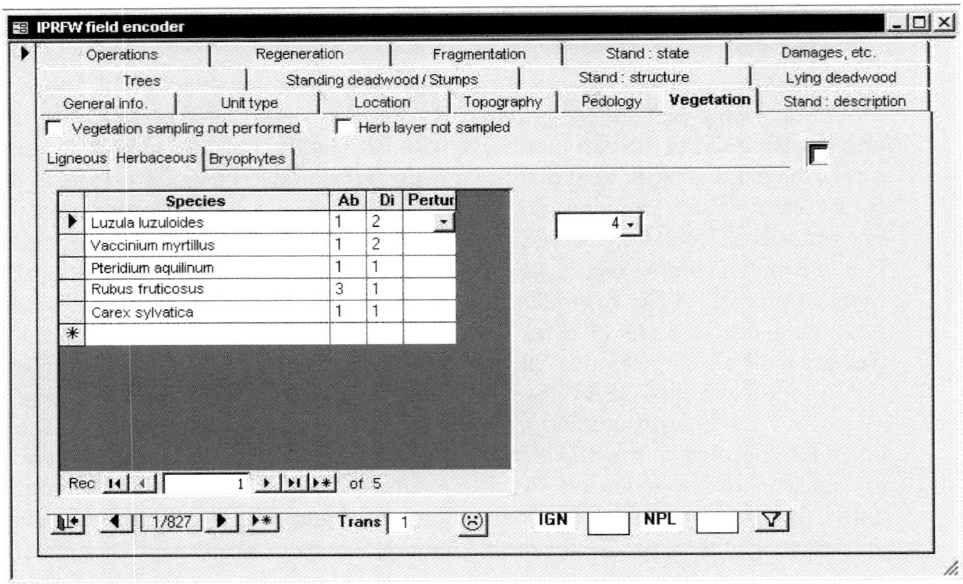

Fig. 13.2. The vegetation form of the field encoding program.

Some Examples of Potential Results

The current method of data collection for the forest inventory provides a strongly consistent floristic database, which can be exploited in many different ways. This section presents examples concerning forest species, forest habitats, vegetation diversity and other specific topics.

Spatial distribution of forest species

Owing to its systematic sampling-plot distribution throughout the whole territory, the Walloon forest inventory data can give an objective picture of the spatial distribution of forest species, as illustrated in Fig. 13.3, which presents two herbaceous species showing two opposite geographical distributions. *Hyacinthoides non-scripta* is an Atlantic species of fresh and mesotrophic soils and *Vaccinium myrtillus* is an acidophilic species essentially present in the Ardenne region.

Information on forest habitats

The vegetation data collected by the inventory enables determination of the vegetation type of the forest stand in which the sample unit is located. The classification used (Noirfalise, 1984) is based on both phytosociological analysis of *relevés* and abiotic characteristics of the site. It is in good correspondence with European classifications, because it was used as a basis of the CORINE European

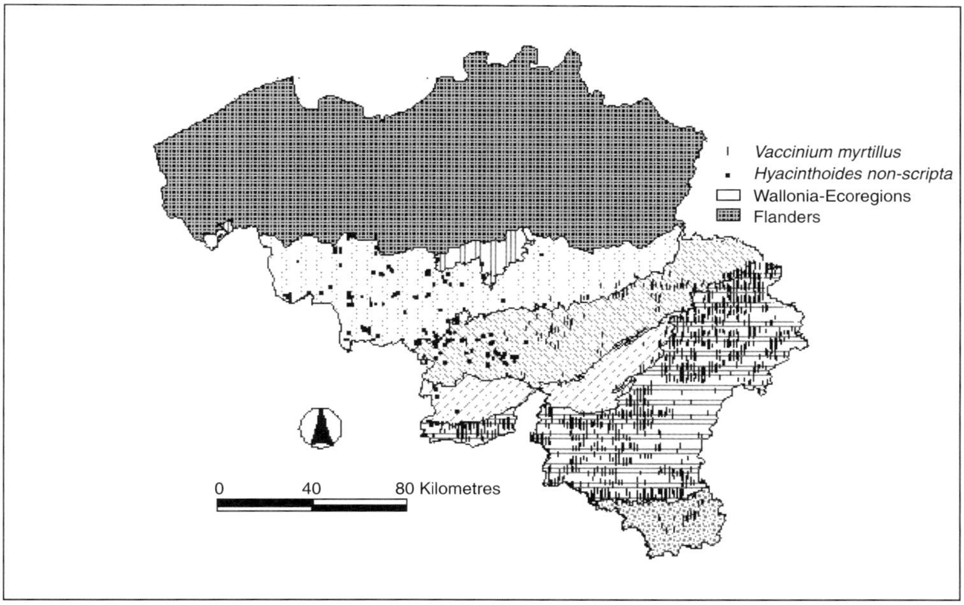

Fig. 13.3. *Hyacinthoides non-scripta* and *Vaccinium myrtillus* distribution map.

Table 13.1. Forest habitat frequency (%) in Wallonia and the Ardenne ecoregion.

Forest types	Frequency (%) in Wallonia	Frequency (%) in Ardenne region
Fagion silvaticae	12	14
Carpinion betuli	23	6
Quercion roboris-petraeae	14	15
Alno-Padion	1	0
Alnion glutinosae	1	0
Betulion pubescentis	0	1
Coniferous stand	38	52
Others	11	12

habitat classification (Coordination de la Recherche de l'Information en Environnement). Some of the abiotic factors collected in the field are also used as additional information for the determination of vegetation type. These factors concern topography, soil properties, aspect, etc. The results of the classification procedure can be used to identify the phytosociological association (e.g. *Luzulo-Fagetum*) and in some cases the sub-association (e.g. *Luzulo-Fagetum vaccinietosum*). This information can lead to further results, such as the distribution of a particular forest vegetation type (drawing points on an existing map) or comparison between the frequency of the main forest habitats in Wallonia or between and within ecological regions (ecoregions) (Table 13.1).

Vegetation diversity assessment and monitoring

Vegetation diversity can be assessed via two components: species richness and species evenness (or equitability). Evenness is an expression of the distribution of the number of individuals among the different species. This diversity can be quantified by means of several indices expressing species richness, evenness or both. The well-known Shannon and Simpson indices combine both species richness and evenness (Standovar, 1996). Table 13.2 lists the different values of Shannon and Simpson indices for ten forest habitats in Wallonia calculated from the spring *relevés* data (Sanchez, 2003). These diversity indices are a useful tool for carrying out an objective comparative analysis of forest vegetation diversity between different land covers in the various ecoregions as well as for all Wallonia. They can also be used within one ecoregion scale or within one forest habitat to compare diversity between sample units. However, these indices are most useful as indicators within the framework of a monitoring project, especially for assessing the long-term evolution of biodiversity.

Specific topics

In addition to the main objectives given to vegetation study (forest biodiversity characterization and monitoring), IPRFW's floristic database gives many other results. Site characteristics are also collected for comparative scientific studies, especially regarding vegetation types and plant indicator values.

An example of these specific analyses, undertaken for a particular purpose, is given by the assessment of the Natura 2000 network in Wallonia (Claessens *et al.*, 2004). In this analysis, the sample units of the inventory were separated

Table 13.2. Values of two diversity indices for each forest habitat. The EUR15 Code refers to the habitat code of 'Community interest habitats' of the European directive 'Habitats' (European Commission, 1999). Priority habitats are asterisked.

Forest habitats		EUR15 Code	Diversity indices[a] Shannon	Simpson
Neutrophile beech forests	*Melico-Fagetum*	9130	3	0.54
Limestone beech forests	*Carici-Fagetum*	9150	3	0.57
Acidophilous beech forests	*Luzulo-Fagetum*	9110	3	0.27
Ravine maple forests	*Tilio-Acerion*	9180*	3	0.54
Oak-hornbeam forests	*Carpinion betuli*	9160	4	0.81
Calciphile hornbeam forests	*Carici-Carpinetum*	–	3	0.52
Acidophilous oak forests	*Luzulo-Quercetum*	–	3	0.32
Alluvial forests of fast-flowing rivers	*Stellario-Alnetum*	91E0*	4	0.71
Alder swamp woods	*Alnion glutinosae*	–	4	0.88
Bog woodlands	*Vaccinio-Betuletum*	91D0*	2	0.18

[a]Formulae extracted from Colwell (2005).

Table 13.3. Proportion of different habitat groups in Wallonia's Natura 2000 (N2000) forest zone.

Habitat groups	Proportion (%) in N2000 perimeter	Selection rate (%)
Coniferous forests	21	15
Priority Habitats	2	50
Habitats of Community Interest	31	51
Other broadleaved forests	35	18

into two groups: the first involves units included in the Natura 2000 network of the Sites of Community Interest (SCI) and the second those that are not. The two groups were then compared in terms of habitats in order to evaluate the extent to which the SCI designation had successfully taken into account the Habitats of Community Interest and especially the Priority Habitats.

The results of the analysis indicate the proportion of each habitat within the SCI perimeter of the forest zone (180,000 ha of the 545,000 ha of the Walloon forest), as shown in Table 13.3. Habitats of Community Interest and Priority Habitats together represent 33% of the Natura 2000 forest area. About 50% of the existing Community Interest and Priority Habitats in the Walloon forest are located in the Natura 2000 forest area. This rate is lower for other deciduous forest habitats, and especially for coniferous forests.

This analysis shows that the inventory data can be used to verify the completion of EU objectives. In this case, the EU objectives are almost attained for the Community Interest Habitats (60%), but not at all for the Priority Habitats (100%).

Other characteristics, such as diversity indices or ancient-forest species frequency, which are considered when defining the Natura 2000 perimeters, should have been taken into account in order to complete the analysis, but several methodological aspects prevented this analysis (see next section).

Methodological Aspects Limiting Vegetation Assessment

A number of methodological aspects, reducing the quality of the vegetation data and hence the resulting analysis, were identified during data processing. Among these, the date on which the *relevés* are performed is considered as the main obstacle for achieving optimal floristic analysis. Other factors, such as habitat classification method or the effect of different field operators on the reliability of the classification procedure, as well as problems encountered in the field, will also be discussed in this section.

Date of observations

Because of the seasonal cycle in temperate forests, vegetation quantity, aspect and composition change constantly throughout the year. However, a precise phytosociological diagnosis can only be completed if an optimum number of species, especially characteristic species that define vegetation types, are present. Unfortunately, many of these species are not visible during winter, especially vernal species such as *Anemone nemorosa* or *Ranunculus ficaria*.

As seen earlier, the organization of the inventory's fieldwork implies that field data are collected all year round. Consequently, the quality of the floristic data varies from sample to sample, according to the date of the flora description. This has three main consequences:

- The number of species present on the sample plot may not reflect the maximum species diversity, which is assessed by diversity indices.
- The quality of the floristic data is geographically unequal. As shown in Fig. 13.4, vernal species, which vary most strongly in relation to *relevé* date, are only detected by the inventory in the regions that are surveyed in spring (the extreme south and centre of Wallonia).
- Because the monitoring is based on two inventory phases, normally separated by a decade, comparisons between floristic data of the two phases become irrelevant, given that the date of observations is not necessarily the same between phases. Therefore, evolution of flora cannot be meaningfully assessed.

Compared with the spatial distribution of *Anemone nemorosa* given by the national floristic survey (Van Rompaey and Delvosalle, 1979), which can be taken as the reference, the data from the Walloon forest inventory reveal a lack of information for entire geographical zones in which data are collected during late summer, autumn and winter.

In order to quantify the impact of the observation date on species number and diversity indices, Sanchez (2003) performed a detailed analysis of the differences between the number of species assessed in winter and in summer for 11 forest habitats (Table 13.4). For this study, two seasons were roughly defined: 'summer', from April to mid-September, and 'winter', from October to February. For each habitat, two *relevés* were performed (one in winter and one in summer) in ten different sample units.

The result of Wilcoxon's test ($n = 11$, $P = 0.003$) clearly confirms the significant difference between the composition of the two samples (winter and summer). This result confirms that there is a significant under-evaluation of vegetation diversity in winter. The relative difference varies according to the different habitats, ranging from 7% (bog woodland) to 38% (neutrophile beech forests).

Vegetation type classification method

Date of observations

As detailed above, the method for classifying vegetation type is based on both floristic composition and abiotic characteristics of the sample unit. Vegetation type is determined at the office using the reference types for Wallonia, i.e. the Noirfalise (1984) forest communities model. The procedure is therefore qualified as a posteriori.

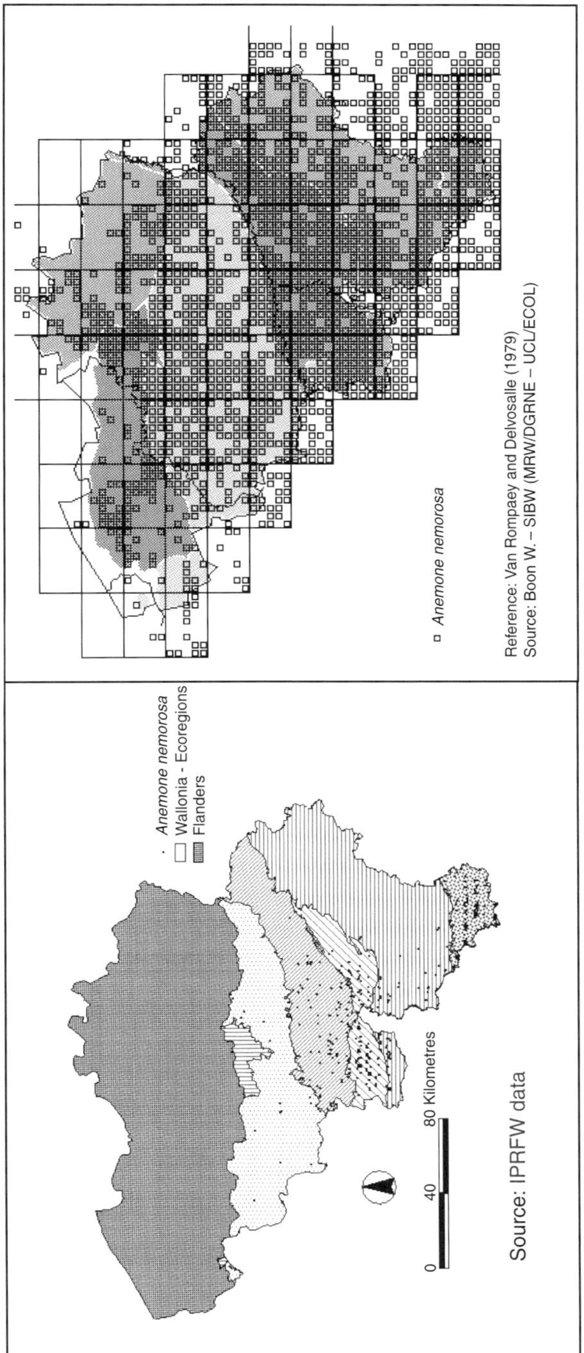

Fig. 13.4. Distribution of *Anemone nemorosa* according to the Walloon Forest Inventory (left) and the national floristic survey (right).

Table 13.4. Average number of herbaceous species per *relevé* and per habitat, in summer and in winter.

	Average herbaceous species number/*relevé*		
Vegetation types[a]	Summer	Winter	Relative difference (%)
Melico-Fagetum	26.9	16.8	38
Carici-Fagetum	25.3	18.0	29
Luzulo-Fagetum	9.2	6.5	29
Acerion pseudoplatani	22.4	16.2	28
Primulo-Carpinetum	28.1	20.1	29
Stellario-Carpinetum	20.0	14.0	30
Carici-Carpinetum	22.2	19.5	12
Luzulo-Quercetum	18.9	16.5	13
Stellario-Alnetum	25.5	20.2	21
Alnion glutinosae	25.4	17.1	33
Betulion pubescentis	13.0	12.1	7
Average	21.5	16.3	24

[a]According to Noirfalise (1984).

In the study cited above (Sanchez, 2003), determination of vegetation types was performed at the office separately and consecutively with winter and summer *relevés* for the same sample units. The results suggested a 19% difference at the syntaxonomic level of the phytosociological association between the winter and summer *relevés*, underlining the important impact of *relevé* date on the quality of the vegetation-type determination, despite the use of certain abiotic factors.

Operator effect

Another source of significant errors is the effect of different operators performing the classification procedure. Sanchez (2003) compared the results of summer *relevé* classifications made by two different operators on the data cited above concerning the 11 different habitats. At the phytosociological association level, 14% of the *relevés* showed differences between operators. Despite the theoretical objectivity of an a posteriori interpretation, the lack of certain important abiotic variables in the Noirfalise classification (1984), or the inaccuracy of these variables, certainly explains part of the differences observed.

Site heterogeneity

The inventory's methodology specifies that sample units must be entirely located within a homogeneous stand. This means that, when a sample unit overlaps an open habitat or a different stand, the sample unit is displaced towards the predominant stand. However, a homogeneous stand does not necessarily mean that ground vegetation and abiotic factors are homogeneous at the scale of a sample unit. For instance, this is often the case in sample units located in valleys, where vegetation changes continuously from marshy to alluvial soils. These special sample units, overlapping two (or more) types of habitat, are difficult to classify.

Proposals for Methodological Adjustments

Date of observations

The vegetation *relevé* method appears to be appropriate with respect to the new objectives assigned to the inventory. These surveys provide information about species, habitats and their diversity. However, the date on which the *relevé* is performed is the main obstacle to an objective interpretation of the floristic data, because, when the *relevé* is not performed during spring or early summer, information on certain species (e.g. vernal species, annual species, etc.) becomes irrelevant, diversity indices are biased and the monitoring of biodiversity no longer gives meaningful results.

It has been proposed that complete *relevés* performed in spring (from late April to early June) would be limited to a restricted number of sample units. These *relevés* would concern one-tenth of the annual sample units that should also be selected for the detailed soil analysis. This would make it possible to conduct global ecological monitoring, and the link with soil analysis opens up possibilities for useful cross-analysis of soil and plant studies. A further advantage is that a great deal of time could be saved on the 90% of remaining plots, where complete *relevés* become less important.

Habitat classification

Two main sources of errors impact negatively on the quality of the habitat classification: the incomplete *relevé*, due to the date of the survey, and the operator effect, due to different interpretations from different operators.

To improve information for forest habitats that are most subject to seasonal changes (Table 13.5), one solution is to first evaluate the potential vegetation in the office, using cartographic criteria. Data from the *relevés* of these particular habitats would be scheduled for a particular date during spring. However, this solution requires a significant adaptation of the field organization.

A second solution would be the use of a field-determination key, based on objective criteria for performing habitat classification. The European Nature Information System (EUNIS) habitat-determination key developed by Wibail *et al.* (2005) is based on both abiotic and floristic criteria, and does not require the identification of all the species present on the site. It can be used to facilitate habitat classification and improve ecological diagnosis due to the abundance of abiotic criteria, and does not require an in-depth knowledge of floristic aspects. Furthermore, this field-determination key obliges all users to follow the same approach, and, because of the abundance of abiotic criteria, it is easier for the operator to deal with sample units located in valleys and other special zones. It includes many secondary habitats (subclimax) not taken into account by the Noirfalise classification (1984), but that remain useful for updating the inventory procedure for habitat classification currently in use. Another advantage of the EUNIS classification is that it is compatible with the CORINE European classification.

Table 13.5. Forest habitats particularly sensitive to winter classification.

Forest habitat	Classification error (%)
Neutrophile Beech forests (*Carici-Fagetum*)	50
Calciphile hornbeam forest (*Carici-Carpinetum*)	30
Alluvial forests of fast-flowing rivers (*Stellario-Alnetum*)	50
Alder swamp woods (*Alnion-Glutinosae*)	25

Table 13.6. List of easily identifiable bryophyte indicator species improving habitat classification.

Scientific name	Indicator value
Plagiomnium undulatum	Nitrophyte hygrophyte
Dicranum scoparium	Acidophilic on moder
Polytrichum commune	Acidophilic hygrophyte
Polytrichum formosum	Large-amplitude acidophilic
Atrichum undulatum	Mesotrophic mull, fresh silts
Leucobryum glaucum	Acidophilic on mor
Sphagnum sp.	Acidophilic hygrophyte on peat

Concerning the plant species recorded, as seen above, bryophytes are optionally noted only if identified by the field operator. This is mainly because of difficulties in bryophyte identification, which can generate errors. Nevertheless, it appears that taking into account several bryophyte species can largely improve the habitat classification. For example, 30% of the sample units classified as bog woodland using the a posteriori classification were later associated to another forest habitat following diagnosis in the field taking into account mosses, especially *Sphagnum* sp. (Sanchez, 2003). Table 13.6 lists some potentially easily identifiable indicator species.

Data control

Errors are always possible at different levels of data collection and analysis: botanical identification, mismatching during data recording, habitat classification, etc.

A computer program (Ecoflore) developed by the French State Forest Service (Office National des Forêts) was designed to perform objective ecological diagnosis and calculate indices that are easily compared between each other and over time, for one particular site (Bruno and Bartoli, 2002). The principle of the program is based on the indicator value and the autecology of each plant species. Each species has particular affinities in terms of moisture and trophic levels of the site. It is therefore possible to illustrate this affinity by representing it in a moisture/trophic diagram, called an 'ecogram'. For each *relevé*, the program calculates a unique 'barycentre' located in the ecogram. Its coordinates are

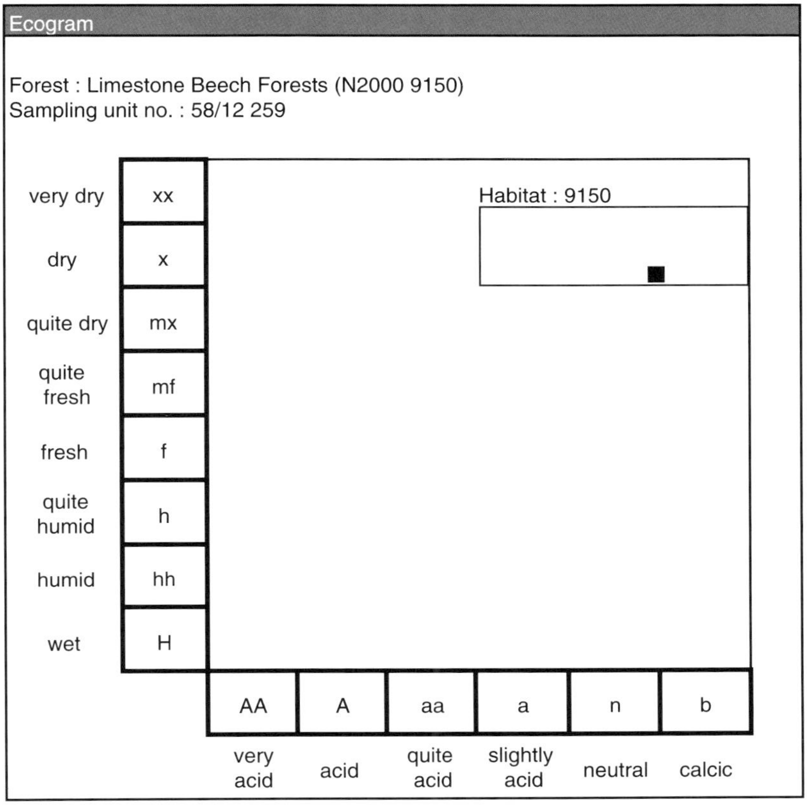

Fig. 13.5. Example of results provided by Ecoflore. The barycentre (small square) corresponds to a *relevé* (no. 58/12 259 of the regional inventory). The large rectangle represents limestone beech forest's edaphic characteristics.

derived from a calculation using the different species' coordinates and their relative abundance. The Ecoflore application also provides the possibility of controlling the consistency of a *relevé* by detecting any errors.

Indeed, if habitats are correctly classified, the barycentre of a *relevé* has to be included in the zone of the ecogram corresponding to the moisture and trophic levels of the relevant habitat, as shown in Fig. 13.5. In the same way, if a species of a *relevé* is not correctly identified, and the error leads to a species with a very different autecology (for example, *Polygonatum odoratum* instead of *Polygonatum multiflorum*), the program automatically identifies this species.

Discussion and Concluding Remarks

The IPRFW has been given an additional assignment of a higher priority for sustainability, and is capable of providing very useful data on the diversity and monitoring of forest vegetation at different scales, as illustrated in this chapter.

The sampling method for vegetation assessment at a regional level has proved its descriptive capacity, but the IPRFW, which has to be performed all year round, is not free from drawbacks. During autumn and winter, *relevés* do not provide a representative description of vegetation diversity and habitats.

These difficulties result from the 'multi-resource' nature of a national or regional forest inventory. The IPRFW's original objective concerned the state of timber resources and, although sustainability and forest biodiversity assessments have been added, only minor methodological adjustments have been made to compensate.

Concerning vegetation assessment in particular, it is clear that *relevés* must be performed during the optimal season. However, data concerning trees and stands are collected preferentially in winter. This highlights a fundamental contradiction undermining the inventory's current methodology.

The proposal suggested in this chapter, i.e. a systematic subsampling of the inventory sample units visited during spring (late April to June), can be considered as a good compromise solution, because it can also guarantee the inventory's role in biodiversity monitoring. At the forest habitat level, the classification method can also be improved by use of a determination key directly in the field. This solution should be less sensitive to incomplete *relevés*, and less vulnerable to the subjectivity of the operator's interpretation.

However, if biodiversity assessment also becomes a main objective of the IPRFW, the internal organization of the inventory will have to be thoroughly adapted. Vegetation assessment and monitoring will have to be performed by a specific team that can operate independently of the constraints of the dendrometric measurements, and which is able to reorganize its fieldwork depending on phenological variations of the vegetation.

Acknowledgements

This study was carried out on behalf of the Walloon Forest Inventory Committee, General Directory of Natural Resources and Environment, and with the financial support of the Walloon Region, Nature and Forest Division (Jambes, Belgium).

References

Braun-Blanquet, J. (1965) *Plant Sociology: the Study of Plant Communities,* edited by Fuller, G.D. and Conard, H.S. Hafner, London. 439 pp.

Bruno, E. and Bartoli, M. (2001) Premiers enseignements de l'utilisation du logiciel Ecoflore pour traiter les relevés botaniques de l'IFN. *Revue forestière française* 53 (3–4), 391–396.

Claessens, H., Lecomte, H. and Rondeux, J. (2004) Quelques données objectives sur la désignation des forêts dans les sites Natura 2000. *Forêt Wallonne* 69–70, 42–47.

Colwell, R.K. (2005) *EstimateS: Statistical Estimation of Species Richness and Shared Species from Samples. Version 7.5.* purl.oclc.org/estimates

European Commission (1999) *Interpretation Manual of European Union Habitats.* EUR15/2, DG Environment, Nature Protection, Coastal Zones and Tourism. Brussels, 121 pp.

Köhl, M. (1996) Assessing and monitoring forest biodiversity in Switzerland and Germany. In: Bachmann, P., Kunsela, K. and Uutera, J. (eds). *Assessment of Biodiversity for Improved Forest Management*. European Forest Institute Proceedings No. 6, Lambinon, Meise, Belgium. pp. 95–104.

Lambinon, J., De Langhe, J.E., Delvosalle, L. and Duvigneaud, J. (1992) *Nouvelle flore de la Belgique, du Grand-Duché de Luxembourg, du Nord de la France et des Régions voisines*, 4th edn. Jardin botanique national de Belgique, Meise, Belgium, 1092 pp.

Mueller-Dombois, D. and Ellenberg, H. (1974) *Aims and Methods of Vegetation Ecology*. Wiley and Sons, New York, 547 pp.

Noirfalise, A. (1984) *Forêts et stations forestières en Belgique*. Les presses agronomiques de Gembloux, Gembloux, Belgium, 229 pp.

Rondeux, J. (1999) Forest inventories and biodiversity. *Unasylva* 196, 35–41.

Rondeux, J. and Lecomte, H. (2001) L'inventaire forestier wallon. *Revue forestière française* 53, 263–267.

Sanchez, C. (2003) Contribution méthodologique à l'analyse de la végétation dans le cadre de l'Inventaire Permanent des Ressources Forestières de Wallonie. MSc thesis, Gembloux Agricultural University, Belgium, 85 pp.

Standovar, T. (1996) Aspects of diversity in forest vegetation. In: Bachmann, P., Kunsela, K. and Uutera, J. (eds) *Assessment of Biodiversity for Improved Forest Management.* European Forest Institute Proceedings No. 6, European Forest Institute, Joensuu, Finland. pp. 17–28.

Tomppo, E. (1996) Biodiversity monitoring in Finnish forest inventories. In: Bachmann, P., Kunsela, K. and Uutera, J. (eds) *Assessment of Biodiversity for Improved Forest Management*. European Forest Institute Proceedings No. 6, European Forest Institute, Joensuu, Finland. pp. 87–94.

Van Rompaey, E. and Delvosalle, L. (1979) *Atlas de la Flore belge et luxembourgeoise*, 2nd edn. Jardin botanique national de Belgique, Meise, Belgium.

Wibail, L., Claessens, H. and Rondeux, J. (2005) *Appui scientifique à la mise en œuvre du réseau Natura 2000 en Wallonie: Finalisation de la typologie Waleunis et synthèse des états de conservation, critères de restauration et mesures de gestion pour les habitats forestiers – Rapport intermédiaire*. Internal Report, Gembloux Agricultural University, Gembloux, Belgium, 40 pp.

14 PractiSFM – an Operational Multi-resource Inventory Protocol for Sustainable Forest Management

F. BARRETT, M.J. SOMERS and M. NIEUWENHUIS

School of Biology and Environmental Science, University College Dublin, Ireland.

Abstract

Following on from the Third Ministerial Conference on the Protection of Forests in Europe in 1998 and the publication of the Irish National Forest Standard (INFS) in 2000, private woodland owners in Ireland are committed, in perpetuity, to sustainable forest management (SFM). With the introduction of SFM, forest certification, increasing regulation and more exacting codes of practice, forest owners and managers require effective and efficient multi-resource inventory, data management and decision-support tools in order to produce management plans that are realistic, practical and sustainable. The PractiSFM (Practical Sustainable Forest Management) research project, initiated in 2001, is addressing these needs. This chapter describes the development and application of the stand-level multi-resource forest inventory component of the PractiSFM system, which has been designed in accordance with the criteria and indicators identified within the INFS.

Introduction

During the last 30 years, traditional forestry objectives of sustainable yield management have been replaced with those of SFM, with a consequent movement away from the traditional single-resource (timber)-based forest inventory towards a multi-resource forest inventory. A multi-resource inventory seeks to engender a more holistic view of the forest ecosystem, whilst establishing a sound knowledge base upon which to make informed forest planning and management decisions and enable monitoring of the progress towards or movement away from sustainable forest management. Countries with historically large forested areas such as the USA and Canada and in Scandinavia have seen significant developments in the area of multi-resource forest inventory (Omule *et al.*, 1996; Zjheng, 1997; Scott *et al.*, 1999).

Following the publication of the Irish National Forest Standard (INFS) in 2000 (Irish Forest Service, 2000a), Irish forest owners and managers have been

 Sustainable Forestry: from Monitoring and Modelling to Knowledge Management and Policy Science (eds K.M. Reynolds, A.J. Thomson, M. Köhl, M.A. Shannon, D. Ray and K. Rennolls)

required to implement SFM principles at forest and stand levels, while at the same time trying to achieve a wide array of objectives, including profit generation. Private woodland owners with limited resources, often lacking in forestry experience, skill and tradition (Ní Dhubháin and Wall, 1998), are facing further challenges and increasing pressure to meet long-term management planning requirements under the new forestry rules, regulations and codes of practice. Furthermore, the need to evaluate forests for sustainability through the quantification and qualification of multiple resources requires data collection methods and skills beyond the traditional, timber-oriented inventory.

While various national forest inventory and monitoring initiatives have been undertaken in Ireland over a number of years, several authors have drawn attention to problems regarding the accuracy, currency, reliability and relevance of the data collected (Anon., 1996, 2004; Gallagher and O'Carroll, 2001). These problems are particularly acute with regard to the assessment and monitoring of SFM in woodlands under private ownership. The new national forest inventory currently being carried out by the Irish Forest Service aims to record a wide array of timber and multi-resource forest variables of relevance to the measurement and monitoring of sustainability in forest management (Department of Agriculture and Food, 2005). However, as estimates will be provided at a regional level only, the data will not serve the needs of private woodland owners at a local stand and forest level.

This chapter describes a successful attempt to address the specific needs of private forest owners and managers in Ireland for a user-friendly, flexible and practical multi-resource forest inventory method, implemented at a stand-level scale. The research project, referred to as PractiSFM (Practical Sustainable Forest Management), has been running since 2001. The PractiSFM system comprises a multi-resource inventory protocol and decision-support system (DSS) to facilitate sustainable forest management of privately owned woodlands in Ireland (Barrett and Nieuwenhuis, 2003). Special attention has been placed in the development of the inventory protocol to criteria and indicators (C & I) and local measures of SFM as identified in the INFS. After introducing the forest areas where the inventory protocol was developed, a brief account of the methodology is made, and its testing and application are presented. Results from the application and field testing are critically discussed. Conclusions concerning the suitability of the PractiSFM system for the evaluation of multiple forest resources and future potential uses of the inventory data and outputs from the system are highlighted.

Material and Methods

Study area

The initial research to develop, apply and test the multi-resource inventory component of PractiSFM took place in a 530-ha privately owned estate in Co. Wicklow, approximately 35 km to the south-west of the capital, Dublin. Ballycurry estate was selected as it incorporates a full range of complex factors that can be encountered within privately owned forests in Ireland. A total 243 ha of the estate is forest, comprising conifer high forest (76%), broadleaved high forest (14%), mixed high forest (6%) and unproductive forest land and open spaces (4%). The forest has been managed commercially since the start of the last century using a clear-fell system, with some areas being recently managed using a continuous cover system. The diverse tree species and age class distributions in some areas have given rise to forest with relatively high biodiversity, while simultaneously other parts of the forest consist of even-aged monocultures. A matrix of aquatic zones and a semi-natural oak woodland National Heritage Area (NHA) located in the estate are noteworthy for their high biological value, biodiversity and conservation functions. In addition to commercial timber production, the forest provides the owners and local residents with an array of goods and services and is located in a relatively densely populated area of high visual amenity and is valued for recreation. The estate is used by local groups for walking and horse riding and by the owners for hunting deer. A local contracting business is responsible for all forest harvesting.

Eight additional forest sites were chosen for assessment and validation of the PractiSFM multi-resource inventory methodologies. The sites, distributed across Ireland, were selected so as to represent typical privately owned forest areas. These sites ranged in size from 6 to 89 ha, with both coniferous and broadleaved species present. All forests were established as even-aged plantations, with plantation age ranging from 14 to 98 years.

Development protocol and criteria

The multi-resource inventory was developed by an iterative process (Fig. 14.1): (i) review and appraisal of stand/forest-level measures of C & I as identified in the INFS; (ii) review of monitoring and multi-resource inventory methodologies at the stand/forest level; (iii) field testing and application at the Ballycurry estate; (iv) consultation to meet the specific needs of private forest owners/managers;

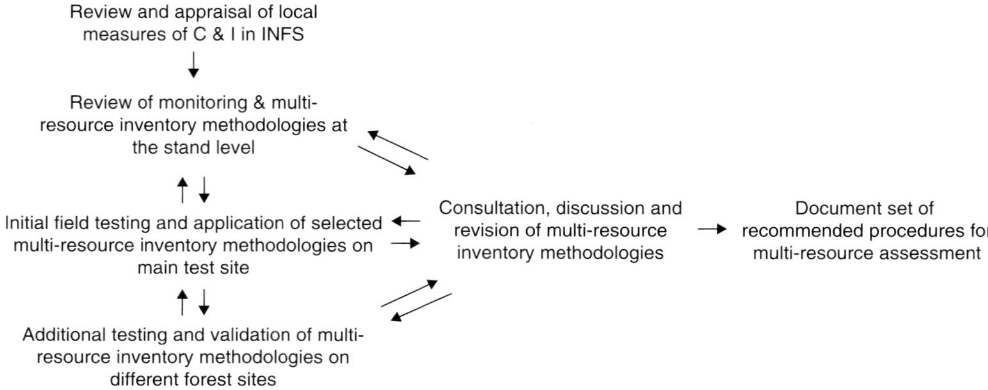

Fig. 14.1. Methods used to develop the multi-resource inventory protocol.

(v) additional field testing on eight sites of varying ownership, size and complexity; and (vi) selection and documentation of the recommended (multi-resource) inventory procedure.

Individuals and organizations charged with inventorying, monitoring and managing private woodlands in Ireland operate with severely limited human, financial and material resources. In this context, the introduction of new methods and tools to improve the efficiency of data collection for inventory and monitoring is of fundamental importance. In order to ensure a practical, relevant and user-friendly focus for the multi-resource inventory, fieldwork was carried out with a leading Irish forest management consultancy company (Purser Tarleton Russell Ltd). A set of screening criteria was adopted where new procedures should: (i) complement and be compatible with traditional and national forest inventory practice; (ii) involve a minimum of extra work; (iii) have a high level of user-friendliness (i.e. variables measured should be easy to detect, interpret and record); and (iv) yield relevant, replicable and credible data.

Initial efforts focused on testing existing multi-resource assessment techniques. Where necessary, the techniques were modified for the needs of private owners and managers in Ireland. Where no existing techniques were found to be practical, new methods were developed.

The multi-resource inventory methodology

Four out of a total of 28 multi-resource assessment criteria tested at Ballycurry are presented and discussed. These are the assessment components of tree and forest health [F, G, H], assessment and evidence of wildlife activity [J], estimation of dead-wood volume [S] and assessment of landscape sensitivity [U] (see Appendix).

Assessment of forest and tree health[1]

To date, forest health has not generally been recorded during stand-level forest inventory conducted in private forests in Ireland. National and international strategic tree health assessment protocols are complex and problems of observer bias can occur (Innes, 1988, 1990; Hanisch and Kilz, 1990; Nicholas *et al.*, 1990; McCarthy, 1993; Bussotti *et al.*, 2002). Field tests showed that these methods were unwieldy and labour-intensive when applied at a stand level. For PractiSFM the stand health assessments developed were confined to a three-level rating of good, moderate or poor, which permitted a rapid visual evaluation in the field. Crown density (crown density reduction classes) and foliage discoloration (needle/leaf colour and extent of needle/leaf discoloration) were used as indicators of stand health. Stand health is considered 'poor' where individual stand trees show more than a 30% reduction in crown density and/or where 30% or more of the crown is observed to have a discoloration of foliage. The symptoms should be described (e.g. pattern of crown defoliation/discoloration) and the symptom distribution estimated in the stand. The chief agents of tree damage or stress within the stand are noted where present i.e. (i) insect; (ii) disease; (iii) wind; (iv) animal; (v) mechanical; (vi) drought; (vii) air pollution; and (viii) other sources. Wind, insect, disease and mechanical damage are important in Irish forests. Air pollution and drought are less important agents of damage and stress. Fire damage is omitted, as a separate survey of damage has to be carried out if fire occurs (Irish Forest Service, 2003). Where more than 5% of the stand area is affected, the area is mapped. Another aspect of forest and tree health is the presence of detrimental or invasive species in the stand (Mrosek and Balsillie, 2001). For example, dense growth of rhododendron and/or laurel will prevent natural regeneration or reforestation (Irish Forest Service, 2002). In this instance recording of the invasive species and its estimated percentage cover within the stand and mapping the location on paper maps are carried out.

Recording evidence of wildlife activity[2]

Comprehensive wildlife inventories are beyond the scope of PractiSFM. A practical assessment was devised based on identification of recent wildlife activity. While the reliability of presence/absence data is highly dependent on species' habitats, this approach is very effective for species that vocalize during their breeding season and species with distinctive signs, such as tracks, faeces, fraying or den building (Gurnell, 1987; Anon., 1995; Pendergast, 2002). A protocol was adopted whereby evidence of activity was described and, where possible, the species named. Live animal observations were also recorded (e.g. deer, rabbits, squirrels). Birds were identified by sight or call where possible. Large wildlife-related features within the stand, such as badger setts or birds' nests, were marked on maps so that, where necessary, they could be protected by exclusion zones.

Habitat suitability indices have been developed for many wildlife species (Bouvier and Howes, 1997; Kuhnke and Watkins, 1999), where modelling habitat requirements for a species aids prediction of the likelihood of its presence. Deer are one of the most important mammal species in Irish forestry, contributing to biodiversity and amenity, providing revenue from hunting, but also having the potential to damage forest and conservation habitats, biodiversity, persons and property (Rooney and Hayden, 2002; Coad, 2004). A deer habitat rating system was developed based on available literature and expert knowledge. Conifer and broadleaved age strata were assigned scores of 1–5 for food and cover. As the age strata information is already part of the traditional timber inventory, there were no additional data collection requirements. Based on the stand-level ratings and the proposed management, a weighted rating for the forest was produced. The habitat ratings can be used to determine the relative impact of proposed forest management strategies (e.g. clear-felling or thinning operations, species selection upon reforestation) on habitat suitability across the forest. Further habitat rating indices, for a range of species, can be developed and integrated into PractiSFM as detailed information on habitat requirements and preferences becomes available.

Estimation of dead-wood volume[3]

Dead wood represents an important resource for biodiversity (Ratcliffe, 1993; Humphrey and Peace, 2003) and affects carbon storage, soil nutrient cycling, energy flows and hydrological processes within the forest. Numerous techniques have been developed for the measurement of dead wood, including fixed-area sample plots, point sampling and line-intersect sampling (Harmon and Sexton,

1996; Kirby *et al.*, 1998). Line-intersect sampling and fixed-area plots were tested at Ballycurry to estimate volume per hectare and stage of decay of coarse woody debris (> 7 cm diameter).

In order to reduce sampling time, a flexible scale that could be implemented visually was devised, i.e. dead-wood quantity is scaled at 0 for 0 m^3/ha, 1 for less than 30 m^3/ha, 2 for 30–150 m^3/ha and 3 for more than 150 m^3/ha. The dead-wood class volume ranges are based on results from recent dead-wood inventories carried out in plantation forests in Britain (Humphrey and Peace, 2003). Estimates of the number of standing dead trees or snags were generated from data recorded during the timber inventory. In order to assess the changes in dead wood over time and to determine the effects of specific thinning or harvesting treatments at forest level, a decay rate function for dead wood, developed by Eriksson and Lindhagen (2001) for Swedish forests, was applied. The dead-wood volume at the end of a planning period was calculated using the starting dead-wood volume estimated at inventory, the dead-wood inputs from harvesting operations (i.e. thinning or clear-fell) scheduled during the planning horizon and the decay rate for the particular species.

Assessment of landscape sensitivity[4]

It was decided that the PractiSFM procedures involved in assessing landscape sensitivity should be easily implementable without a need for specialized GIS or landscape visualization software. Recommendations regarding forest design, aesthetic planning and mitigation of adverse visual impacts resulting from forest operations have been published for the four distinct landscape types commonly found in Ireland (Irish Forest Service, 2000b). Based on these recommendations and following consultation with Irish forest design landscape experts, a simple stand-level landscape sensitivity classification system was adopted. The assessment divides the landscape into 'far-view scenery' or external stand landscape sensitivity and 'within-stand scenery' or internal stand landscape sensitivity (see also Alho and Kangas, 1997). Stands were classified on a three-level scale (0–2) in terms of internal, within-stand landscape sensitivity and also on a three-level scale (0–2) for external landscape sensitivity. The following factors were used to derive the internal landscape sensitivity class: (i) access (by forest/public road, forest trail); (ii) diversity of tree species; (iii) relative stand area; (iv) presence of within-stand open spaces; (v) number of tree storeys; (vi) within-stand visibility; and (vii) presence of features of historical or cultural significance (stone walls, dwellings). The external landscape sensitivity classification was assigned from (up to) three significant viewpoints outside the forest. Stands were classified as externally visually sensitive if visible from designated or recognized tourist routes, amenity areas or population centres (towns, villages) while at the same time taking into account the landscape type as identified in the Irish forest design guidelines (Irish Forest Service, 2000b).

Validation of the multi-resource inventory

Subsequent testing and validation of the PractiSFM inventory protocol on a range of other sites typical of privately owned forests in Ireland were carried out by an independent assessor. This further testing permitted the: (i) analysis of whether the results of the inventory were representative of the different sites; (ii) assessment of the time and resources required; (iii) evaluation of the effectiveness and completeness of the documented protocol; and (iv) correction and revision of the assessment protocols and associated documentation where problems occurred in the interpretation of the methodology. The PractiSFM multi-resource inventory, including a plot-based timber inventory, was carried out for each of the stands at the eight sites. Where forest mapping data were not available, stand boundaries and other physical and environmental features were mapped using a Global Positioning System(GPS).

Results and Discussion

Completeness and representativeness of multi-resource inventory

The multi-resource inventory protocol was capable of accommodating the wide array of biological, physical, cultural and socio-economic characteristics encountered at the other eight validation sites. The testing and validation process permitted the revision of the protocol where inadequacies were identified. These revisions included: (i) a methodology for recording and mapping within-stand open spaces (excluding roads); (ii) a revision of dead-wood volume classes; (iii) a record of forest wind zone (Miller, 1986); (iv) a record of the forest soil type (Horgan *et al.*, 2003); and (v) a record of any proposed statutory or non-statutory designations within and/or adjacent to the forest being assessed. A combination of expert knowledge and additional information describing landscape quality and type, sourced from county landscape character maps (Anon., 2000b), facilitated allocation of internal and external landscape sensitivity scores.

Time and resource requirements for the multi-resource inventory protocol

Experience in the field showed that the PractiSFM methodologies complemented the traditional timber inventory well. Only at the early stages in the assessment of Ballycurry and the first of the eight additional sites was time spent referring to the field notes to clarify the protocol. The time required to carry out the PractiSFM inventory varied according to the physical complexity of the different sites, the age and mix of tree species present and the amount of mapping and other physical and environmental data already available for the stands being assessed. Productivity was found to range from 0.6 to 8.2 ha/h, with most time spent collecting data related to timber, i.e. dbh measurements, tree heights. For both the timber and the non-timber PractiSFM inventory components, stands with complex structures required more time for assessment than single-storeyed, even-aged, monospecific crops. However, younger coniferous stands (e.g. Sitka spruce stands less than 10 years old) proved especially difficult to assess where inspection or brash paths were not present.

While the use of electronic aids for data recording (i.e. electronic calipers, GIS, data logger and GPS) increased speed and efficiency in the field and reduced the need for paper records and subsequent manual data entry, they are not essential. Mapping represents an important component of the PractiSFM methodologies and, instead of using a GIS and GPS, photocopied Ordnance Survey maps can be used in the field or later within the office to record relevant information. Length of stone walls, hedgerows and rivers within the forest were some of the features quantified and recorded in this way.

Precision and accuracy of multi-resource methodologies

Multi-resource inventories and associated variables and measurement techniques should match individual informational needs, resources, budgets, data-processing capabilities, forms of analysis to be employed and tabulations to be reported (Whyte, 1999). In the PractiSFM multi-resource inventory protocols, a useful trade-off is achieved between the investment in sampling and the level of accuracy and precision obtained. The system was designed to facilitate collection of timber data according to point, line and plot sampling strategies typically used in Irish forestry (Purser, 2000), facilitating statistical analysis and error estimation. The timber inventory provides an opportunity for reconnaissance and a rapid visual assessment of various non-timber (multi-resource) attributes of the stand. Where appropriate, additional time is spent gathering more detailed information on specific stand and forest multi-resource attributes.

Quantitative continuous data (e.g. length of hedgerows, adjacency to streams), quantitative categorical data (e.g. dead wood, natural regeneration) and qualitative categorical data (e.g. internal and external landscape sensitivity) were collected as part of the protocols. The quantitative continuous data provide an opportunity for statistical analysis and trend analysis. The quantitative categorical data allow trend analysis at a stand and, especially, at a forest level, as the classes are clearly defined. However, the attribution of categorical score classes can be problematic as the assessor can have a tendency to classify a limited vegetation component more extensively and a large vegetation component less extensively (Jukola-Solunen and Salemaa, 1985). This issue was encountered in the assessment of the stand ground vegetation layer (see Appendix, component [R]). The adoption of classes defined by specific quantitative minimum and maximum values (Lund, 1998) and experience gained through applying and validating the assessment in the field reduce the possibility for making this type of error. In cases where classification is essential (e.g. if the stand management objective is to achieve reforestation through natural regeneration) or difficult (e.g. where the variable being assessed is thought to occur at class threshold values), plots or transects should be used.

Utilizing PractiSFM data for sustainable forest management

The adoption of a multi-resource inventory protocol by private Irish forest owners and managers to plan, conduct and evaluate sustainable forest management represents a huge commitment to collecting, storage, retrieval, analysis and aggregation of data. To this end, the PractiSFM DSS was developed to aid in the production of tactical/operational sustainable management plans for 10-year planning periods. The PractiSFM DSS provides a means to integrate, map and analyse timber and multi-resource inventory data using the database created by the PractiSFM multi-resource inventory (Barrett and Nieuwenhuis, 2003). The DSS can be used to formulate planning scenarios and generate information at a forest and stand level, such as timber volume/value assortments, age-class strata areas, hectares of visually sensitive forest area affected by thinning and clear-fell operations, cumulative dead-wood volume production, productive man-hours and forest-level wildlife food/cover habitat indices.

The PractiSFM multi-resource inventory protocol assumes some basic competencies in the areas of traditional forest inventory on the part of the assessor. Thus, some training will be required by non-experts such as forest owners in the areas of timber inventory (e.g. dbh, height measurement) and basic statistical concepts (e.g. stratified random sampling, number of plots vs. crop variability, plot layout/measurement, etc.) in order to implement the procedures successfully. The PractiSFM field sheet has been designed to include adequate descriptive notes and key information (e.g. description of types of evidence demonstrating specific wildlife activity, descriptions of tree canopy density classes/foliage discoloration classes) regarding the procedures involved in recording each of the non-timber inventory variables. Development of photographic keys to facilitate the classification and identification of the following variables described in the field sheet is planned: stand health, wildlife activity, dead-wood volume, natural regeneration, endangered species and landscape sensitivity. Documentation listing the steps taken to arrive at the final set of methodologies was created in addition to the field sheet. This was provided: to explain the reasoning behind and justification for the various multi-resource inventory procedures; and to show the relevance and importance of each of the parameters measured against C & I in the INFS. An instruction manual for the PractiSFM DSS has also been produced as additional background and training material.

The PractiSFM system can fulfil a number of additional data and planning needs at local and national levels by: (i) facilitating the standardization of management plan reporting to the Irish Forest Service; (ii) feeding into the national forest inventory and providing statistics on a range of forest parameters at a local level; and (iii) facilitating regional and national timber forecasting and strategic planning for the forest industry.

A national forest certification standard has yet to be agreed for Ireland (Little, 2005). However, an analysis of the data requirements for small woodland owners in the United Kingdom Woodland Assurance Scheme (UKWAS) (Anon., 2000a) showed that the PractiSFM multi-resource inventory data requirements are very similar to those required by UKWAS (e.g. requirements for statistics comprising forest area, species, present/future growth and yield, identification of special habitats, amenities and records of public use of the forest). Use of the PractiSFM system (i.e. the combination of the multi-resource inventory and DSS components) will allow private woodland owners to demonstrate their commitment to SFM and achieve certification in any country with a certification standard similar to UKWAS.

Conclusions

Recent years have seen the widespread adoption and implementation of C & I for SFM. However, little has been done to provide forest managers and owners with practical and operational tools to assess the measures of sustainability in the forest at a stand level. The PractiSFM system provides flexible, efficient and cost-effective methods for data collection, without the need for highly specialized knowledge, while requiring only limited training. Further refinements based on additional research and field experience will be required, but the inventory protocol is flexible, allowing the incorporation of new research findings. The cooperative participation of resource management and assessment specialists, in addition to the consultant foresters involved in the project, contributed to identifying and achieving the proper scope, effectiveness and efficiency of the multi-resource inventory protocol. The results of a PractiSFM multi-resource inventory provide a benchmark of the current physical, social and biological conditions within the forest. Future work on the PractiSFM DSS will focus on the analysis of changes occurring in the forest over successive inventories, allowing assessment of the impact of particular management regimes on the overall sustainability of the forest resource.

Acknowledgements

The project is funded by COFORD (National Council for Forest Research and Development in Ireland) and Ballycurry Estate, Co. Wicklow, Ireland, and is carried out in cooperation with forestry consultants Purser Tarleton Russell Ltd. and the owners of Ballycurry Estate, Geoffry and Charles Tottenham. The authors would like to thank the private forest management companies who provided the test sites and Henry Philips, Gerhardt Gallagher, the Irish Forest Service and the Regional Foresters Consultants Group of the Society of Irish Foresters, who provided useful comments and suggestions regarding the approaches taken in the PractiSFM multi-resource inventory methodologies and in the design and implementation of the PractiSFM DSS.

Notes

1. Assessment of forest and tree health relates to Criterion 2 of the INFS (Maintenance of forest ecosystem health and vitality).
2. Recording of evidence of wildlife activity relates to Criterion 4 of the INFS (Maintenance, conservation and appropriate enhancement of biological diversity in forest ecosystems).
3. Estimation of stand-level dead-wood volume relates to Criterion 4 of the INFS (Maintenance, conservation and appropriate enhancement of biological diversity in forest ecosystems).
4. Assessment of landscape sensitivity relates to Criterion 6 of the INFS (Maintenance of other socio-economic and cultural conditions).

Appendix

Irish National Forest Standard (INFS) Criteria and Indicators (C & I), multi-resource components of PractiSFM and variables measured/recorded/mapped for each stand using the PractiSFM multi-resource inventory protocol.

INFS C & I	Multi-resource components of PractiSFM	Actual variables measured/recorded for each stand	Mapped
3.2, 4.2, 4.3, 4.4, 4.5, 4.6	Timber inventory [A]	Stand name, area, plot area, tree species, provenance, diameter distribution (<, > 7 cm live/dead) mean dbh, mean height, top height	Yes
3.2, 4.5	Tree quality class [B]	Straightness (1–3 scale) Branchiness (1–3 scale)	No
1.2, 4.3, 4.4, 6.6	Management [C]	Thin year, thin type/intensity clear-fell year, prune year, vegetation control year, grant potential, official designations (e.g. NHA, ASSI)	Yes
6.3, 6.6	Road access [D]	Access presence/absence	Yes
4.3	Open space [E]	% area of stand	
2.2, 2.3	Health class [F]	Health (1–3 scale)	Yes
2.2, 2.3	Damage and stress in stand [G]	Insect, disease, wind, browsing, stripping, fraying, mechanical, drought, air pollution, other, wind zone	Yes
2.2, 2.3	Invasive species [H]	Species name, distribution in stand	Yes
	Physical properties [I]	Elevation (m), aspect, grid reference	No
4.3	Evidence of wildlife [J]	Species name (feeding, nesting, fraying, denning, marking, tracks, other), deer food/cover habitat ratings	Yes
2.5, 5.3	Forest soil type [K]	Soil group (podzol, peat, gley, etc.)	Yes
2.5, 5.3	Soil damage [L]	Severity (0–3 scale), % area of stand	Yes
4.3, 4.4	Aquatic zones [M]	Length major/minor rivers (m)	Yes
4.2, 4.3	Hedgerows [N]	Length of hedgerows (m)	Yes
4.3, 6.6	Stone walls [O]	Length of stone walls (m)	Yes
3.2, 4.5	Natural regeneration [P]	Scale (0–3 scale), species, stems/ha	Yes
4.2	Vertical structure [Q]	Structure class (0–1 scale)	Yes
3.5, 4.2	Horizontal structure (Vegetation layer) [R]	% Area of stand (moss/lichen (0–2 scale), grass/herb (0–2 scale), shrub (0–2 scale), bracken (0–2 scale))	Yes
4.3	Dead-wood volume [S]	Volume per ha (0–3 scale)	No
4.5	Superior trees in stand [T]	Species name, dbh, % area of stand	Yes
4.6	Landscape sensitivity [U]	Internal sensitivity rating (0–2 scale) External sensitivity rating (0–2 scale)	Yes
3.2, 3.5, 6.5	Amenity sensitivity [V]	Amenity (0–1 scale) List primary/secondary uses	Yes

Continued

Continued.

6.6	Cultural sensitivity [W]	Cultural (0–1 scale) Archaeological/ historic buildings, etc.	Yes
6.5, 6.6	Statutory designation(s) [X]	Designation type and proximity to/ distance from designated area	Yes
6.5	Adjacency to dwellings [Y]	Number of dwellings	Yes
6.5	Dumping [Z]	Dumping rating (0–3 scale) Nature, location of waste	Yes
6.7	Terrain classification [AA]	Ground conditions (1–5 scale), roughness (1–5 scale), slope (1–5 scale)	Yes
6.7	Safety [AB]	Safety rating (0–1 scale) Nature, location of hazard	Yes

References

Alho, J.M. and Kangas, J. (1997) Analyzing uncertainties in experts' opinions of forest plan performance. *Forest Science* 43(4), 521–528.

Anon. (1995) *Forest Operations and Badger Setts*. Forestry Commission Practice Guide 9, Forestry Commission, Edinburgh, UK.

Anon. (1996) *Growing for the Future: A Strategic Plan for the Development of the Forestry Sector in Ireland*. Department of Agriculture, Food and Forestry, Stationery Office, Dublin, Ireland.

Anon. (2000a) *Certification Standard for the UK Woodland Assurance Scheme*. Forestry Commission, Edinburgh, UK.

Anon. (2000b) *Landscape and Landscape Assessment. Consultation Draft of Guidelines for Planning Authorities*. Department of the Environment and Local Government, Dublin, Ireland.

Anon. (2004) *A Review and Appraisal of Ireland's Forestry Development Strategy*. Final Report, Peter Bacon and Associates, Economic Consultants, in association with Deloitte, Killinick, Wexford, Ireland.

Barrett, F. and Nieuwenhuis, M. (2003) A computer-based decision support system for sustainable forest management. *Irish Timber and Forestry* 12(3), 20–23.

Bouvier, J. and Howes, L. (1997) *Eastern Ontario Matrices Linking Wildlife to Habitat: A Biodiversity Management Tool*. Eastern Ontario Model Forest Information Report No. 47. Kemptville, Ontario

Bussotti, F., Cozzi, A., Ferretti, M., Cenni, E., Bettini, D. and Nibbi, R. (2002) Crown condition assessment at the CONECOFOR Permanent Monitoring Plots. *Journal of Limnology* 61 (Suppl. 1), 12–18.

Coad, B. (2004) Deer management – integrated control policies and practices. In: *Proceedings: Deer Management and Control – Policy and Practice, 26 November 2004, Tullamore, Co. Offaly, Ireland*, http://www.coford.ie/mammals04/03-BarryCoad.pdf, last accessed 18 May 2005.

Department of Agriculture and Food (2005) *Annual Review and Outlook for Agriculture and Food 2004/2005*. Economics and Planning Division, Department of Agriculture and Food, Dublin, Ireland.

Eriksson, L. and Lindhagen, A. (2001) A model indicating effects of multipurpose use of forestry on stand level. In: Franc, A., Laroussinie, O. and Karjalainen, T. (eds) *Criteria and Indicators for Sustainable Forest Management at the Forest Management Unit Level*. Proceedings No. 38,

European Forest Institute, Joensuu, Finland, pp. 83–98. http://www.efi.fi/attachment/f5d80ba3clb89242106f2f97ae8e3894/e35db8f2ac985535312a956le70769f4/Proc_38.pdf

Gallagher, G. and O'Carroll, J. (2001) *Forecast of Roundwood Production from the Forests of Ireland 2001–2015*. COFORD (National Council for Forest Research and Development in Ireland), Dublin, Ireland.

Gurnell, J. (1987) *The Natural History of Squirrels*. Christopher Helm, Bromley, UK.

Hanisch, B. and Kilz, E. (1990) *Monitoring of Forest Damage – Spruce and Pine*. Verlag Eugen Ulmer, Stuttgart/A. and C. Black, London/Arts Graphiques Européens, Le Plessis.

Harmon, M.E. and Sexton, J. (1996) *Guidelines for Measurements of Woody Detritus in Forest Ecosystems*. Publication No. 20, US Long Term Ecological Research Network Office, University of Washington, Seattle, Washington.

Horgan, T., Keane, M., McCarthy, R., Lally, M. and Thompson, D. (2003) *A Guide to Forest Tree Species Selection and Silviculture in Ireland*. COFORD (National Council for Forest Research and Development in Ireland), Dublin, Ireland.

Humphrey, J.W. and Peace, A. (2003) Deadwood. In: Humphrey, J.W., Ferris F. and Quine, C.P. (eds) *Biodiversity in Britain's Planted Forests*. Forestry Commission, Edinburgh, pp. 41–50.

Innes, J.L. (1988) Forest health surveys: problems in assessing observer objectivity. *Canadian Journal of Forest Research* 18, 560–565.

Innes, J.L. (1990) *Assessment of Tree Condition*. Forestry Commission Field Book 12, HMSO, London.

Irish Forest Service (2000a) *Irish National Forest Standard*. Forest Service, Department of Communications, Marine and Natural Resources, Dublin, Ireland.

Irish Forest Service (2000b) *Forestry and the Landscape Guidelines*. Forest Service, Department of Communications, Marine and Natural Resources, Dublin, Ireland.

Irish Forest Service (2002) *Forest Protection Guidelines*. Forest Service, Department of Communications, Marine and Natural Resources, Dublin, Ireland.

Irish Forest Service (2003) *Forestry Schemes Manual*. Forest Service, Department of Communications, Marine and Natural Resources, Dublin, Ireland.

Jukola-Solunen, E.L. and Salemaa, M. (1985) A comparison of different methods of quantitative vegetation analysis. *Silva Fennica* 19(3), 325–337.

Kirby, K.J., Reed, C.M., Thomas, R.C. and Goldsmith, F.B. (1998) Preliminary estimates of fallen dead wood and standing dead wood in managed and unmanaged forests in Britain. *Journal of Applied Ecology* 35, 148–155.

Kuhnke, D.H. and Watkins, W. (1999) *Selecting Wildlife Species for Integrating Habitat Supply Models into Forest Management Planning in Manitoba*. Northern Forestry Centre Information Report NOR-X-357, Natural Resources Canada, Edmonton, Alberta.

Little, D. (2005) Forest certification in Ireland, Irish Forestry Certification Initiative Ltd (IFCI) – progress to date. *Irish Timber and Forestry* 14(5), 27–28.

Lund, G.H. (1998) *IUFRO Guidelines for Designing Multipurpose Resource Inventories*. IUFRO World Series Vol. 8. IUFRO, Vienna, Austria.

McCarthy, R. (1993) Monitoring forest condition in Ireland (1988–1991). *Irish Forestry* 50(1), 21–34.

Miller, K.F. (1986) Windthrow hazard in conifer plantations. *Irish Forestry* 43, 66–78.

Mrosek, T. and Balsillie, D. (2001) Development and testing of a criteria and indicators system for sustainable forest management at the forest management unit level: case study at the Haliburton Forest & Wild Life Reserve Ltd., Ontario, Canada. In: Franc, A., Laroussinie, O. and Karjalainen, T. (eds) *Criteria and Indicators for Sustainable Forest Management at the Forest Management Unit Level*. Proceedings No. 38, European Forest Institute, Joensuu, Finland, pp. 215–231. http://www.efi.fi/attachment/f5d80ba3clb89242106f2f97ae8e3894/e35db8f2ac985535312a956le70769f4/Proc_38.pdf

Nicholas, N., Gregoire, T. and Zedaker, S. (1990) The reliability of tree crown position classification. *Canadian Journal of Forest Research* 21(5), 698–701.

Ní Dhubháin, A. and Wall, S. (1998) *Management Requirements for Farm Woodlands*. COFORD (National Council for Forest Research and Development in Ireland), Dublin, Ireland.

Omule, A.Y., Munro, D.D. and Gilbert, D.E. (1996) Designing multiple resource inventories: a British Columbia experience. In: *Proceedings of the AIFM International Conference, 21–24 November 1994*. ASEAN Institute of Forest Management, Seremban, Malaysia, Kuala Lumpur, Malaysia, pp. 145–159.

Pendergast, B. (2002) Wildlife and habitat inventory for a results-based forest practices code. BC *Journal of Ecosystems and Management* 2(2), 2–8. http://www.forrex.org/publications/JEM/jem.asp

Purser, P. (2000) *Timber Measurement Manual, Standard Procedures for Measurement of Round Timber for Sales Purposes in Ireland*. COFORD (Council for Forest Research and Development in Ireland), Dublin, Ireland.

Ratcliffe, R.P. (1993) *Biodiversity in Britain's Forests*. Forest Authority, Edinburgh, UK.

Rooney, S. and Hayden, T. (2002) *Forest Mammals – Management and Control*. COFORD (Council for Forest Research and Development in Ireland), Dublin, Ireland.

Scott, C.T., Tyrrell, L.E., Smith, M.L. and Funk, D.T. (1999) A monitoring system for natural areas in the northeastern and midwestern United States. In: Aguirre-Bravo, C. and Franco, C.F. (eds) *Proceedings of the North American Science Symposium: Towards a Unified Framework for Inventorying and Monitoring Forest Ecosystem Resources, Guadalajara, Mexico, 2–6 November 1998*. Proceedings RMRS-P-12, USDA Forest Service, Rocky Mountain Research Station, Fort Collins, Colorado, pp. 315–318.

Whyte, G. (1999) Inventory, modelling and auditing systems for planning and controlling forestry operations. *International Journal of Forest Engineering* 10(2), 2–31.

Zjheng, X. (1997) Toward multi-resource forest environment monitoring in Switzerland, Sweden and Finland. *World Forestry Research* 10(2), 58–64.

15 The Importance of Forest Stand-level Inventory to Sustain Multiple Forest Values in the Presence of Endangered Species

DEBORA L. JOHNSON,[1] K. NORMAN JOHNSON[2] AND DAVID W. HANN[2]

[1]*Oregon State University, College of Forestry, College Forests, Corvallis, Oregon, USA;* [2]*Oregon State University, College of Forestry, Department of Forest Resources, Corvallis, Oregon, USA.*

Abstract

Over the last decade, Oregon State University (OSU) College Forests has successfully implemented a plan for its 4600-ha forest in the Willamette Valley of Oregon that called for significant timber harvest under a range of silvicultural systems, including clear-cutting, while providing teaching, research and extension opportunities. Despite discovery of endangered species, increasing recreational use and demands from OSU faculty for research and field teaching uses, the college has been able to implement the plan and produce the expected volume and revenue. Central to this success was the forest inventory installed in the mid-1980s, combined with a growth and yield model calibrated to the forest. Rather than using class averages like most public forests, the College Forests use projected stand estimates, which enable accurate aggregation and disaggregation of the forest stands for both strategic planning and detailed harvest scheduling. This approach both enables accurate initial estimates of the harvest schedule and facilitates the inevitable shifts in the timings and location of the harvest as unforeseen events occur. One such event was the arrival of northern spotted owls (*Strix occidentalis*), a federally listed threatened species. Utilizing this inventory and projection system, the college identified stands that functioned as spotted owl habitat, adjusted the location of harvests to conserve that habitat during the plan period (10 years) and still met its harvest and revenue targets. In the recent plan revision, the college has taken the next step to dynamically schedule investment and harvest in ways that will maintain desired amounts of the spotted owl habitat in the long run.

Introduction

The College of Forestry at Oregon State University manages the 4600-ha McDonald–Dunn teaching, demonstration and research forest on the western

Fig. 15.1. McDonald–Dunn Forest is located in Oregon's Willamette Valley.

edge of Oregon's Willamette Valley, near the city of Corvallis (Fig. 15.1). Researchers and teachers throughout the university use the forest, and it receives at least 150,000 recreational visits each year. Timber harvested from the forest produces revenue to maintain and operate the forest, and to meet special college instructional and research needs.

History

McDonald–Dunn Forest is part of the historical homeland of the Kalapuya Indians. Evidence of their use of the land dates back over 10,000 years. Epidemics in the late 1700s and early 1800s decimated Indian tribes in the Willamette Valley (Mackey, 2004). The survivors ceded most of their lands to the US government and became part of what are now the Confederated Tribes of Grand Ronde and the Confederated Tribes of Siletz.

Because the Kalapuya used regular burning to favour subsistence plant and animal species, early Euro-Americans arriving in the Willamette Valley found an

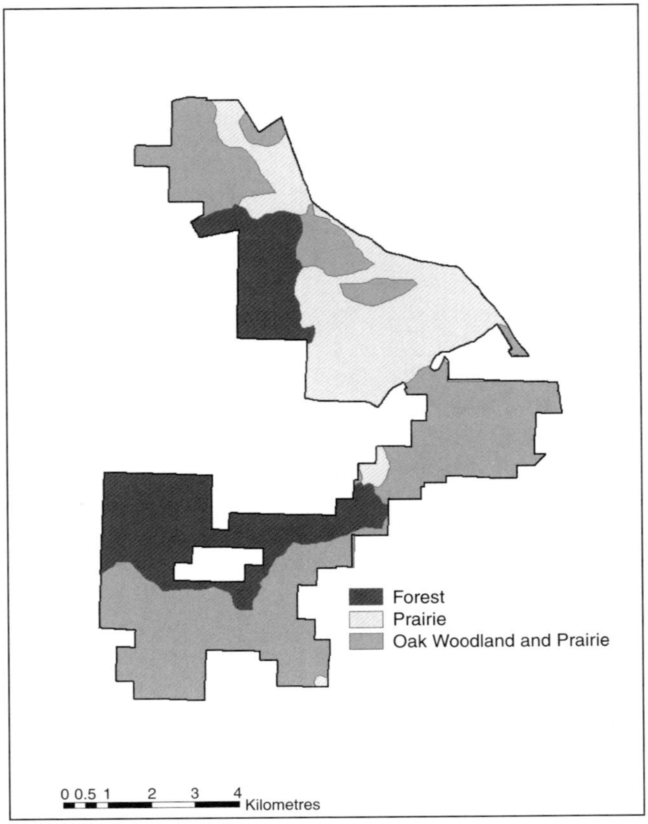

Fig. 15.2. Vegetation in the mid-1800s.

open landscape dominated by prairie and oak savannah (*Quercus garryana*) (Fig. 15.2). The end of Indian burning practices brought encroachment of the conifer forest (Fig. 15.3).

By 1850, most of the area that was to become the McDonald–Dunn Forest was occupied by homesteaders, who raised grains and animals, but, as the encroaching forest matured, sawmills were constructed.

The OSU College of Forestry acquired the McDonald portion of the forest through gifts and purchases between 1925 and 1962. The adjacent Dunn Forest was used by the military for training during the Second World War and was acquired after the war ended. Early management activities were largely focused on reforestation and road building until 1962, when the 'Columbus Day storm' brought sustained winds over 160 km per hour and catastrophic windthrow to the McDonald–Dunn Forest. This event directed management of the forest for at least the next decade, as the forest manager salvaged and worked through stands damaged by the storm. Reliance on timber revenue by the college gradually increased, and by the mid-1980s harvest shifted from thinning and small clear-cuts in the central and northern part of the forest to clear-cuts of older

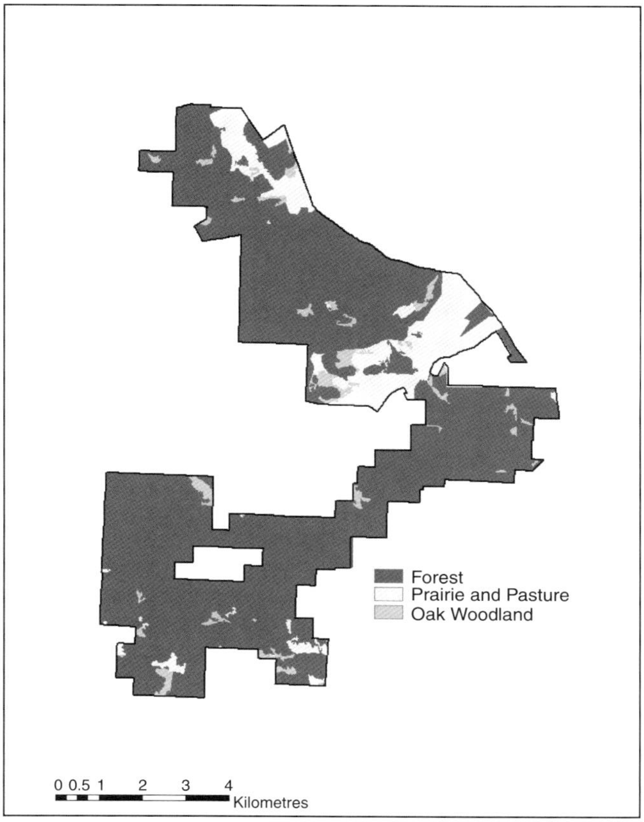

Fig. 15.3. Current vegetation.

timber in the southern part of the forest – the area that is closest to and most heavily used by the residents of Corvallis.

Corvallis, in the meantime, had become a politically liberal city of about 50,000 with a large number of affluent and highly educated residents, who (based on voting patterns) tended to focus heavily on public values. When the dean of the college was confronted by neighbours opposed to clear-cut harvests in their recreational areas and adjacent to their homes, he had no long-term plan for the forest that he could use to defend himself.

Creating the Forest Plan

The primary mission of the College Forests is to support teaching, research and extension. In addition, the college has many other goals for this forest, including revenue generation, accommodation of recreation use and conservation of biodiversity. A plan was needed that would integrate this mission and additional goals, demonstrate a long-term vision and serve as a guide for action. In 1993,

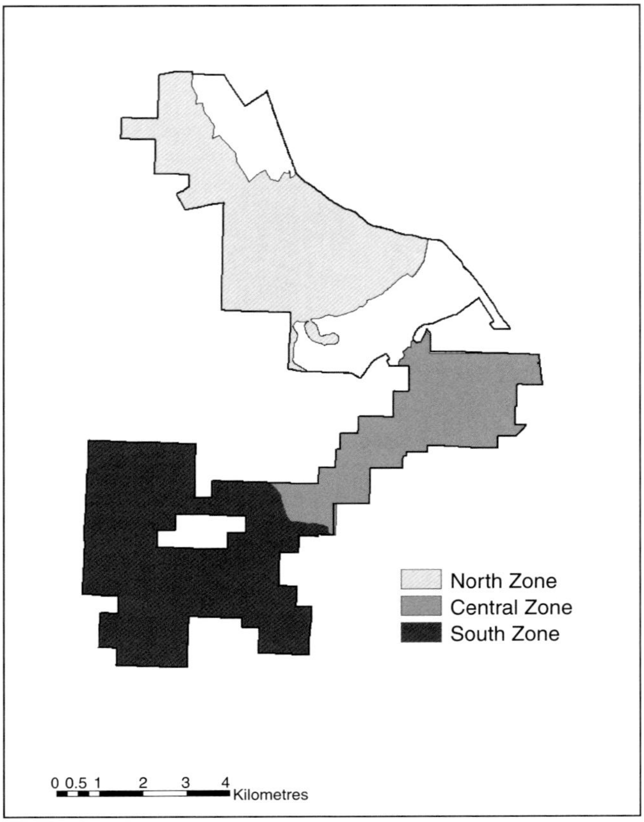

Fig. 15.4. Forest zones.

an interdisciplinary group of faculty from the university was convened. They developed a plan that divided the forest into three zones, each with a different management theme (Fig. 15.4).

The northern part of the forest was allocated to a short-rotation, even-aged regime. Even-aged, two-storey regimes designed to protect sensitive parts of the viewshed were put in place in the central zone, and the southernmost part of the forest, most heavily used by both recreationists and university instruction, because of its proximity to Corvallis, was to be managed with uneven-aged treatments designed to accelerate the development of mid- to late-seral forest conditions. Unique areas that have special teaching uses and old-growth conifer forests were identified and withdrawn from timber harvest (Fig. 15.5).

Forest Inventory

Among public forests in the Pacific Northwest, the College Forest is unique in its ability to accurately portray the potential outcomes from an analysis of harvest

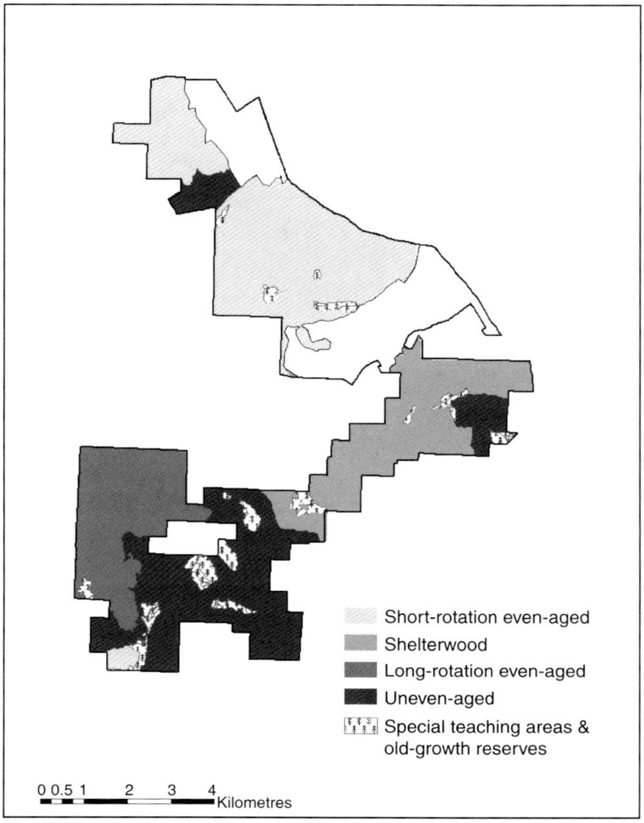

Fig. 15.5. Land allocation.

scheduling, because of the forest's stand-level inventory. Most public forests base their inventory estimates on class averages, which introduces two problems: (i) it is difficult to apply these estimates to individual stands; and (ii) when actual volumes harvested differ from estimated inventory volumes (for the same prescription), there is no way to distinguish whether the difference is due to an atypical stand or to bias in the inventory itself. The OSU stand-level inventory overcomes these problems.

A grid of permanent inventory plots was originally installed on the forest in the early 1980s, and a subset of plots have been remeasured in most subsequent years. The sampling intensity varies from 4.95 sample points per hectare in regeneration to 0.62 sample points per hectare in old-growth forest (Fig. 15.6). As regeneration stands mature, the number of plots measured drops down to what is needed for the desired sampling error. The sample-plot design consists of three nested subplots installed at each point: (i) a 1/566-ha (1/229-acre) fixed-area plot for trees ranging from 15.2 cm. tall to 10.2 cm dbh (6 in. tall to 4 in. dbh); (ii) a 1/141-ha (1/57-acre) fixed-area plot for trees ranging from 10.2 to 20.3 cm (4.1 to 8 in. dbh); and (iii) a 4.592-BAF (basal area factor) (20-BAF)

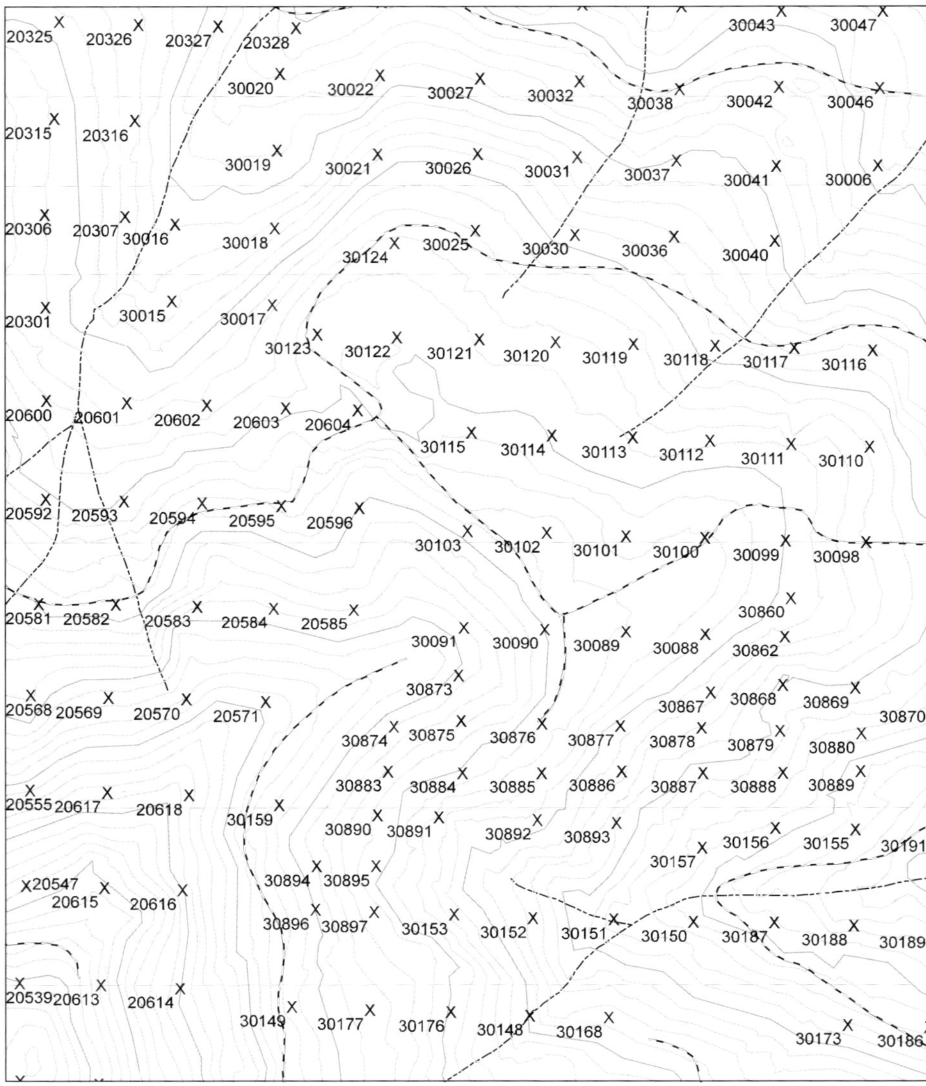

Fig. 15.6. Example of plot layout.

variable-radius plot for trees larger than 20.3 cm (8 in.). Each sample tree is measured for diameter, total height and height to live crown. Many of these stand-level design features reflect common features of stand-level inventories used throughout the region.

While the basic plot design of this inventory is similar to that used on both public and private industrial lands throughout the state, private land managers tend to use many temporary plots to create estimates of stand-level averages, while public agencies use a sparse number of permanent plots to develop general estimates of forest-wide averages. Although private landowners often collect data at the stand level, they do not always maintain individual tree information in their databases, since their traditional stand-level growth-and-yield models don't require this information. Without original sample tree data it is difficult to re-aggregate these data over time as stand boundaries change, and it is difficult to regenerate tree lists that are needed for other types of models, particularly habitat models.

Permanent plots provide the foundation for monitoring traditional and non-traditional silvicultural treatments as well as changes in ecosystem structure and function. They also provide data sets for validating and, if necessary, recalibrating the stand development model (Marshall *et al.*, 1997), and can be augmented with temporary plots to get higher levels of precision. Although an inventory built from permanent plots costs more to install than one based on temporary plots, after the initial plot installation costs have been absorbed, remeasurement costs per plot decrease, and subsets of the plots are useful for monitoring other plant and animal species or communities.

Stand development model

When the inventory was initially installed, past growth rates were reconstructed for 136 stands to calibrate the ORGANON (Oregon Growth Analysis and Projection) stand development model (Hann, 2003). The stands used for calibration had not been treated for the previous 5 years, had significant basal area in the two target species of Douglas fir (*Pseudotsuga menziesii* (Mirb.) Franco) and grand fir (*Abies grandis* (Dougl.) Lindl.) and covered as wide a range in site index, age and density as possible.

ORGANON is a single-tree, distance-independent model (Munro, 1974) that incorporates crown attributes in many of its functions. It can predict future development and resulting wood-quality attributes of even-aged and uneven-aged stands with pure or mixed-species composition. In addition to a large number of cutting alternatives, treatments can include fertilization and pruning. During simulation, new trees can be added through an in-growth routine (Marshall *et al.*, 1997).

Applied to an existing stand, ORGANON takes the initial sample of trees from the inventory, a stand age and site index (King, 1966) and predicts how each sample tree's diameter, total height, height to crown base and expansion factor (number of trees per unit area that the tree represents) will change, given a species-specific diameter growth rate (Zumrawi and Hann, 1993), height growth

Table 15.1. ORGANON growth-and-yield model validation results using data on total stem cubic foot volume per acre from the control plots on four installations near the McDonald–Dunn Forest. (From Marshall *et al.*, 1997.)

	Number of control plots	Size of control plots (acres)	Initial breast height age	King's site index (feet)	Difference (%) at end of 30-year projection
Black Rock A	1	1.0	39	116	–4.2
Black Rock B	1	1.0	40	113	–1.6
Burnt Woods	2	0.1	18	136	+3.7
Hoskins	3	0.2	13	136	–2.0
Weighted average					–0.6

rate (Hann *et al.*, 2003), height to crown base (Zumrawi and Hann, 1989) and mortality-rate equations (Hann *et al.*, 2003), respectively. Individual tree records are aggregated to provide stand-level information. The resulting model has been validated with 30 years of independent remeasurement data from four local growth-and-yield installations (Table 15.1).

Harvest scheduling

A harvest schedule, based on the different management themes and special areas, was developed for the plan. In this analysis, each stand was simulated, using every treatment that might occur for a 100-year period, with the ORGANON growth-and-yield model (Hann, 2003).

The simulations were then passed to the FORPLAN linear programming harvest scheduling model (Johnson *et al.*, 1986), which created solutions linked to the GIS stands layer. After working with the forest planning team, a tentative harvest schedule was produced for the forest staff to implement (Fig. 15.7).

Plan Implementation

After the plan was completed, the forest staff began laying out timber sales, based on the new harvest schedule, and the college convened a Forest Advisory Committee (FAC) made up of an interdisciplinary group of faculty as well as stakeholders from the community. The FAC was charged with providing advice on tactical adjustments to the plan. When implementation issues surfaced, the FAC either met and worked on the issues themselves or convened subcommittees that brought in other types of expert help as needed. The FAC developed many important refinements to the forest plan, and also provided timely advice to the dean about how to proceed when unforeseen events challenged plan implementation.

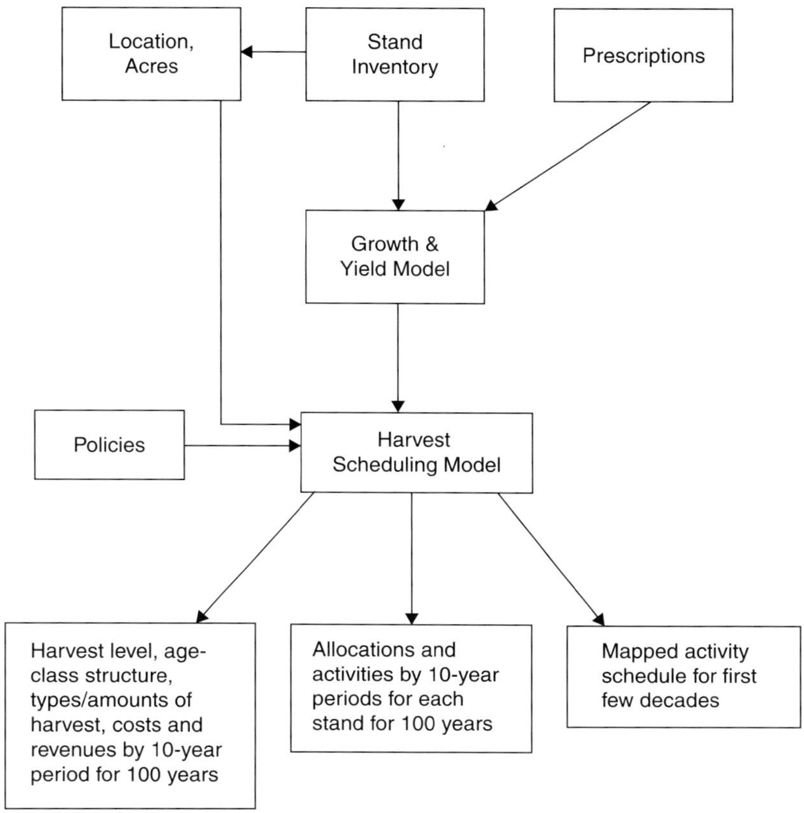

Fig. 15.7. Harvest scheduling analysis process.

Spotted owls arrive

Shortly after this plan was finalized and implementation began, a pair of northern spotted owls (*Strix occidentalis*, Fig. 15.8) began nesting in the south zone of the forest. Because northern spotted owls are listed as a threatened species under the US Endangered Species Act, implementation of the plan could not occur if it meant harming the owls or adversely affecting their habitat. According to guidelines issued by the US Fish and Wildlife Service (USFW), northern spotted owls need at least 40% suitable nesting, roosting and foraging (NRF) habitat within 2.4 km (the home range) of their nest site.

The evaluation of spotted owl habitat within the forest around the nest site required the development of a definition that fitted our forest type, based on the known ecology of the spotted owl. The main components of this habitat are nesting, roosting, foraging and dispersal. Optimal nesting and roosting habitat typically includes a fairly high canopy closure, a multilayered, multi-species canopy with large overstorey trees, large accumulations of fallen trees and other debris and sufficient open space below the canopy for owls to fly (Thomas *et al.*, 1990).

Fig. 15.8. Juvenile northern spotted owl. Photo by Dan Schafer.

Nest trees are often large trees with broken tops or broom-like structures that serve as nesting platforms. Good foraging habitat is believed to be closed-canopy forests with adequate flying space under the crowns. Roosting habitat includes patches of mid-storey shade-tolerant conifers or hardwoods. Although definitions around these components can be found in the spotted owl literature, none of them fitted the type of owl habitat found in the Willamette Valley ecoregion. Based on inventory data from stands that the owls appeared to be using, a definition of NRF habitat was crafted around three main variables: (i) crown closure must be greater than 55%; (ii) the quadratic mean diameter of trees over 15.2 cm (6 in.) must be greater than or equal to 30.5 cm (12 in.); and (iii) there must be at least 14.8 trees per hectare (6 trees per acre) greater than 76.2 cm (30 in.) diameter at breast height (Bettinger *et al.*, 2003).

When this definition of NRF was applied back to the stand data for the south zone, there was less than the 40% required under the USFW guidelines within the recommended home range (Fig. 15.9). To compare this estimate to the amount of habitat in the actual home range for this owl pair, radio telemetry data were collected for one full year. The actual home range was then constructed from the telemetry observations, using kernel-density estimators (Fig. 15.10). This confirmed that our definition was primarily identifying stands that the owls were using, and that the actual home range was similar to the USFW 2.4 km home-range circle. This definition guided the first decade of plan implementation and efforts to develop alternative plans for suitable owl habitat.

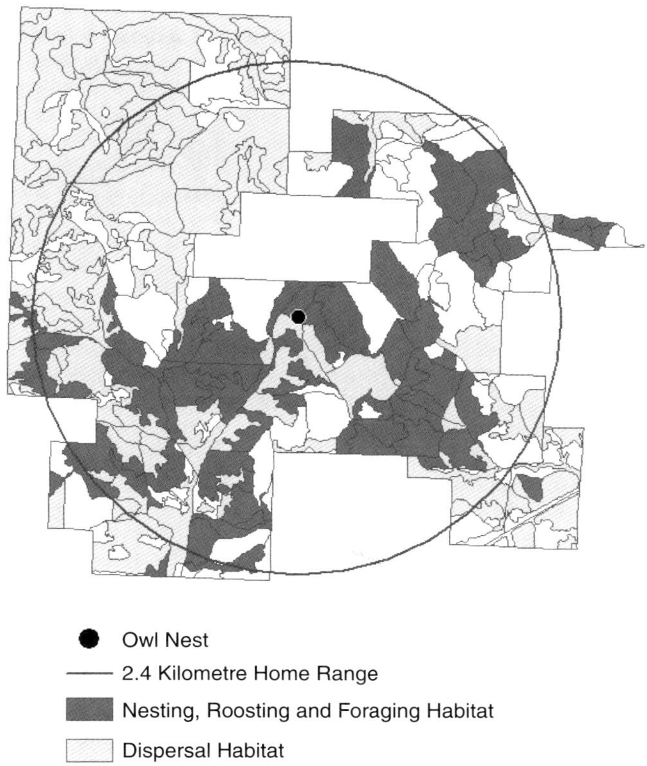

Fig. 15.9. USFW theoretical home range for the northern spotted owl.

Other surprises

Other things also happened that influenced the harvest schedule. A research project on uneven-aged treatments was superimposed on stands in the central zone that had originally been allocated to two-storey regimes. Management of forested lands in two agricultural farms was transferred from the College of Forestry to the College of Agriculture. Field foresters sometimes chose different stands or prescriptions from those scheduled because of logistical problems. Costs associated with some types of logging were higher than anticipated. Finally, an increase in revenue was required towards the end of the 10-year planning period because of a financial emergency in the College of Forestry.

Comparison of planned and actual harvest schedule

How well did College Forest staff implement the planned harvest schedule on the 10-year life of the plan? On a year-to-year basis, the harvest showed significant

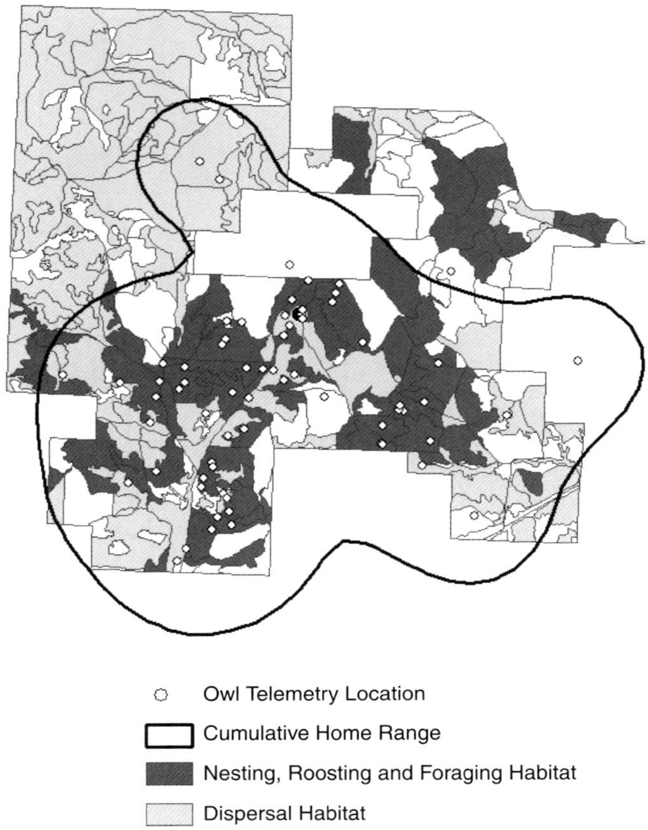

Fig. 15.10. Telemetry locations were used to calculate the actual home range.

oscillation as the forest staff coped with changing conditions, surprises and markets (years 1994–2003 in Fig. 15.11). Overall, though, the staff did surprisingly well in meeting plan targets:

- Harvest level. The scheduled annual harvest was 4.1 MMBF (million board feet) while the actual average annual harvest for the decade was 4.5 MMBF, due to the large increase in harvest during the last year of the period (2003).
- Acres harvested by treatment type (Fig. 15.12). Clear-cut acres for the decade were slightly higher than scheduled, because of the need to generate additional revenue towards the end of the period, and thinning and uneven-aged treatments were slightly below those scheduled, because of the influence of the owls and high logging costs associated with thinning.
- Volume harvested by treatment type (Fig. 15.13). The volume harvested by treatment type for the decade was close to plan estimates, although some individual stand estimates were better than others.

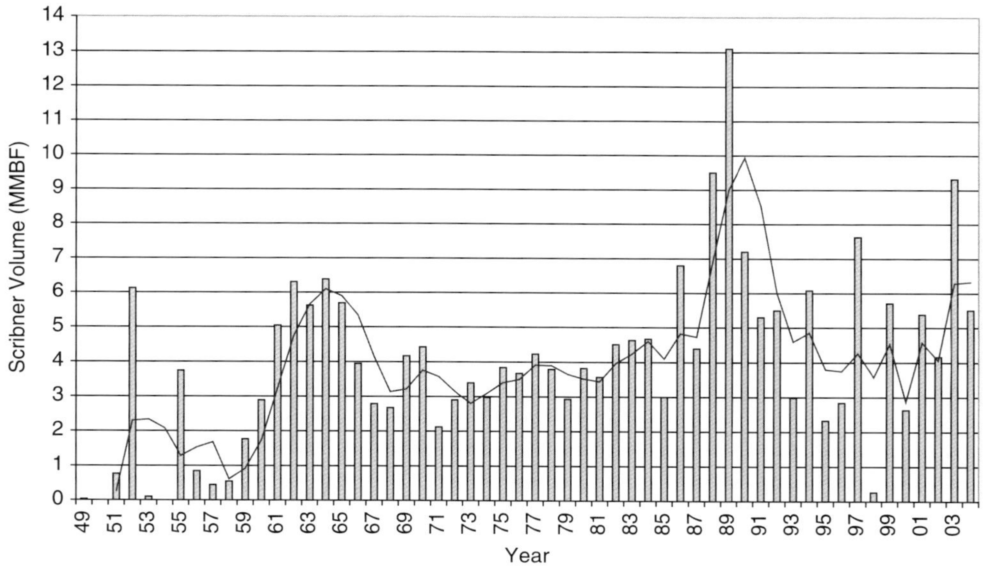

Fig. 15.11. Volume harvested between 1949 and 2004 on McDonald–Dunn Forest with a 3-year moving trend line. From 1994 to 2003 timber harvest fluctuated as forest staff coped with changing conditions, surprises and markets.

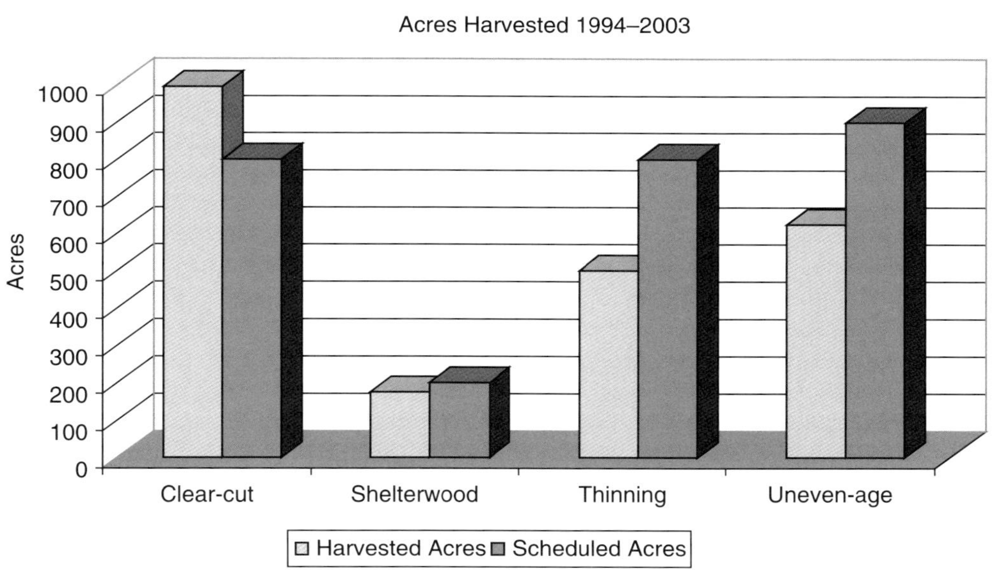

Fig. 15.12. Acres scheduled for harvest and actual acres harvested by treatment type.

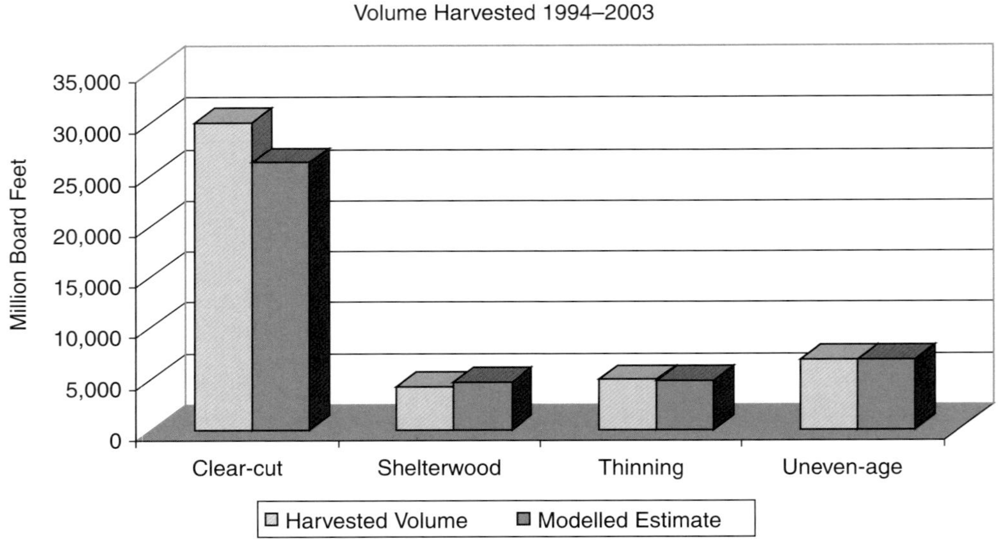

Fig. 15.13. Cut-out comparisons for stands that were actually harvested.

- Stands selected for harvest. Comparing stands originally scheduled for harvest (Fig. 15.14) with stands actually harvested (Fig. 15.15), we see that shifts occurred in many places in response to events and field information, but these changes maintained the general amount and distribution of harvest among treatment types (Figs 15.12 and 15.13).

Despite the many unforeseen events that occurred during the plan decade, College Forest staff were able to meet the plan targets for harvest, which included a significant amount of clear-cutting. How was this possible while other public forest plans floundered in the 1990s? Three major factors made this possible. First, the College Forest plan, as written, was feasible to implement, in part due to the stand-level inventory, which enabled a confident statement of the timber outputs that would be associated with various activities. Other public forest plans, based on forest-level inventories, could not identify the spatial location where their harvest volume would come from, making the plans difficult to adjust. Secondly, the ability of the public to challenge activities was much more limited on the College Forest as compared with federal forests. Citizen class-action lawsuits, the backbone of environmental challenge on federal forests, would be difficult to bring on the College Forest, which is legally treated similarly to private land. Thirdly, and related to the second factor, the forest manager and staff had wide latitude to quickly adjust the schedule as they saw fit, as long as they abided by the goals and standards for the different themes and allocations outlined in the plan. The environmental analysis mandated for federal forests under the National Environmental Policy Act was not an issue in managing the College Forest.

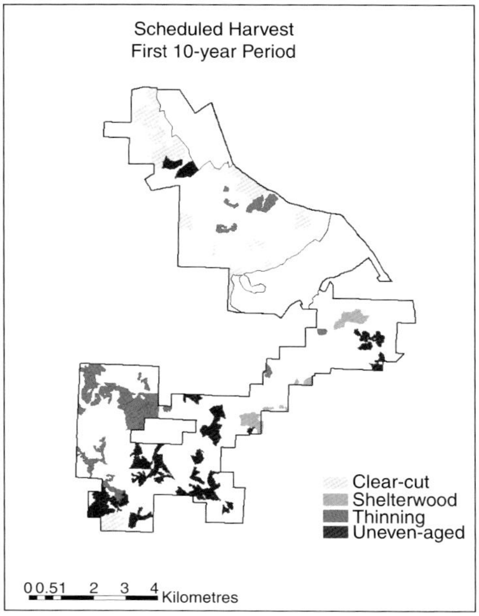

Fig. 15.14. Harvest schedule for the first 10-year period.

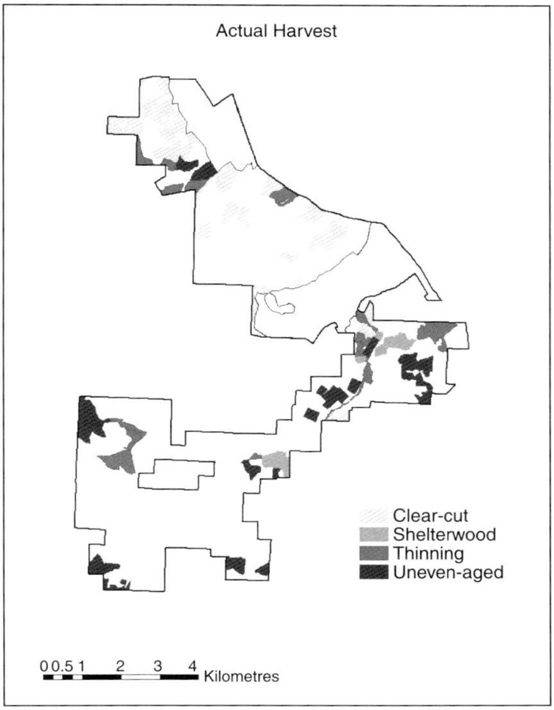

Fig. 15.15. Actual harvest that occurred during the first 10-year period.

Plan revision

The revision process for the forest plan began in 2003. Between 1994 and 2004, three different pairs of spotted owls had occupied five different circles (Fig. 15.16). The changes in adult owls and nesting sites are apparently the result of pressure from the larger and more aggressive barred owls (*Strix varia*), which have expanded their range into the Pacific Northwest. A key decision in the revised plan was to maintain the current level of NRF habitat for the northern spotted owls throughout the south zone of the forest, because shifting owl circles make it very difficult to manage for habitat within a particular circle.

Using our definition for suitable habitat, we created a dynamic harvest scheduling model, which maintains the current level of NRF in the south zone of the forest (as shown in Fig. 15.9) but allows it to shift within the south zone as new habitat develops.

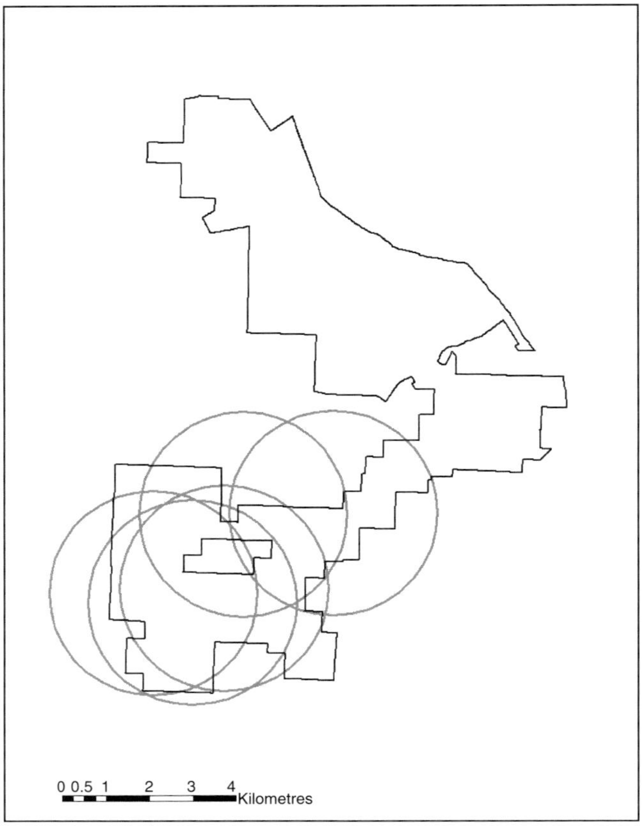

Fig. 15.16. Home range circles associated with spotted owls between 1994 and 2004.

Other efforts

The College Forests has also used the stand-level inventory data for two other large planning efforts. A non-reserve-based landscape plan was developed for the College's Blodgett Forest in the North Coast of Oregon, which created mature forest habitat, using active management, while also generating revenue for teaching and research (Sessions *et al.*, 2000). The spatially explicit harvest schedule for this plan was built around a complex set of yields and spatial habitat goals.

Secondly, a plan was built for the south zone of McDonald–Dunn Forest around a replicated set of treatments, designed to test different types of uneven-aged management. The spatial harvest scheduling model developed for this effort allowed decision makers to evaluate the effects of different methods of implementing the plan on spotted owl habitat and revenue (Bettinger *et al.*, 2003).

Contributions of the inventory to planning and implementation

The stand-level inventory linked to the ORGANON growth-and-yield model has served as an invaluable data source for managing the College of Forestry's research properties. Over the past decade, these tools have been critical for:

- Calculating and displaying a harvest schedule.
- Evaluating the current and future conditions of wildlife habitat.
- Providing credible growth-and-yield estimates for forest decision makers and forest users.
- Quickly evaluating changes in harvest volume due to modifications of the harvest schedule.
- Dampening the 'we are out of places to cut' reaction that can affect plan implementation.

The habitat model that we built for northern spotted owls, using the stand-level inventory, not only helped us understand where and what we could harvest under our old plan; it was also very important for constraining harvest levels under our new plan.

One of the most important beneficiaries of the inventory system and the continuing evolution of the forest plans has been students in the college, who have had the opportunity to routinely visit this forest, study the simulations and plans and then design their own solutions to real-life problems.

References

Bettinger, P., Johnson, D.L. and Johnson, K.N. (2003) Spatial forest plan development with ecological and economic goals. *Ecological Modelling* 169, 215–236.

Hann, D.W. (2003) *ORGANON User's Manual*, 7th edn. Department of Forest Resources, Oregon State University, Corvallis, Oregon.

Hann, D.W., Marshall, D.D. and Hanus, M.L. (2003) *Equations for Predicting Height-to-crown-base, 5-Year-diameter-growth Rate, 5-Year-height-growth Rate, 5-Year-mortality Rate, and Maximum Size–Density Trajectory for Douglas-fir and Western Hemlock in the Coastal Region of the Pacific Northwest*. Research Contribution 40, Oregon State University, Forest Research Laboratory, Corvallis, Oregon.

Johnson, K.N., Stuart, T. and Crim, S. (1986) *FORPLAN (Version II) – an Overview*. USDA Forest Service, Land Management Planning Systems Section, Washington, DC.

King, J.E. (1966) *Site Index Curves for Douglas-fir in the Pacific Northwest.* Weyerhaeuser Forestry Paper 8, Weyerhaeuser, Centralia, Washington State.

Mackey, H. (2004) *The Kalapuyans: A Sourcebook on the Indians of the Willamette Valley*. Mission Mill Museum Association, Salem, Oregon, and the Confederated Tribes of Grand Ronde, Grand Ronde, Oregon.

Marshall, D.D., Johnson, D.L. and Hann, D.W. (1997) A forest management information system for education, research and operations. *Journal of Forestry* 95(10), 27–30.

Munro, D.D. (1974) Forest growth models – a prognosis. In: Fries, J. (ed.) *Growth Models for Tree and Stand Simulation.* Research Note 30, Royal College of Forestry, Stockholm, Sweden, pp. 7–21.

Sessions, J., Johnson, D., Ross, J. and Sharer, B. (2000) The Blodgett plan: an active-management approach to developing mature forest habitat. *Journal of Forestry* 98(12), 29–33.

Thomas, J.W., Forsman, E.D., Lint, J.B., Meslow, E.C., Noon, B.R. and Verner, J. (1990) *A Conservation Strategy for the Northern Spotted Owl: a Report of the Interagency Scientific Committee to Address the Conservation of the Northern Spotted Owl.* USDA Forest Service, USDI Bureau of Land Management, Fish and Wildlife Service, and National Park Service, Portland, Oregon.

Zumrawi, A.A. and Hann, D.W. (1989) *Equations for Predicting Height to Crown Base for Six Tree Species in the Central Western Willamette Valley of Oregon.* Research Paper 52, Oregon State University, Forest Research Laboratory, Corvallis, Oregon.

Zumrawi, A.A. and Hann, D.W. (1993) *Diameter Growth Equations for Douglas-fir and Grand Fir in the Western Willamette Valley of Oregon.* Research Contribution 4, Oregon State University, Forest Research Laboratory, Corvallis, Oregon.

16 Forest Land Change Assessment by Continuous Inventory

P. CORONA,[1] E. POMPEI[2] AND G. SCARASCIA MUGNOZZA[1]

[1]*Dipartimento di Scienze dell'Ambiente Forestale e delle Sue Risorse, University of Tuscia, Viterbo, Italy;* [2]*National Forest Service, Ministry of Agricultural and Forest Policies, Rome, Italy.*

Abstract

Forest expansion is one of the main factors characterizing the landscape dynamics in many European countries, and inventory and management of the areas recolonized by forests are focal issues for sustainable forestry. However, reliable updated information on such a process is often not available for wide territories (e.g. on a national or regional scale). This chapter shows a probabilistic estimation approach based on land-use classification, repeated on the same sampling points for two successive occasions. The experimental approach was tested in the Abruzzo region (a region bigger than 1 million ha in central Italy), where the rate of forest expansion was assessed by multi-temporal classification of permanent sampling points on ortho-corrected aerial photographs. The forest area in the year 1990, a figure of great interest under the Kyoto protocol, was also assessed. An unaligned systematic sampling design was adopted, with a grid of 1 km^2. The results show a pronounced increase in the cover of forests and other wooded land in recent decades, at an average annual rate of 0.23% with respect to the total land area, and corresponding relevant losses of grassland and cropland. The estimation of forest area in the year 1990 was affected by a standard error around 1%. The assessment procedure was relatively easy to implement and is repeatable. The adopted set of estimators are straightforwardly applicable, and in the examined case has proved to be rather efficient.

Introduction

Forest expansion is one of the major factors currently characterizing the landscape dynamics in many European countries (MCPFE, 2003), and also in many other temperate countries of the northern hemisphere (FAO, 2003). Such a phenomenon is mainly due to forest recolonization on abandoned farmland. It may have a critical (either positive or negative) role for biodiversity and landscape conservation and enhancement, depending upon local environmental conditions and social perception, while it always implies an increase in atmospheric

 Sustainable Forestry: from Monitoring and Modelling to Knowledge Management and Policy Science (eds K.M. Reynolds, A.J. Thomson, M. Köhl, M.A. Shannon, D. Ray and K. Rennolls)

carbon sequestration potential. In such a perspective, inventory and management of the areas recolonized by forests are focal issues for sustainable forestry. However, reliable updated statistical data on forest expansion are often not available for large areas (e.g. on a national or regional scale), and this is distinctively true for Italy.

The aim of this chapter is to present and discuss, as a case study, a probabilistic estimation approach to land-use classification repeated on the same sampling units for two successive occasions. The experimental approach was tested in the Abruzzo region, a main administrative region in central Italy (over 1 million ha – extending from the Apennine mountain range down to the Adriatic Sea (Fig. 16.1)), where the rate of forest expansion was assessed by multi-temporal classification of permanent sampling points on ortho-corrected aerial photos.

Developing specific estimators, the approach was also used to assess the forest area in 1990: this figure is of great interest in view of reporting requirements by the Kyoto Protocol (United Nations Framework Convention on Climate Changes, UN-FCCC), since 1990 is taken as the base year for assessing reductions and emissions of greenhouse gases due to afforestation, reforestation, deforestation and revegetation during the first committment period (2008–2012). Afforestation includes the natural expansion of forests, provided that it can be demonstrated that such an expansion is driven by targeted agricultural/environmental planning measures (e.g., Common Agricultural Policy measures by the European Union).

Materials and Methods

The sampling design in the time domain is a pure panel in which the same sampling units are observed at every point in time (Fuller, 1999). The number and geographical locations of the sampling points were the same as those used for the first phase of the current Italian National Forest Inventory (INFC): the territory of the Abruzzo region was divided into adjacent 1 km^2 squares and a sampling point was randomly chosen within each square (Fig. 16.1). Fattorini *et al.* (2004) have proved the superiority of such a design, called unaligned systematic sampling design, over the more common aligned systematic sampling.

Ortho-corrected panchromatic aerial photos, taken in 2002 and during the 1980s, were used for the multi-temporal assessment of the forest area. Both series of orthophotos have 1 m^2 pixel resolution (nominal scale: 1 : 10000). For the year 2002, 322 digital orthophotos were available fully covering all the examined region, while three different non-overlapping areas of the region were covered by 185, 86 and 61 orthophotos taken, respectively, in the years 1981, 1985 and 1987.

The sampling points were classified into six land-use categories according to a classification system similar to that officially adopted for the first phase of INFC (2003) (Table 16.1, Fig. 16.2). Due to the good quality (brightness and contrast) of both the series of orthophotos, no problems were encountered in sampling point classification (Figs 16.3 and 16.4): such a classification was taken as the ground truth.

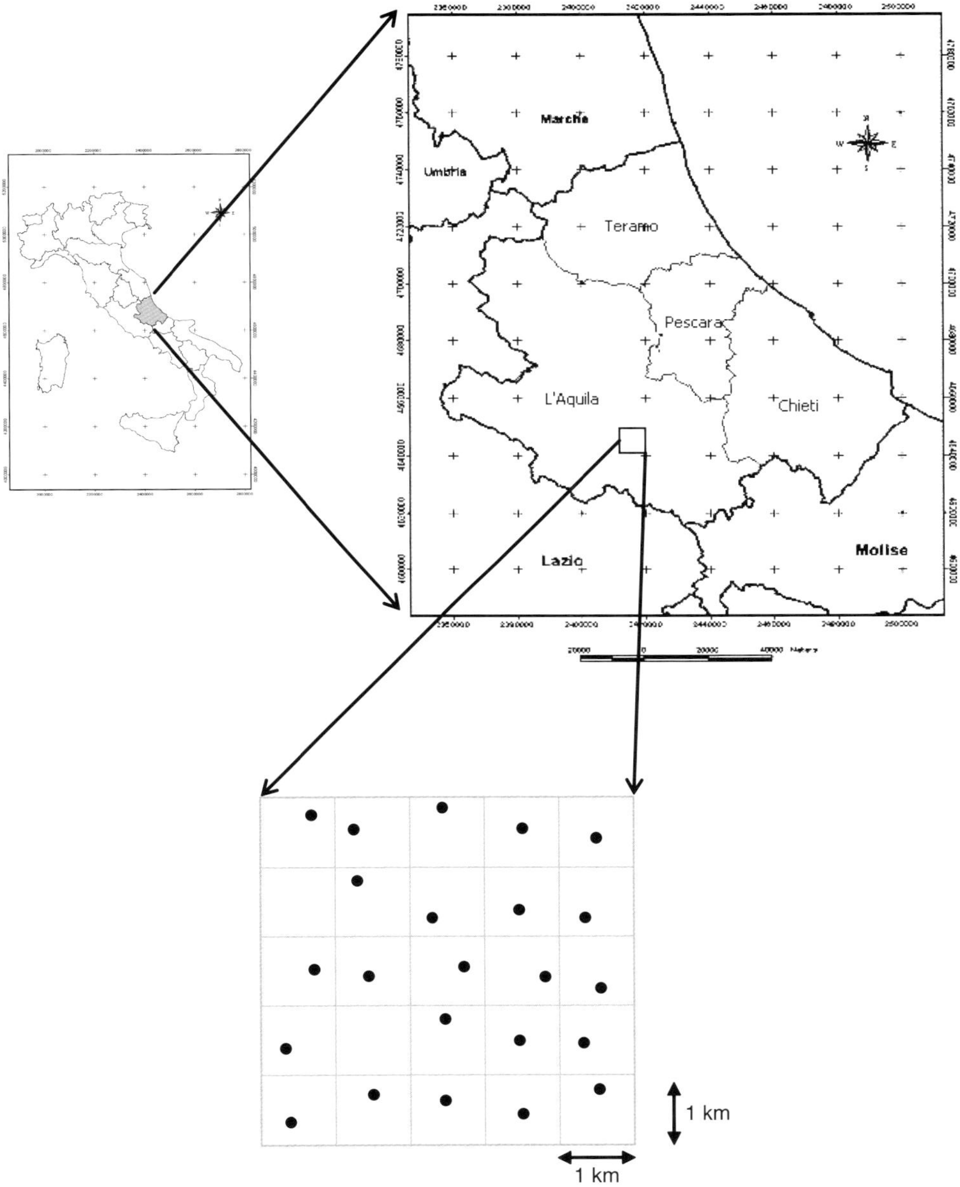

Fig. 16.1. Geographical location of the Abruzzo region and excerpt of the unaligned systematic sampling design adopted for allocating the sampling points.

Table 16.1. Land-use nomenclature used in the study. Names of land categories are a mixture of land-cover (e.g. forest, grassland) and land-use (e.g. cropland, settlements) classes, following IPCC (2003). The definition of 'forest' is analogous to that of the Forest Resource Assessment of UN-FAO (UN-ECE/FAO, 1997), except for the parameter 'potential height at maturity' which is so far undetectable by remote sensing.

Land category	Definition
Settlements	Built-up areas, including urban outskirts, villages and rural agglomerations, with a minimum size of 5000 m^2; industrial, handicraft and commercial settlements, big plants of public and private services with annexed artificial spaces, quarries, dumps, transportation infrastructures; roads and railways larger than 20 m are also included
Cropland	Cultivated areas, with a minimum size of 5000 m^2: arable and tillage land, dry or irrigated land, nurseries (not forest nurseries), horticulture cultivations, greenhouse or plastic tunnel cultivations, vineyards, olive groves and all cultivated arboreal and shrubbery species that bear fruit (even hazel groves), and permanent cultivated meadows are included
Forest	Land covered by forest trees and characterized by: minimum area of 5000 m^2, crown cover higher than 10%, area width larger than 20 m
Other wooded land	Land covered by forest trees and characterized by: minimum area of 5000 m^2, crown cover between 5% and 10%, area width larger than 20 m
Grassland	Rangeland, pastures, fallow areas and all the natural vegetation land covered by herbs (with a minimum coverage higher than 40%), with a minimum area of 5000 m^2
Other land	All other land not included in the above-mentioned categories

Estimating the Annual Area Change of a Given Land-use Category

The area change of the *i*-th land-use category between the first and the second forest inventories can be estimated as:

$$\hat{Z}_i = A\hat{z}_i$$

where

$\hat{z}_i = \hat{p}_{i2} - \hat{p}_{i1}$;

$\hat{p}_{i2}$ = estimated proportion of sampling points of the *i*th land-use category at the time of the second inventory;
$\hat{p}_{i1}$ = estimated proportion of sampling points of the *i*th land-use category at the time of the first inventory;
A = land area (known without error).

(A)

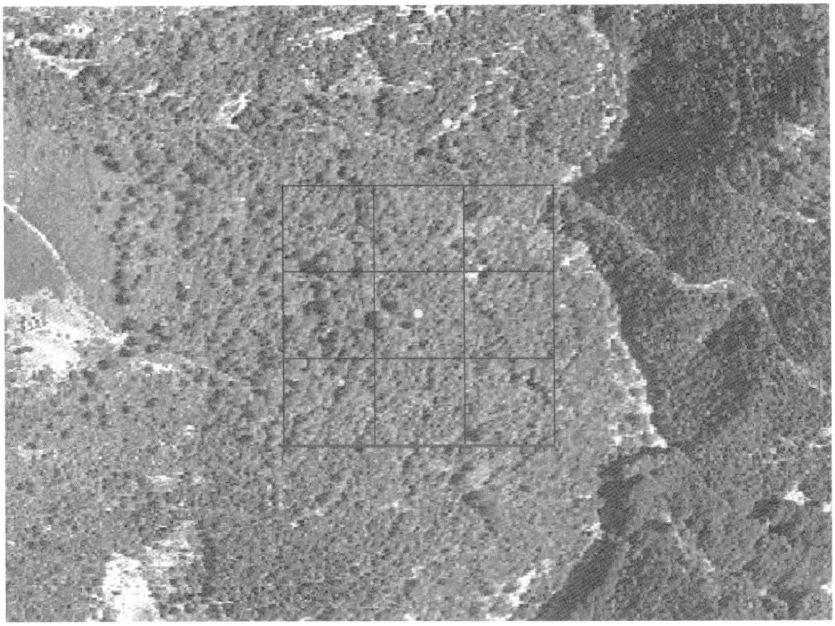

(B)

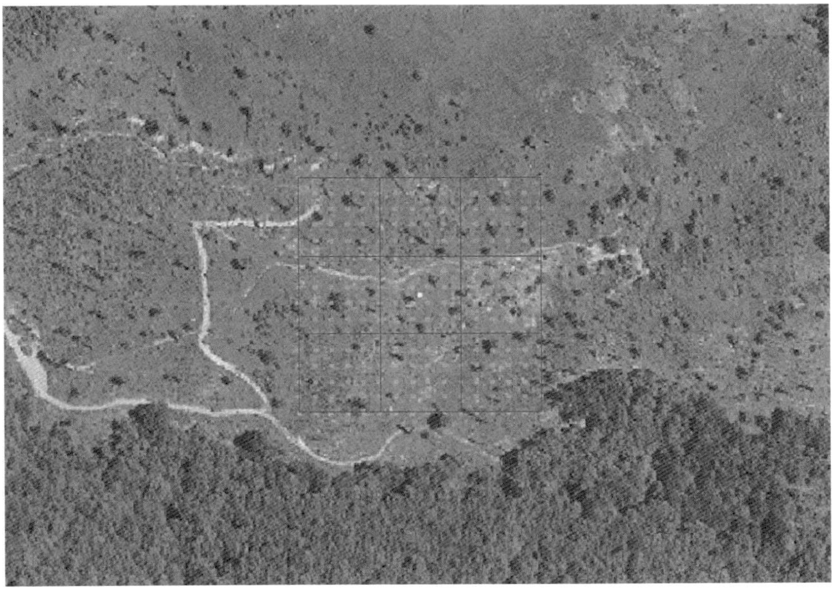

Fig. 16.2. Example of sampling photo-point classification: (A) forest, (B) other wooded land (from INFC, 2003).

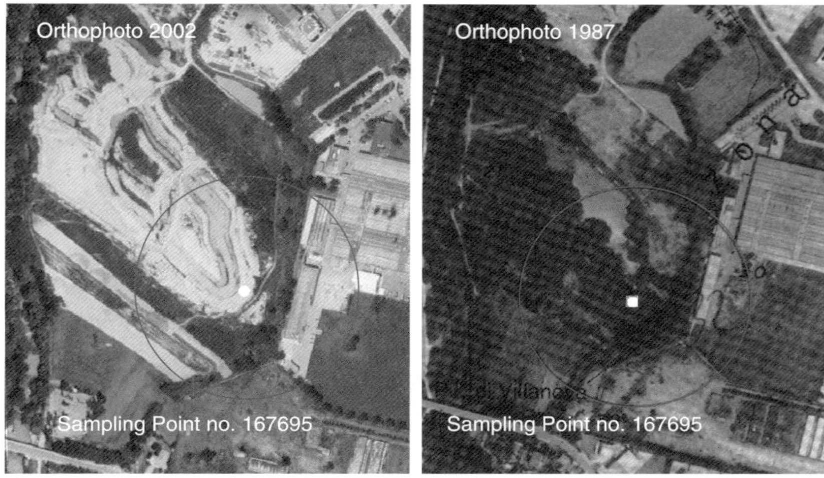

Fig. 16.3. Example of land-use change detected by multi-temporal orthophoto analysis.

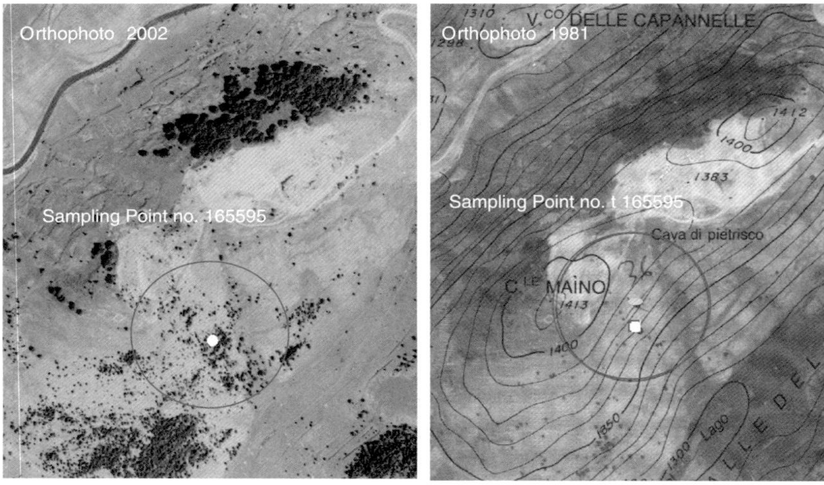

Fig. 16.4. Examples of land-use change detected by multi-temporal orthophoto analysis.

The estimated standard error of $\hat{Z}_i$ is:

$$\hat{s}_{\hat{Z}_i} = A\hat{s}_{\hat{Z}_i}$$

where:

$$\hat{s}_{\hat{Z}_i} = \sqrt{\hat{s}^2_{\hat{p}_{i2}} + \hat{s}^2_{\hat{p}_{i1}} - 2\hat{s}_{\hat{p}_{i2}\hat{p}_{i1}}}$$

$$\hat{s}^2_{\hat{p}_{i2}} = \frac{\hat{p}_{i2}(1-\hat{p}_{i2})}{n-1} = \text{conservative estimator of the variance of } \hat{p}_{i2};$$

$$\hat{s}^2_{\hat{p}_{i1}} = \frac{\hat{p}_{i1}(1-\hat{p}_{i1})}{n-1} = \text{conservative estimator of the variance of } \hat{p}_{i1};$$

$$\hat{s}_{p_{i2}p_{i1}} = \frac{\hat{p}_{i1=2} - \hat{p}_{i2}\hat{p}_{i1}}{n-1} = \text{conservative estimator of the covariance } \hat{p}_{i2} \text{ and } \hat{p}_{i1};$$

$\hat{p}_{i1=2}$ = proportion of the sampling points classified in the *i*th land-use category at both the first and second inventories.
n = number of sampling points.

The average annual area change of the *i*th land-use category can be estimated by:

$$\hat{\upsilon}_i = \hat{Z}_i / l$$

where l = number of years between the first and second inventory occasion. The estimator of the standard error of $\hat{\upsilon}_i$ is:

$$\hat{s}_{\upsilon_i} = \hat{s}_{\hat{Z}_i} / l$$

Estimating the Area of a Given Land-use Category at a Year Between Two Successive Occasions

The area of the *i*th land-use category at an intermediate year between the two successive inventory occasions, for example, the base year 1990, can be straightforwardly estimated assuming that the annual area change between the two occasions is constant.

Thus, the estimator $\hat{A}_{i(2-x)}$ of the area of the *i*th land-use category *x* years before the second occasion is:

$$\hat{A}_{i(2-x)} = \hat{A}_{i2} - x\hat{\upsilon}_i = \left(1 - \frac{x}{l}\right)\hat{A}_{i2} + \frac{x}{l}\hat{A}_{i1}$$

where

$\hat{A}_{i2} = \hat{p}_{i2} A$ = estimated area of the *i*th land-use category at the second inventory occasion;

$\hat{A}_{i1} = \hat{p}_{i1} A$ = estimated area of the *i*th land-use category at the first inventory occasion.

The estimator of the standard error of $\hat{A}_{i(2-x)}$ is:

$$\hat{S}_{\hat{A}_{i(2-x)}} = \sqrt{\left(1 - \frac{x}{l}\right)^2 \hat{S}^2_{\hat{A}_{i2}} + \left(\frac{x}{l}\right)^2 \hat{S}^2_{\hat{A}_{i1}} + 2\frac{lx - x^2}{l^2}\hat{S}_{\hat{A}_{i2}\hat{A}_{i1}}}$$

where:

$$\hat{S}^2_{\hat{A}_{i2}} = A^2 \hat{S}^2_{p_{i2}} = \text{estimated variance of } \hat{A}_{i2};$$

$$\hat{S}^2_{\hat{A}_{i1}} = A^2 \hat{S}^2_{p_{i1}} = \text{estimated variance of } \hat{A}_{i1};$$

$$\hat{S}_{\hat{A}_{i2}\hat{A}_{i1}} = A^2 \hat{S}_{p_{i2}p_{i1}} = \text{estimated covariance of } \hat{A}_{i2} \text{ and } \hat{A}_{i1}.$$

Results

In the whole Abruzzo region, the average annual area change of the observed land-use categories is estimated to be (confidence intervals at 0.05 probability level are reported in parentheses):

settlements: + 323 (± 83) ha/year;

cropland: – 720 (± 124) ha/year;

forest: + 2439 (± 219) ha/year;

other wooded land: – 1053 (± 186) ha/year;

grassland: – 989 (± 144) ha/year;

other land: 0 (± 0) ha/year.

The multi-temporal assessment carried out shows significant changes in the landscape mosaic. In particular, the development of the social and economic conditions and of industrial and commercial activities near the urban centres and along the coast has led to a migration of the population away from rural areas and to a related forest expansion due to the progressive abandonment of farmland: such a trend is emphasized in the interior mountain areas, and even in the hilly zone (Fig. 16.5). At the same time, artificial areas, such as those occupied by urban settlements and infrastructures, have enlarged, mainly to the detriment of cropland.

The land-use category 'other land' did not show any change during the assessment period due to the types of landscape elements included within this category (rocky peaks in the highest mountains, wetland, lakes and rivers).

The relevant increase of forest land was mainly due to the transition from other wooded land, and secondly, from rangeland and farmland (Table 16.2). However, even leaving out the internal changes between other wooded land and forest land, a significant increase of wooded areas (forest + other wooded land) can still be evidenced: 1387 (± 166) ha/year.

The forest area of the Abruzzo region estimated in the base year 1990 was 403,426 ha (± 9295 ha), while the area of other wooded land covered 38,635 ha (± 3242 ha). Overall, in 1990, wooded areas amounted to 442,061 ha (± 9844 ha).

Discussion and Conclusion

Forest expansion is the most relevant and important factor driving landscape dynamics in the study region of Abruzzo. During the last two decades forest area has increased at an average annual rate of 0.23% and 0.60%, if calculated with

Fig. 16.5. Changed sampling points (classified as non-forest in the first inventory analysis and as forest in the second inventory analysis) within the Abruzzo region.

Table 16.2. Land-use change matrix (expressed in terms of proportions) for the Abruzzo region between the first (initial) and the second (final) inventory occasions.

	Initial land-use						
Final land-use	Settlements	Cropland	Forest	Other wooded land	Grassland	Other land	Total
Settlements	0.030524	0.004810	0.000370	0	0.000277	0	0.035981
Cropland	0	0.378966	0.000277	0	0	0	0.379243
Forest	0	0.007030	0.355472	0.024420	0.013320	0	0.400242
Other wooded land	0	0.000277	0.000000	0.018223	0.005550	0	0.02405
Grassland	0	0.000647	0.000185	0	0.138470	0	0.139302
Other land	0	0	0	0	0	0.021182	0.021182
Total	0.030524	0.391730	0.356304	0.042643	0.157617	0.021182	1

respect to the total land area of the region or to the forest area in 1990. Such a significant trend has a direct impact on the role of the forests, and connected social expectations, especially in terms of the functions of landscape management, nature conservation, recreation activities and global climatic changes.

According to the results, the multi-temporal assessment procedure tested has proved to be relatively easy to implement, and the estimation functions as well as the estimators of the corresponding variances are straightforwardly applicable. Moreover, they have turned out to be satisfyingly efficient, considering the adopted sampling intensity (1 sampling point per km^2, the same as in the first phase of the current National Forest Inventory, INFC). The case study tested a repeated photo-point classification on ortho-corrected aerial panchromatic images over a territory of 1,081,070 ha where forest covers around 40% of the land. The sampling effort corresponded to 200 man-days (at an approximate cost of 30,000 euros, according to standard labour costs in Italy) and provided a standard error below 5% for the estimation of forest annual expansion and a standard error around 1% for the estimation of forest area in the year 1990.

This latter estimate is a reliable assessment of the 1990 forest inventory baseline, which complies with the Kyoto Protocol reporting requirements.

Methods for the *ex post* assessment of forest area have to be based on historical remotely sensed imagery. It is imperative that the images have an appropriate spatial resolution to grasp the definition of 'forest'. For instance, the Kyoto Protocol definition involves a minimum forest size not larger than 1 ha, and units as small as 1 ha can be identified by sensor systems with a spatial resolution not lower than 30 m (IPCC, 2003). In Italy, minimum forest size is 0.5 ha (i.e. the same size threshold as defined by the Forest Resource Assessment of UN-FAO; see Table 16.1). Consequently, remotely sensed images must have a geometric resolution higher than 7–8 m, as in the present study.

Two main approaches can be adopted to assess the total area of each land-use category: the considered territory can be completely mapped by polygon delineation and classification (wall-to-wall mapping), so that such areas can be straightforwardly calculated by summing up the area of each polygon belonging to a given land-use category; otherwise, the estimation of such areas can be carried out by sampling and classification of geographically located points, as reported in this study. A combination of sampling and mapping is also possible (e.g. area frame sampling; see Gallego, 1995).

In the case of large areas, point sampling is faster and cheaper and, above all, tends to provide more accurate results than mapping. In fact, there are important risks in using land maps produced through interpretation of satellite images or aerial photographs as a direct tool to estimate spatial variables. When mapping, the interpretation errors tend to be systematic and, in general, there is no compensation between commission and omission errors (i.e. the areas of a land-use type A mapped as land-use type B are not compensated by the areas of land-use type B mapped as A). The outcome is that the sum of sampling and non-sampling errors in sample surveys for land statistics estimation is very often smaller than non-sampling errors by mapping for land statistics (Carfagna and Gallego, 1999). This is particularly true for forest area change estimation, as described here, which is characterized by a pronounced change from cropland

and grassland to forest and, above all, from other wooded land to forest. In this type of analysis the classification is differentiated with respect to the threshold of forest tree canopy cover (non-forest $< 5\%$, other wooded land $\geq 5\%$, forest $> 10\%$): omission and commission errors by mapping with continuous delineation and classification of polygons having forest tree canopy cover less or greater than the 5% or 10% threshold are usually much more relevant than those made by the easier and operatively more objective classification of sampling points.

To conclude, the assessment of forest dynamics across large areas (and the associated carbon figures) obtained by a probabilistic estimation like that tested (land-use classification by permanent sampling points), is a viable approach, more easily and more objectively repeatable than conventional repeated mapping. This is most relevant from a practical consideration, since it supports the feasibility of the methodological approach illustrated in the revised Good Practice Guidance by the Intergovernmental Panel on Climate Change (IPCC, 2003). The possibility of deriving uncertainty estimates based on formal statistical procedures is an important advantage of this sampling procedure in comparison to other forest area assessment methods. The reliability of the information produced can be quantitatively assessed from the data, and this can be very useful when estimating the amount (and error) of change in the carbon stock of forests.

Acknowledgements

This chapter was carried out by the Authors in equal parts. The Authors would like to thank the Italian National Forest Service (Corpo Forestale dello Stato) for supporting the PhD curriculum of Enrico Pompei and for providing basic data from the current Italian National Forest Inventory and technical support. The work was partially carried out under a grant provided by the CARB-IUS (Italy–USA) bilateral project (Research Unit: University of Tuscia, Italy).

References

Carfagna, E. and Gallego, F.J. (1999) Thematic maps and statistics. In: *Land Cover and Land Use Information Systems for European Union Policy Needs*. Office for Official Publications of the European Communities, Luxembourg, pp. 219–228.

FAO (2003) *State of World's forests 2003*. Food and Agriculture Organization of the United Nations, Rome, Italy.

Fattorini, L., Marcheselli, M. and Pisani, C. (2004) Two-phase estimation of coverages with second-phase corrections. *Environmetrics* 14, 1–12.

Fuller, W.A. (1999) Environmental surveys over time. *Journal of Agricultural, Biological and Environmental Statistics* 4, 331–345.

Gallego, F.J. (1995) *Sampling Frames of Square Segments*. EUR 16317, Office for Official Publications of the European Communities, Luxembourg.

INFC (2003) *Manuale di fotointerpretazione per la classificazione delle unità di campionamento di prima fase. Inventario Nazionale delle Foreste e dei Serbatoi Forestali di Carbonio*. MiPAF – Direzione Generale per le Risorse Forestali Montane e Idriche, Corpo Forestale dello Stato. Documento a cura dell'Istituto Sperimentale per l'Assestamento Forestale e per l'Alpicoltura, Trento, Italy.

IPCC (2003) *Good Practice Guidance for Land Use, Land-Use Change and Forestry*. Intergovernmental Panel on Climate Change, National Greenhouse Gas Inventories Programme, Institute for Global Environmental Strategies, Kanagawa, Japan.

MCPFE (2003) *State of Europe's Forests 2003*. Ministerial Conference on the Protection of Forests in Europe, Liaison Unit Vienna, Vienna, Austria.

UN-ECE/FAO (1997) *UN-ECE/FAO Temperate and Boreal Forest Resources Assessment 2000. Terms and Definitions*. United Nations, Geneva, Switzerland.

17 A Study on Tree Colonization of Abandoned Land in the Italian Alps: Extent and Some Characteristics of New Forest Stands in Trentino

F. DE NATALE,[1] P. GASPARINI[1] AND A. CARRIERO[2]

[1]*Forest and Range Management Research Institute, ISAFA – CRA p.zza Nicolini 6, 38100, Trento, Italy;* [2]*Forest and Wildlife Service, Provincial Council of Trento, via Trener 3, 38100 Trento, Italy.*

Abstract

This chapter reports on a study concerning the tree colonization on abandoned land in Trentino, a region of 600,000 ha in the eastern Alps of Italy. The study refers to a period of about 30 years and was carried out through a two-stage sampling design. In the first stage some area units were selected according to a probability depending on the value of the auxiliary variable 'forest edge length'. In the second stage a set of sample points were randomly distributed over each drawn area unit and surveyed. The total extent of new forest area was estimated by comparing the forest–non-forest classification of the sample points on two orthophotos, respectively of years 1973 and 1999. Sample points intercepting new stand patches were surveyed further in the field and on orthophotos to analyse ecological and landscape features. Details on sampling method and surveys are given in the first part of the chapter. In the second part, the report refers to some first results of data collection focusing on factors characterizing the distribution of tree colonization and on the effects of land abandonment on the landscape pattern.

Introduction

During the last century, the socio-economic conditions of European mountain areas have completely changed, particularly in locations far from industrial areas. The change is a consequence of the migration of workers towards more densely inhabited valleys and close-to-mountains lowland areas, where there are better income opportunities. As a result the population of many small mountain villages has been declining and ageing. Actually, this process began during the first part of the 19th century, increased during the second half of the 19th century through the first three decades of the 20th century, and intensified during the

 Sustainable Forestry: from Monitoring and Modelling to Knowledge Management and Policy Science (eds K.M. Reynolds, A.J. Thomson, M. Köhl, M.A. Shannon, D. Ray and K. Rennolls)

1950s and 1960s (Piussi, 2000). In the Alps, despite an increase of population from 7 million up to 11 million in the period 1870–1990, the proportion of people living at higher altitudes and in remote valleys has generally decreased (EEA, 1999). This trend has led to a different spatial distribution of the population and to a change in the relationship between the population distribution and the areas of forest. Up to the middle of the 19th century, an inverse relationship was observed: an increase in population corresponded to a decrease in forest area. However, in the last century, both forest area and population have continued to increase, as described for France by Mather *et al.* (1999) and for Switzerland by Mather and Fairbairn (2000). Actually, the latter authors have observed that in Switzerland an inverse relationship between forest cover area and population size still exists, but it has to be referred to the agricultural population. Besides the migration from mountains to valleys, a shift in mountain workers from the primary to the tertiary sector has affected the Alpine region in the last decades, thanks to the development of mountain tourism and outdoor recreation.

As a consequence of demographic and economic changes, the traditional agricultural practices have been progressively abandoned and significant land-use changes have occurred. Extensive agriculture, pastoralism and dairy farming have been drastically reduced and forests have expanded on abandoned land. MacDonald *et al.* (2000) reported that this agricultural decline has had a particularly high environmental impact in mountain areas, where the semi-natural habitats, originated by low-intensity farming, support high species richness. Besides the loss of the traditional cultural landscape, the results of this process are a reduction of human-made landscape diversity and the loss of open habitats such as meadows, pastures and glades.

In the mountain areas, forest expansion seems to have started on the poorest sites near the treeline. Tree colonization is often associated with a reduction in grazing or browsing pressure (French *et al.*, 1997; Piussi, 2000; Motta and Nola, 2001). However, an analysis of the causes of forest expansion at the treeline is fairly difficult, because of the unknown impact of climate change (Didier, 2001; Motta and Nola, 2001). As described by Piussi (2000), reforestation depends on the interaction of basic factors such as the availability of propagules, the characteristics of dissemination, the competition between tree seedlings and herbaceous vegetation and the former land use. Other authors have also remarked that the distance from the forest boundary is an important factor (French *et al.*, 1997; Pelleri and Sulli, 1997; Kettle *et al.*, 2000; Endress and Chinea, 2001). The consequences of forest expansion are also dependent on the extent of remaining rural areas and on the impact of industrialization and tourism in neighbouring mountain areas. Indeed, the threat to cultural landscapes and natural and semi-natural habitats created by traditional human activities has become more severe as demand for land in the mountain valleys has dramatically increased because of the recent expansion of urban areas and the development of transport infrastructures and recreation facilities.

In the Italian Alps, the decline of populations of several species, such as black grouse (*Tetrao tetrix*, L.), rock partridge – Alpine subspecies (*Alectoris graeca saxatilis*), corncrake (*Crex crex*) and brown hare (*Lepus europaeaus*), was caused by the abandonment of traditional agriculture and a reduction of semi-natural habitats maintained by traditional agricultural practices (Odasso *et al.*, 2002). For the same area, most species included in the red lists of endangered species live in non-forested habitats derived from traditional Alpine land-use systems (Provincia Autonoma di Bolzano, 1994; Prosser, 2001). The effects of land-use change on Alpine ecosystems were recently analysed by the research project Ecological Effects of Land-use Changes on European Terrestrial Mountain Ecosystems (ECOMONT) funded by the European Union within the IV Framework Programme (Cernusca *et al.*, 1999). However, depending on new forest types, local conditions and existing biotopes, changes may positively or negatively affect species diversity and landscape attractiveness (Hunziker, 1995; EEA, 1999; Höchtl *et al.*, 2005). Some studies in the Italian Alps have shown that the loss of semi-natural habitats, caused by abandonment, has led to a decrease in floristic diversity (Tasser and Tappeiner, 2002), while at the landscape scale a higher ecosystem dynamic has been observed (Höchtl *et al.*, 2005).

From a different point of view, the expansion of forest area over large territories in the temperate regions can be seen as a positive process to help counter forest destruction in other parts of the world and to sequester atmospheric carbon. With this in mind, the scientific community has developed an increasing interest in the forest area expansion process in recent years as a consequence of the rising importance of land-use change in relation to the carbon cycle and the Kyoto Protocol. In many countries, methods are being tested to quantify the rate of forest expansion with reference to the 1990 baseline, using remotely sensed data. On the other hand, historical forest inventory data are frequently compared and different sampling techniques are used in order to produce reliable estimates of afforestation and reforestation processes (Brassel and Brändli, 1999; Corona *et al.*, 2005).

In 2002, the Forest and Wildlife Service of the province of Trento (Italy) launched the research project 'Monitoring secondary forests', aimed at deriving information to promote the sustainable management of new forest stands and to direct tree colonization dynamics through management practices. The project involved three partners: the University of Padua for the development of a typology of secondary forest plant communities; the centre for Scientific and Technological Research (ITC-IRST, Trento) for mapping new forest stands by automatic classification of ortho-rectified aerial photographs; and the Forest and Range Management Research Institute (ISAFA-CRA, Trento) for the assessment and description of new forests through the collection of sample data. This chapter reports on methods used by the third group for the inventory of secondary forests, and presents some results of the data collection, focusing on the estimate of the new forest extent and of its distribution according to the main environmental factors observed.

Study Area and Period

The province of Trento, or Trentino, is a mountainous region located in the north-east of Italy, in the Alpine region (Fig. 17.1). The area covers about 600,000 ha, of which forests and shrubs cover more than 60% of the area.

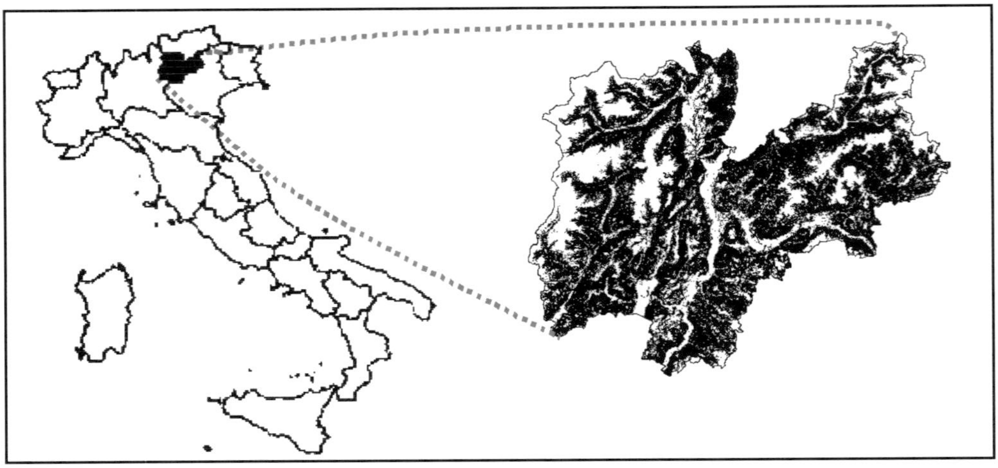

Fig. 17.1. Location and forest cover (shaded area of inset map) of the study area.

Tourism, agriculture (fruit orchards and vineyards) and forestry are the main resources of this region. Most human activities are concentrated in valley flats, where the main urban and industrial areas are located.

The study was based on the comparison of two different years, 1973 and 1999, to assess changes of forest cover during this period of nearly 30 years. The first year 1973 was chosen because it was in this year that complete high-quality aerial photograph coverage was available. The high-resolution photographs were necessary to ensure good detection of the phenomenon, which occurs sparsely and in small patches. In any case, a study period of almost 30 years was considered long enough to identify significant land-use changes. Moreover the observed period is interesting in relation to major demographic changes: the total population increased by 10% – from 433,600 in 1973 to 477,859 in 2000 (ISTAT, 1973, 1999) – but, as reported by the local statistics (Provincia Autonoma di Trento, Servizio Statistica), the employees in agriculture decreased by more than 50% (from 21,378 in 1971 to 10,223 in 2001).

Material and Methods

Secondary forest assessment in the study area was performed by comparing the forest–non-forest classification at sample points observed by photo-interpretation for the two years 1973 and 1999. Forest was defined using the following criteria: minimum area of 1000 m^2, minimum width of 10 m, canopy cover of at least 20% and minimum height of 2 m. The forest definition used has been adopted by the province of Trento and differs from the national definition of the second National Forest Inventory. Finally, a secondary forest is defined as a stand established after 1973 on land that had, at that time, a different land use.

Data

For the year 1999, digital, colour, ortho-rectified, aerial photographs (orthophotos) produced by Compagnia Generale Ripreseaeree (CGR) were used. These had been taken originally at an average scale of 1 : 40,000, and were re-sampled at the nominal scale 1 : 10,000 with a ground resolution of 1 m. The data for 1973 came from digital black and white aerial photos, which were ortho-rectified by the project partner ITC-IRST, with a Root Mean Square Error (RMSE) of 6 m. All 1973 photos were scanned at a resolution of 1 m to be compatible with the 1999 orthophotos. A land-use map (LUM) derived from photo-interpretation of the 1999 colour orthophotos and a 40 m resolution digital elevation model (DEM) covering the whole area were used both to improve the classification and to collect data for the sampling units. All materials used were supplied by the Forest and Wildlife Service of Trento. The analysis of the geographical data was carried out in ArcView 3.2.

Sampling design

The sampling scheme was designed to address the issue of ensuring the randomness of point selection and, at the same time, minimize the number of 1973 photos used. For 1999, a complete orthophoto cover of the study area was available, while, for the year 1973, the aerial photos had to be ortho-rectified to be overlaid on the 1999 data. In addition, a number of secondary forest points large enough to obtain reliable information from ground surveys had to be identified by photo-interpretation.

As a consequence of these requirements, an area sampling procedure with a two-stage design was adopted. The first stage involved choosing areas, and the second stage involved selecting sample points. However, area sampling might be unsuitable for studying rare phenomena (Fabbris, 1989), such as tree colonization on abandoned land. In order to ensure a more efficient sample, an auxiliary variable was chosen for which a correlation with the studied phenomenon was very likely. Since distance to the forest edge is an important factor affecting tree colonization, the auxilary variable used for the sampling area selection was the length of the forest edge, derived from the digital land-use map of the province of Trento.

For the first stage of sampling, the whole study area was subdivided into n square units (n = 472) with side length of 4000 m (primary sampling units or PSUs). Each PSU was assigned a weight proportional to the total length of forest edges intercepted by its area. To determine the sample size, a pilot study was carried out in two areas (the first in the north-west and the second in the south-east of the province) to derive a preliminary assessment of the extent of tree colonization. Finally, 100 PSUs were randomly selected (with replacement), with probability of selection proportional to the weight. Figure 17.2 shows the distribution of PSUs and the forest/non-forest boundaries. Altogether, 89 different square units were drawn, with 11 units selected twice. Formula (1) was used to calculate the selection probability P_i of each PSU:

$$P_i = \frac{X_i}{X} \tag{1}$$

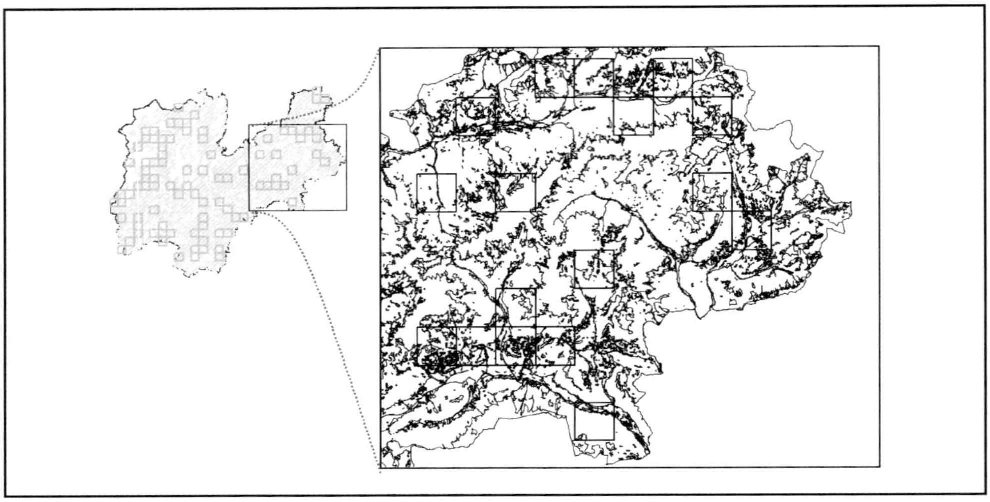

Fig. 17.2. Forest edge layer (auxiliary variable) intercepted by primary sample unit areas (PSUs).

where X_i is the value of the auxiliary variable in the primary unit i and $X = \sum_{i=1}^{N} X_i$ is the total value of the auxiliary variable in the study area. The selection probability of each PSU therefore depends on the value of the auxiliary variable and is equal to the ratio between the forest edge length in the unit considered and the total edge length in the whole area.

In the second stage, a total sample of 10,000 points (secondary sampling units or SSUs) was made by selecting randomly 100 points in each PSU drawn in the first stage. The selection probability of each SSU P_{ij} ($j = 1 \div 100$) in a PSU depends on the selection probability of the ith PSU (P_i) and on the selection probability of the jth SSU within the PSU (m_i/M_i):

$$P_{ij} = P_i \times \frac{m_i}{M_i} \tag{2}$$

where M_i is the number of all SSUs that form the ith primary unit and m_i is the number of SSUs sampled in the ith primary unit. The value M_i was defined by considering each SSU equal to the pixel resolution of 1 m^2, and is 16,000,000 for most units, except for the PSUs intercepted by the province boundaries, which are smaller than 1600 ha.

The photo-interpretation of 1973 and 1999 orthophotos made it possible to classify each SSU as forest or non-forest for both dates. In order to label an SSU as forest, the forest polygon in which the point was located had to be identified by connecting the trees on the forest boundary with a distance between them of less than 10 m. The three criteria stand area, crown cover and width, checked by photo-interpretation, also had to be satisfied. Since the 'minimum height' could not be assessed from orthophotos, different ancillary data (LUM, forest planning maps, topographic maps) were used to provide evidence of the height criterion for the photo-interpretation. Sample points that intercepted secondary stands were identified by comparing the results of this classification. To make the comparison more reliable, the classifications for 1973 and 1999 were carried out at the same time, opening separate windows to view each simultaneously, in order to detect differences due to spatial accuracy problems.

The new forest extent estimate ($\hat{Y}$) is given by

$$\hat{Y} = \frac{1}{n}\sum_{i=1}^{n} T_i \tag{3}$$

where n is the number of PSUs selected and T_i is given by

$$T_i = \sum_{j=1}^{m_i} \frac{y_{ij}}{P_i \frac{m_i}{M_i}} \tag{4}$$

where y_{ij} is the value of variable y (1 in the case of a point intercepting a new forest stand, 0 otherwise) in the jth unit (SSU) of the ith area (PSU). The variable y_{ij} was weighted on the basis of the selection probability of the unit considered. The variance of $\hat{Y}$ is estimated by

$$v\left(\hat{Y}\right) = \frac{1}{n(n-1)}\sum_{i=1}^{n}\left(T_i - \hat{Y}\right) \tag{5}$$

The estimators of secondary forest features are given by the ratios between two estimated values, the value of the feature observed and, as denominator, the new forest area. In order to have a good estimate of the variance, we adopted a Taylor expansion, which performs a linearization of the ratios (Cicchitelli *et al.*, 1992).

Data collection

The main objectives of the study were to estimate the extent of secondary forests in Trentino, and provide information with which to characterize secondary forest stands. In the sample points classed as secondary forest, many attributes were assessed by GIS and field survey. Although in the first stage 334 SSUs were classed as secondary forest, a survey was planned in only 297 of them, i.e. the SSUs located in stands with an area of more than 1000 m^2. This was considered to be the minimum area extent to observe the selected attributes. In two SSUs the field survey showed that a land-use change had occurred after the end of the study period (1999), while two other sample points were inaccessible. For these reasons, the field data were reduced to 293 sample units. Several other measurements were made from orthophotos, including the size of new forest patches, the distance to the old forest boundary (by comparing the two orthophotos) and the crown cover of wooded areas. Elevation, aspect and slope were derived from the DEM, and the category of surrounding forest was derived from forest planning maps. Other attributes, such as origin (natural, artificial or mixed), development stage (early colonization, well established or intermediate) and forest category of the new stands (coniferous, broadleaved and shrubs), were assessed in the field. The previous land use (in 1973) and the main current land use of the surroundings were assessed by using both the orthophotos and the data collected

Table 17.1. Attributes surveyed in the new forest stands, by information source.

GIS and photo-interpretation	Field surveys
• Elevation	• Main surrounding land use
• Aspect	• Previous land use
• Slope	• Origin
• Size of each new forest patch	• Development stage
• Crown cover	• Forest category
• Distance from the previous forest boundary	
• Category of surrounding forests	

in the field. Among the attributes surveyed in the field, a list of those discussed in the present chapter is given in Table 17.1. The 297 SSUs were surveyed in the field during summer 2004 and the work required was approximately 100 man-days, while the photo-interpretation took 1 man-day for each PSU.

Results and Discussion

Here we focus on the extent of new forest expansion and on the distribution of tree colonization according to the main environmental factors. Some consideration is also given to the efficiency of the sampling design and its operational applicability to land-cover change detection.

Estimate of the new forest area

As a result of the photo-interpretation, 334 of 10,000 SSUs were classed as secondary forest and their distribution is shown in Fig. 17.3. It should also be noted that 48 of the 10,000 SSUs have been affected by deforestation.

The total extent of new forest estimated by formula (3) is 18,218 ha, which means that 2.9% of the study area (or 4.5% of the forest area in 1999) is covered by secondary forest stands that have established since 1973. The standard error of the estimated value is 0.3% and the confidence interval is between 2.4% and 3.5%.

The estimated average annual rate of tree colonization is 0.11% of the total extent of the study area, but the net forest expansion, including deforestation, is 0.10%. This value is about one-half of that assessed by Corona *et al.* (2005) in the Abruzzo region (central Italy) for a shorter comparison period (1990–2002). The Abruzzo study estimated an average annual rate of forest expansion of 0.23%. A recent estimate of forest expansion in the Alpine region in the last decades is given by the comparison of the two most recent National Forest Inventory (NFI) reports for Switzerland, which showed an increase in the forest extent of 1.1% of the country area between 1982–86 and 1993–95 (Brassel and Brändli, 1999). However, the estimates vary with the subregion being considered; for example, in the southern Alps, a much higher increase of forest cover

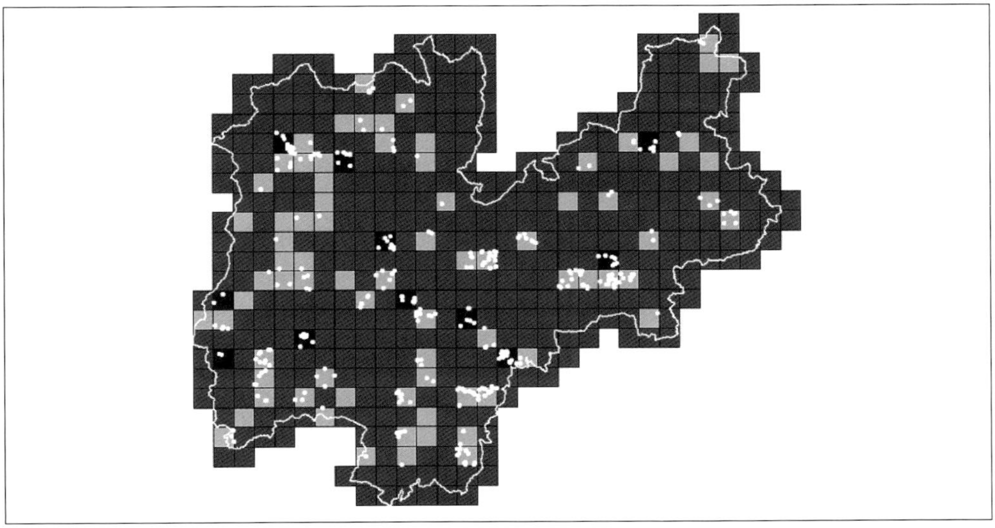

Fig. 17.3. The location of the secondary sampling areas (SSUs) classed as new forest (white dots) in the primary sampling areas (PSUs) of the whole study area. The PSUs sampled twice are coloured in black.

Table 17.2. Estimated area, standard error, coefficient of variation and confidence interval given separately for the first 50 PSUs and the second 50 PSUs and for the whole sample.

					Confidence interval	
Number of PSUs	Number of SSUs	Estimate of new forest area (ha)	SE	Coefficient of variation	Lower bound (*P* 0.05)	Upper bound (*P* 0.05)
50 (I)	5,000	15,897	1,960	0.123	12,054	19,740
50 (II)	5,000	20,885	2,796	0,134	15,405	26,366
100	10,000	18,218	1,667	0.091	14,884	21,551

(2.6%) was reported. In Trentino the new forest expansion is not uniformly distributed. The rate of colonization varied significantly among the selected PSUs, ranging from 0 to 17%. Indeed, in some valleys land-use change was extremely evident, since in 1999 the landscape pattern was quite different from the pattern evident in 1973.

The auxiliary variable used for sample selection was correlated with new forest expansion, as shown by the coefficient of correlation, which is quite low (equal to 0.27) but very significant (P value = 0.0059).

Regarding the sampling efficiency, in Table 17.2 the estimate derived from the first selection of 50 PSUs is compared with the one derived from 50 PSUs selected later, and with the estimate obtained from the whole sample. The three estimates do not differ significantly and therefore a 50 PSU sample is sufficient to obtain a reliable and representative estimate of the new forest area.

Nevertheless, since secondary forest expansion is quite rare, a first-stage sample of 100 units was necessary to detect a sufficient number of SSUs intercepting new forest patches in order to obtain an SSU sample large enough to derive reliable information on the features of secondary forest stands.

Attributes of the new forests

The data collection involved a wide range of assessments concerning: physiography, stand site, vegetation composition, dendrometric attributes, stand management factors and landscape features linked to tree colonization dynamics. Some of the results pertaining to site features and colonization dynamics are reported and discussed in the following paragraphs.

Site features

One of the main tasks of the survey was to investigate the role of site features on the distribution of secondary forests in the study area. Figure 17.4a shows the distribution of the new forest area by different elevation classes: the phenomenon is concentrated in the mid-elevation classes, typically marginal agricultural areas, with a peak between 800 and 1000 m. A second peak, less evident, occurs at a higher elevation, between 1600 m and 1800 m, and is due to the progressive abandonment of high-elevation pastures and the consequent increase in the elevation of the treeline. This trend is confirmed in Fig. 17.4b, which illustrates the distribution of the colonization coefficient by elevation class. By 'colonization coefficient' we mean the ratio of the new forest area to the total area calculated in each elevation class. The values suggest that the new forest expansion occurs mostly at middle and higher elevations. The first peak, ranging from 400 to 1000 m, may be wider than that observed in Fig. 17.4a, as the elevation range between 400 and 800 m occurs at a low frequency in the study area.

As regards the other observed site features, Fig. 17.4c and d illustrates the distribution of the new forest area, respectively, per aspect and per slope class. The graphs show that the tree colonization has mainly affected the warmer sites, with a south aspect, and moderately sloped areas, that is, the sites more suitable to agriculture. These results suggest that less productive and less accessible fields had already been abandoned before the monitoring period and that the more recent tree colonization has occurred mainly in marginal areas.

Colonization dynamics

In order to analyse the tree colonization process, some attributes concerning the new stand establishment were observed. In terms of the current surrounding land use, the secondary forest is located mainly in glades and on meadows and pastures (Fig. 17.5a), while the distribution by previous land use (Fig. 17.5b) highlights the fact that tree colonization has occurred principally on abandoned meadows and pastures. Important characteristics of newly established stands

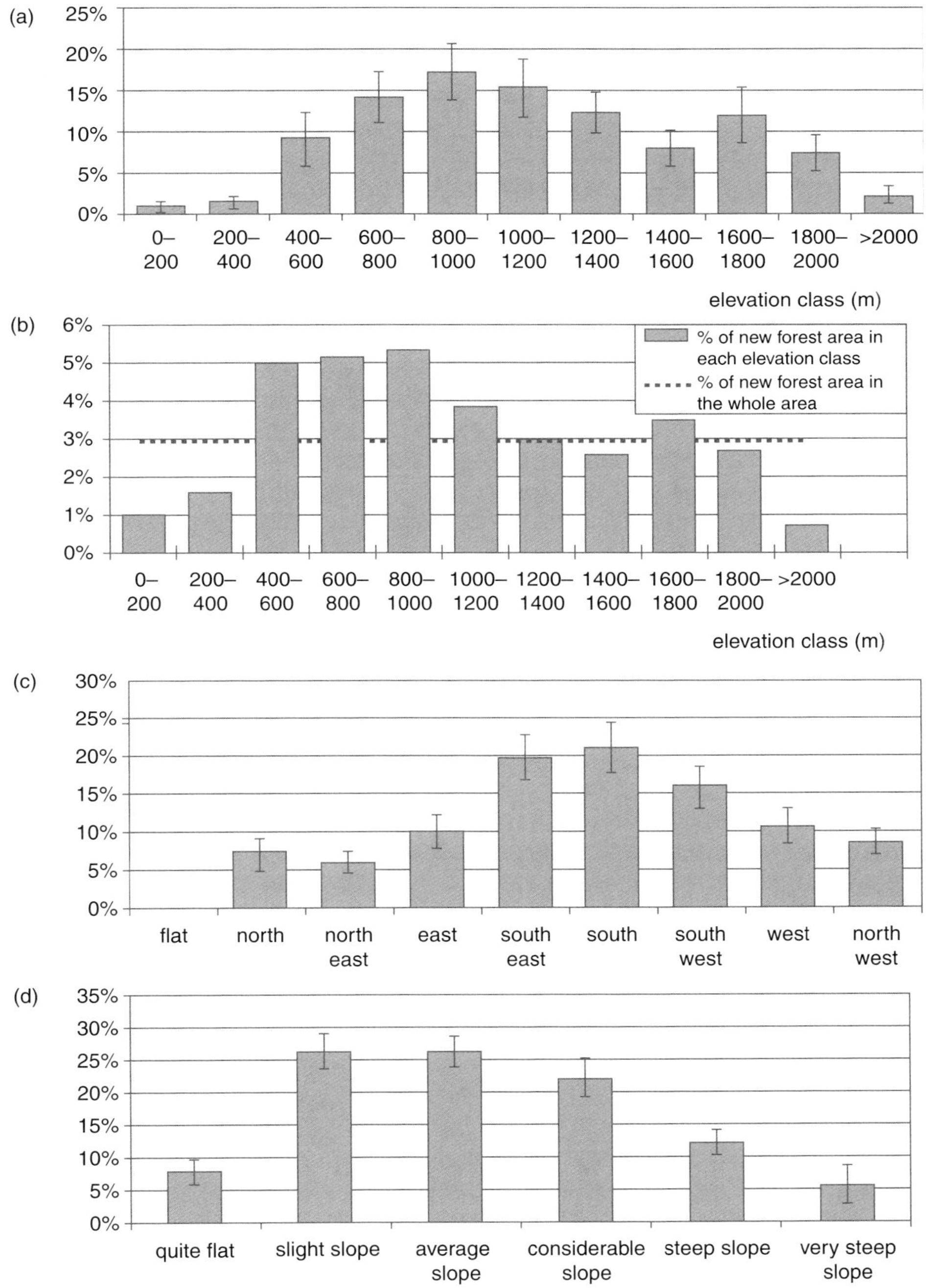

Fig. 17.4. Site features of new forest stands: (a) distribution of new forest stands per elevation class; (b) distribution of the colonization coefficient per elevation class; (c) distribution of new forest stands per aspect class; (d) distribution of new forest stands per slope class. The lines over the bars represent the standard errors of the estimates.

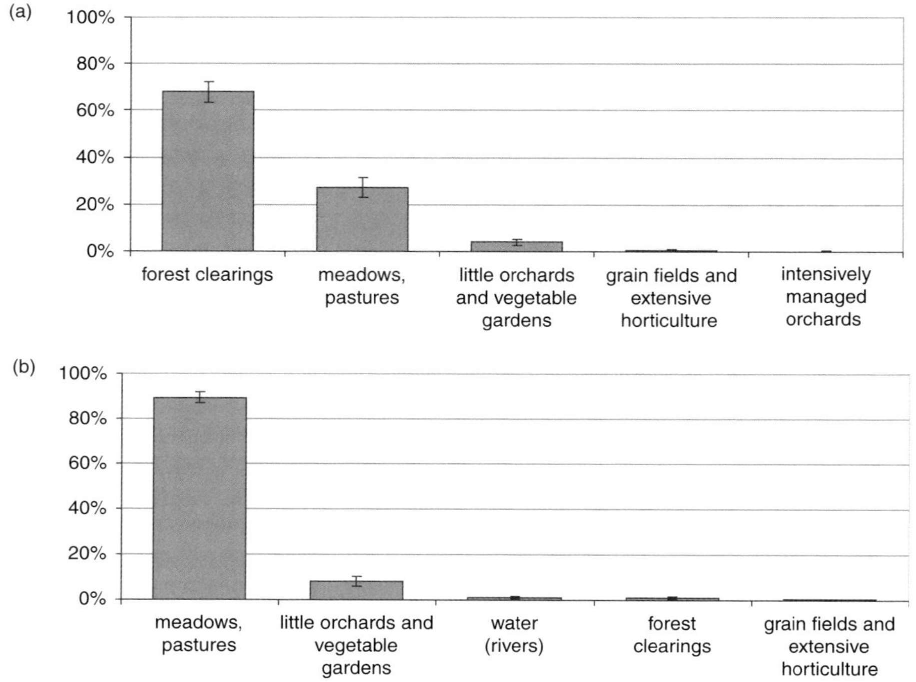

Fig. 17.5. New forest stands in relation to land use: (a) distribution of new forest stands by surrounding land use; (b) distribution of new forest stands by previous land use. The lines over the bars represent the standard errors of the estimates.

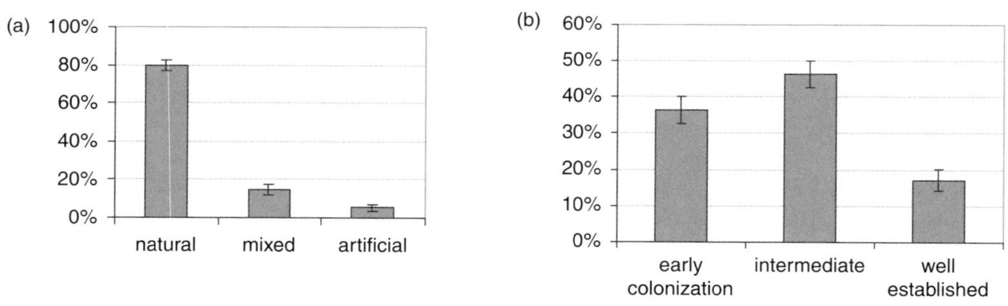

Fig. 17.6. Origin and development of the new stands: (a) distribution of new forest stands per origin source; (b) distribution of new forest stands per development stage. The lines over the bars represent the standard errors of the estimates.

include the following: most have a natural origin (Fig. 17.6a), only a few are well established (Fig. 17.6b) and more than 30% are very young while about 50% are at an intermediate stage. This distribution suggests that tree colonization has continued throughout the whole observation period, even though some of the

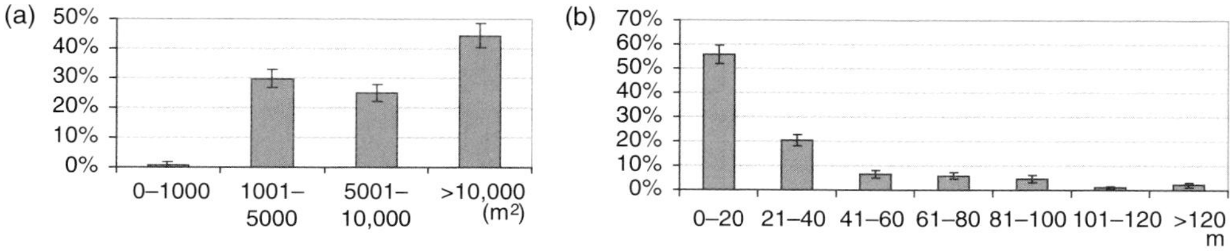

Fig. 17.7. Some spatial properties of the new stands: (a) distribution of new forest stands per patch size; (b) distribution of new forest stands per distance of the new forest boundary from the old one. The lines over the bars represent the standard errors of the estimates.

new stands identified by photo-interpretation had been removed in order to re-establish an agricultural use (two sampling plots) before the follow-up field survey.

Figure 17.7 illustrates some spatial properties of the new forest patches. Most patches are very small and their extent is less than 1 ha (Fig. 17.7a). The distribution of new stands classed by distance to the former forest edge (Fig. 17.7b) suggests that they generally occur near old forests.

The influence of canopy cover was also considered. Most new forest stands have closed canopies: 79.3% of the SSUs were assigned to the highest crown cover class (80–100%), with a standard error of 3% ($P = 0.05$). However it should be remembered that many of these forests were formerly very open stands that closed canopy in recent years. Finally, the new stands were grouped into broad forest categories: coniferous, broadleaved and shrubs. More than 50% of the surveyed SSUs were assigned to the broadleaved class, while the frequency of coniferous and shrub class SSUs were, respectively, 43.6% and 5.9%. The survey also classified the surrounding forest type, and this was mainly coniferous (64.3% of the SSUs), while the broadleaved forest class occurred in only 29.3% of the SSUs. Therefore, it seems that the tree colonization process has increased the species diversity, even though it might be a temporary result of land-use changes.

In terms of species composition, it should also be noticed that most species observed in new stands are natural (indigenous). Among the regenerating species, *Robinia pseudoacacia* was the only significant invasive species noted and it occurred in 7.9% (SE = 2.7%) of new stands.

Conclusions

In Italy, the tree colonization of abandoned lands is an important issue for most mountain areas. Many authors have analysed this process (Salbitano, 1987; Sulli, 1996; Fontana, 1997; Aceto *et al.*, 2000) but information about the overall extent of forest expansion is still lacking. However, this information is important in order to provide a solid quantitative foundation for forest policy oriented to sustainable management of the landscape.

In this study, a procedure to detect new forest stands based on a two-stage sampling design has been set up. In the first stage some area units were selected according to a probability depending on the value of the auxiliary variable 'forest edge length'. This variable was chosen on the basis of its suitability described in the literature, and of the availability of continuous data in the study area, and, although it had a fairly low correlation with the phenomenon, this was highly statistically significant. The sampling units were surveyed by photo-interpretation at two reference dates. Field surveys were limited to the sampling units classed as new forest. The data collection involved a wide range of assessments concerning physiography, stand site, vegetation composition, dendrometric attributes, stand management and landscape features linked to tree colonization dynamics. The sampling procedure was efficient and helped the organization of both photo-interpretation and field surveys.

This study provides a quantitative estimate of the shift in land use from non-forest uses to forest areas that occurred between 1973 and 1999 in the province of Trento and leads to the conclusion that the phenomenon has been rather limited (it has involved about 3% of the whole land area) but is continuous and concentrated in some particular environments. The new forest patches have established mainly in abandoned meadows and pastures at mountain (1600–1800 m) and hill (800–1000) elevation, with warmer aspect and moderate slope. Generally the new forest has originated from natural expansion along the previous forest edge. Most new stands are very small with an elongated shape, but have a relatively closed canopy.

The tree colonization process has involved a loss of diversity in the landscape, because it has generally occurred in meadows and pastures and is progressively closing the forest clearings. At the same time, the tree colonization may increase the forest tree species diversity, at least temporarily. Generally, the composition of new stands is indeed very different from that of the surrounding forest, with a remarkable increase of broadleaved stands in a conifer-species-dominated landscape.

Acknowledgements

Financial support for this work was provided by the province of Trento. We wish to thank our statistical consultants D. Filipponi and M. Picci for their valuable assistance in the definition of the sampling design and in data processing. Our thanks are also expressed to: S. Agnoloni, A. Anderle and S. Morelli for their assistance in several work phases and their collaboration in the photo-interpretation; and A. De Francesco, V. Menghini, R. Pasquazzo, D. Manzoni, D. Fedel, C. Rizzi, A. Gomiero and A. Penasa, trainees of the University of Padua in the Trento Forest Service, for their contribution to the fieldwork. The authors finally want to thank three anonymous referees for their useful comments on the original manuscript.

References

Aceto, P., Pividori, M. and Siniscalco, C. (2000) Dinamica evolutiva di popolamenti forestali di neoformazione nel piano montano. *Monti e Boschi* 1, 4–12.

Brassel, P. and Brändli U.-B. (1999) *Schweizerisches Landesforstinventar. Ergebnisse der Zweitaufnahme 1993–1995*. WSL Birmensdorf, Haupt Bern, 444 pp.

Cernusca, A., Tappeiner, U. and Bayfield, N. (1999) *Land-use Changes in European Mountain Ecosystems. ECOMONT – Concept and Results*. Blackwell Wissenschafts-Verlag, Berlin.

Cicchitelli, G., Herzel, A. and Montanari, G.E. (1992) *Il campionamento statistico*. Il Mulino, Bologna, Italy, 571 pp.

Corona, P., Pompei, E. and Scarascia Mugnozza, G. (2005) Stima probabilistica del tasso di espansione annua e del valore al 1990 della superficie forestale nella Regione Abruzzo. *Forest* 2(2), 178–184.

Didier, L. (2001) Invasion patterns of European larch and Swiss stone pine in subalpine pastures in the French Alps. *Forest Ecology and Management* 145, 67–77.

EEA (1999) *Environment in the European Union at the Turn of the Century*. Environmental Assessment Report No. 2, European Environment Agency, Copenhagen, Denmark.

Endress, B.A. and Chinea, G.D. (2001) Landscape patterns of tropical forest recovery in the Republic of Palau. *Biotropica* 33(4), 555–565.

Fabbris, L., (1989) *L'indagine campionaria, metodi disegni e tecniche di campionamento*. NIS, Rome, 261 pp.

Fontana, S. (1997) Boschi di neoformazione: un caso nelle prealpi venete. *Sherwood Foreste ed Alberi Oggi* 23, 13–17.

French, D.D., Miller, G.R. and Cummins, R.P. (1997) Recent development of high-altitude *Pinus sylvestris* scrub in the northern Cairngorm mountains, Scotland. *Biological Conservation*, 79, 133–144.

Höchtl, F., Lehringer, S. and Konold, W. (2005) 'Wilderness': what it means when it becomes a reality – a case study from the southwestern Alps. *Landscape and Urban Planning* 70, 85–95.

Hunziker, M. (1995) The spontaneous reafforestation in abandoned agricultural lands: perception and aesthetic assessment by locals and tourists. *Landscape and Urban Planning* 31, 399–410.

ISTAT (1973) *Popolazione e movimento anagrafico dei comuni. Anno 1973*. Istituto Nazionale di Statistica, Rome.

ISTAT (1999) *Popolazione e movimento anagrafico dei comuni. Anno 1999*. Istituto Nazionale di Statistica, Rome.

Kettle, W.D., Rich, P.M., Kindscher, K., Pittman, G.L. and Fu, P.D. (2000) Land-use history in ecosystem restoration: a 40 year study in prairie–forest ecotone. *Restoration Ecology* 8(3), 307–317.

MacDonald, D., Crabtree, J.R., Wiesinger, G., Dax, T., Stamou, N., Fleury, P., Gutierrez Lazpita, J. and Gibon, A. (2000) Agricultural abandonment in mountain areas of Europe: environmental consequences and policy response. *Journal of Environmental Management* 59, 47–69.

Mather, A.S. and Fairbairn, J. (2000) From floods to reforestation: the forest transition in Switzerland. *Environment and History* 6, 399–421.

Mather, A.S., Fairbairn, J. and Needle, C.L. (1999) The course and drivers of the forest transition: the case of France. *Journal of Rural Studies* 15(1), 65–90.

Motta, R. and Nola, P. (2001) Growth trends and dynamics in sub-alpine forest stands in the Varaita Valley (Piedmont, Italy) and their relationships with human activities and global change. *Journal of Vegetation Science* 12, 219–230.

Odasso, M., Mayr, S., De Franceschi, P., Zorzi, S. and Mattedi, S. (2002) *Miglioramenti ambientali a fini faunistici*. Provincia Autonoma di Trento, Assessorato all'Agricoltura e alla Montagna, Servizio Faunistico, Trento, Italy, 167 pp.

Pelleri, F. and Sulli, M. (1997) Campi abbandonati e avanzamento del bosco. Un caso di studio nelle Prealpi lombarde (Comune di Brinzio, Provincia di Varese). *Annali ISSA* 28, 89–115.

Piussi, P. (2000) *Expansion of European Mountain Forests*. In: Price, M.F. and Butt, N. (eds) *Forests in Sustainable Mountain Development*. CAB International, Wallingford, UK, pp. 19–25.

Prosser, F. (2001) *Lista Rossa della Flora del Trentino. Pteridofite e Fanerogame*. Museo Civico di Rovereto, Edizioni Osiride, Rovereto, Italy, 107 pp.

Provincia Autonoma di Bolzano (1994) *Lista Rossa delle specie minacciate in Alto Adige*. Provincia Autonoma di Bolzano/Alto Adige, Ripartizione Tutela del paesaggio e della natura, Bolzano, Italy, 409 pp.

Salbitano, F. (1987) Vegetazione forestale ed insediamento del bosco in campi abbandonati in un settore delle prealpi Giulie (Taipana, Udine). *Gortania* 9, 83–143.

Sulli, M. (1996) Campi abbandonati e avanzamento del bosco: temi di ricerca ecologico forestale e priorità tecnico-economiche Sherwood. *Foreste ed Alberi Oggi*, 8(1/96), 7–9.

Tasser, E. and Tappeiner, U. (2002) Impact of land-use changes on mountain vegetation. *Applied Vegetation Science* 5, 173–184.

18 Using Spatial Statistics to Improve the Primary Forestry Supply Chain

J.D. HAMANN AND K. BOSTON

Department of Forest Engineering, College of Forestry, Oregon State University, Corvallis, Oregon USA.

Abstract

The benefits of spatial inventories are numerous and allow for the direct development of precision harvest planning. By better identifying areas containing desired products within forest harvest blocks, customer demands can be processed to meet specific orders while minimizing handling, processing and storage costs.

Simulations were performed to examine the potential benefits of Kriging to estimate product volumes in unsampled areas within a harvest block. Using a stem map, we sampled the area using 0.01 ha fixed-radius plots on a 22.2 m × 16.6 m grid to obtain saw-log volume estimates with and without spatial information. The actual total saw-log volume for the area was 1243.3 m^3 while the estimated saw-log volumes for the non-spatial and spatial sampling method were 1185.5 m^3 and 1196.2 m^3, respectively. The predicted variance of the sample estimated with Kriging (1067.8) was less than 2% of the variance estimate for the non-spatial sample (74,403.4).

To demonstrate an advantage of a spatially explicit inventory, we optimized a harvest operation, using a simulated annealing algorithm. We optimized a fictitious operation so that the optimal harvest pattern would minimize the sum of the squared differences between a demand function and the predicted production for the harvest unit.

Introduction

An aggregative manufacturing process (e.g. consumer electronics, automobiles and conference proceedings) is not typically considered a stochastic process where component delivery timing or the quality of raw materials is delivered with little or no uncertainty. Unlike this more common process type, log manufacturing is a disaggregative manufacturing process (e.g. butchering and mineral extraction), which is almost always associated with many types of variation. The quality and quantity of the raw material within a given order can vary within certain predefined levels, such as diameter ranges, lengths or surface characteristics.

 Sustainable Forestry: from Monitoring and Modelling to Knowledge Management and Policy Science (eds K.M. Reynolds, A.J. Thomson, M. Köhl, M.A. Shannon, D. Ray and K. Rennolls)

Unlike finished panel products or boards, which have a minimal variation, the production of the raw material is subject to variation within the stem, stand and season. To reduce this variation, forest product companies have sought to optimize the sample size for a desired level of precision (Zeide, 1980; Gambill *et al.*, 1985; Oderwald and Jones, 1992; Brooks and Wiant, 2004) and increase the level of detail in sampling procedures by measuring additional attributes (Mandallez and Ye, 1999) and by including additional data sources in the inventory (Kilkki and Päivinen, 1987; Korhonen and Kangas, 1997; Holmström, 2002). While obtaining detailed information on operational harvest units can be costly, it can lead to significant improvements in the financial performance of an integrated forest products firm, as log mix can be optimally matched with processing facilities (Wagner *et al.*, 1996; Uusitalo, 1997).

An examination of previous optimization studies reveals a continuous increase in the volume, resolution and complexity of data required for operations planning and optimization. Hehnen *et al.* (1984) demonstrated the benefits of including a merchandizing decision support simulator in a real-time optimization process, so that the simulator could examine the influence of changes in operational parameters and economic conditions on the current merchandizing operation. By monitoring the production requirements against the actual production levels, the simulator was used to adjust the operation to maximize value recovery for the entire operation. The authors concluded they were able to increase the return on capital investment by 40% resulting from the increase in the amount of data made available for making decisions.

In addition to increasing the volume of data, Hay and Dahl (1984) discarded general descriptions of the stem in favour of utilizing the true shape of the stem to determine log and lumber recovery at each potential conversion facility. The authors cited a lack of resolution in the data to be the key factor in developing a log allocation system. They then used more detailed information to assess the optimal operation of two competing sawmills, concluding that the optimal configuration required that both facilities operate with log diets mutually beneficial to the overall operation. This differed from their original finding, based on the less detailed information.

While obtaining more and higher-resolution information on stems as they enter the mill for processing seems advantageous, the problem of handling the stem before that point has rarely been considered. Most optimization studies have focused on the decisions made after the log has reached the conversion facility. Wagner *et al.* (1996) used an approach similar to the one presented by Hay and Dahl (1984) to determine log acquisition policies, stating as the rationale for their study that delivered log costs accounted for 70–80% of operating costs. The authors developed a system of simulations and forecasts to determine which timber sales would be optimal from a package of sales available. They used pre-sale sample information to determine which sales would have the greatest increase in profit margin. Using regression equations to estimate lumber grade recovery and production rates, they were able to reduce the lead time of timber purchase to stem processing to 3 months and, in doing so, they were able to reduce storage requirements, handling costs and losses due to damage.

It would seem that variance reduction is the most important task for primary forest supply chain management and optimization, and, since the spatial component of forest accounts for the most variation regarding the production of logs, it seems appropriate to examine the spatial aspect of log production (Korhonen and Kangas, 1997; Uusitalo, 1997; Rasinmäki and Melkas, 2005). Until recently, there has been little work that examines the addition of spatial data in sampling to reduce the variation in production estimates. Murphy *et al.* (2004) found that they could achieve a 17 to 22% increase in stand value from a 3% sample of a radiata pine (*Pinus radiata* D. Don) stand in New Zealand. The authors concluded that, with the harvest manager's ability to capitalize on within-stand variation (harvest pattern selection) and inability to control order book requirements (constraints), it might be possible to reduce variation in production estimates by simply modifying the harvest pattern.

Choosing an operational pattern that reduces variation in the delivery of raw materials is not new to other disciplines. Mining geology (Journel and Huijbregts, 1978) and, to a lesser extent, soil science (Kravchenko, 2003) have used spatial prediction and mapping for short-term planning, where, unlike commonly practised sampling methods in forestry, variance estimates are generated for unsampled areas. While mining geology has accepted the high cost of inventory and data collection, forest managers in North America have emphasized cost reduction using a variety of methods, such as reducing sampling intensity, utilizing remote sensing data and decreasing inventory remeasurement periods. In contrast, Scandinavian forest managers have begun to examine the additional benefits of high-resolution data from spatial prediction methods (Korhonen and Kangas, 1997; Culvenor, 2002; Holmström, 2002) or from stem map data obtained during processing in harvester operations (Stendahl and Dahlin, 2002; Rasinmäki and Melkas, 2005).

The papers mentioned above reveal the potential improvements in the forestry supply chain when improved data are utilized in the log manufacturing process. This chapter examines how spatial information can be added to traditional sampling procedures to increase the usefulness of the sample while maintaining or possibly increasing precision in the context of the primary forestry supply chain. A simple scheduling model that demonstrates the additional gains from spatial information is presented.

Data and Methods

Study site and stem map

The stem map used for this analysis was obtained from the National Aeronautics and Space Administration (NASA) Forest Ecosystem Dynamics (FED) project (Walthall *et al.*, 1993). The study site is a 3 ha rectangle (150 m × 200 m) located 56 km north of Bangor, Maine, in Penobscot County (45°12′N, 68°44′W). This data set was desirable because of the flat to gently rolling topography, which was important for our fictitious ground-based harvest machine. An example with a cable operation would have been inappropriate due to the relaxation that we allowed the harvester to relocate to any harvest block within the study area rather than restricting the harvester to a single or continuous path.

The original data file contained 5967 tree records. Each tree record contained: *x* and *y* coordinates, species, diameter at breast height (dbh), total height (tht), canopy position and an indicator variable to represent a dead stem. We selected only those trees that were living, over 8 cm dbh and standing for our study, and for simplification assumed all trees were a single species. The final number of stems within the 3-ha stem map was 4390.

The stems were then cut into log lengths of 4 m to a 2 cm top, using the taper equation presented by Kozak *et al.* (1968). If the stem could not be cut into a round number of standard log lengths, any remaining stem length was cut into a short log no less than 2 m. Stems were merchandized with a stump height of 0.3 m and each log included 0.2 m of trim. For each log in the stem, the start and end height of the log, the nominal length and the actual length (nominal length plus trim), the small- and large-end diameters and the Smalian volume were recorded. The volume of all logs with a small-end diameter over 10 cm was totalled and assigned as the saw-log volume for that stem record. The log volumes for those logs with a small-end diameter less than 10 cm were tallied as pulp volume. The resulting stand summaries are presented in Table 18.1 and Fig. 18.1.

Non-spatial sampling

To determine the baseline for comparing the traditional (non-spatial) and spatial sampling methods, we simulated a commonly applied square grid sample design of 0.01 ha fixed-radius plots on a 22.2 m × 16.6 m grid. The distance of each stem to the plot centre was computed to determine total saw-log volume per plot. Figure 18.2 presents the sample trees and the associated plot centres with the sampled stems.

Spatial prediction

To determine the benefits of adding spatial information, we used the plot locations from the grid generated during the non-spatial sampling phase to then predict the saw-log volume in unsampled locations. Initially, we had great difficulty fitting variograms to the rectangular sampling grid. We then attempted to fit a variogram for a sampling grid where the distance between the plots was one-half of the distance between lines of plots. Again, this configuration gave unsatisfactory results. Citing Dalenius *et al.* (1961), Webster and Oliver (2001) suggested that a triangular pattern is better than a square grid for estimating the mean of a region if the variogram is exponential. Following this suggestion, we altered the sampling grid so that every other plot in the sample was removed, yielding the pattern presented in Fig. 18.2.

After visual examination of the stem map, we decided to use universal Kriging (Kriging with a trend model) to predict the saw-log volumes over the

Table 18.1. Stand summaries by diameter class.

Dbh class (cm)	Stems	Logs	Saw-log volume (m^3)	Pulp volume (m^3)
5–10	1425	1228	0.0	66.1
10–15	1294	2952	39.7	244.6
15–20	771	2163	246.0	119.8
20–30	746	2528	655.7	115.6
30–40	154	627	302.0	27.4
Total	4390	9498	1243.3	573.5

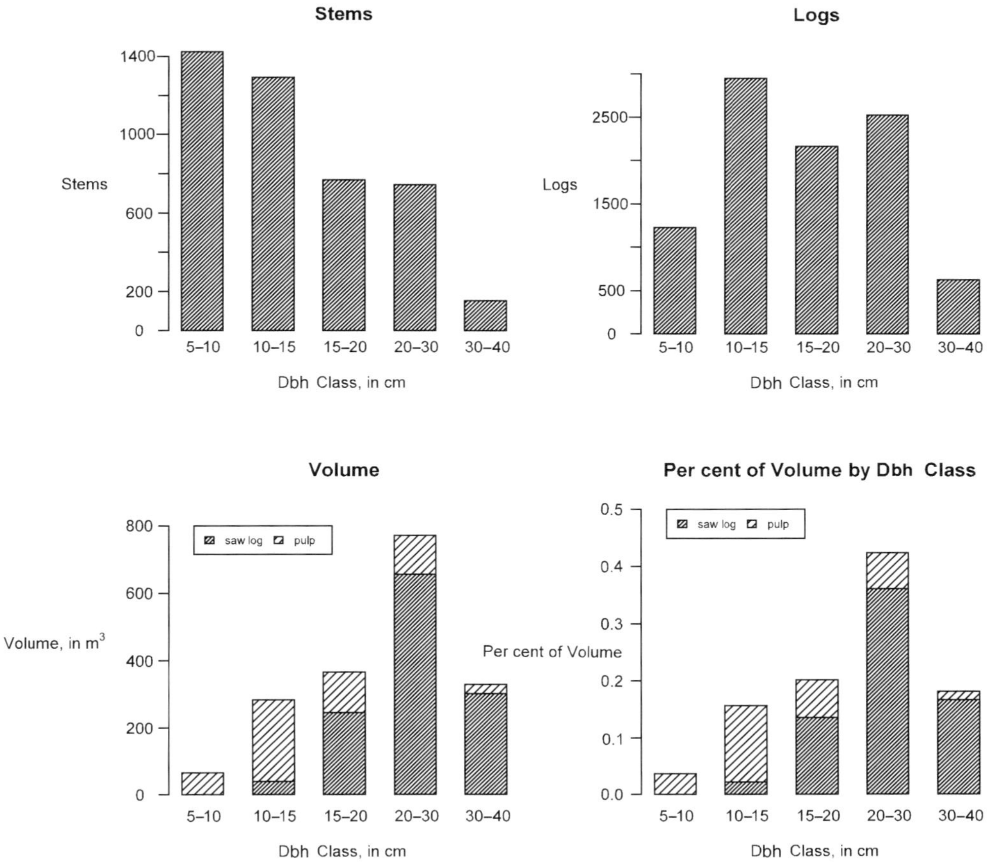

Fig. 18.1. Stem, log and volume distributions by diameter class for the stem map.

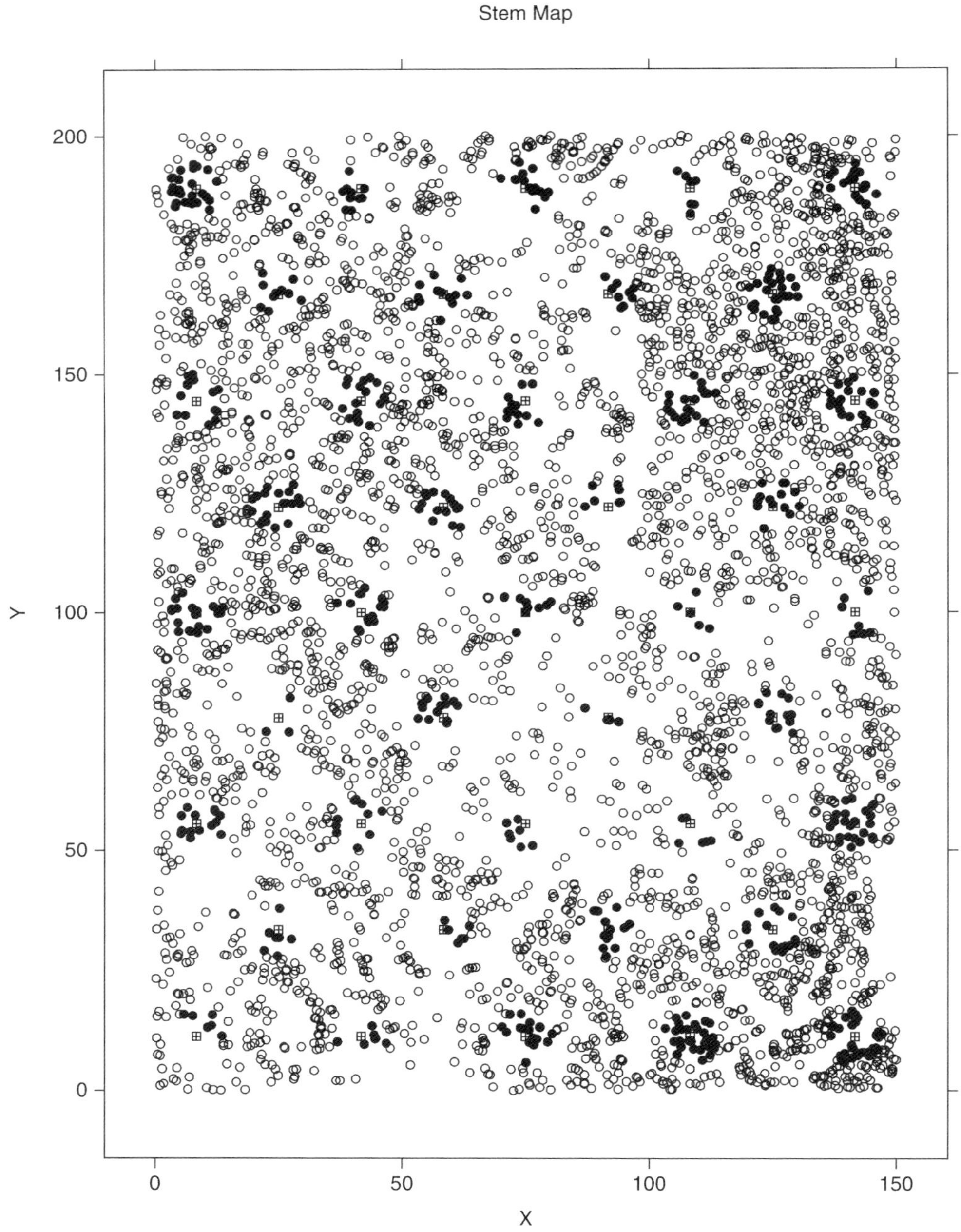

Fig. 18.2. Plot locations with sampled stems where the units are in metres. The circles represent individuals not sampled and the dark circles represent those trees included in the fixed-radius plots. The squares are located at the fixed-area plot centres.

entire area. The model fitting and prediction were performed using the gstat package within R (Ihaka and Gentleman, 1996; Pebesma, 2004). We fitted an exponential model for the variogram, which gave adequate results upon visual inspection.

Since we had few data points to estimate a variogram, we did not attempt to refine our variogram by combining models by examining periodicity or anisotropy. The variogram was used to estimate the saw-log volumes in 10 m × 10 m cells for those cells that did not contain sample points. To compare the estimated variance for the total predicted saw-log volume, we used the methods described by Journel and Huijbregts (1978) and Kim and Baafi (1984) to combine the variance estimates from all the individual predicted cells.

Saw-log demand

Since the goal of the study was to determine if the inclusion of spatial estimates in a tactical harvest operation was beneficial, we decided to evaluate the Kriging estimates. We compared the deviations between a generated random demand function for a 12-day production period and the predicted saw-log volumes from various harvest patterns. The demand levels represented the saw-log volume required by a fictitious sawmill and were developed so that the sum of the daily demand levels was equal to the total volume produced from the entire study area. To make our demand levels as realistic as possible, the average production required was the sum of the total volume for the area divided by 12 production days. We varied the production requirements from a minimum of 50 m^3 to 140 m^3 per day, with a standard deviation of 25 m^3 per day. The daily demand levels are presented in Table 18.2.

Harvest path optimization

Once we had obtained volume estimates for all unsampled areas, we divided the area into 12 daily production blocks (50 m × 50 m) to simulate a harvest operation. To obtain both the combined predicted saw-log volume and the associated variance estimates for each harvest block, we again used the methods as described by Journel and Huijbregts (1978) and Kim and Baafi (1984) to combine the variance estimates from the individual predicted cells for the 12 harvest blocks.

To determine the daily production without the benefits of spatial information, a serpentine path was placed over the 12 blocks and the daily production levels were obtained by summing the saw-log volume for all cells within each cutting block. Then, to obtain an optimal harvest pattern, simulated annealing (Metropolis *et al.*, 1953) was used to find the harvest pattern that minimized the sum of the squared differences between the demand curve and the estimated daily production. Using two-opt moves and a logarithmic cooling schedule, 50,000 iterations were performed for each level. Once the optimal harvest pattern was determined, the daily harvest areas were labelled

Table 18.2. Predicted and actual production levels for non-spatial (serpentine harvest pattern) and spatial (optimal harvest pattern) sampling methods. The difference values are the customer demands minus the predicted values.

Production period	Customer demand	Non-spatial (serpentine) pattern			Spatial (optimal) pattern		
		Predicted	Actual	Difference	Predicted	Actual	Difference
1	115.0	103.6	70.9	11.4	99.8	110.3	15.2
2	115.0	103.6	126.9	11.4	99.3	62.0	15.7
3	115.0	103.6	106.4	11.4	92.9	106.4	22.1
4	50.0	103.6	113.0	– 53.6	69.3	53.6	– 19.3
5	60.0	103.6	62.0	– 43.6	87.0	113.0	– 27.0
6	100.0	103.6	110.3	– 3.6	87.0	70.9	13.0
7	105.0	103.6	91.1	1.4	88.2	105.5	16.8
8	120.0	103.6	130.4	16.4	129.5	154.0	– 9.5
9	140.0	103.6	154.0	36.4	134.5	119.1	5.5
10	120.0	103.6	119.1	16.4	127.7	130.4	– 7.7
11	110.0	103.6	105.5	6.4	88.5	91.1	21.5
12	110.0	103.6	53.6	6.4	92.5	126.9	17.5

with a number from 1 to 12 for visual examination and a plot of the demand, predicted supply and actual supply was generated, to examine the results for any anomalies.

Results

Volume estimates

The total predicted saw-log volumes for both the non-spatial and the spatial sampling were similar to the actual saw-log volume for the area. The actual total saw-log volume for the area was 1243.3 m^3, with the estimated saw-log volumes for the non-spatial and spatial sampling methods being 1185.5 m^3 and 1196.2 m^3, respectively. The variance estimate for the total predicted saw-log volume from the spatial method was 1067.8 m^3. This was less than 2% of the estimated variance of 74,403.4 m^3 for the non-spatial method. We generated the 95% confidence limits for the total predicted saw-log volume for both the non-spatial and spatial sampling methods and present them graphically in Fig. 18.3.

Production

The resulting estimated daily saw-log volume production for the two sampling methods differed as well. Since there was no spatial information for the non-spatial sampling method, the predicted daily production was simply the total estimated

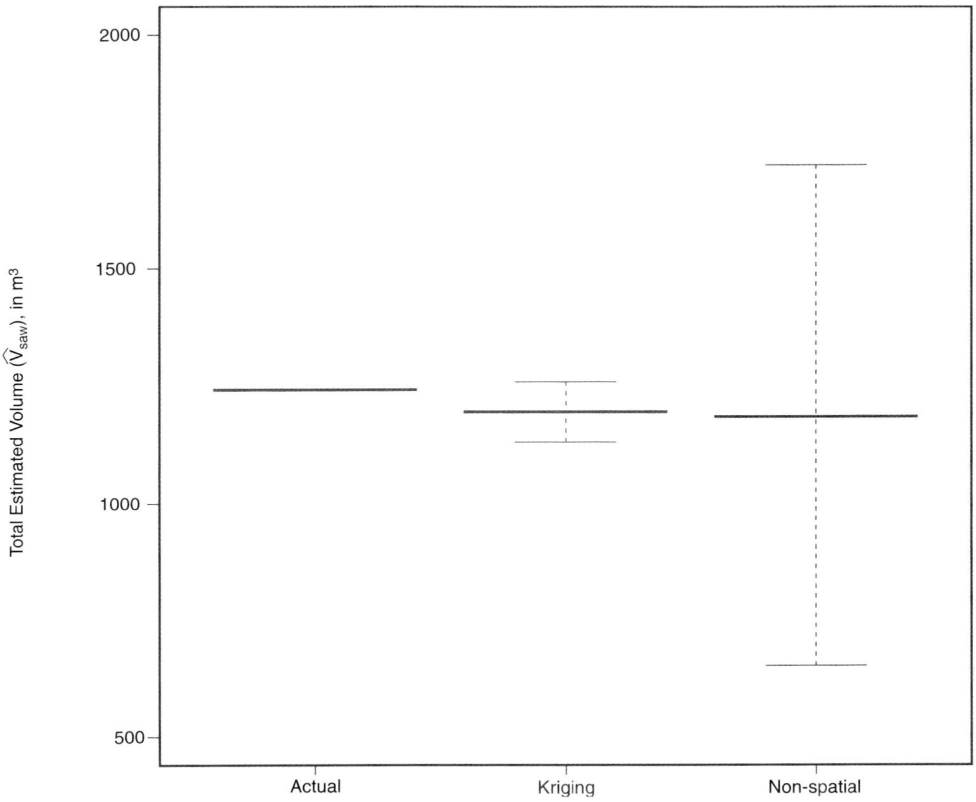

Fig. 18.3. Total estimated saw-log volumes for Kriging and non-spatial sampling methods with 95% confidence limits.

saw-log volume divided by the 12 operating days. The estimated average daily production from the non-spatial sampling method was 103.61 m^3. The estimated daily production for the spatially explicit sampling method ranged from a minimum of 69.26 m^3 to a maximum of 134.51 m^3 and the actual daily production from the spatial sampling method ranged from 53.61 m^3 to a maximum of 154.01 m^3.

The sum of the squared deviations between required and produced volumes for the non-spatial harvest pattern was 7123.27, or more than twice the sum of the squared differences for the spatially explicit harvest pattern of 3469.15. A table of the predicted and actual production for the initial harvest pattern and the optimal harvest pattern is presented in Table 18.2. Figure 18.4 presents a chart of the harvest production for the non-spatial and spatial (optimal) harvest

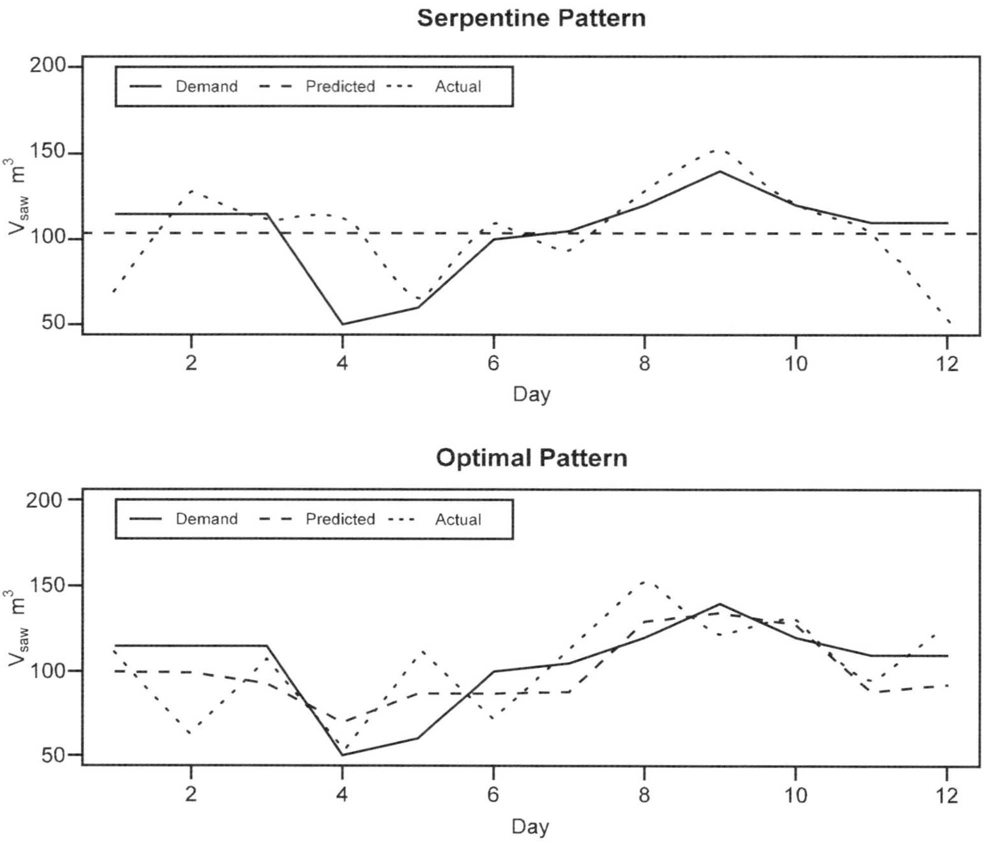

Fig. 18.4. Predicted and actual production levels for non-spatial (serpentine harvest pattern) and spatial (optimal harvest pattern) sampling methods.

production and Fig. 18.5 graphically displays the non-spatial and optimal harvest patterns.

Discussion

While the ability to minimize the differences between the consumer's demand and the supplier's production is an important part of any attempt to manage a supply chain, there are a multitude of issues that prevent the development of a method to minimize the deviation between the demand and production of log products.

Initially, it was assumed a rectangular sampling grid would yield a sufficient variogram for little effort to be needed to obtain the Kriging results. In retrospect, the major task for this study was simply obtaining an adequate variogram. Our sample of 41 plots for the 3 ha area was well above what would be considered standard for an operational forest inventory environment. Once a

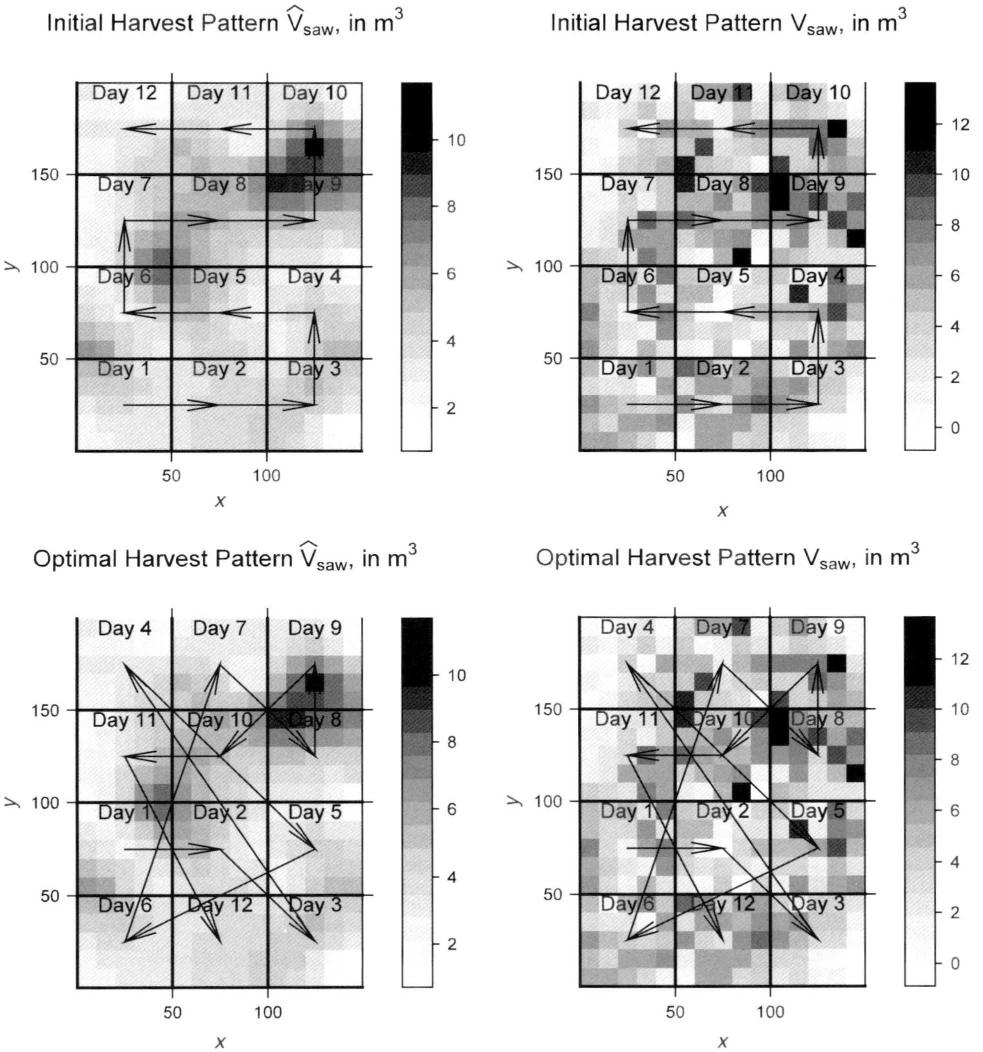

Fig. 18.5. Harvest patterns for the non-spatial sample (top) and the spatially explicit sample with the optimal harvest pattern (below). The left images contain the predicted saw-log volumes in the study area and on the right are the actual saw-log volumes. The numbers represent the harvest order.

variogram was obtained, producing a map of the saw-log volume and resulting variance estimates for the daily cutting blocks was routine with the software we used.

The harvester path jumped around as the demand fluctuated, and, as the purpose of the research was to examine if spatially explicit information was valuable in tactical planning, we ignored many operational costs and constraints. Since the additional processing costs (harvester movement, forwarding and additional processing) were ignored as well as the inclusion of a measure of uncertainty within the objective function, we speculate that, should these costs be included, the path of the harvester would become more continuous, yielding a more practical path.

Conclusions

The goal of this study was to determine if spatially predicting saw-log volume from traditional non-spatial sampling methods could reduce the deviation between a customer's demand and a producer's supply in a simple harvest operation. We demonstrated that, while the sample intensity increased above what would be considered a traditional non-spatial sampling intensity, there were potential benefits associated with being able to control harvest operations with better precision.

Our examination only contained size breaks for the saw-log products and, in an operational setting, there would be more than one product. Twenty or more log products, with surface attributes such as knot size and defect indicators, are typical in the Pacific Northwest. In that case, alternative Kriging procedures, such as coKriging or indicator Kriging could have been used. Now that it appears that the addition of spatially explicit information is advantageous in forest operations, we can examine additional aspects of spatial sampling, such as optimal sample design and precision tactical planning and its effects on the operation of the firm.

References

Brooks, J.R. and Wiant, H.V., Jr (2004) Efficient sampling grids for timber cruises. *Northern Journal of Applied Forestry* 21(2), 80–82.

Culvenor, D.S. (2002) TIDA: an algorithm for the delineation of tree crowns in high spatial resolution remotely sensed imagery. *Computers and Geosciences* 28, 33–44.

Dalenius, T., Hjáek, J. and Zubrzycki, S. (1961) On plane sampling and related geometrical problems. In: Neyman, J. (ed.) *Proceedings of the Fourth Berkeley Symposium on Mathematical Statistics and Probability*, Vol. 1. University of California Press, Berkeley, pp. 164–177.

Gambill, C.W., Wiant, H.V., Jr. and Yandle, D.O. (1985) Optimum plot size and BAF. *Forest Science* 31(3), 587–594.

Hay, D.A. and Dahl, P.N. (1984) Strategic and midterm planning of forest-to-product flows. *Interfaces* 14(5), 33–43.

Hehnen, M.T., Chou, S.C., Scheurman, H.L., Robinson, G.J., Luken, T.P. and Baker, D.W. (1984) An integrated decision support and manufacturing control system. *Interfaces* 14(5), 44–52.

Holmström, H. (2002) Estimation of single-tree characteristics using the kNN method and plotwise aerial photograph interpretations. *Forest Ecology and Management* 167, 303–314.

Ihaka, R. and Gentleman, R. (1996) R: a language for data analysis and graphics. *Journal of Computational and Graphical Statistics* 5(3), 299–314.

Journel, A. and Huijbregts, C.J. (1978) *Mining Geostatistics*, Chapter 5. Academic Press, London, p. 600.

Kilkki, P. and Päivinen, R. (1987) Reference sample plots to combine field measurements and satellite data in forest inventory. *Helsinginyliopiston Metsänarvioimistieteen laitoksentiedonantoja* 19, 209–215.

Kim, Y.C. and Baafi, E.Y. (1984) Combining local Kringing variances for short-term mine planning. In: Verly, G., David, M., Journel, A. and Marechal, A. (eds) *Geostatistics for Natural Resources Characterization*. Vol. 122 of Series C: Mathematical and Physical Sciences, NATO Advanced Study Institute on Geostatistics for Natural Resources Characterization, Reidel Publishing, Dordrecht, the Netherlands, pp. 185–199.

Korhonen, K. and Kangas, A. (1997) Application of nearest-neighbor regression for generating sample tree information. *Scandinavian Journal of Forest Research* 12, 97–101.

Kozak, A., Munro, D.D. and Smith, J. (1968) Taper functions and their application in forest inventory. *Forestry Chronicle* 45, 278–283.

Kravchenko, A.N. (2003) Influence of spatial structure on accuracy of interpolation methods. *Soil Science of America Journal* 67, 1564–1571.

Mandallez, D. and Ye, R. (1999) Forest inventory with optimal two-phase, two-stage sampling schemes based on the anticipated variance. *Canadian Journal of Forest Research* 29, 1691–1708.

Metropolis, N., Rosenbluth, A., Rosebluth, M., Teller, A. and Teller, E. (1953) Equation of state calculations by fast computing machines. *Journal of Chemical Physics* 21, 1087–1092.

Murphy, G., Marshall, H. and Bolding, M.C. (2004) Adaptive control of bucking on harvesters to meet order book constraints. *Forest Products Journal* 54(12), 114–121.

Oderwald, R.G. and Jones, E. (1992) Sample sizes for point, double sampling. *Canadian Journal of Forest Research* 22, 980–983.

Pebesma, E. (2004) Multivariable geostatistics in s: the gstat package. *Computers and Geosciences* 30, 683–691.

Rasinmäki, J. and Melkas, T. (2005) A method for estimating tree composition and volume using harvester data. *Scandinavian Journal of Forest Research* 20, 85–95.

Stendahl, J. and Dahlin, B. (2002) Possibilities for harvester-based forest inventory in thinnings. *Scandinavian Journal of Forest Research* 17, 548–555.

Uusitalo, J. (1997) Pre-harvest measurement of pine stands for sawing production planning. *Acta Forestalia Fennica* 259, 56.

Wagner, F.G., Brody, J.A., Ladd, D.S. and Beard, J.S. (1996) Sawtimber valuation and sawlog allocation through simulation of Temple-Inland sawmills. *Interfaces* 26(6), 3–8.

Walthall, C., Kim, M., Williams, D., Meeson, B., Agbu, P., Newcomer, J. and Levine, E. (1993) Data sets for modeling: a retrospective collection of bidirectional reflectance and forest ecosystem dynamics multisensor aircraft campaign data sets. *Remote Sensing of Environment* 46, 340–346.

Webster, R. and Oliver, M. (2001) *Geostatistics for Environmental Scientists*, Chapter 8. John Wiley and Sons, Chichester, UK, p. 271.

Zeide, B. (1980) Plot size optimization. *Forest Science* 26(2), 251–257.

19 Evaluation of Commercial Airborne LiDAR and SAR Products to Estimate Top Height and Associated Parameters in Production Forests in Britain

E.D. WALLINGTON[1] AND J.C. SUÁREZ[2]

[1]*Forestry Commission, Silvan House, 231 Corstorphine Road, Edinburgh UK;*
[2]*Forest Research, Northern Research Station, Roslin, Midlothian, UK.*

Abstract

The theory of sustainable forestry has always existed, but has only recently been defined as a management concept, and, as such, monitoring and information reporting have become an important part of forest management. This chapter demonstrates the use of two commercially available remote-sensing systems for estimating forest top height and related parameters, and discusses the operational use of such systems with regard to monitoring sustainable forestry indicators.

Airborne LiDAR and X-band Interferometric Synthetic Aperture Radar (InSAR) systems are used to retrieve forest stand top height. Top-height estimation was subsequently used to estimate stand mean, dominant and subdominant heights and mean stand diameter. Top-height estimation was achieved with errors of around 2% from LiDAR and 8% from SAR. Mean height estimation was achieved with errors of around 1% from LiDAR and 7% from SAR. Dominant and subdominant height estimation was achieved with errors of around 2% from LiDAR and 8% from SAR. Mean dbh estimation was achieved with errors of around 10.5% from LiDAR and 13.5% from SAR.

Potential reasons for errors are discussed, and the use of remote sensing is discussed in relation to operational sustainable forest management at a range of scales.

Keywords: LiDAR, InSAR, forest parameter retrieval, top height, mean height, dominant height, mean dbh.

Introduction

The goal of forest management is the maximization of a range of benefits for a wide collection of stakeholders. Nowadays, these benefits are not necessarily aligned with forest production, because environmental and recreation factors are gaining wider interest. As such, there is a required need for the continued flow of benefits from forests to satisfy present and future human needs (Franklin, 2001), and, if managed responsibly, forests will continue to enhance our quality of life and that of future generations (FC, 2002). The theory of sustainable forestry has always existed, but has only recently been defined as a management concept, i.e. maintaining forest ecosystems in a sustainable condition through the use of specific forest management objectives (Franklin, 2001). Monitoring and information reporting to promote adaptive and sustainable forest management have emerged as important components in the field of forest management (Hickey *et al.*, 2005). Sustainable forest management is a complex process, requiring input and collaboration from a number of forest sectors (e.g. management for timber, biodiversity, conservation, recreation, health, etc.) and related organizations, using a wide range of data sources and inventory techniques. Management objectives will vary given certain situations, desired end products or results and available data. Thus, a set of key indicators and guidelines have been developed in the UK, as in other countries, to help guide sustainable forestry management practices (FC, 2002, 2004).

Earth observation (EO) through the use of remote-sensing systems has the capability to provide valuable information for sustainable forestry management (Hall, 2000; Rosenqvist *et al.*, 2003).This is achieved through input into management decisions, and continued monitoring towards meeting, assessing and maintaining the indicators of sustainable forestry. As an example (with UK indicators in parentheses; see FC, 2002), EO systems are capable of monitoring woodland area through change-detection algorithms. This allows the quantification of woodland increase (A1), the creation of new wooded areas (A2) and the assessment of woodland losses (A3). Thus, EO allows the monitoring of forest changes in each country, and forms the basis for the implementation of forest policies in relation to increasing woodland cover. The ability of EO to identify species (A4) is essential to monitor the composition and diversity of woodlands, and to assess whether conversions of conifer plantations to more mixed and broadleaf forests are occurring and meeting targets. Remote sensing can map variability (in terms of forest parameters), and therefore contribute to the design of better retention schemes in a cost-effective way, by assisting forest planners to adjust economic cycles and returns. Added to this is the ability to contribute to the assessment and monitoring of the visual appearance of the landscape (A5). This is achieved by mapping the percentage of woodland cover and its location, density and proximity to other forests. This information can then be used to simulate the real world and forecast changes (FA, 1992). When integrated to computer-based models, the system can be used to model scenarios of future landscape appearance based on certain management interventions and objectives. As an example, the use of three-dimensional representations, integrated in the Forestry Commission Forester extension to ArcGIS, constructs semi-realistic landscapes that are easily interpreted by planners. This approach provides a system for forest planners to use variability in the landscape for design plans (e.g. height and diameter distributions). This information can be used for planning of forest operations, like long-term retentions, transition to continuous-cover forestry (CCF), felling plans and thinning regimes, all of which contribute to sustainable forestry. Through the above, EO can provide information that can be used towards certification to the UK Woodland Assurance Standard (UKWAS; A6) and monitoring management practices (A7).

The methodology presented in this chapter helps to improve the location and estimation of standing volume for wood-processing industries. This information enhances the field-based inventories described in the *National Inventory of Woodland and Trees* (FC, 2003; D1). Height estimates can be used to ensure that the gross annual increment in growing stock exceeds the volume being harvested or lost (e.g. by windthrow) (D2/3). Likewise, these height estimates can be used to estimate forest biomass and carbon content (D5). This information can be input into global climate models, carbon trading and assessing the UK's ability to meet targets such as those set out in the Kyoto Protocol (Rosenqvist *et al.*, 2003).

This chapter demonstrates the use of commercially available airborne remote-sensing systems in the UK to provide estimates of forest top height (defined in this study as the average height of the 100 tallest trees/ha, H_{100}; Philip, 1994). This supplements traditional field inventory, and has direct relevance to many of the aforementioned indicators. The work presented is a comparative study between commercial LiDAR (Light Detection and Ranging; Aronoff and Petrie, 2005) and Synthetic Aperture Radar (SAR; Henderson and Lewis, 1998; Woodhouse, 2005) data validated with ground-reference data. The perceived advantages of these two airborne systems are the high resolution of the data and their versatility to capture data over a large geographical area at will. Additional advantages are the increasingly competitive costs, and the possibility to process the data into a product compatible with many GIS systems (Longley *et al.*, 2005).

This study aims at the development of a cost-effective method for the estimation of the height of forest stands. Previous studies have estimated forest height to within 7–30% for LiDAR (e.g. Nilsson, 1996; Zimble *et al.*, 2003) and 4–90% for SAR (e.g. Yong *et al.*, 2003; Walker *et al.*, 2004) dependent on system and wavelength used. Top height is an important parameter in traditional production forestry (West, 2004) and can also be used as an indicator in studies looking at, for example, forest stand structure, growth dynamics, biofuel estimations or carbon sequestration – all of which need to be assessed and monitored as contributions towards sustainable forest management. Likewise, top height is used to estimate structural parameters within the stand, such as dominance and mean diameter. Additionally, the method seeks to enhance traditional field routines for forest inventory, and to explore operative issues for its implementation in the forest industry.

Study Areas and Data Sets

Study sites

Three study sites were considered: Coed Y Brenin Forest District in north Wales (52°49′12″N, 3°53′27″W, lat./long.), Kielder Forest District in Northumberland,

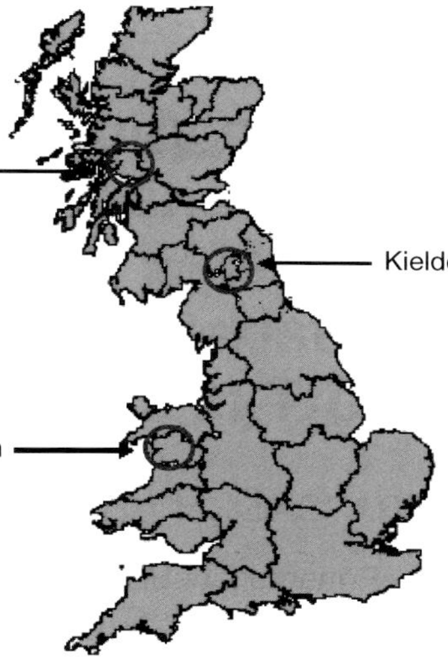

Fig. 19.1. Location of test sites.

northern England (55°11′44″N, 2°32′11″W, lat./long.), and Aberfoyle in south-west Scotland (56°10′00″N, 4°22′00″ W, lat./long.) (see Fig. 19.1). Forest stands consisted of Sitka spruce (*Picea sitchensis*) plantation. The SAR data were analysed over the three study sites, Kielder, Coed Y Brenin and Aberfoyle, whilst the LiDAR were analysed over selected stands in Kielder and Aberfoyle. Direct comparison of the two techniques was possible in Kielder and Aberfoyle, where both data sets had comparable coverage, and were subsequently used for analysis.

Data sets

Intermap Technologies has recently captured elevation data for the entire UK under their Nextmap Britain campaign (Intermap, 2005). This was achieved through using X-band SAR Interferometry (InSAR) from their airborne Star-3i sensor. Such data are rapidly being acquired for a large number of countries, so these data sets are becoming available to industrial users.

The UK Environment Agency (EA, 2005) recently acquired the capability of collecting LiDAR data for the production of high-resolution digital terrain models. This government agency also works as a contractor for other agencies throughout the UK.

InSAR digital surface model

The InSAR (Henderson and Lewis, 1998) derived Digital Surface Model (DSM) was supplied by Intermap Technologies (Intermap, 2005), and was geo-referenced to the UK national grid (OSGB36) prior to delivery. The DSM represents the first surface the signal came into contact with, whether ground or vegetation canopy. The DSM is presented with a pixel size of 5 m × 5 m and a vertical RMSE of between 0.5 and 2.0 m dependent on flying height. These accuracies are quoted for moderately sloped, unobstructed terrain (Intermap, 2003). Penetration and attenuation of the radar signal will occur, even at X-band, so the DSM height represents the height of the scattering phase centre, and not the actual canopy surface, as is sometimes assumed in reference to optical systems.

Ordnance survey DEM

The Ordnance Survey (OS) Profile 10 m Digital Elevation Model (OSDEM) was used as a ground-reference surface. The OSDEM has a pixel size of 10 m × 10 m (re-sampled to 5 m × 5 m to match the DSM; this was done purely to ease comparison and not as an attempt to improve resolution). The OSDEM has a stated accuracy of ± 5 m (OS, 2001), but this is expected to be significantly improved over small areas. As a product derived from aerial photogrammetry, there are potential limitations with the DEM, for example where dense vegetation exists. These issues are inherent in any photogrammetrically derived product. As this is the only high-resolution wide-area coverage of ground surface for the UK, it is assumed to be a representation of the true ground surface for the purposes of this study.

LiDAR

Airborne LiDAR was obtained by the Environment Agency using an Optech ALTM2033 scanner (Optech, 2005). The first survey was undertaken in Aberfoyle in September 2002 at a flying altitude of 1000 m a.s.l. The sampling intensity was 3–4 returns per m^2 and a beam divergence of 10 cm. The scanning angle was 20°. A second area in Kielder forest was surveyed in April 2003. Sampling intensity varied from 6 to 23 returns per m^2. Beam divergence was up to 1 m and the scanning angle was 10°. Data were distributed in ASCII XYZ format, where first and last returns, with their corresponding intensities, were located to the OS national grid. RMSE were ± 40 cm in X and Y and ± 9–15 cm in Z.

Ground reference data

Standard forest inventory techniques (Hamilton, 1975; Phillp, 1994; Husch *et al.*, 2003) were used to establish the top height of the sample stands. This height was then used as the true top height of the stand. Individual tree locations were mapped using differential GPS and a total station, with x/y positional accuracy of < 1 m; these locations were subsequently used for locating individual trees in the LiDAR imagery for analysis.

Top-height Retrieval from Remote Sensing

The technique for retrieval of top height used in this study was similar for the InSAR and LiDAR data sets, but a number of different steps were taken in each. The generic algorithm for height retrieval was:

$$H_{Tree} = H_{Canopy} - H_{Ground} \quad (1)$$

InSAR

In all the stands, height values from the DSM and OSDEM were retrieved from 50 m × 50 m plots. Care was taken to extract values away from potential error sources, for example edge effects, which have a significant impact on height retrieval (Woodhouse *et al.*, 2006). Following equation (1), subtraction of the OSDEM from the DSM was performed to recover the height per pixel within the plot (Wallington *et al.*, 2004). Retrieved top height of the stand (Hr_{100}) was estimated by averaging the highest 25 pixel heights. This technique follows standard forest practice of taking the tallest 100 trees/ha (Philip, 1994). Retrieved top height was then compared with the measured top height per stand (Fig. 19.2).

Retrieval of top height for the three test sites gave consistent results (Fig. 19.2), with underestimations of 18% (Aberfoyle), 21% (Wales) and 24% (Kielder). The average underestimation for the three data sets combined was 23% ($R^2 = 0.91$). The retrieved height was an underestimation, as expected. This was attributed to signal interaction with scatterers within the canopy of a similar size to the signal wavelength, and the resultant effect on penetration and attenuation through the canopy. At shorter wavelengths (X-band), penetration is limited to the upper canopy, and as such the resultant height of the scattering phase centre is predominantly dominated by scattering from the smaller scatterers in the canopy. The results show that X-band does not easily penetrate the dense canopy of Sitka spruce stands, and this is attributed to the compactness of their canopies. In contrast, other species such as Japanese larch and Scots pine present less dense canopies, and will allow more penetration of the signal through them (Wallington *et al.*, 2005).

Further to the above-mentioned penetration, a number of other factors may also be affecting height retrieval, thus producing underestimations. The aforementioned

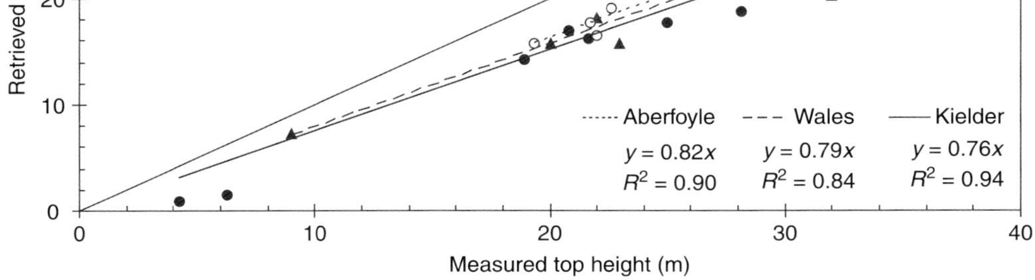

Fig. 19.2. Retrieved top height from SAR in Wales (Coed Y Brenin), Kielder and Aberfoyle. 1 : 1 line is shown.

edge effects have been shown to significantly reduce height retrieval (Woodhouse *et al.*, 2006). This is due to the relative contributions of canopy and ground scattering to the resulting scattering phase centre. Increased ground contribution results in a lower height. Edge effects occur at the edges of forest stands, but may also occur within a stand with emergent trees or canopy gaps. Other sources of potential error include the density of the stand, the height of the trees, canopy shape and the slope of the underlying ground surface in relation to the sensor position (Izzawati *et al.*, 2004). The OSDEM has an accuracy of ± 5 m, and it is reasonable to assume that an error in this ground surface may be contributing to the height underestimation.

LiDAR

LiDAR instruments can generate accurate models of canopy height that can be used to estimate other forest parameters such as canopy heights, stand volume and the vertical structure of the forest canopy. A normalized model of forest canopy is obtained by subtracting bare ground values from the canopy layer. In commercial airborne systems, the canopy layer is retrieved from the first laser return that measures the intensity of the signal as it first encounters an object on the ground. The last return will provide information about the location and height of the midpoint of the last strong waveform that is normally associated with the terrain.

In order to approximate a model of the ground surface, the last returns were filtered to eliminate those hits being intercepted by the forest canopy and therefore not reaching the underlying terrain. The method involved an initial selection of points within a kernel of 10 m × 10 m according to the local minima (Suárez *et al.*, 2005a). Then, a basic DTM model was constructed using an interpolation method based on Kriging with no anisotropy. After, this a second selection of ground hits was performed by comparing LiDAR hits with this model within an empirically defined threshold of ± 30 cm. This operation densified the initial selection of points, allowing the construction of a higher-resolution DTM using the same interpolation method (Fig. 19.3).

A normalized model for canopy height was calculated for every laser hit as the difference between each first return and the resultant terrain model. Individual tree heights were accurately predicted in 73% of the cases within ± 1 m and 96% within ± 2 m. The largest under-predictions were observed in subdominants. Suppressed and dead trees were missed completely. Generally, individual tree heights were 7–8% shorter than observed due to the low number of laser hits intercepted by the apices. In Kielder, the higher density of returns per m^2 (6 to 23) reduced underestimations to less than 2% (Fig. 19.4).

Correction of retrieved height

Comparison of height retrieval for LiDAR and SAR was possible in Kielder and Aberfoyle study sites, where there was comparable data coverage. Once the

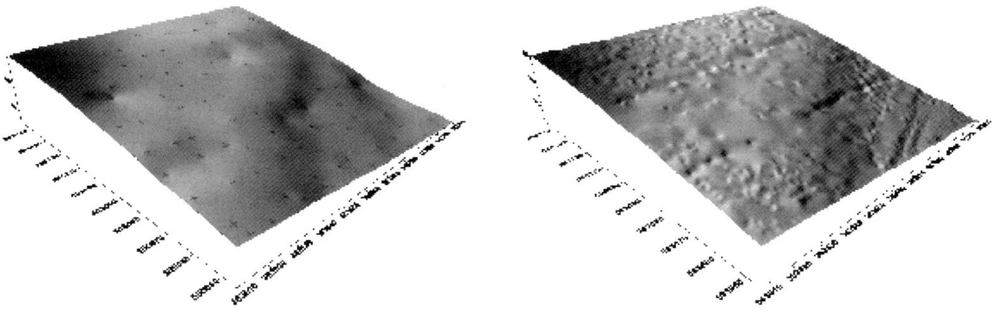

Fig. 19.3. Basic DTM produced using LiDAR (left), and the final higher-resolution DTM (right). A lack of LiDAR penetration can be identified by the loss of detail in the centre of the lower drainage channel (right).

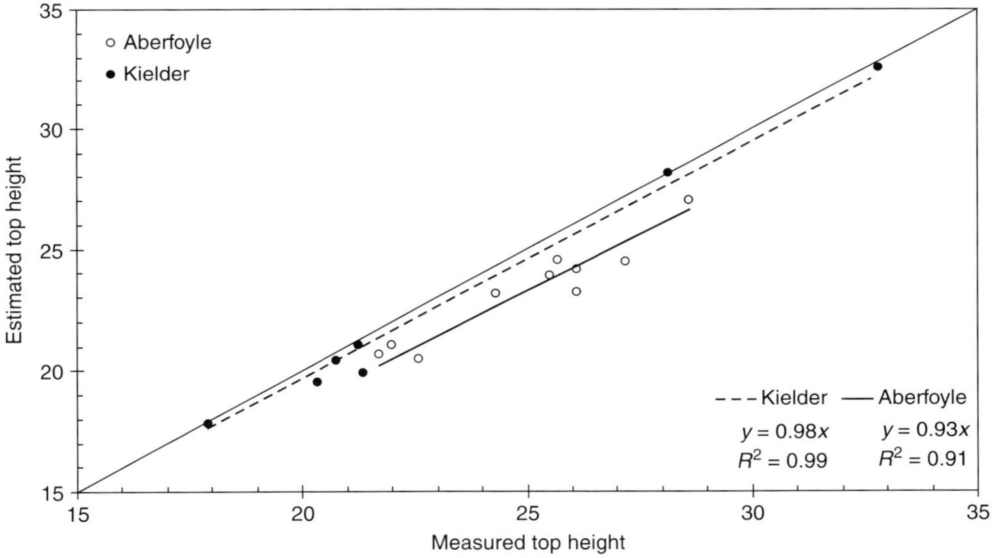

Fig. 19.4. Retrieved top height from LiDAR in Aberfoyle and Kielder. 1 : 1 line is shown.

retrieved top height had been estimated by LiDAR and SAR, these values were corrected by the amount of overall underestimation to obtain a recovered top height (Table 19.1). Retrieved height estimates with an error greater than 2% were corrected. It was found that correction of errors below the 2% threshold resulted in an overall increase in error due to overcompensation. Through correction, top-height estimation errors for LiDAR were improved from 4.95% to 1.75% and from 22.94% to 7.68% for SAR. These errors are within generally accepted tolerances.

Table 19.1. Summary of errors in estimations for retrieved and recovered top height (H_{100}).

	Stand	Measured top height (m)	LiDAR top height (% error)		SAR top height (% error)	
			Retrieved	Recovered	Retrieved	Recovered
Kielder	1	21.7	–	–	24.92	8.40
	2	20.8	–	–	18.76	0.89
	3	18.9	–	–	24.40	7.77
	4	28.8	–	–	21.04	3.67
	5	25.0	–	–	28.81	13.14
	6	32.8	0.96	0.96	13.88	5.06
	7	28.2	0.01	0.01	33.04	18.31
	8	20.3	4.19	0.60	–	–
	9	21.4	6.80	2.14	–	–
	10	17.9	0.56	0.56	–	–
	11	20.8	1.73	1.73	–	–
	12	21.2	1.08	1.08	–	–
Aberfoyle	1	27.2	9.97	5.47	16.11	2.34
	2	22.6	9.47	4.94	–	–
	3	22.6	–	–	15.79	2.74
	4	25.7	4.74	0.02	–	–
	5	21.7	4.89	0.13	18.27	0.28
	6	22.0	4.54	0.23	25.16	8.69
	7	25.5	6.39	1.71	–	–
	8	28.6	5.32	0.59	–	–
	9	26.1	7.40	2.77	–	–
	10	26.1	11.16	6.72	–	–
	11	23.8	–	–	–	–
	12	24.3	4.86	0.10	35.14	20.88
Average			4.95	1.75	22.94	7.68

Height Estimation Based on Retrieved Top Height

Using the corrected top height discussed above, it is possible to estimate other stand parameters that are useful for forest structure studies and forest management decisions. Stand mean, dominant and subdominant heights can be estimated using equations (2)–(4), which have been developed for Sitka spruce stands. These equations were derived from measurements based on 11,571 trees across Great Britain and used within the UK Forest Research Environmental Database (FRED).

$$\text{Mean height} = \text{Top height} \times 1.0467 - 2.1452 \quad (2)$$

$$\text{Dominant height} = \text{Mean height} \times 1.05 \quad (3)$$

$$\text{Subdominant height} = \text{Mean height}/1.05 \quad (4)$$

The three heights were calculated using the measured top heights, and, from the retrieved top-height estimates from LiDAR and SAR, comparisons were made to assess accuracy (Figs 19.5–19.7).

Mean height estimation

Mean height estimates for LiDAR were found to have average errors of 2% (Table 19.2), with an overall trend of 2% underestimation from regression (Fig. 19.5). Average SAR errors were 23%, with an overall underestimation of 6%.

Dominant height estimations

Dominant height estimates for LiDAR were found to have average errors of 6% (Table 19.2), with an overall trend of 6% underestimation from regression (Fig 19.6). Average SAR errors were 11%, with an overall underestimation of 10%.

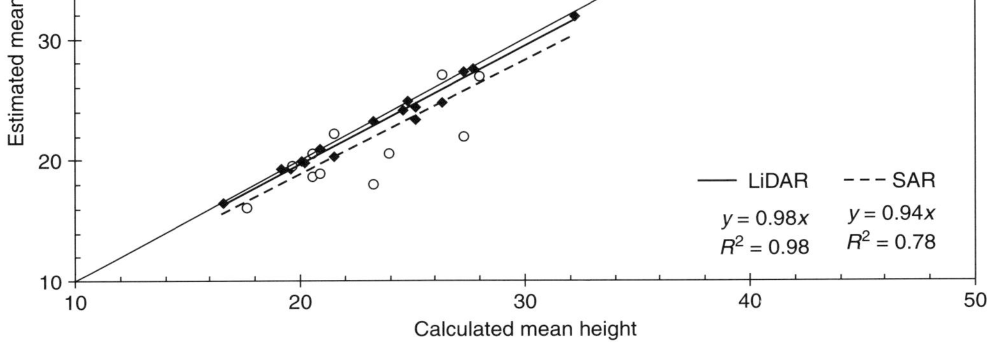

Fig. 19.5. Retrieved mean height from LiDAR and InSAR in Aberfoyle and Kielder. 1 : 1 line is shown.

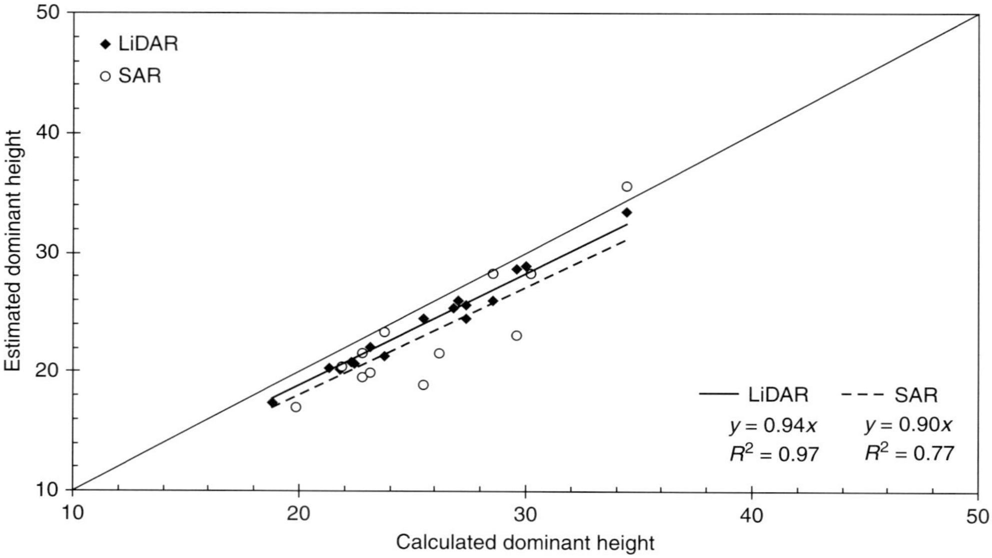

Fig. 19.6. Retrieved dominant height from LiDAR and InSAR in Aberfoyle and Kielder. 1 : 1 line is shown.

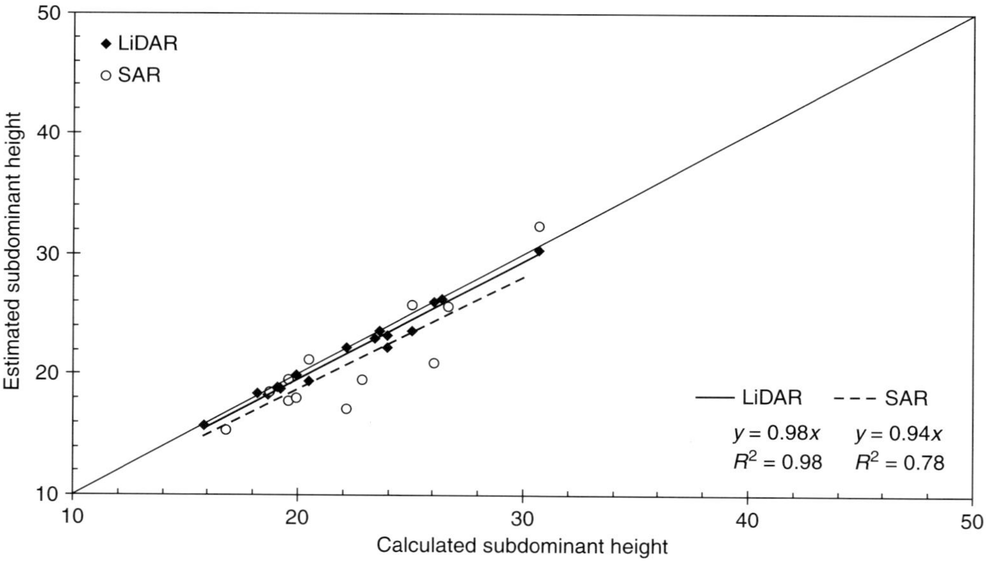

Fig. 19.7. Retrieved subdominant height from LiDAR and InSAR in Aberfoyle and Kielder. 1 : 1 line is shown.

Table 19.2. Summary of retrieved and recovered height estimation errors.

	LiDAR (% error)		SAR (% error)	
	Retrieved	Recovered	Retrieved	Recovered
Top height	4.95	1.75	22.94	7.68
Mean height	1.91	1.05	8.39	7.18
Dominant height	6.10	2.06	11.14	7.83
Subdominant height	1.91	1.35	8.39	7.18

Subdominant height estimations

Subdominant height estimates for LiDAR were found to have average errors of 2% (Table 19.2), with an overall trend of 2% underestimation from regression (Fig. 19.7). Average SAR errors were 8%, with an overall underestimation of 6%.

These results were promising, but, through correction by the amount of underestimation, the recovered heights show reduced errors (Table 19.2). These heights are all based on corrected retrieved top height, and it should be noted that any errors in estimation of top height are carried through to subsequent height estimations.

Mean Diameter Estimation

Another key forest parameter is mean diameter of a stand. Mean diameter can be used to assess forest growth, input into yield predictions, and also as an indicator of diameter distributions within a stand (Hamilton, 1975) and across a forested landscape. Using measurements from 11,571 trees across Great Britain, contained within the UK Forest Research Environmental Database (FRED), a linear relationship between diameter at breast height (1.3 m; dbh) and the total height of individual trees was derived (Fig. 19.8).

$$\text{Individual dbh} = \text{Total height} \times 1.3213 \qquad (5)$$

Mean height refers to the mean total height of a stand (Hamilton, 1975), and for the purposes of this study it was assumed that mean height is equivalent to total height. Given this assumption and a linear relationship, mean height can be substituted in equation (5) to give:

$$\text{Mean dbh} = \text{Mean height} \times 1.3213 \qquad (6)$$

Equation (6) was used to calculate stand mean dbh using the retrieved mean heights from LiDAR and SAR. Mean height estimates for LiDAR were found to have average errors of 15.5% (Table 19.3), with an overall trend of 12% overestimation from regression (Fig. 19.9). Average SAR errors were 28%, with an overall overestimation of 21%. The retrieved mean dbh estimates were then corrected, given the amount of overestimation to produce recovered

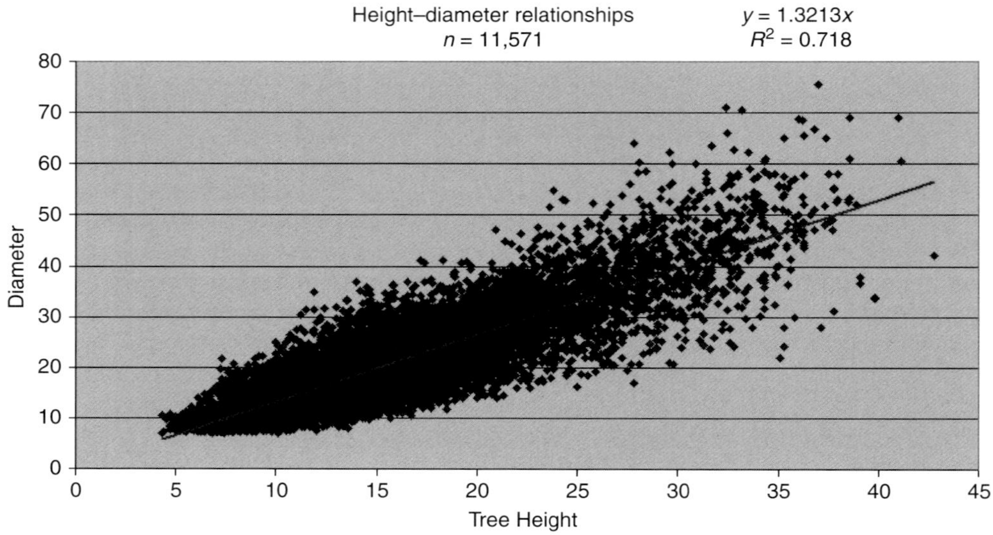

Fig. 19.8. Relationship of individual tree total height to diameter at breast height (dbh).

Table 19.3. Summary of retrieved and recovered mean diameter estimation errors.

	LiDAR (% error)		SAR (% error)	
	Retrieved	Recovered	Retrieved	Recovered
Mean diameter	15.51	10.50	28.07	13.47

values, and the amount of error was reduced to 10.5% and 13.5%, respectively (Table 19.3).

Summary and Conclusion

This chapter has demonstrated the use of commercially available remote-sensing systems to estimate key forest parameters. Airborne LiDAR and SAR were used to initially estimate top height of a stand, from which mean, dominant and subdominant heights and mean dbh were estimated. The following results were found:

- Top-height estimation from LiDAR gave an overall underestimation of between 2 and 6%, resulting in errors of 4.95%, which were reduced to 1.75% after correction. Likewise, SAR gave an overall underestimation of between 2 and 7%, with errors of 22.94%, which were reduced to 7.68% after correction.

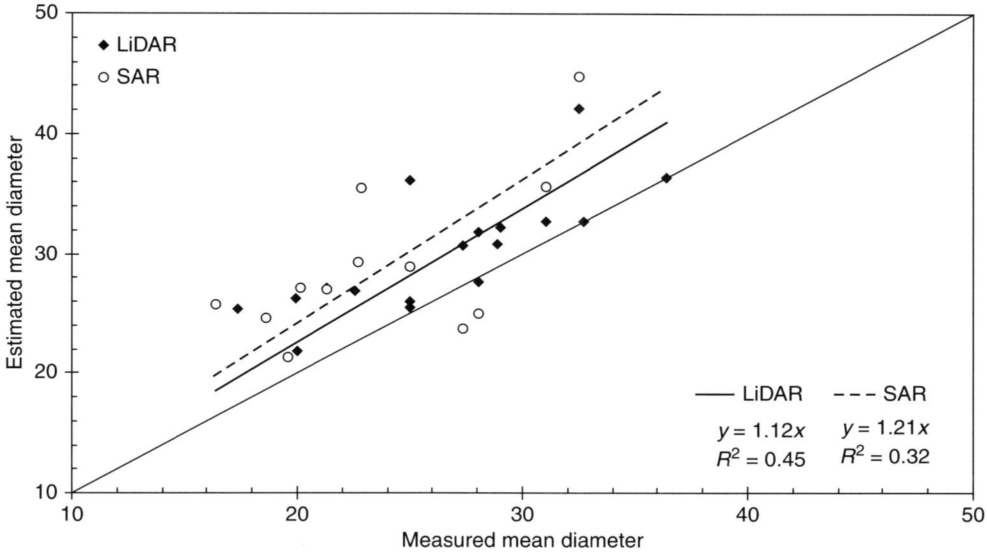

Fig. 19.9. Measured mean diameter against estimated mean diameter estimated from LiDAR and InSAR data. 1 : 1 line is shown.

- Mean height estimation from LiDAR gave an overall underestimation of 2%, resulting in errors of 1.91%, which were reduced to 1.05% after correction. Likewise, SAR gave an overall underestimation of 6%, with errors of 8.39%, which were reduced to 7.18% after correction.
- Estimation of dominant height from LiDAR gave an overall underestimation of 6%, resulting in errors of 6.10%, which were reduced to 2.06% after correction. Likewise, SAR gave an overall underestimation of 10%, with errors of 11.14%, which were reduced to 7.83% after correction.
- Estimation of subdominant height from LiDAR gave an overall underestimation of 2%, resulting in errors of 1.91%, which were reduced to 1.35% after correction. Likewise, SAR gave an overall underestimation of 6%, with errors of 8.39%, which were reduced to 7.18% after correction.
- Mean dbh estimation from LiDAR gave an overall overestimation of 12%, resulting in errors of 15.51%, which were reduced to 10.50% after correction. Likewise, SAR gave an overall overestimation of 21%, with errors of 18.07%, which were reduced to 13.47% after correction.

These results indicate the capabilities of both airborne LiDAR and short-wavelength SAR interferometry to provide accurate estimates of key forest parameters. Having a sample of ground reference data allows correction of retrieved heights to gain better accuracies; this correction can then be applied over large areas. The parameters discussed are of use to a wide range of end-users, and complement current inventory techniques. The ability of remote sensing products to cover large areas rapidly and to provide a sampling intensity that approaches full coverage, as opposed to traditional sampling strategies

(Suárez *et al.*, 2005b), will allow more detailed assessment and monitoring for sustainable forest management and related industries. Both of the airborne data sets discussed in this chapter are available commercially, and, as such, the techniques described are nearing operational status. Further potential is foreseen with the ever-increasing availability and use of satellite systems for data capture, allowing even greater areas to be rapidly mapped.

The operational capability of remote sensing to provide data for the forest industry shows significant potential for local, regional and global forest mapping. However, the ability to map at an appropriate resolution and with continued data availability has always been an issue. With ongoing airborne high-resolution data acquisition now available commercially, and with new satellites being launched in the near future (e.g. TerraSAR-X), the issue of data availability and updating is being resolved. The ability to map large areas of land has frequently led to a cost-effective method, which can be used to support and supplement field inventories.

Acknowledgements

The authors would like to thank Intermap Technologies and the Environment Agency for kindly supplying the data for the test sites. Thanks are also due to Stuart Snape (UK Forest Enterprise) for supplying ground-reference data for Coed Y Brenin; to Daniel Donoghue and Pete Watt of Durham University for supplementary ground data for Kielder, and to Iain Woodhouse and Izzawati of Edinburgh University for assistance with the interpretation of radar modelling results. This study was funded by the National Environment Research Council (NERC) and the UK Forestry Commission.

References

Aronoff, S. and Petrie, G. (2005) Active sensors: radar and Lidar. In: Aronoff, S. (ed.) *Remote Sensing for GIS Managers.* ESRI Press, New York.

EA (2005) http://www.environment-agency.gov.uk/ UK Environment Agency, last accessed 14 June 2005.

FA (1992) *Lowland Landscape Design Guidelines*. Forest Authority, Edinburgh, UK.

FC (2002) *UK Indicators of Sustainable Forestry*. Forestry Commission, Edinburgh, UK.

FC (2003) *National Inventory of Woodland and Trees – Great Britain*. Inventory Report, Forestry Commission, Edinburgh, UK.

FC (2004) The UK *Forestry Standard: The Government's Approach to Sustainable Forestry*, 2nd edn. Forestry Commission, Edinburgh, UK.

Franklin, S.E. (2001) *Remote Sensing for Sustainable Forest Management*. CRC Press, Boca Raton, Florida.

Hall, R.T. (2000) Applications of remote sensing to forestry – current and future. *Forestry Chronicle*, 76(6), 855–857.

Hamilton, G.J. (1975) *Forest Mensuration,* 5th impression. Forestry Commission Booklet 39, Forestry Commission, Edinburgh, UK.

Henderson, F.M. and Lewis, A.J. (eds) (1998) *Principles and Applications of Imaging Radar, Manual of Remote Sensing*, 3rd edn, Vol. 2. John Wiley and Sons, New York.
Hickey, G.M., Innes, J.L., Kozak, R.A., Bull, G.Q. and Vertinsky, I. (2005) Monitoring and information reporting for sustainable forest management: an international multiple case study analysis. *Forest Ecology and Management* 209, 237–259.
Husch, B., Beers, T.W. and Kershaw, J.A. (2003) *Forest Mensuration,* 4th edn. John Wiley and Sons, New Jersey.
Intermap (2003) *Product Handbook and Quick Start Guide*. Intermap Technologies Corporation, Denver, Colorado.
Intermap (2005) http://www.intermap.com. Intermap Technologies. Last accessed 14 June 2005.
Izzawati, Woodhouse, I.H. and Wallington, E.D. (2004) The impact of forest heterogeneity on the height retrieval using X-band interferometry. In: *Proceedings of IEEE IGARSS 2004, 20–24 September 2004, Anchorage, Alaska.* IEEE GRSS, Piscataway, New Jersey.
Longley, P.A., Goodchild, M.F., Maguire, D.J. and Rhind, D.W. (2005) *Geographic Information Systems and Science,* 2nd edn. John Wiley and Sons, Chichester, UK.
Nilsson, M. (1996) Estimation of tree heights and stand volume using an airborne Lidar system. *Remote Sensing of Environment* 56, 1–7.
Optech (2005) Optech Incorporated, Toronto, Ontario, Canada.
OS (2001) *Land-form Profile User Guide.* UK Ordnance Survey, Southampton, UK.
Philip, M.S. (1994) *Measuring Trees and Forests,* 2nd edn. CAB International, Wallingford, UK.
Rosenqvist, A., Milne, A., Lucas, R., Imhoff, M. and Dobson, C. (2003) A review of remote sensing technology in support of the Kyoto Protocol. *Environmental Science and Policy* 6, 441–455.
Suárez, J.C., Ontiveros, C., Smith, S. and Snape, S. (2005a) Use of airborne LiDAR and aerial photography in the estimation of individual tree heights in forestry. *Computers and Geosciences* 31(2), 253–262.
Suárez, J.C., Smith, S., Bull, G., Malthus, T., Donoghue, D. and Knox, D. (2005b) The use of remote sensing techniques in operational forestry. *Quarterly Journal of Forestry* 99(1), 31–42.
Walker, W.S., Pierce, L.E., Kellndorfer, J.M., Dobson, M.C., Hunsaker, C.T. and Fites, J.A. (2004) A comparison of forest canopy height estimates derived from SRTM and TOPSAR in the Sierra Nevada of California. In: *Proceedings of IEEE IGARSS 2004,* 20–24 September 2004, *Anchorage, Alaska.* IEEE GRSS, Piscataway, New Jersey.
Wallington, E.D., Izzawati and Woodhouse, I.H. (2004) Forest height estimation from X-band SAR. In: *Proceedings of IEEE IGARSS 2004, 20–24 September 2004, Anchorage, Alaska.* IEEE GRSS, Piscataway, New Jersey.
Wallington, E.D., Woodhouse, I.H. and Izzawati (2005) Commercial airborne X-band Synthetic Aperture Radar for forest height inventory. In: Olsson, H. (ed.) *Proceedings of ForestSAT Conference 2005, 30 May–3 June, Boras, Sweden.* Swedish National Board of Forestry.
West, P.W. (2004) *Tree and Forest Measurement.* Springer-Verlag, Berlin.
Woodhouse, I.H. (2005) *Introduction to Microwave Remote Sensing.* Taylor and Francis, Abingdon, UK.
Woodhouse, I.H., Izzawati, Wallington, E.D. and Turner, D. (2006) Edge effects on tree height retrieval using X-band interferometry. *IEEE Geoscience and Remote Sensing Letters* 3(3), 344–348.
Yong, P., Zengyuan, L., Sun, G., Erxue, C. and Xuejian, C. (2003) Comparison of tree height estimations from C and L-band InSAR data. In: *Proceedings of IEEE IGARSS 2003, 21–25 July 2003, Toulouse, France.* IEEE GRSS, Piscataway, New Jersey.
Zimble, D.A., Evans, D.L., Carlson, G.C., Parker, R.C., Grado, S.C. and Gerard, P.D. (2003) Characterizing vertical forest structure using small-footprint airborne LiDAR. *Remote Sensing of Environment* 87, 171–182.

20 Potential Contributions of Statistics and Modelling to Sustainable Forest Management: Review and Synthesis*

KEITH RENNOLLS,[1] MARGARIDA TOMÉ,[2] RONALD E. MCROBERTS,[3] JEROME K. VANCLAY,[4] VALERIE LEMAY,[5] BIING T. GUAN[6] AND GEORGE Z. GERTNER[7]

[1]University of Greenwich, UK; [2]Technical University of Lisbon, Portugal; [3]USDA Forest Service, St Paul, USA; [4]Southern Cross University, Australia; [5]University of British Columbia, Vancouver, Canada; [6]National Taiwan University, Taipei, Taiwan; [7]University of Illinois, Urbana, USA.

Abstract

This chapter provides a review of the statistical and modelling disciplines, their techniques and potential contribution to sustainable forest management (SFM). The main topics covered are:

- Mensuration and models for SFM
- Inventory and monitoring for forest sustainability: criteria and indicators
- Models of tropical forests for the conservation of biodiversity
- Integrating information and models across spatial and temporal scales for SFM
- Climate and carbon models in relation to sustainability
- New techniques for the statistical analysis of sustainability data
- Uncertainty analysis in modelling and monitoring for SFM
- Forest data, information and model archives

There are major contributions to be made, in particular in the areas of information and model integration, where a synthesis of information and models across both spatial and temporal scales is required. There is a great need for international collaboration on the development of open and shared forest data and model repositories/archives, as well as continued development of forest information systems.

* All the authors are, and have been, officers of: IUFRO Div. 4 Forest Assessment, Modelling and Management; IUFRO 4.01.00 Forest Mensuration and Modelling; and IUFRO 4.03.00 Informatics, Modelling and Statistics. See http://www.iufro.org/

Introduction

First we give a brief introduction to the topic of sustainable (multifunctional) forest management (SFM) and some of its obvious challenges. We then give descriptions of 'statistics' and 'modelling' (S & M) and discuss briefly their relationship to SFM. The main body of the chapter focuses on a number of important areas in which there are further challenges to achieving the aims of SFM. The ways in which S & M have potential to meet these challenges and so contribute to research, understanding and practice of SFM are reviewed and discussed.

Sustainable (multifunctional) forest management (SFM)

Forests have always had multiple uses and have always provided multiple services for society, including timber production. However, during the last two decades, and in particular since the Earth Summit in Rio de Janeiro in 1992, forest management has become increasingly complex. This has most significantly been so for the tropical forests of developing nations, where there have been large human populations directly dependent on non-timber products for their livelihood and survival. Hence the social and economic values of non-timber forest products to such forest-dependent communities must be included as part of the forest management envelope. The increasing scientific evidence that unique ecosystems and irreplaceable biodiversity resources are being lost through deforestation of tropical forest regions and by climate change (whatever its cause) has heightened awareness of the need for forest management planning to include biodiversity conservation considerations. It is also of major importance to take into account the carbon sequestration role of forests and the potential they have to ameliorate global warming trends. Of course, the traditional role of forest management, to ensure the efficient production of timber products, remains of high economic importance for all the world's forested regions.

We have arrived at the concept, clearly enunciated in the Rio Earth Summit resolutions, of 'sustainable multifunctional forest management', the objective of which is the optimization of the production of multiple products – timber, wood for fuel, cork, berries, game, medicinal plants, etc. – while maintaining the equilibrium of the forest ecosystem and satisfying society's current and future needs and demands for recreation, biodiversity, landscape and environmental conservation. This is not an easy task. Strictly speaking, optimization of all functions of multifunctional forestry is impossible, because the various functions compete for the same limited resources. Hence, optimization can only be in terms of a composite measure of those multiple functions (Chapter 6, this volume).

Statistics

Statistics, as a discipline, is concerned with the detection, display and representation of patterns and relationships that exist within data sets but which are often obscured by noise in the data. Clearly, statistics is relevant to SFM wherever there are data to be collected or analysed, wherever conclusions need to be drawn and whenever statements or predictions and decisions about sustainability need to be made.

Statistics is also necessarily concerned with the way in which data are collected, because this can influence how the data should be analysed, modelled and interpreted. Hence statistical design of the data-collection process is a crucial, though difficult, task. Experimental design and sample survey and inventory design are two areas in which data-collection design has been developed to a high standard. Design of data collection for model building is less formalized and less well developed, but is probably an area that needs considerable future attention in relation to SFM (Rennolls, 1999a).

The different types of data that arise in SFM (e.g. species labels, risk classes, biodiversity indices, etc.), together with a range of aims, invoke the use of a wide range of established techniques, based on assumed models. For analysis of structure in attribute space there is association/correlation analysis and latent-variable models (such as principal components and factor analysis). Examining dependencies and prediction models calls for regression analysis (including, for example, mixed models and the multilayer-perception neural network). In case-space the methods of cluster analysis and classification (e.g. multiple discriminant analysis, decision trees, etc.) are available. Simultaneous analysis in attribute-space and case-space has ordination methods such as canonical correlation analysis and correspondence analysis, which have been much used in population ecology studies. Methods are known as supervised or unsupervised according to whether they involve fitting procedures (Hastie *et al.*, 2001). All these techniques are used fairly routinely in SFM-related activities, from the analysis and use of satellite imagery, through modelling stand growth, to finding appropriate SFM indicators and indicator species, to mention just a few examples. However, there is much scope and need for the development and application of new models and methods appropriate to new problems as they arise in SFM.

Noise in data may arise from stochasticity, or from sampling and from measurement error, and leads to uncertainty in conclusions and risks in decision making. Characterizing these sources of error, their nature, properties and detailed structure, and how they percolate through an SFM system is important if decision making in SFM is to be evidence-based.

Statistics is the scientific methodology of the soft sciences (Rennolls, 1995b), and is often concerned with clarifying definitions of what is meant by entities and constructs, their attributes, measures and derived measures (indices, indicators, etc.). Hence, statistics in SFM asks for clear definitions of biodiversity and sustainability. It asks about the meaningfulness of propositions/models (i.e. that they should be invariant under a change of measurement scale), and if those propositions/models can be demonstrated to be true in some sense, and identifiable and estimable from the data that might be available. For example, is it meaningful to aim for SFM in a climate-change scenario? Answers depend not only on the scales of measurement (and the resolution of measurements), but more importantly on the spatial and temporal scales within which each of the component features of SFM is meaningfully placed. It seems that, in much SFM activity to date, basic definitions are not clear, and issues of meaningfulness and achievability are assumed (Chapter 6, this volume).

Statistics is closely related to machine-learning-based data mining and knowledge discovery in databases (KDD). Statistics, data mining and KDD depend heavily on data-visualization techniques (Andrienko and Andrienko, 2005). These range from the simple scatter plot and histogram to multidimensional techniques, kernel-smoothing and other non-parametric summarization methods. Tabular techniques range from the two-way contingency table up to the data hypercube used in KDD. It is probably the case that in an area as complex as SFM, with multiple data sources over a range of spatial and temporal scales, rational evidence-based policy and decision making will only be possible with extensive visualization support, including the use of geographic information systems (GIS). The ultimate use of discovered information and knowledge is in its deployment within information and decision-support systems. It is probably in the designing of such systems that there is most need for interdisciplinary interaction and collaboration.

Modelling

The statistical models mentioned in the last section are data-driven. They merely attempt to describe the patterns and relationships inherent in a data set, and hence are often called descriptive or empirical models. Statistical modelling adopts the principle of parsimony: that the simplest model that adequately describes the data is the best model (Baker, 2004). However, this principle can be a liability when science insists that the real world is complex, and that correspondingly complex models are required. Scientists are by definition modellers, and can and should make use of all of the models and techniques of statistics when they are appropriate. Statistical modellers become scientific modellers as soon as they replace their parsimonious statistical models by models that come from the scientific theories of the domain from within which the data were collected. They should do so whenever possible. Hence, there should be no essential dichotomy between data-driven statistical modelling and science-driven modelling.

The main topics we cover in the following sections of this chapter are:

Mensuration and models for SFM
Inventory and monitoring for forest sustainability: criteria and indicators
Models of tropical forests for the conservation of biodiversity
Integrating information and models across spatial and temporal scales for SFM
Climate and carbon models in relation to sustainability
New techniques for the statistical analysis of sustainability data
Uncertainty analysis in modelling and monitoring for SFM
Forest data, information and model archives

Mensuration and Models for SFM

Forest mensuration and sustainability

The measurement of a single forest stand, and of a single tree, must be fundamental to any sustainability monitoring programme that attempts to compare forests over time and space. Traditional forest mensuration has used sample plots, randomly or systematically located, to measure trees and hence to estimate stand state variables, such as mean number stocking, mean diameter at breast height (dbh) and stand top height. Sample plots are usually of a fixed size but plots with a fixed number of trees can also be used (Paulo *et al.*, 2005). The variable-plot method, using a relascope/angle gauge, provides an estimate of basal area stocking (Bitterlich, 1947; Holgate, 1967) and top height (Garcia, 1998), and variations in its use have much potential in relation to plot mensuration for SFM (Gregoire *et al.*, 1995; Jordan *et al.*, 2004). All mensurational terms have specific meanings: for example, mean basal area is usually meant to be the quadratic mean diameter calculated from the mean basal area per tree; stocking is usually interpreted as the relascope measure consisting of the fraction that the stem basal area (at 1.3 m) is of ground area, rather than mean basal area per tree or number of trees per unit area. For stands consisting of simple mixtures of species, the measurement of each of the species on each plot should provide no problem. However, if the stand contains many species, mostly rare, and occurring in different layers within the canopy, as in the case of a tropical forest, then some form of nested sampling might be required. The gradsect of Gillison and Brewer (1985), a transect that is oriented along the line of maximum gradient, while it reveals maximal sample species abundance, introduces a sampling bias for other stand variables. LiDAR offers ultra-high-resolution three-dimensional structural measurement of canopy surfaces and within-stand structure, which offers much in terms of accuracy of stand measurement but poses considerable analytical and processing challenges (Magnussen, 1999; Hill and Thompson, 2005). Integration of all forms of mensurational and sensing data, for SFM purposes, using data-fusion techniques, is the vision of 'digital forestry' (Zhao, G.A. *et al.*, 2005).

The validity of all mensurational measures depends implicitly on assumptions about how the tree or sample plot is selected. Appropriate use of mensurational data depends on how it is defined/obtained/sampled, and this information should always be captured in measurement metadata. For example, the usual measurement of stand top height often involves selecting the largest-girth trees in a plot for height measurement. This means that the bivariate height–diameter relationship and its spatial structure influence the way that top height should be estimated from the collected data: simple proportionality approaches are erroneous (Matérn, 1976; Rennolls, 1978, 1979; Garcia, 1998; Magnussen *et al.*, 1999). The use of spatial models in the analysis of forest data was pioneered by Matérn (1960).

A problem arises for the monitoring of sustainability over space and time because different countries, and even different regions in the same country, have differing mensurational practices. This problem of mensurational incomparability has been long known by organizations involved in attempting to combine forestry information from different countries (Köhl *et al.*, 1997, 2000). Clear metadata for all mensurational data sets are essential. Models for interconversion between differing standards need to be developed (Rennolls *et al.*, 2004b) if valid analyses of wide-scale spatio-temporal sustainability are to be possible: the administrative alternative of harmonization is prohibitively expensive.

Forest growth-and-yield models

Models of forest growth and dynamics, used to predict the effects of different management practices on forest ecosystem behaviour over time, increasingly play a key role in supporting sustainable forest management (Munro, 1974). Models have evolved with the evolution of forest management objectives, and in so doing they have made use of increased knowledge about ecosystem functioning and technology development. Forest models were originally empirical growth-and-yield models, but today have evolved into a spectrum of models that ranges from state-space stand-level models (Garcia, 2003), through distribution-based models and individual-tree models to complex process-based ecophysiological models (see Thornley, Chapter 21, this volume). Each of these models has an area of application for which it remains the best model for management purposes.

Empirical forest models for SFM

Empirical forest models may be developed from any data collected from a forest – not necessarily growth-and-yield data. The data and models could relate to tree total biomass or biomass by components (Reed and Tomé, 1998); tree health or mortality (Rennolls and Peace, 1986); tree species (and hence biodiversity: Rennolls and Laumonier, 1999a, 2000); seed, insect, pathogen or fire dispersal; movement of animals around the forest, and the way in which they damage trees or tree stability (Blackwell *et al.*, 1990). Such models can get very complex and become much more than merely empirical models. If the forest data are measurements on trees, over a stand and over time, then the models might be either stand-structure models or growth models. All features of a forest and their associated models would seem to be of relevance in relation to SFM, because each feature might conceivably be a means by which the desired sustainability is undermined.

Most empirical growth models use site index (stand height at a predefined age) to characterize site productivity, though there are important exceptions. Such empirical models are adequate for describing growth and yield for the range of silvicultural practices and site conditions that were used to produce the sample-plot data to which the models were fitted. Some of the main contributions to the methodology of site-index and growth modelling are Pienaar and Turnbull (1973), Bailey and Clutter (1974), Garcia (1983, 1993, 2005), Borders *et al.*, (1988), Rennolls (1995a), Amaro *et al.* (1998) and Cieszewski and Bailey (2000), though there still seem to be some open questions concerning appropriate parameter characterizations for multivariate site-index and growth-model development.

Empirical growth-and-yield models exist for almost all of the most important temperate and boreal forest types, but they are not appropriate for yield predictions under conditions of changing climate, and are not valid for new silvicultural management treatments (Amaro and Tomé, 1999; Amaro *et al.*, 2003).

There are many empirical and descriptive stem models available, but process-based stem models are needed to relate stem form to SFM drivers (e.g. pipe models that relate canopy structure and stem form: Shinozaki *et al.*, 1964; Valentine, 1985; Rennolls, 1994; Mäkelä and Vanninen, 1998; Mäkelä, 2002).

Distributional models of dbh are particularly important in empirical growth models because they lead to the assortment outputs. Stage (2002) applied the important Forest Vegetation Simulator (FVS) model, which uses an empirical (non-parametric) diameter distribution model, in the context of climate uncertainty, and Zumrawi *et al.* (2002) considered the scaling properties of this model. Parametric distributional models have been given considerable attention over the last few decades, with the Weibull, the Beta and the S_B being most popular (Hafley and Schreuder, 1977). A recent addition to the toolkit is the logit–logistic (LL) distribution (Wang and Rennolls, 2005). Each of these is extendable into bivariate and higher-dimensionality multivariate models (Schreuder and Hafley, 1977). However, such models have limited usage in SFM unless the model parameters can be related to local stand and environmental variables (Rennolls *et al.*, 1985).

Individual-tree models for SFM

Individual-tree models are most suitable for the simulation of complex forests and the effects of novel management interventions on them. In temperate and boreal forests, mixed and uneven-aged stands and/or individual-tree-based management strategies, such as continuous-cover forestry, are most appropriately modelled using individual-tree models. Mitchell (1975) presented the earliest complete individual-tree forest model, Tree and Stand Simulator (TASS), which is still used as a practical forest management tool in British Columbia (Mitchell and Cameron, 1985). However, calibration of individual-tree models usually requires complete spatio-temporal data on large sets of contiguous trees. Such data sets are very expensive to collect and are not generally available. Further issues often regarded as problematic are that individual-tree models are computationally demanding, that their predictions are sensitive to initial configurations of trees and that they might need to be smoothed for use in stand-level management. Liu and Ashton (1995) argue that individual-tree models, hybridized with appropriate stand-level models or constraints, would be the ideal way of developing models suitable for use in tropical forests. Gap models are spatially explicit, but the modelling unit could in principle be either a tree or an area unit (Koehler and Huth, 1998; Koehler *et al.*, 2000). The Hubbell (2001) 'neutral' individual-tree dispersal-based (gap) model (i.e. all trees of all species labels have identical dynamics) is distance-independent. Hubbell's model has led to a recent breakthrough in forest biodiversity modelling, with the random drift phenomenon being central to this characterization. Chave (2004) provided a recent review of this ecological research on forest biodiversity. By using nodes on a lattice as the modelling unit in Hubbell's model, a simple gap model is obtained. It is clear that there is much scope for the convergence of forest-biodiversity modelling and forest-dynamics modelling, which will be of much use in support of future efforts towards sustainable forest management.

Process-based forest models for SFM

Process-based models (Dixon *et al.*, 1990; Mäkelä, 2002, 2003; see also Thornley, Chapter 21, this volume) have been developed to aid understanding of forest behaviour through the capture of knowledge of plant/soil and carbon/nutrient/water processes and interactions in the form of an integrated model.

Such models are useful for long-term predictions, especially under changing management and climate. Practical problems with this type of model are their requirement for either very detailed data or extensive prior knowledge of process parameters of the component sub-models (Gertner *et al.*, 1999). Process-based models also require detailed climate and soil conditions to be specified: but usually neither detailed current soil condition nor future climate is known. Furthermore, the detail on the timber outputs of such models is generally inadequate. For example, the assessment of wood quality or the choice of harvesting procedures and the estimation of their costs depend on high-resolution information on spatial structure of a forest stand – such information is not usable in most process-based models – and features such as wood quality cannot be produced as outputs by (existing) stand-level process-based models.

Models of the 3PG type (Landsberg and Waring, 1997; Sands and Landsberg, 2002) are intermediate between very detailed process-based models and growth-and-yield models, in terms of their soil and climate data input requirements. This may be one of the reasons why the 3PG model has been calibrated for several species and regions (Almeida *et al.*, 2004; Tomé *et al.*, 2004).

Models for the conservation of biodiversity in tropical forests

Plant and faunal biodiversity depend on initial status (e.g. whether a biodiversity 'hot spot' or remote island), site conditions (e.g. climate, natural disturbance regime, site productivity) and anthropogenic effects (e.g. fragmentation, harvesting, pollution). Sustaining this biodiversity requires knowledge of the surviving species and of their requirements for survival and proliferation, and involves managing the environment to enhance survival and moderate unwanted proliferation. Given the extent and rate of anthropogenic change (conversion of natural vegetation to agricultural and urban use, pollution, climate change), it may be argued that it is more important to provide for the process of speciation, rather than to provide for individual species. In either case, the key issues for biometricians and modellers include inventory of species and habitats and monitoring and assessment of natural and anthropogenic changes, including climatic trends, natural disturbance regimes, site productivity, fragmentation (by broad-scale land conversion as well as by intrusion of linear barriers such as pipelines and roads), harvesting and hunting, and pollution, such as nitrogen deposition. Modellers have dealt with these components rather unevenly, and it appears that no model has attempted to integrate all these aspects.

Floral and faunal inventory of terrestrial biodiversity is uneven in the tropics (and elsewhere). Known 'hot spots' and charismatic taxonomic groups tend to be over-sampled, while large areas and many taxa remain under-represented. Maps indicating 'hot spots' and advocating greater conservation efforts rarely draw on all the information available (i.e. they rely on samples and expert opinion, rather than species distribution models; Guisan and Thuiller, 2005) and tend not to indicate precision. Biometricians could contribute to bio-inventory efforts by helping to highlight areas of potential high biodiversity that remain under-sampled and by drawing attention to areas where the variance of current estimates is high. An inverse relationship between productivity and biodiversity has been reported for several habitats (Mittelbach *et al.*, 2001; Venterink *et al.*, 2003), and remote sensing is now routinely used to monitor and map carbon fluxes, but the relationship between productivity and richness does not appear to have been applied in bio-inventory. Site productivity has been neglected in the sustainability debate in other ways. Site index is a standard yardstick for sustainable plantation management, but site productivity assessment in tropical forests remains neglected and warrants further research (Vanclay, 1992; Vanclay *et al.*, 1997), with important implications for sustainable harvests and for expected species richness.

Much attention has been devoted to anthropogenic impacts on tropical forests: the consequences of fragmentation on species richness (Köhler, *et al.*, 2003), time to recovery following harvesting (e.g. Vanclay, 1990), estimation of allowable cut (e.g. Vanclay, 1996) and (in temperate forests) the consequences of nitrogen deposition (e.g. Spiecker *et al.*, 1996). However, there have been few attempts to consider the interactions between these various anthropogenic disruptions. If fragmentation leads to a gradual decline in species richness in tropical forests, what is the implication for allowable cut and for other environmental services? Can nutrient removal through harvesting timber (and other non-wood harvests) help to mitigate the impact of nitrogen deposition (or other pollution)? Modellers have supported studies of fragmentation (Turner, 1996), sustainable harvesting (Vanclay, 1994), pollution (Spiecker *et al.*, 1996), and subsistence land use (Prabhu *et al.*, 2003), but such studies remain to be integrated so that holistic system-wide impacts and opportunities can be examined.

Hybrid forest models

One approach that has been identified as a valuable way forward in sustainable forest modelling is the development of methodologies to combine a stand-level growth-and-yield model with a stand-level process-based model. In so doing, it is hoped that the users of such ‘hybrid’ models might benefit from the strengths of both model types, while the weaknesses of one model might be compensated by the strengths of the other (Monserud, 2003). Most of the attempts at hybridization of process-based and empirical models base the linkage on the site index/mean annual increment in volume relationship (e.g. Almeida *et al.*, 2004). Tomé *et al.* (2005) used a link between basal area and total above-ground biomass.

However, following Liu and Ashton (1995), it is suggested that a forest model that is able to predict detailed spatial stand structure and able to make detailed biodiversity predictions under a wide range of climatic, environmental and management scenarios will need to be a ‘super-hybrid’, combining the strengths of stand-level empirical growth-and-yield models and stand-level process-based models and an individual-tree model representation of stand-structure dynamics. The integrated models of Daniels and Burkhart (1988) and Rennolls and Blackwell (1988) represent early work in developing such super-hybrid models: they use a nested range of forest models, from individual-tree models through stand-level models to process-based model structures.

Requirements specification for an integrated SFM model

At present, there is no single model that fully meets all the requirements for evaluating the sustainability of multifunctional forest management. Currently there are models each of which addresses one or more of the requirements. The modelling requirements envelope of an integrated SFM model would include the following:

Take forest inventory data as its input
Give 'good' predictions under climate-change scenarios
Take into account the full species list and genetics of the plant material
Provide information on the effect of different silvicultural alternatives on all forest outputs – not only on tree growth but also on other forest products and services
Provide reliable predictions of mature stand structure and wood quality
Take account of the effects of several different types of forest damage event – insect defoliations, acidic deposition, fire, windthrow, etc.
This 'wish list' at least serves the purpose of providing a target towards which forest modellers might aspire.

Decision-support tools for SFM

The integration of forest growth models into decision-support systems (DSS) is a means of transferring new knowledge from scientists to policymakers and forest managers. Decision-support tools for optimal sustainable multi-purpose forest management need to be tailored for different user groups and stakeholders.

The type of model or models used in a DSS will depend on the domain in question (e.g. timber yield, sustainability of soil nutrient status, biodiversity conservation) and the data available. Of course, one cannot answer questions concerning sustainability of soil nutrients or biodiversity conservation if one only has data suitable for an empirical growth-and-yield model (e.g. stocking, age and site index).

Inventory and Monitoring for Forest Sustainability: Criteria and Indicators

Forest sustainability assessments at broad geographical scales require data of the scope and breadth that only national forest inventories are currently capable of providing. Traditionally, these inventories have been based on the design-based sampling paradigm (Schreuder *et al.*, 1993), and have focused on individual tree attributes, such as species, age, diameter, height, mortality, removals and regeneration, and collective tree attributes, such as forest cover type and proportion crown cover. This traditional approach is well suited for estimating some indicators associated with the Montreal Process and the Ministerial Conference on the Protection of Forests in Europe (MCPFE), such as forest area and timber supply. However, to address the entire set of criteria and indicators, new approaches are required. First, the traditional (*T*) variables are inadequate as indicators for criteria such as ecosystem health, water and soil resources. Secondly, the additional (*A*) variables often require different sampling designs. Thirdly, new approaches to estimation, particularly for addressing issues of uncertainty, are required.

The statistical challenges include construction of suitable sampling designs, integration of them with traditional sample designs, compensation for inadequate sample sizes, integration of data in multiple forms and from multiple sources and the estimation of uncertainty.

For purposes of addressing the Montreal Process and MCPFE criteria and indicators, national inventory programmes have augmented traditional sampling regimes to include information on *A* variables, such as tree crown condition, lichen diversity, air pollutants, understorey vegetation, soil chemistry and erosion and down woody material (dead wood). These attributes are sufficiently different in kind from the *T* variables for separate sampling designs to be required (McRoberts, 2004). For example, information on down woody material is collected from line transects, soil information is collected from soil cores, while tree-crown condition and air-pollutant injury are usually visually estimated. Because the greatest portion of costs for a forest inventory is attributed to travel to and from sample locations, the sampling designs for *A* variables are often integrated with the traditional sampling designs. This integration often takes the form of sampling the *A* variables at, or in close proximity to, the sampling locations for the *T* variables. Several undesirable effects may result. First, while the sampling design may be appropriate for the *T* variables, it may be inadequate for the *A* variables. As a first example, the strata constructed for stratified sampling of volume may be of virtually no use for stratified sampling of biodiversity or soil resources. As a second example, air-pollutant injury usually requires the presence of sensitive bioindicator species such as lichens. The presence of these species tends to be highly spatially clustered, a condition for which traditional inventory sampling designs are often inadequate. This approach often does not produce estimates of the *A* variables with adequate precision. One cause is that sampling the *A* variables is usually substantially more costly than sampling the *T* variables. As a result, the *A*-variable sampling design is often less intense. For example, the ratio of *A*-variable plots to *T*-variable plots in the USA is 1 : 16 (Bechtold and Patterson, 2005). A second cause is that natural variability for the *A* variables is often much greater than for the *T* variables, meaning that, even with comparable sample sizes, the precision of estimates for the *A* variables will be less.

Model-based approaches merit consideration as solutions to some of these difficulties (Smith, 1976). Rennolls (1981, 1989) provides an early example of the use of the model-based paradigm. The detrimental effects of inadequate sample sizes for particular *A* variables may be partially alleviated by developing models to predict them from observations of *T* variables. This approach has been used successfully in the USA for predicting down woody material from *T* variables (Woodall *et al.*, 2004). As a second example, the probability of the presence of pollutant-sensitive bioindicator species may be predicted from models based on *T* variables and used as ancillary information for adaptive cluster sampling, as outlined by Ståhl *et al.* (2001).

New approaches to estimation are also required for some C & I variables (Requardt *et al.*, 2004). For example, tree-species richness is a measure of biodiversity for criteria of both the Montreal Process and the MCPFE. Because the total number of species occurring in any area will always be as great as or greater than the number observed on sample plots, traditional sample-based estimators of species richness are inherently biased (Dorazio and Royle, 2005), and new non-parametric estimators have been developed (Rennolls and Laumonier, 1999b, 2006; Chao and Shen, 2004).

Policymakers and public administrators have historically asked the question of forest inventories, 'How much?' and have been satisfied with sample-based estimates for large geographical areas such as counties, provinces or regions. Forest managers and private landowners have always demanded results with high spatial resolution, the ideal requirement being a spatially explicit map. Increasingly, policymakers and public administrators are also asking the question, 'Where?' and are demanding maps. This demand is particularly acute for C & I variables. Some indicators, such as area of productive timberland, are oriented towards the contributions of forest land, while other indicators, such as fragmentation or invasive species, are oriented towards threats to the contributions. Estimates of the amount of productive forest land per county threatened by an invasive insect are no longer sufficient; land managers and planners now require spatially explicit estimates so that they can manage or eliminate the threat. The design of forest biodiversity reserves in large transnational regions also requires spatially explicit maps.

It is important that map users realize that maps produced from forest inventories are estimates and have associated errors and uncertainty. The analyses necessary to produce the spatially explicit map estimates for sustainability analyses must often integrate or combine spatial data for multiple indicators (Stein *et al.*, 2005). There are several important steps to the process. First, a sample of ground-based field inventory data for both *T* and *A* variables is required, though it may be possible to replace ground inventory by high-resolution aerial photography or the use of new remotely sensed/scanned data, for example from LiDAR. These data are used to establish a predictive regression model between the *A* and *T* variables. This fitted model may then be used in conjunction with the more extensive data on *T* variables to obtain model-based estimates of contributions and threats, where *T* variables are measured but *A* variables are not. One may similarly make use of remotely sensed covariate data that are available over wide spatial regions, usually from satellite imagery, to obtain estimates over the whole domain covered by the imagery (McRoberts *et al.*, 2002; Tomppo *et al.*, 2002; LeMay and Temesgen, 2005). Note that even interpolation approaches such as Kriging are essentially model-based, because they require models of spatial semi-variograms to be chosen and fitted. Alternative (non-parametric) approaches include nearest neighbour methods (Tomppo, 1991; McRoberts *et al.*, 2002), most similar neighbour (LeMay and Temesgen, 2005) and gradient nearest neighbour (Ohmann and Gregory, 2002). Geographic information system techniques may then be used to integrate the maps to produce spatially explicit map estimates and predictions of where contributing forest lands are threatened. Estimates of total areas of contribution or threat are obtained by aggregation. The uncertainty associated with the predictions for individual mapping units (e.g. pixels) needs to be estimated as a measure of map reliability. This can be a difficult task, because there are several confounding sources of map error (Rennolls, 2002), and the map uncertainty will not be spatially homogeneous

(Rennolls, 1999b; Gertner *et al.*, 2002). Estimation of the uncertainty associated with the estimated total area at threat has its theoretical and computational difficulties, particularly if small mapping units, such as the pixels of even moderate-resolution satellite imagery, are used.

Integrating Information and Models across Spatial and Temporal Scales for SFM

The shift towards SFM has prompted demand for information and supporting models across a wide range of spatial scales. This information and the supporting models (ideally) must be compatible and consistent as we scale operations through stand, forest, regional and ultimately to the global scale. Because forests cover about 30% of the earth's land areas (FAO, 2005), this task of acquiring the necessary information at local and broader scales and consistently integrating this information and models across spatial scales is challenging, to say the least.

Timescales for forest models also vary vastly. Photosynthesis sub-models of leaves in forest process models might have seconds as their natural time unit. However, daily and seasonal timescales are clearly appropriate at lower levels of temporal resolution. The annual time unit is the common basis for most stand-level models. Rotation time is a natural time unit for economic and sustainability analysis. Timescales for climatic and geologically based models are very long. Consistency of models over timescales is an important requirement. Hybridization is a possible route to building models that span more than one timescale.

The need for forest models that scale from stand to landscape

Traditional forest models and inventories have focused on one stand, with stand measures (such as timber volume) that are additive: estimates or forecasts for a forest, composed of several stands, is simply the sum of the estimates or forecasts for each stand. For many non-timber forest products (such as scenic beauty, species or landscape biodiversity, water quantity and quality, forest fire susceptibility, wildlife needs) the forest attribute or measure used is not simply additive over its sub-compartments. Assessment of such attributes needs to involve both temporal and spatial interactions of management options at the landscape level. Measures or indicators of scenic beauty and recreation might use multiple-criteria analysis techniques (Mendoza and Prabhu, 2003), based on expert evaluation, or contingent valuation (Li and Mattsson, 1995). Non-timber benefits might be assessed through stand characteristics such as area of old forest or dead-wood volume, through structural biodiversity indices or through landscape metrics such as habitat suitability indices.

Hence, we need to add to the requirements specifications of the previous section the rather demanding requirement that the model should provide pertinent output measures or indicators of multifunctional sustainable forest management at the landscape level: that is, that our model should scale from forest compartment or stand through the forest level to the landscape scale.

Models, mapping and GIS

Some of the challenges of estimating maps from multi-source information at a single time, together with uncertainty maps of such estimated maps, have been indicated above. Combining such map estimates into a sequence of map estimates in order to monitor changes over short periods of time is required for SFM, but is a complex process involving many statistical methods and modelling stages. As the information needs increase to longer time periods and to increased detail at all levels, models and methods used to link these sources of information and provide consistency across scales will necessarily become more complex. Linkages between series of remotely sensed images and spatially explicit growth-forecast models are required, and these need to be integrated in a GIS framework. Such linkage between the GIS data and forecast models (e.g. Laumonier *et al.*, 1999; Coops and Waring, 2001) can improve the ability to estimate impacts of a variety of forest policies on forest attributes and economic returns.

Many methods to integrate across scales are in development, but have not been thoroughly evaluated in terms of their accuracy. There are issues of positional accuracy of all data sources. Sometimes it is not possible to estimate accuracy, since the analysis, depending on a mixture of scales in time and space, is intractable. Stage (2003) noted that long-term field installations are commonly used to assess accuracy. However, often only limited circumstances are covered by these installations, they only cover prior circumstances and they are limited in both space and time.

Compromises must be made in terms of scale and the detail needed. For example, it is not possible to obtain habitat information for every species, nor is it possible to test every management intervention that might be considered. For some interventions, there will be few data, and models will not be designed to extend to very extreme cases. Nelson (2003) noted that the level of details needed is likely to vary with the scale of analysis; using one large model for all scales and information needs is not likely to be feasible.

Climate and Carbon Models in Relation to Sustainability

There has been much modelling activity over many years at many centres throughout the world in the area of climate change. Most national forest authorities have policies that include the conduct of a national carbon inventory, so that compliance with the Kyoto protocol of the CBD may be evaluated. Most national forest authorities make projections or predictions, but the level of sophistication of the underlying model base varies considerably.

There seems to be a partial dichotomy in the approaches being used. North American and Australian efforts have focused on the development of holistic global models, whereas the European approach has been (with some exceptions) more oriented towards a carbon-inventory audit. Much relevant recent climate and carbon work on both modelling and metrics at the global scale may be found on the US National Academy of Sciences website (http://dels.nas.edu/basc/reports.shtml).

The NASA Goddard Institute for Space Studies has been working on global climate change models for 25 years (Hansen *et al.*, 1983). The Massachusetts Institute of Technology (MIT) Joint Program on the Science and Policy of Global Change was founded in 1991 as an interdisciplinary organization, and focuses on the integration of natural and social science aspects of the climate issue. The Australian government has a greenhouse programme that is developing a sophisticated integrated greenhouse-gas model adapted to Australian conditions and requirements. The Lawrence Livermore Laboratory has taken a rather less detailed approach and considered extreme scenarios.

At the European level, the aim is to develop a 'quality (information) system for greenhouse gas (GHG) emissions and sinks'. Ray *et al.* (2002) are rather tentative about the use of mathematical computer models, and apply the Ecological Site Classification (ESC) model to the UKCIP98 scenarios. Milne *et al.* (2005) describe an inventory-audit approach that makes use of the model C-Flow as well as a range of the Rothamsted models for carbon capture and release. Broadmeadow *et al.* (2005) report new sampling designs for monitoring UK forest carbon.

The Hadley Centre of the Meteorological Office in the UK adopts a model-based approach, and runs a sophisticated programme in weather research, based on the use of powerful computational facilities and detailed physics-based models. Betts (2005) advocates the need for the development of integrated global modelling of all interacting systems (20 years after the Americans started doing it). The Hadley approach is a very high-resolution approach requiring large data sets and powerful computational solutions. It is not clear if this level of resolution is appropriate to make meaningful global projections and predictions.

Lenton and van Oijen (2002) use cellular automata to develop global models of Gaia as a self-adaptive system. Addition of local interactions into the model introduces instabilities, which are in turn quelled by extra variability and natural selection. Rial *et al.* (2004) in considering the non-linearities inherent in multi-component climate models, gives examples, over a range of timescales, of sudden and rapid transitions of climate that we do not adequately understand. Such examples warn us not to be too confident in our well-developed models. There are many uncertainties.

New Techniques for the Statistical Analysis of Sustainability Data

Many new management methods for sustainable forestry are based on the results from broad-scale, long-term experiments. A good example is the use of thinning as a primary tool to increase the structural complexity of forests to promote forest biodiversity in the Pacific Northwest region of North America (Monserud, 2002). Because of the inherent spatial heterogeneity of forests and resource constraints on data collection, the nature and structure of the data available from broad-scale, long-term experiments and observational studies pose considerable analytical challenges. For example, some experiments might have only a few true replications, some might only have pseudo-replications in space or in time (Hurlbert, 1984) and some might not have any replication at all. Over the past two decades, many new statistical techniques have been developed that can help us to analyse data from complicated experiments. Here, we discuss two approaches that are particularly useful in the analysis of broad-scale, long-term forest management experiments. Monserud (2002) discussed additional ways in which maximal information can be extracted from broad-scale experiments.

The mixed-effects model

Though the mixed-effects model is the same in form as earlier variance-components and multi-level models, the routine use of mixed-effects models was only realized with the relatively recent arrival of inexpensive computing power and suitable software (e.g. Pinheiro and Bates, 2000). Heterogeneous and autocorrelated observations (in space and in time) are often primary problems in analysing experimental data. On the other hand, such features are hallmarks of broad-scale long-term forest experimental data. In fact, it is true to say that the whole of forest research in this area is built on such problematic data.

With the use of mixed-effects models, we can accommodate within a unified framework several random effects (e.g. the effects due to individual experimental units) as correlated sources of variance (Gregoire *et al.*, 1997; Peek *et al.*, 2002). Use of parametric mixed-effects models is now common in forest research (McRoberts, 1996; Zhao, D. *et al.*, 2005). The approach is particularly useful for experiments with pseudo-replications (i.e. replications correlated in space, time or both). By accounting for correlations among the data, we can make proper inferences from such experimental data. The mixed model is a descriptive parametric model, with the functional dependence on the parameters being either linear or non-linear. Statistical inference from the mixed-effects model framework still depends on the normality assumption for the error distributions. Such statistical inference is termed distribution-dependent and parametric (the parameters in this case being the means and variances of the assumed normal distributions).

Non-parametric statistical approaches

The assumption of a normal distribution for model errors has concerned researchers for a long time. In real life, data that meet the normality assumption on the model errors are uncommon, and often the response is not a continuous measure (a requirement of the normality assumption if interpreted strictly). An important approach that alleviates the need for the normality assumption is the randomization method. These ideas go back to the randomization distribution, used by R.A. Fisher (1936) in his classic, *The Design of Experiments*. Again, it is only with the arrival of inexpensive computing power that randomization techniques can be adopted practically (Efron, 1979a). Methods within this framework include randomization tests, Tukey's jackknifing, Efron's bootstrapping (Efron, 1979b) and Monte Carlo tests. Manly (1997) provided an excellent overview on the use of randomization methods in biology. Theoretical approximations of the randomization process provide non-parametric estimators of population parameters such as species richness (Chao and Shen, 2004). The use of Monte Carlo approaches to estimate the parameters of complex Bayesian models is the foundation of modern spatial point-pattern analysis (e.g. Diggle, 2003), an area that has an important potential contribution to make to research related to sustainable forest management.

Another issue in the analysis of many complex forest experimental data is that it is often either not possible, easy or convenient to formulate an explicit parametric trend or response model (either linear or non-linear in its parameters). It is often sufficient to approximate response trends using non-parametric data-smoothing approaches. Several modern statistical methods are based on the approximation of the true response function by a series expansion in terms of a set of basis functions (Hastie *et al.*, 2001). Examples include smoothing splines (Wahba, 1990), generalized additive models (GAM) (Hastie and Tibshirani, 1986, 1990), artificial neural networks and wavelet representations. Guan *et al.* (1997) used neural networks to approximate complex uncertainty functions.

Mixed-effects models and non-parametric modelling approaches can also be merged to produce semi-parametric and non-parametric mixed-effects models. Verbyla *et al.* (1999) used smoothing splines to analyse longitudinal experiments, and Guan *et al.* (2006) applied a semi-parametric mixed-effects model based on smoothing splines to analyse effects of thinning on microclimates. Such techniques have the potential to contribute significantly to the way in which sustainability data are analysed in the future.

Uncertainty Analysis in Modelling and Monitoring for SFM

Model users often use models as if they were a true representation of a deterministic reality. However, reality and data obtained from it are often stochastic in nature or so complex that they may be regarded as so. Even in simple parametric trend models the response variable, conditional on the regressor variables, has an error distribution, and hence parameter estimates are uncertain also. More general error-in-regressor models introduce more uncertainties into parameter estimation. Prediction from fitted models is therefore uncertain, and the sensitivity of predictions can depend critically on how the sample data are distributed over time and space.

Assessing the quality of a modelling system that is to be used to assess sustainability is a difficult task. This is particularly true for multi-component systems, whose prediction qualities are determined by the interactions of those components as well as by their inputs. In such a system, the outputs from one component are used as inputs for other components. Errors from components propagate and accumulate throughout the entire system, and the effects will be evident in the final predictions.

Error and uncertainty analysis constitutes a fundamental aspect of simulation studies of models (e.g. Gertner and O'Neill, 1981; Dale and Hemstrom, 1984; Recknagel, 1984; Gertner, 1987; Hannes *et al.*, 1991; Summers *et al.*, 1993;

McRoberts, 1996). Important objectives of this type of analysis include identification and quantification of different sources of input uncertainty, which propagate through the model to produce error in model predictions and projections. One framework for conducting error–uncertainty analysis is with an error budget (Gelb *et al.*, 1974; Gertner and Köhl, 1992; Gertner *et al.*, 1995, 1996, 2004; Guan *et al.*, 1997).

An error budget can be considered to be a catalogue of the different error sources that allow an (additive) partition of the prediction variance (and bias) in terms of their sources. Additivity of the error budget is ensured by the inclusion of interaction terms in the budget model to take account of non-additive effects. An error budget may be regarded as a form of mixed-effects model and this framework can be used to formulate a model of the effects at different temporal and spatial scales (Wang *et al.*, 2005).

All major sources of error may be accounted for by building analysis of variance (ANOVA)-like tables that show how prediction variance originates from the propagation of different sources of model-input error. The components that contribute the most towards the final prediction variance or bias can be identified, and thus the absolute and relative contributions of the various sources of errors may be examined, partitioned, compared and assessed in different ways. For example, errors may be partitioned as:

Sampling errors, measurement errors, grouping errors and expert-opinion error, etc.
Errors contributed by different inputs, groups of inputs and modules, etc.
Errors associated with stochastic and non-stochastic input variables
Controllable and uncontrollable sources of uncertainty, etc.

Error budgets allow optimization of both data collection and field experimentation in terms of error management (Köhl, 2001). When a model is used for formulating and testing sustainability hypotheses, the estimated uncertainty can and should be accounted for as if the hypotheses were tested with real-world data (Parysow and Gertner, 1999).

The primary benefit of the use of an error budget is acknowledgement that error and uncertainty exist and that they are reported. When using the SFM model, the model user will know that there is uncertainty, and will use the model with full knowledge of the uncertainty and risk (probability and cost) of making incorrect predictions. The SFM model user can assess the probable costs of making mistakes due to SFM-model uncertainties versus the costs for improving the models.

Forest Data, Information and Model Archives

Forest mensurationists have been collecting and storing forest and tree data for centuries so that eventually an understanding of how forests grow and yield timber can be developed. This empirical research has been paralleled by long-term data-collection exercises of forest ecologists and environmental scientists. Such long-term data have been regarded in the forest research sector as a crucial and valuable resource to be conserved for eventual use. In recent decades, the relatively cheap availability of large computer storage resources, combined with the ability to rapidly manipulate, search, analyse and reanalyse these forest data, has meant that such data have become even more useful than in the pre-computer era. Also, the automatic capture of forest data by various remote sensing techniques has seen an explosion in the amount of forest data that is now stored in computer databases. The importance of conserving such vast information resources has been well recognized in many national and international projects, e.g. the EU European forest information system (EFIS) project (Schuck *et al.*, 2005), the FAO FRA2000 project, and the IUFRO GFIS Task Force (Päivinen *et al.*, 2001). Other projects catalogue information about long-standing forest experiments from permanent and temporary sample plots (e.g. TROPIS: Vanclay, 1999; Baker and Wright, 2001; Miles, 2001, in Rennolls, 2001). However, there has been little progress in developing international collaboration on systems to implement open-access data archives of historical data for wider SFM analysis. In the context of the European NEFIS project, Rennolls *et al.* (2004b) make recommendations on the need for using modelling methods to satisfy harmonization requirements, while Rennolls *et al.* (2004a) and Rennolls (2005a, b) outline the architectural and ontological features required in the construction of international forest data repositories. Rennolls (2004b) suggested that data archives need to store historical satellite imagery to facilitate future SFM growth and change modelling. Rennolls *et al.* (2001) argue that a Forest Model Archive (FMA) is also needed, and some prototypes do now exist (www.forestmodelarchive.info). Such open, shared archives of forest models would encourage international SFM collaboration and help to support and enable the building of hybrid models that scale across the spatial and temporal scales that are involved in SFM.

Conclusions and Recommendations

In the last two decades, there have been substantial developments in both statistical techniques and forest modelling. The needs of multifunctional SFM call for consistent and openly accessible information that spans a range of spatial and temporal scales. It seems likely that this will only occur by the development of underpinning models that are consistent and scale appropriately over the levels of the space and time hierarchy.

The challenge of developing such a scalable model base is very substantial. The quantity of data available from automatic sensing devices is increasing very rapidly, and to deal with such data volume, in conjunction with the increasing complexity of models needed, will mean that expanding computing resources will need to be fully used. It is likely that grid technology will need to be integrated in the programme. It is clear that there will be much interdisciplinary collaboration required between statisticians and modellers, on the one hand, and software, information and knowledge engineers, on the other, if the needs of policy and decision makers in SFM are to be met.

Having stated this need for interdisciplinary collaboration, it may also be noted that there is some danger that professionally trained statisticians and mathematical modellers might be dropped from team efforts. With increasing computing power and easy access to statistical and modelling software packages, it is possible that forest scientists might regard it as advantageous to take over these central modelling activities themselves. Rennolls (2004a) has argued that this would not be a satisfactory approach, because the current challenges are of such magnitude that continued involvement of statisticians and mathematical modellers in the development of new techniques is of crucial importance.

In the area of sustainable multifunctional forestry, it is clear that the most desperate situation continues to exist in the developing world. Economic income from timber trade is crucial to the economic survival of many developing countries, and rapid clearance of the remaining tropical forests of the world continues. Tropical forest biodiversity resources are bound to be lost. Conservation activities, including research into SFM modelling and information management, can play an important role. However, the scientists and biometricians of the developing nations do not have the time or the resources to deal with the issues before it is too late. The richer developed nations of the world, all of which have temperate or boreal forests, do not have the biodiversity threats that face the developing nations of the tropics. It should be the duty of global science to address these issues, and to develop effective means for the sharing of scientific and technical expertise, and to support financially as well as scientifically the solution of these major global forest sustainability problems.

Acknowledgements

We are grateful to Keith Reynolds, Duncan Ray and Christine Chang for the support they provided for us in the preparation of this chapter.

References

Almeida, A., Landsberg, J.J. and Sands, P.J. (2004) Parameterisation of 3-PG model for fast-growing *Eucalyptus grandis* plantations. *Forest Ecology and Management* 193, 179–195.

Amaro, A. and Tomé, M. (eds) (1999) *Empirical and Process Based Models for Forest Tree and Stand Growth Simulation*. Publisan, Lisbon, Portugal.

Amaro, A., Reed, D., Tomé, M. and Themido, I. (1998) Modelling dominant height growth: eucalyptus plantations in Portugal. *Forest Science* 44, 37–46.

Amaro, A., Reed, D. and Soares, P. (eds) (2003) *Modelling Forest Systems*. CAB International, Wallingford, UK.

Andrienko, N. and Andrienko, G. (2005) *Exploratory Analysis of Spatial and Temporal Data: A Systematic Approach*. Springer-Verlag, New York.

Bailey, R.L. and Clutter, J.L. (1974) Base-age invariant polymorphic site curves. *Forest Science* 20, 155–159.

Baker, A. (2004) Simplicity. In: Zalta, E.N. (ed.) *The Stanford Encyclopedia of Philosophy*, Winter 2004 edn. http://plato.stanford.edu/archives/win2004/entries/ simplicity/

Baker, N. and Wright, H. (2001) The UK archive of tropical forest inventory: conserving the data. In: Rennolls, K. (ed.) *Proceedings of IUFRO 4.11 Conference on Forest Biometry, Modelling and Information Sciences*. University of Greenwich, London. http://cms1.gre.ac.uk/conferences/iufro/proceedingsBakerWright1.pdf

Bechtold, W.A. and Patterson, P.L. (eds) (2005) *Forest Inventory and Analysis National Sample Design and Estimation Procedures*. General Technical Report SRS-80, US Department of Agriculture, Forest Service, Southern Research Station, Asheville, North Carolina.

Betts, R.A. (2005) Integrated approaches to climate–crop modelling: needs and challenges. *Philosophical Transactions of the Royal Society: Biological Sciences* 360, 2049–2065.

Bitterlich, W. (1947) Die Winkelzahlmessung. *Allgemeine Forst-und Holtzwirtschaftliche Zeitung* 58, 94–96.

Blackwell, P.G., Rennolls, K. and Coutts, M.P. (1990) A root anchorage model for shallowly rooted Sitka spruce. *Forestry* 63, 73–91.

Borders, B.E., Bailey, R.L. and Clutter, M.L. (1988) Forest growth models: parametric estimation using real growth series. In: Ek, A.R., Shirley, S.R. and Burk, T.E. (eds) *Proceedings of the IUFRO Conference on Forest Growth Modelling and Prediction, 23–27 August 1987, Minneapolis, Minnesota,* Vol. 2. General Technical Report NC-120, US Department of Agriculture, Forest Service, North Central Forest Experiment Station, St Paul, Minnesota, pp. 660–667.

Broadmeadow, M. (ed.) (2002) *Climate Change: Impacts on UK Forests*. Forestry Commission Bulletin 125, Forestry Commission, Edinburgh. http://www.forestresearch.gov.uk/fr/infd-5zyhmc

Broadmeadow, M.S.J., Matthews, R.W., Mackie, E., Wilkinson, M., Benham, S. and Harris, K. (2005) Survey methods for Kyoto Protocol monitoring and verification of UK forest carbon stocks. In: Milne, R. and Mobbs, D.C. (eds) *UK Emissions by Sources and Removals by Sinks due to Land Use, Land Use Change and Forestry Activities*. http://www.edinburgh.ceh.ac.uk/ukcarbon/docs/Defra_Report_2005_Section6.pdf

Chao, A. and Shen, T.J. (2004) Non-parametric prediction in species sampling. *Journal of Agricultural, Biological, and Environmental Statistics* 9, 253–269.

Chave, J. (2004) Neutral theory and community ecology. *Ecology Letters* 7, 241–253.

Cieszewski, C.J. and Bailey, R.L. (2000) Generalized algebraic difference approach: theory based derivation of dynamic site equations with polymorphism and variable asymptotes. *Forest Science* 46, 116–126.

Coops, N.C. and Waring, R.H. (2001) The use of multi-scale remote sensing imagery to derive regional estimates of forest growth capacity using 3-PGS. *Remote Sensing of Environment* 75, 324–334.

Dale, V.H. and Hemstrom, M. (1984) *CLIMACS: A Computer Model of Forest Stand Development for Western Oregon and Washington*. Research Paper PNW-327, US Department of Agriculture, Forest Service, Pacific Northwest Forest and Range Experiment Station, Portland, Oregon.

Daniels, R.F. and Burkhart, H.E. (1988) An integrated system of forest stand models. *Forest Ecology and Management* 23, 159–177.

Diggle, P.J. (2003) *Statistical Analysis of Spatial Point Patterns*. Arnold, New York.

Dixon, R.K., Meldahl, R.S., Ruark, G.A. and Warren, W.G. (eds) (1990) *Process Modelling of Forest Growth Responses to Environmental Stress.* Timber Press, Portland, Oregon.

Dorazio, R.M. and Royle, A.J. (2005) Estimating size and composition of biological communities by modelling the occurrence of species. *Journal of the American Statistical Association* 100, 389–398.

Efron, B. (1979a) Computers and the theory of statistics: thinking the unthinkable. *Society of Industrial and Applied Mathematics Review* 21, 460–480.

Efron, B. (1979b) Bootstrap methods: another look at the jackknife. *Annals of Statistics* 7, 1–26.

FAO (2005) Website on forest statistics: www.fao.org/forestry accessed on 31 October 2005.

Garcia, O. (1983) A stochastic differential equation model for the height growth of forest stands. *Biometrics* 39, 1059–1072.

Garcia, O. (1993) Stand growth models: theory and practice. In: *Advancement in Forest Inventory and Forest Management Sciences – Proceedings of the IUFRO Seoul Conference*. Korea Forestry Research Institute, Seoul, Korea, pp. 22–45.

Garcia, O. (1998) Estimating top height with variable plot sizes. *Canadian Journal of Forest Research* 28, 1509–1517.

Garcia, O. (2003) Dimensionality reduction in growth models: an example. *Forest Biometry, Modelling and Information Sciences* 1, 1–15.

Garcia, O. (2005) Unifying sigmoid univariate growth equations. *Forest Biometry, Modelling and Information Sciences* 1, 63–68. www. fbmis.info

Gelb, A., Kasper, J., Jr, Nash, R., Jr, Price, C. and Sutherland, A., Jr (1974) *Applied Optimal Estimation*. MIT Press, Cambridge, Massachusetts.

Gertner, G.Z. (1987) Approximating precision in simulated projections: an efficient alternative to Monte Carlo methods. *Forest Science* 33, 230–239.

Gertner, G.Z. and Köhl, M. (1992) An assessment of some nonsampling errors in a national survey using an error budget. *Forest Science* 38, 525–538.

Gertner, G.Z. and O'Neill, R.V. (1981) A comparison of sensitivity and error analysis based on a stream ecosystem model. *Ecological Modeling* 12, 173–190.

Gertner, G.Z., Cao, X. and Zhu, H. (1995) A quality assessment of a Weibull based growth projection system. *Forest Ecology and Management* 71, 235–250.

Gertner, G.Z., Parysow, P. and Guan, B. (1996) Projection variance partitioning of a conceptual forest growth model with orthogonal polynomials. *Forest Science* 42, 474–486.

Gertner, G.Z., Fang, S. and Skovsgaard, J.P. (1999) A Bayesian approach for estimating the parameters of a forest process model based on long-term growth data. *Ecological Modeling* 119, 249–265.

Gertner, G., Wang, G.X., Fang, S.F. and Anderson, A.B. (2002) Mapping and uncertainty of predictions based on multiple primary variables from joint co-simulation with Landsat TM image and polynomial regression. *Remote Sensing of Environment* 83, 498–510.

Gertner, G., Fang, S., Wang, G. and Anderson, A. (2004) Partitioning spatial model uncertainty when inputs are from joint simulations of correlated multiple attributes. *Transactions in GIS* 8, 441–458.

Gillison, A.N. and Brewer, K.R.W. (1985) The use of gradient directed transects or gradsects in natural resource surveys. *Journal of Environmental Management* 20, 103–127.

Gregoire, T.G., Valentine, H.T. and Furnival, G.M. (1995) Sampling methods to estimate foliage and other characteristics of individual trees. *Ecology* 76, 1181–1194.

Gregoire, T.G., Brillinger, D.R., Diggle, P.J., Russek-Cohen, E., Warren, W.G. and Wolfinger, R.D. (1997) *Modelling Longitudinal and Spatially Correlated Data*. Springer-Verlag, New York.

Guan, B.T., Gertner, G.Z. and Parysow, P. (1997) A framework for uncertainty assessment of mechanistic forest growth models: a neural network example. *Ecological Modeling* 95, 47–58.

Guan, B.T., Weng, S.H., Kuo, S.R., Chang, T.Y., Hsu, H.W. and Shen, J.W. (2006) Analyzing the effects of stand thinning on microclimates with semiparametric smoothing splines. *Canadian Journal of Forest Research* 36, 1641–1648.

Guisan, A. and Thuiller, W. (2005) Predicting species distribution: offering more than simple habitat models. *Ecology Letters* 8, 993–1009.

Hafley, W.L. and Schreuder, H.T. (1977) Statistical distributions for fitting diameter and height data in even-aged stands. *Canadian Journal of Forest Research* 7, 481–487.

Hannes, S.J., Roberts, L.A., Warwick, J.J. and Cale, W.G. (1991) Testing the utility of the first order uncertainty analysis. *Ecological Modeling* 58, 1–23.

Hansen, J., Russell, G., Rind, D., Stone, P., Lacis, A., Lebedeff, S., Ruedy, R. and Travis, L. (1983) Efficient three-dimensional global models for climate studies: Models I and II. *Monthly Weather Review* 111, 609–662.

Hastie, T.J. and Tibshirani, R.J. (1986) Generalized additive models [with discussion]. *Statistical Science* 1, 297–318.

Hastie, T.J. and Tibshirani, R.J. (1990) *Generalized Additive Models.* Chapman and Hall, New York.

Hastie, T.J., Tibshirani, R.J. and Friedman, J.H. (2001) *The Elements of Statistical Learning.* Springer, New York.

Hill, R.A. and Thomson, A.G. (2005) Mapping woodland species composition and structure using airborne spectral and LiDAR data. *International Journal of Remote Sensing* 26, 3763–3779.

Holgate, P. (1967) The angle-count method. *Biometrika* 54, 615–623.

Hubbell, S.P. (2001) *The Unified Neutral Theory of Biodiversity and Biogeography.* Princeton University Press, Princeton, New Jersey.

Hurlbert, S.H. (1984) Pseudoreplication and the design of ecological field experiments. *Ecological Monographs* 54, 187–211.

Jordan, G.J., Ducey, M.J. and Gove, J.H. (2004) Comparing line-intersect, fixed-area, and point relascope sampling for dead and downed coarse woody material in a managed northern hardwood forest. *Canadian Journal of Forest Research* 34, 1766–1775.

Koehler, P. and Huth, A. (1998) The effects of tree species grouping in tropical rain forest modelling – simulations with the individual based model FORMIND. *Ecological Modeling* 109, 301–321.

Koehler, P., Ditzer, T. and Huth, A. (2000) Concepts for the aggregation of tropical tree species into functional types and the application on Sabah's lowland rain forests. *Journal of Tropical Ecology* 16, 591–602.

Köhl, M. (2001) Error sources and their influence on NFI inventory results. In: Brassel, P. and Lischke, H. (eds) *Swiss National Forest Inventory: Methods and Models of Second Assessment.* Swiss Federal Research Institute WSL, Birmensdorf, Switzerland, pp. 297–305.

Köhl, M., Päivinen, R., Traub, B. and Miina, S. (1997) Comparative study. In: *Study on European Forestry Information and Communication System: Report on Forest Inventory and Survey Systems.* European Commission, Luxembourg, pp. 1267–1322.

Köhl, M., Traub, B. and Päivinen, R. (2000) Harmonisation and standardisation in multinational environmental statistics – mission impossible? *Environmental Monitoring and Assessment* 63, 361–380.

Köhler, P., Chave, J., Riéra, B. and Huth, A. (2003) Simulating the long-term response of tropical wet forests to fragmentation. *Ecosystems* 6, 114–128.

Landsberg, J. (2003a) Modelling forest ecosystems: state of the art, challenges, and future directions. *Canadian Journal of Forest Research* 33, 385–397.

Landsberg, J. (2003b) Physiology in forest models: history and the furture. *Forest Biometry, Modelling and Information Systems* 1, 49–63. www.fbmis.info

Landsberg, J.J. and Waring, R.H. (1997) A generalized model of forest productivity using simplified concepts of radiation-use efficiency, carbon balance and partitioning. *Forest Ecology and Management* 95, 209–228.

Laumonier, Y., King, B., Legg, C. and Rennolls, K. (eds) (1999) *Data Management and Modelling Using Remote Sensing and GIS for Tropical Forest Land Inventory.* Rodeo International Publishers, Jakarta, Indonesia.

LeMay, V. and Temesgen, H. (2005) Comparison of nearest neighbor methods for estimating basal area and stems per hectare using aerial auxiliary variables. *Forest Science* 51, 109–119.

Lenton, T.M. and van Oijen, M. (2002) Gaia as a complex adaptive system. *Philosophical Transactions of the Royal Society: Biological Sciences* 357, 683–695.

Li, C.-Z. and Mattsson, L. (1995) Discrete choice under preference uncertainty: an improved structural model for contingent valuation. *Journal of Environmental Economics and Management* 28, 256–269.

Liu, J. and Ashton, P.S. (1995) Individual-based simulation models for forest succession and management. *Forest Ecology and Management* 73, 157–175.

Magnussen, S. (1999) Effect of plot size on estimates of top height in Douglas-fir. *Western Journal of Applied Forestry* 14, 17–27.
Magnussen, S., Eggermont, P. and LaRiccia, V.N. (1999) Recovering tree heights from airborne laser scanner data. *Forest Science* 45, 407–422.
Mäkelä, A. (2002) Derivation of stem taper from the pipe theory in a carbon balance framework. *Tree Physiology* 22, 891–905.
Mäkelä, A. (2003) Process-based modelling of tree and stand growth: towards a hierarchical treatment of multiscale processes. *Canadian Journal of Forest Research* 33, 398–409.
Mäkelä, A. and Vanninen, P. (1998) Impacts of size and competition on tree form and distribution of aboveground biomass in Scots pine. *Canadian Journal of Forest Research* 28, 216–227.
Manly, B.F.J. (1997) *Randomization, Bootstrap and Monte Carlo Methods in Biology*, 2nd edn. Chapman and Hall, New York.
McRoberts, R.E. (1996) Estimating variation in field crew estimates of site index. *Canadian Journal of Forest Research* 26, 560–565.
McRoberts, R.E. (2004) Biometric research. In: Burley, J., Evans, J. and Youngquist, J. (eds) *Encyclopedia of Forest Science*. Elsevier, Oxford, UK, pp. 152–158.
McRoberts, R.E., Nelson, M.D. and Wendt, D.G. (2002) Stratified estimation of forest area using satellite imagery, inventory data, and the k-nearest neighbors technique. *Remote Sensing of Environment* 82, 457–468.
Matérn, B. (1960) *Spatial Variation*. Meddelanden fran Statens Skogsforskningsinstitut 49, No. 5, 144 pp. [Second edition (1986), *Lecture Notes in Statistics*, No. 36. Springer, New York.]
Matérn, B. (1976) Om skattning av övre höjden. *Sveriges Skogsvårdsforbunds Tidskift* 74, 51–53.
Mendoza, G.A. and Prabhu, R. (2003) Qualitative multi-criteria approaches to assessing indicators of sustainable forest resource management. *Forest Ecology and Management* 174, 329–343.
Miles, P.D. (2001) Information systems for the forest resource analyst. In: Rennolls, K. http://cms1.gre.ac.uk/conferences/infro/proceedings/miles3.pdf
Milne, R. and Mobbs, D.C. (eds) (2005) *UK Emissions by Sources and Removals by Sinks due to Land Use, Land Use Change and Forestry Activities*. http://www.edinburgh.ceh.ac.uk/ukcarbon/docs/Defra_Report_2005.pdf
Milne, R., Mobbs, D.C. and Thomson, A.M. (2005) Land use change and forestry: the 2003 UK greenhouse gas inventory and projections to 2020. In: Milne, R. and Mobbs, D.C. (eds) *UK Emissions by Sources and Removals by Sinks due to Land Use, Land Use Change and Forestry Activities*. http://www.edinburgh.ceh.ac.uk/ukcarbon/docs/Defra_Report_2005_Section2.pdf
Mitchell, K.J. (1975) Dynamics and simulated yield of Douglas-fir. *Forest Science Monograph* 17, 1–39.
Mitchell, K.J. and Cameron, I.R. (1985) *Managed Stand Yield Tables for Coastal Douglas-fir: Initial Density and Precommercial Thinning*. Land Management Report 31, B.C. Ministry of Forests, Research Branch, Victoria, British Columbia.
Mittelbach, G.G., Steiner, C.F., Scheiner, S.M., Gross, K.L., Reynolds, H.L., Waide, R.B., Willig, M.R., Dodson, S.I. and Gough, L. (2001) What is the observed relationship between species richness and productivity? *Ecology* 82, 2381–2396.
Monserud, R.A. (2002) Large-scale management experiments in the moist maritime forests of the Pacific Northwest. *Landscape and Urban Planning* 30, 159–180.
Monserud, R.A. (2003) Evaluating forest models in a sustainable forest management context. *Forest Biometry, Modelling and Information* Sciences 1, 35–47.
Munro, D.D. (1974) Forest growth models: a prognosis. In: Fries, J. (ed.) *Growth Models for Tree and Stand Simulation*. Research Note 30, Royal College of Forestry, Stockholm, pp. 7–21.
Nelson, J. (2003) Forest-level models and challenges for their successful application. *Canadian Journal of Forest Research* 33, 422–429.

Ohmann, J.L. and Gregory, M.J. (2002) Predictive mapping of forest composition and structure with direct gradient analysis and nearest neighbor imputation in coastal Oregon, U.S.A. *Canadian Journal of Forest Research* 32, 725–741.

Päivinen, R., Richards, T., Binh, N.T., Wood, R. and Margherita Sini (2001) The IUFRO global forest information service: towards a common resource. In: Rennolls, http://cms1.gre.ac.uk/conferences/infro/proceedings/paivinenRichards.pdf

Parysow, P. and Gertner, G.Z. (1999) The role of interactions in hypothesis testing of ecological scenarios with process models. *Ecological Modeling* 116, 107–124.

Paulo, M.J., Tomé, M., Otten, A. and Stein, A. (2005) Comparison of three sampling methods in the characterisation of cork oak stands for management purposes. *Canadian Journal of Forest Research* 35, 2295–2303.

Peek, M.S., Russek-Cohen, E., Wait, D.A. and Forseth, I.N. (2002) Physiological response curve analysis using nonlinear mixed models. *Oecologia* 132, 175–180.

Pienaar, L.V. and Turnbull, K.J. (1973) The Chapman–Richards generalisation of von Bertalanffy's growth model for basal area growth and yield in even-aged stands. *Forest Science* 19, 2–22.

Pinheiro, J.C. and Bates, D.M. (2000) *Mixed-effects Models in S and S-Plus*. Springer-Verlag, New York.

Prabhu, R., Haggith, M., Mudavanhu, H., Muetzelfeldt, R., Standa-Gunda, W. and Vanclay, J.K. (2003) ZimFlores: a model to advise co-management of the Mafungautsi Forest in Zimbabwe. *Small-scale Forest Management, Economics and Policy* 2, 185–210.

Ray, D., Pyatt, G. and Broadmeadow, M. (2002) Modelling the future climatic suitability of plantation forest tree species. In: Broadmeadow, M. (ed.) *Climate Change: Impacts on UK Forests.* Bulletin 125, Forestry Commission, Edinburgh, pp. 151–169. http://www.forestresearch.gov.uk/fr/INFD-5ZYJPT

Recknagel, F. (1984) A comprehensive sensitivity analysis for an ecological simulation model. *Ecological Modeling* 26, 77–96.

Reed, D.D. and Tomé, M. (1998) Total aboveground biomass and net dry matter accumulation by plant component in young *Eucalyptus globulus* in response to irrigation. *Forest Ecology and Management* 103, 21–32.

Rennolls, K. (1978) 'Top height': its definition and estimation. *Commonwealth Forestry Review* 57, 215–219.

Rennolls, K. (1979) A contribution to the mathematics of top height. *Mitteilungen der Forstlichen Versuchs- und Forschungsanstalt Baden-Württemberg* 91, 187–195. In: Schopfer, W. (ed.) *Proceedings of IUFRO Subject Group S6.02*. Freiburg, Germany.

Rennolls, K. (1981) The total area of woodland in Berkshire is. . . *The Statistician* 30, 275–287.

Rennolls, K. (1989) *Design of the Census of Woodlands and Trees 1979–1982*. Forestry Commission Occasional Paper 18, Forestry Commission, Edinburgh, UK.

Rennolls, K. (1994) Pipe-model theory of stem-profile development. *Forest Ecology and Management* 69, 41–55.

Rennolls, K. (1995a) Forest height growth modelling. *Forest Ecology and Management* 71, 217–225.

Rennolls, K. (1995b) *Statistics in Science and Society*. Inaugural lecture, CASSM Research and Development Paper No. 4, CASSM Publishing, University of Greenwich, London.

Rennolls, K. (1999a) Data requirements for forest modelling. In: Amaro, A. and Tomé, M. (eds) *Empirical and Process-based Models for Forest Tree and Stand Growth Simulation*. Edicoes Salamandra, Lisbon.

Rennolls, K. (1999b) Algorithms and software for Landsat image classification with a case study on tropical rain forests in Sumatra. In: Zawila-Niedzwiecki, T. and Brach, M. (eds) *Proceedings of IUFRO Conference on Remote Sensing and Forest Monitoring.* Warsaw Agricultural University, Rogow, Poland. http://rogow99.sggw.waw.pl/09_session_5/08/

Rennolls, K. (ed.) (2001) *Proceedings of IUFRO 4.11 Conference on Forest Biometry, Modelling and Information Sciences*. University of Greenwich, London. http://cms1.gre.ac.uk/conferences/iufro/proceedings/

Rennolls, K. (2002) Components of error of maps from remotely sensed images. In: Hunter, G. and Lowell, K. (eds) *Proceedings of the 5th International Symposium on Spatial Accuracy Assessment in Natural Resources and Environmental Sciences, Melbourne, Australia*. Delft University Press, Delft, Netherlands, pp. 223–230.

Rennolls, K. (2004a) Towards the sustainable use of Europe's forests: a centre for European forest science! – some missing catalytic elements? In: Anderson, F., Birot, Y. and Päivinen, R. (eds) *Towards the Sustainable Use of Europe's Forests – Forest Ecosystem and Landscape Research: Scientific Challenges and Opportunities*. EFI Proceedings No. 49. European Forest Institute, Joensuu, Finland, pp. 317–322.

Rennolls, K. (2004b) Satellite 'image-banks' in forest information systems? Prospects for accurate estimation of forest change and growth from data-fusion models. In: Mowrer, H.T., McRoberts, R. and VanDeusen, P.C. (eds) *Proceedings of the Joint Meeting of the Sixth International Symposium on Spatial Accuracy Assessment in Natural Resources and Environmental Sciences and the Fifteenth Annual Conference of TIES, The International Environmetrics Society*. http://www.spatial-accuracy.org/PDF/Rennolls.pdf

Rennolls, K. (2005a) A partial ontology for forest inventory and mensuration. In: *Proceedings of the Second International Workshop on Forest and Environmental Information and Decision Support Systems*. IEEE Computer Society Press, Washington, DC, pp. 679–683.

Rennolls, K. (2005b) An architectural design for a forest information and communications portal (FICP). In: *Proceedings of the Second International Workshop on Forest and Environmental Information and Decision Support Systems*. IEEE Computer Society Press, Washington, DC, pp. 684–690.

Rennolls, K. and Blackwell, P. (1988) An integrated forest process model: its calibration and its predictive performance. *Forest Ecology and Management* 25, 31–58.

Rennolls, K. and Laumonier, Y. (1999a) Species–area and species–diameter curves for three forest sites in Sumatra. *Journal of Tropical Forest Science* 11, 784–800.

Rennolls, K. and Laumonier, Y. (1999b) Analysis of species hyper-diversity in the tropical rain forests of Indonesia: the problem of non-observance. In: Sassa, K. (ed.) *Proceedings of the IUFRO Division 8 Conference on Environmental Forest Science*. Kyoto University, Kyoto, Japan, pp. 355–362.

Rennolls, K. and Laumonier, Y. (2000) Species diversity structure analysis at two sites in the tropical rain forest of Sumatra. *Journal of Tropical Ecology* 16, 253–270.

Rennolls, K. and Laumonier, Y. (2006) A new local estimator of regional species diversity, in terms of 'shadow species', with a case study from Sumatra. *Journal of Tropical Forest Ecology* 22, 321–329.

Rennolls, K. and Peace, A. (1986) Flow models of mortality and yield for unthinned forest stands. *Forestry* 59, 47–58.

Rennolls, K., Geary, D.N. and Rollinson, T.J.D. (1985) Characterising diameter distributions by the use of the Weibull distribution. *Forestry* 58, 57–66.

Rennolls, K., Blackwell, P. and Coutts, M. (1990) A root anchorage model for shallowly rooted Sitka spruce. *Forestry* 63(1), 73–91.

Rennolls, K., Ibrahim, M.T. and Smith, P.A. (2001) A forest model archive? In: Rennolls, K. (ed.) *Proceedings of IUFRO 4.11 Conference on Forest Biometry, Modelling and Information Sciences*. University of Greenwich, London. http://cms1.gre.ac.uk/conferences/iufro/proceedings/RennIbrSmithFMA.pdf

Rennolls, K., Richards, T., Fedorec, A.M., Ibrahim, M.T., McManus, K. and Butler, A. (2004a) Requirements and design of an integrated European environmental information communication system (IEEICS). In: *Proceedings of the First International Workshop on Forest and Environmental Information and Decision Support Systems*. IEEE Computer Society Press, Washington, DC, pp. 610–614.

Rennolls, K., Richards, T., Fedorec, A.M., Ibrahim, M.T., McManus, K. and Butler, A. (2004b) Models and tools for an integrated European environmental management and decision support system (IEEMDSS). In: *Proceedings of the First International Workshop on Forest and Environmental Information and Decision Support Systems*. IEEE Computer Society Press, Washington, DC, pp. 615–619.

Requardt, A., Köhl, M. and Näscher, F. (2004) A challenge for the joint European criteria and indicators: can sustainability be measured and documented? *Allgemeine Forst Zeitschrift für Waldwirtschaft und Umweltvorsorge* 59, 494–498.

Rial, J., Pielke, R.A., Sr, Benison, M., Claussen, M., Canadell, J., Cox P., Held, H., de Noblet-Ducoudre, N., Prinn, R., Reynolds, J. and Salas, J.D. (2004) Nonlinearities, feedbacks and critical thresholds within the Earth's climate system. *Climatic Change* 65, 11–38.

Sands, P.J. and Landsberg, J.J. (2002) Parameterisation of 3-PG for plantation grown *Eucalyptus globulus*. *Forest Ecological Management* 163, 273–292.

Schreuder, H.T. and Hafley, W.L. (1977) A useful bivariate distribution for describing stand structure of tree heights and diameters. *Biometrics* 33, 471–478.

Schreuder, H.T., Gregoire, T.G. and Wood, G.B. (1993) *Sampling Methods for Multiresource Forest Inventory*. Wiley, New York.

Schuck, A., Andrienko, G., Andrienko, N., Folving, S., Köhl, M., Miina, S., Päivinen, R., Richards, T. and Voss, H. (2005) The European forest information system – an Internet based interface between information providers and the user community. *Computers and Electronics in Agriculture* 47, 185–206.

Shinozaki, K.K., Yoda, K., Hozumi, K. and Kira, T. (1964) A quantitative analysis of plant form – the pipe model theory. I. Basic analyses. *Japanese Journal of Ecology* 14, 97–105.

Smith, T.M.F. (1976) The foundations of survey sampling: a review. *Journal of the Royal Statistical Society, Series A* 139, 183–204.

Spiecker, H., Mielikäinen, K., Köhl, M. and Skovsgaard, J.P. (eds) (1996) *Growth Trends in European Forests-Studies from 12 Countries*. EFI Research Reports No. 5, Springer-Verlag, Heidelberg, Germany.

Stage, A.R. (2002) Using FVS and its fire and fuels extension in the context of uncertain climate. In: Crookston, N.L. and Havis, R.N. (eds) *Proceedings of the Second Forest Vegetation Simulator Conference*. RMRS-P-25, US Department of Agriculture, Forest Service, Rocky Mountain Research Station, Ogden, Utah, pp. 104–107.

Stage, A.R. (2003) How forest models are connected to reality: evaluation criteria for their use in decision support. *Canadian Journal of Forest Research* 33, 410–421.

Ståhl, G., Ringvall, A. and Fridman, J. (2001) Assessment of coarse woody debris – a methodological overview. *Ecological Bulletins* 49, 57–70.

Stein, S.M., McRoberts, R.E. and Alig, R.J. (2005) *Forest on the Edge: Housing Development on America's Private Forests*. General Technical Report GTR-PNW-636, US Department of Agriculture, Forest Service, Pacific Northwest Research Station, Portland, Oregon.

Summers, J.K., Wilson, H.T. and Kou, J. (1993) A method for quantifying the prediction uncertainties associated with water quality models. *Ecological Modeling* 65, 161–176.

Tomé, J., Tomé, M., Fontes, L., Soares, P., Pacheco, C.A. and Araújo, C. (2004) Testing 3PG with irrigated and fertilized plots established in *Eucalyptus globulus* plantations in Portugal. In: Hasenauer, H. and Mäkelä, A. (eds) *Modelling Forest Production – Scientific Tools, Data Needs and Sources, Validation and Application. Proceedings of the International Conference on Modelling Forest Production, 19–22 April 2004, Vienna, Austria*. Department of Forest and Soil Sciences, Boku University of Natural Resources and Applied Life Sciences, Vienna, Austria, pp. 382–390.

Tomé, M., Faias, S.P., Tomé, J., Cortiçada, A., Soares, P. and Araújo, C. (2005) Hybridizing a stand level process-based model with growth and yield models for *Eucalyptus globulus* plantations in Portugal. In: Borralho, N.M.G., Pereira, J.S., Marques, C., Coutinho, J., Madeira, M. and Tomé, M. (eds) *Proceedings of IUFRO Conference – Eucalyptus in a Changing World*.

Aveiro, Portugal, 11–15 October 2004. RAIZ, Instituto de Investigação da Floresta e do Papel, Lisbon, Portugal, pp. 290–297.

Tomppo, E. (1991) Satellite image-based national forest inventory of Finland. *International Archives of Photogrammetry and Remote Sensing* 28, 419–424.

Tomppo, E., Nilsson, M., Rosengren, M., Aalto, P. and Kennedy, P. (2002) Simultaneous use of Landsat-TM and IRS-1C WiFS data in estimating large area tree stem volume and aboveground biomass. *Remote Sensing of Environment* 82, 156–177.

Turner, I.M. (1996) Species loss in fragments of tropical rainforest: a review of the evidence. *Journal of Applied Ecology* 33, 200–209.

Valentine, H.T. (1985) Tree-growth models – derivations employing the pipe-model theory. *Journal of Theoretical Biology* 117, 579–585.

Vanclay, J.K. (1990) Effects of selection logging on rainforest productivity. *Australian Forestry* 53, 200–214.

Vanclay, J.K. (1992) Assessing site productivity in tropical moist forests. *Forest Ecology and Management* 54, 257–287.

Vanclay, J.K. (1994) Sustainable timber harvesting: simulation studies in the tropical rainforests of north Queensland. *Forest Ecology and Management* 69, 299–320.

Vanclay, J.K. (1996) *Estimating Sustainable Timber Production from Tropical Forests*. CIFOR Working Paper No. 11, Bogor, Indonesia.

Vanclay, J.K. (1999) TROPIS: tree growth and permanent plot information system. *Journal of Forest Planning* 4, 1–5.

Vanclay, J.K., Gillison, A.N. and Keenan, R.J. (1997) Using plant functional attributes to quantify stand dynamics in mixed forests for growth modelling and yield prediction. *Forest Ecology and Management* 94, 149–163.

Venterink, H.O., Wassen, M.J., Verkroost, A.W.M. and de Ruiter, P.C. (2003) Species richness–productivity patterns differ between N-, P-, and K-limited wetlands. *Ecology* 84, 2191–2199.

Verbyla, A.P., Cullis, B.R., Kenward, M.G. and Welham, S.J. (1999) The analysis of designed experiments and longitudinal data by using smoothing splines [with discussion]. *Applied Statistics* 48, 269–311.

Wahba, G. (1990) *Spline Models for Observational Data*. Society for Industrial and Applied Mathematics, Philadelphia, Pennsylvania.

Wang, G., Gertner, G.Z., Fang, S. and Anderson, A.B. (2005) A methodology for spatial uncertainty analysis of remote sensing and GIS products. *Photogrammetric Engineering and Remote Sensing* 71, 1423–1432.

Wang, M. and Rennolls, R. (2005) Tree diameter distribution modelling: introducing the logit–logistic distribution. *Canadian Journal of Forest Research* 35, 1305–1313.

Woodall, C.W., Holden, G.R. and Vissage, J.S. (2004) Fuel mapping for the future. *Fire Management Notes* 64, 19–21.

Zhao, D., Wilson, M. and Borders, B.E. (2005) Modelling response curves and testing treatment effects in repeated measures experiments: a multilevel nonlinear mixed-effects model approach. *Canadian Journal of Forest Research* 35, 122–132.

Zhao, G.A., Shao, G.F., Reynolds, K.M., Wimberly, M.C., Warner, T., Moser, J.W., Rennolls, K., Magnussen, S., Köhl, M., Anderson, H.E., Mendoza, G.A., Dai, L.M., Huth, A., Zhang, L.J., Brey, J., Sun, Y.J., Ye, R.H., Martin, B.A. and Li, F.R. (2005) Digital forestry: a white paper. *Journal of Forestry* 103, 47–50.

Zumrawi, A.A., Stage, A.R. and Snowdon, B. (2002) Stand-level scaling of a single-tree, distance-independent, diameter-growth model: interim calibration of prognosis for the southeastern interior of British Columbia. In: Crookston, N.L. and Havis, R.N. (eds) *Proceedings of the Second Forest Vegetation Simulator Conference*. RMRS-P-25, US Department of Agriculture, Forest Service, Rocky Mountain Research Station, Ogden, Utah, pp. 151–157.

21 Modelling Forest Ecosystems: the Edinburgh Forest Model

J.H.M. THORNLEY

Centre for Ecology and Hydrology, Edinburgh, UK.

Abstract

First, possible reasons for building a large ecosystem model are discussed, stressing the importance of modelling objectives. It is suggested that a model is de rigeur for any research programme that aims to take a firm grasp of the responses of forests. Then, an outline account of the Edinburgh Forest Model (EFM) is given. The EFM is a standard model of the genre, taking account of pools and fluxes of carbon (C), nitrogen (N) and water. The core model comprises tree, soil and water sub-models. The tree sub-model has a phenology sub-model. Evergreen or deciduous forests can be simulated. These can be even-aged stands of trees as in a plantation or, using stand averages with self-thinning and regeneration, natural forests. There is also a substantial aphid sub-model for optional use in evergreen mode: many consider that the most important effects of climate change will be via pest and disease vectors, rather than via physiology. Finally there are environmental and management drivers. The former accepts monthly, daily or diurnal data, or provides internally generated sinusoidal data. The latter permit the simulation of fertilizer application, pruning, thinning, clear-felling and fire. Last, some typical simulations and applications of the model are presented. They relate to conifers and beech in plantations or natural forest, fertilizer application, aphid dynamics and climate change. These illustrate the scope of the model. Mechanistic models provide a framework for attempting to unify many variable results, understanding why they arise and making predictions about the future time course of these ecosystems. There seems to be no other way of doing this work. A short conclusions section closes the chapter.

Keywords: Ecosystem, forest, model, plantation, natural forest, aphid.

Introduction

Many of the ideas and general principles of modelling in biology, agriculture and plant ecology can be traced back to de Wit (1970), who pioneered crop simulation.

In any modelling project, the most important initial exercise is to define objectives. People build models for many different reasons. Much of the controversy that sometimes emerges in modelling discussions is rooted in differing objectives. Ecosystem modelling is essentially a long-term commitment of at least 10 years and more. Short-term work is, almost inevitably, rather superficial. The learning curve is too long and covers too many different areas (biology, mathematics, computing) for it to be otherwise. Because of the requirement for expertise in several different areas, teamwork and collaboration are important. An agreed and coherent set of objectives provides motivation and direction for researchers. Some of the reasons why people build models are therefore considered.

Models are not necessarily, or even usually, mathematical, but here our concern is with mathematical models (for a wider discussion see Thornley and Johnson, 2000; Thornley and France, 2005). The need for mechanistic mathematical models is driven by the increasingly quantitative nature of many biological data, the widening knowledge base about parts of the system, the requirement for integrating the behaviour of different parts of a complex system when seeking understanding and the imperative of being able to predict possible futures reliably, and is facilitated by the rapid advances in computer technology.

Models are now accepted as de rigeur for any research programme that aims to take a firm grasp of the responses of forests, which can be unexpected. Mechanistic models are required to provide the understanding needed for appropriate and flexible management of forests, whatever the prevailing environmental or economic objectives. The models are necessarily large, reflecting the complexity of the forest ecosystem. In one sense, such work is 'big' science, requiring a long-term commitment, although the resources needed are comparatively modest. The challenge is to develop models of 'engineering strength'. This requires a suitable research environment: this should be reasonably stable, multidisciplinary and well connected to experimental programmes; it must also permit adequate support for the four essential legs of an ecosystem model: research, development, documentation and application. Some modelling researchers are dismayed by the wasteful fragmentation of much plant ecosystem modelling research. If science is about giving an account of work that can be understood and built upon by others, much work in this area hardly qualifies as science.

Although biologists and foresters are sometimes wary of modelling, it has to be emphasized that modelling and mathematics are the servants of science. The hypotheses expressed in mathematics and computer programs are derived from biological concepts. The model and its associated computer program provide a framework for representing, integrating, exploring and applying ideas about how we think the system works.

All models are wrong: that is, they provide a limited view of reality. Hopefully, the limitations do not lead to serious errors, but that can never be guaranteed. Because all models are wrong, they are easily criticized. Most models do some things well and others poorly or not at all. A model that does a few things well, perhaps breaking new ground, can be of great value, and should not be discarded just because of the dubious bits that have to be tacked on to make it all work.

The EFM is deterministic, although occasionally its predictions are suggestive of chaos. There are no random number generators within the model. It makes definite predictions for quantities such as plant dry mass without any associated probability distribution. However, some driving variables, such as rainfall and migration of diseases, pests or predators, may be rather capricious. An apparent stochasticity may merely reflect ignorance about what is going on. A deterministic simulator such as the EFM is driven by specified inputs for rainfall, wind, etc. These inputs can be given values that occur in actual weather, thus allowing unambiguous connections to be made between outputs (predictions) and inputs. A range of inputs can be used to represent (say) the variability of rainfall.

The Edinburgh Forest Model (EFM)

Most plant ecosystem models are now quite similar at the qualitative level, and few would dispute the statement that a reasonable level of consensus is emerging. The EFM is mechanistic, dynamic and deterministic. The principal objective has been to obtain an understanding of the important factors determining forest ecosystem responses in Britain. The model is written on approximately two to three levels of description. The top level of the model is drawn in Fig. 21.1.

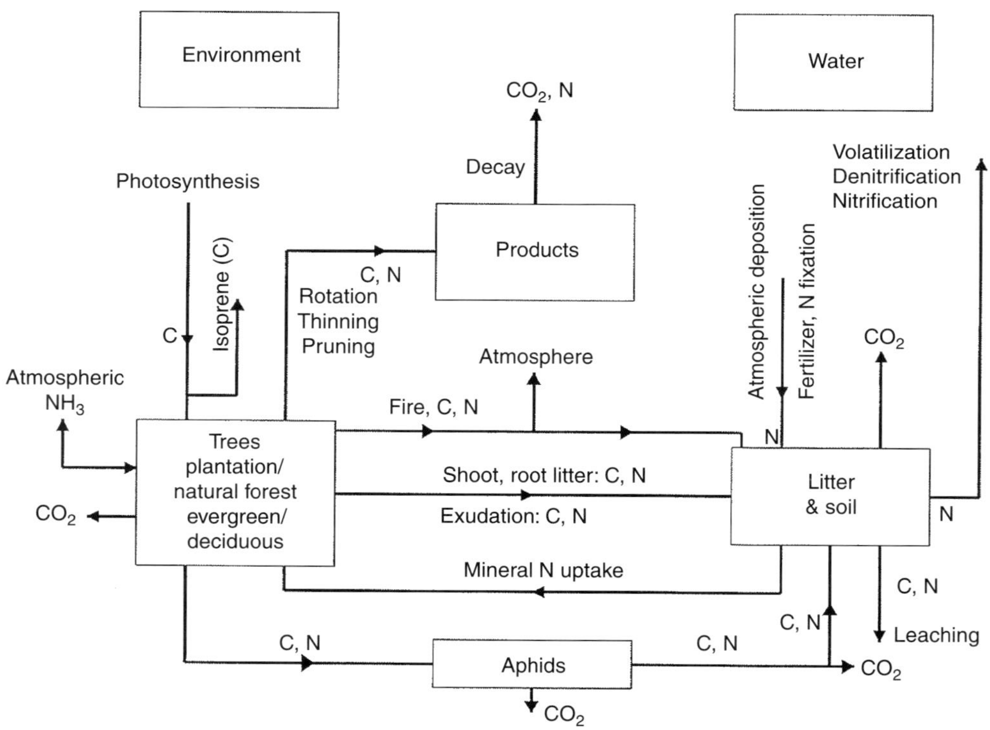

Fig. 21.1. Edinburgh Forest Model (EFM): overview.

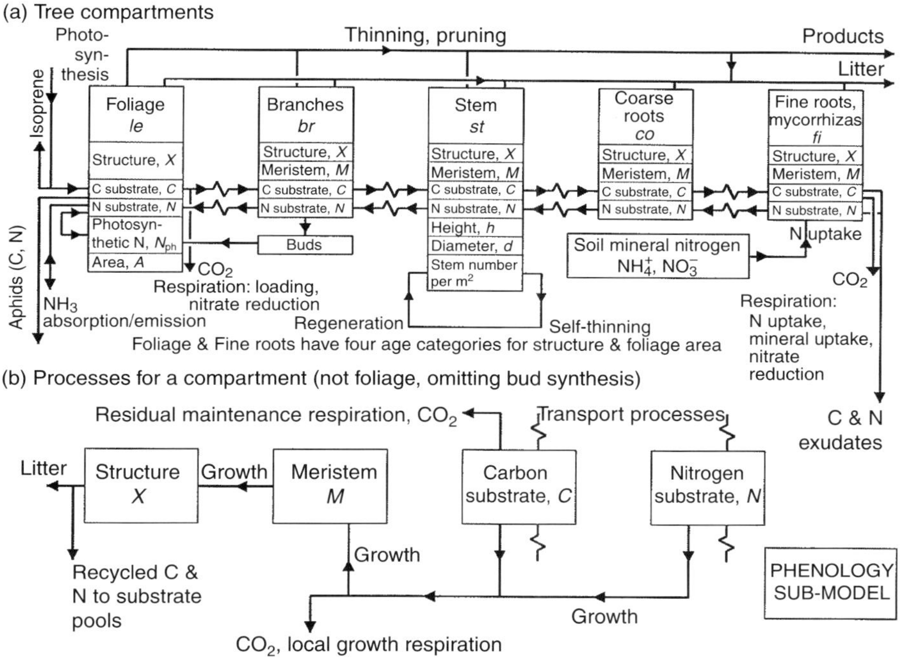

Fig. 21.2. Tree sub-model.

The model accounts for the pools and fluxes of carbon (C), nitrogen (N) and water in the forest. There are sub-models for the tree (Fig. 21.2), litter and soil (Fig. 21.3), water (Fig. 21.4), aphids (optional) (Fig. 21.5) and phenology (used in the deciduous forest option) (Fig. 21.6). Table 21.1 lists the environmental and management inputs that drive the system.

The model is formulated employing standard methods for modelling continuous dynamic systems. The rate–state approach is used. That is, there is a set of state variables defining the system. State variables represent quantitites of substance (where 'substance' can include geometrical aspects such as area) and are shown in the boxes in Figs 21.2 to 21.5. Rate of change of state variable Y is calculated by

$$\frac{dy}{dt} = I_Y - O_Y$$

I_Y and O_Y denote inputs and outputs of Y. Input and output fluxes to a box (state variable) are determined using some assumed mathematical equations. For example, photosynthesis in Fig. 21.2, which is an input to the foliage C substrate box (labile carbohydrates), is computed assuming that rate of photosynthesis depends upon light, CO_2 and air temperature, using an empirical hyperbolic relationship that is known to describe measured leaf photosynthetic responses rather accurately (Cannell and Thornley, 1998b). The processes in Fig. 21.2 (and in Figs 21.3 to 21.5) and their equations can

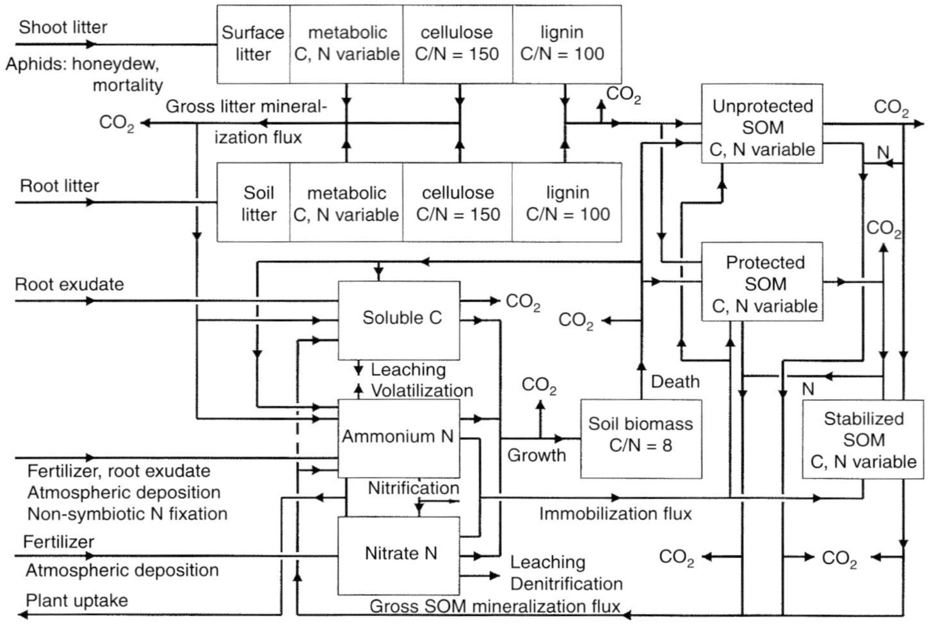

Fig. 21.3. Litter and soil sub-model.

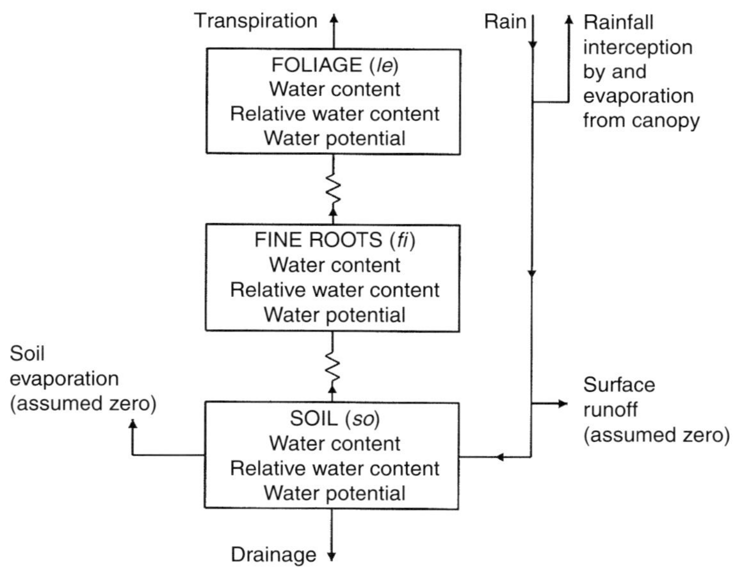

Fig. 21.4. Water sub-model.

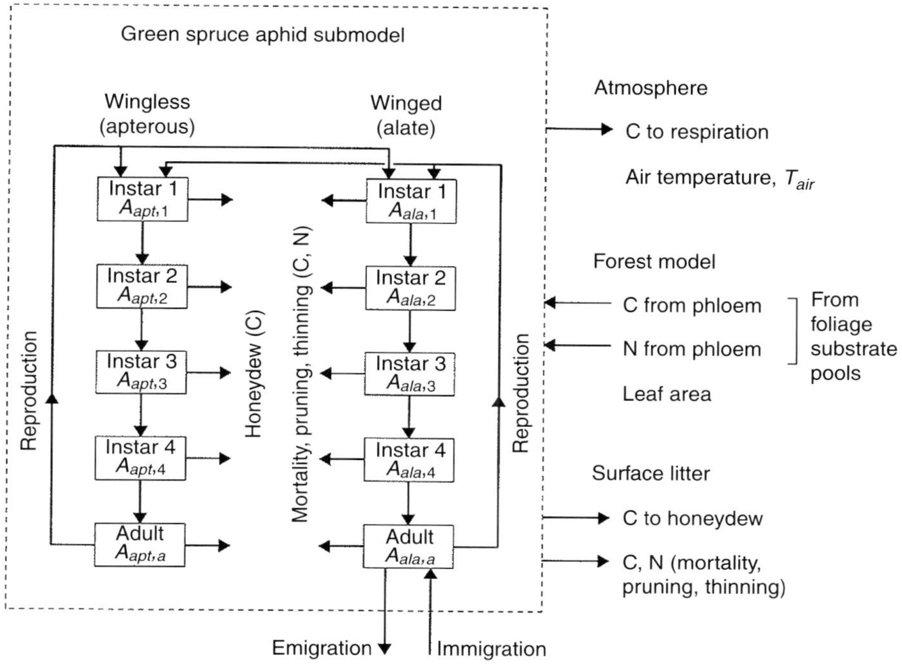

Fig. 21.5. Green spruce aphid sub-model. All morphs produce honeydew and all are subject to mortality, pruning and thinning.

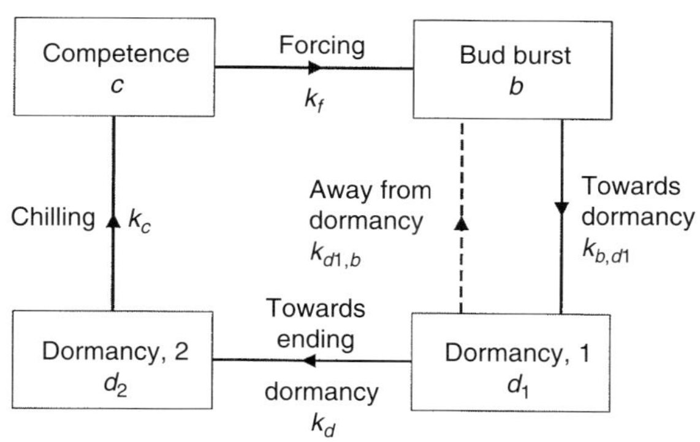

Morphogen, $m = c + b + d_1 + d_2 = 100$
Bud burst occurs when $b = 100$
Dormancy begins when $d_1 = 100$
Dormancy ends when $d_2 = 100$
Chilling & forcing can occur simultaneously
Other processes are exclusive
$k_{d1,b} = 0$ pro tem
Initialization: 0 h, 1 Jan, $c = 100$, $b = d_1 = d_2 = 0$

Fig. 21.6. Phenology sub-model. This is used to trigger various processes in the foliage part of the tree sub-model (Fig. 21.2) concerned with bud burst, leaf growth and leaf fall. It only operates when the EFM is used in deciduous mode.

Table 21.1. Environmental and management variables used in the Edinburgh Forest Model. An internal time step of 11.25 min (1/128 day) is generally used for all dynamic components of the model. The model can be driven by diurnal, daily or monthly data, or can generate its own environmental data assuming specified sinusoids.

Quantity	Units
Environment	
Atmospheric CO_2 concentration	vpm (μmol/mol)
Environmental N input to soil ammonium pool	kg ammonium $N/m^2/day$
Environmental N input to soil nitrate pool	kg nitrate $N/m^2/day$
Photosynthetically active radiation	$J/m^2/day$
Air temperature	°C
Soil temperature	°C
Rainfall	$mm(kg/m^2)/day$
Wind speed at specified reference height	m/s
Relative humidity	
Reference height for meteorological measurements	m
Management	
Fertilizer	
Amount of N applied at each fertilizer application	kg N/m^2
Fractions of N fertilizer entering soil ammonium and nitrate pools	
Number of fertilizer applications	
Times of fertilizer applications	day
Rotation length	year
Thinning	
Time of thinning	year
Fraction of stems removed	
Pruning: fraction removed per year	per year
Fire: time of fire and parameters specifying fire severity	

be regarded as empirical relationships that supply the mechanisms used to interpret and provide an understanding of the level 1 description in Fig. 21.1. The empirical leaf photosynthesis equation provides one of the contributing mechanisms when viewed from above, the ecosystem/plant level. Alternatively, in a different model, the leaf level could be regarded as the top level of a mechanistic model of leaf photosynthesis. This would have lower levels comprising perhaps light reactions and Calvin cycle processes, all of which are described by essentially empirical equations, providing a mechanistic explanation for the leaf level.

Tree sub-model (Fig. 21.2)

Key assumptions

- Tree dry matter is regarded as being 'structural' (e.g. cellulose, hemicellulose, lignin) and 'storage' or substrates (mono- and disaccharides, starch, fructans for C substrates; amino acids, nitrate, some labile proteins for N substrates). This assumption partially decouples sources from sinks, which have different environmental dependencies. It permits a more meaningful calculation of growth, and also allows realistic within-plant allocation patterns to be calculated mechanistically.
- Within-plant allocation (shoot:root ratio) is calculated using the mechanistic transport-resistance approach (Thornley, 1991, 1998b). This method allows the great plasticity observed in allocation patterns to be realistically simulated.
- Using a 'photosynthetic' protein pool, a phenomenological algorithm simulates photosynthetic acclimatization to light, nitrogen, CO_2 (down-regulation) and temperature (Thornley, 1998c).
- Photosynthetic contributions of sunlit and shade leaves in the canopy are separately calculated (Thornley, 2002).
- Foliage and fine root structural material are assigned an age structure. Tissue ageing and turnover can be an important dynamic feature of forests, allowing trees to withstand drought, pruning, fire and other catastrophes.
- Ammonia fluxes through the stomates. Although these fluxes are small, they can be significant in low N deposition ecosystems close to equilibrium.

Weaknesses

- No explicit representation of plant proteins, except for the 'photosynthetic' protein pool (see above). The explicit consideration of plant proteins would allow variable plant nitrogen:carbon relations to be more realistically incorporated, including a further refinement of the way in which 'maintenance' respiration is calculated (Cannell and Thornley, 2000; Thornley and Cannell, 2000c).
- No direct representation of development in relation to reproductive growth, e.g. seed production, soil seed banks and seed germination. This limits the EFM for some ecosystem applications.

Soil sub-model (Fig. 21.3)

Key assumptions

The soil sub-model is standard in most respects. Where the present formulation departs from many other models is:

- Ammonium N and nitrate N pools are separately represented. This allows loss processes that act on these different substrates to be more realistically accounted for – e.g. volatilization and nitrification occur from the ammonium pool, whereas nitrate is the substrate for denitrification and leaching.

Also, plant and microbes do not respond equally to the two substrates. Although these two pools are small, only their explicit representation allows N ecosystem losses to be sensibly calculated (N losses are a key factor in many forest ecosystems).
- Non-symbiotic N fixation is calculated. Forests often grow under very low N inputs. Albeit small, non-symbiotic N fixation in the soil is then a crucial input, determining long-term responses.
- A soluble carbon pool represents C substrates in the soil such as carbohydrates and organic acids. This pool drives C leaching, microbial growth and non-symbiotic N fixation.

Just as in the tree sub-model, it seems that representing relatively small but labile substrate pools allows the model to simulate alternative uses and changing priorities in a natural and flexible manner. The alternative, of omitting substrate pools, is to use unsatisfactory ad hoc devices that block off further model development.

Weaknesses

- Only a single soil horizon is considered.
- Only a single microbial pool is included. Perhaps as a result of this assumption, it was necessary to use linear kinetics for soil organic matter transformations rather than more realistic non-linear expressions, such as the Michaelis–Menten equation, which has been used extensively elsewhere.

Water sub-model (Fig. 21.4)

Key assumptions

- A mechanistic physicochemical basis is chosen for this sub-model.
- Plant water potential and its components are represented.
- Soil water potential and soil hydraulic conductivity are calculated from soil relative water content.
- The plant cells are elastic. Osmotic effects are included.
- Water fluxes are driven by water potential gradients (but see Roderick, 2001).

Weaknesses

- A single horizon for soil water is used.
- The soil is assumed to be homogeneous.
- Simple and quite widely used equations are used to calculate soil water potential and soil hydraulic conductivity from soil water content, which, arguably, rarely apply to real soils with their complex pore size distributions, wormholes and cracks.
- Water fluxes depend on water potential gradients, as assumed by many others. However, this approach has been reasonably disputed (Roderick, 2001).

A dilemma for both the soil and water sub-models is that a more empirical approach might give greater accuracy for specific calibrated applications, but could close off possibilities for general mechanistic model development.

Aphid sub-model (Fig. 21.5)

Key assumptions

This model is based on that of Newman *et al.* (2003), who considered cereal and grass aphids.

- Aphid growth and development is linked to the quantity and quality of C and N substrates available in the phloem, in this case the C and N substrate concentrations in the foliage (Fig. 21.2).
- In addition to the adult morph (wingless and winged), four developmental morphs (instars) are represented in order to give an adequate description of aphid dynamics.
- Effects of temperature, nutritional status and aphid number density on development, fecundity, mortality and allocation of offspring to wingless or winged are considered.

Weaknesses

- No representation of sexual processes or overwintering eggs.
- Possible effects of day length and tree phenology are not included.
- Parameterization is general, and is possibly more applicable to cereal and grass aphids than to the spruce aphid.
- Predicted aphid dynamics appear unrealistic, but the reasons for this are not yet understood.

Phenology sub-model (Fig. 21.6)

Key assumptions

- This functions essentially as a seasonal clock, driven by temperature.
- A cyclical scheme allows one season to affect the next.

Weaknesses

- No effects of tree size or substrate status on phenology.
- Doubts about the stability of the scheme for particular environments.

Environmental data (Table 21.1)

The model accepts a variety of environmental data inputs: diurnal data (e.g. every 30 minutes), daily data and monthly data, on all of which can be imposed a slow year-to-year drift to represent climate change (Fig. 21.16a). There are also options by which components of the environment can be modified by multiplication or addition. Our experience is that the environmental input sections of the program usually require some simple modification for particular application.

Computer program

The source program of the EFM, efm.csl, is much annotated, and obtainable from www.nbu.ac.uk/efm/ or from the author. The program is written in ACSL (Advanced Computer Simulation Language), a software package designed for dynamic modelling problems (ACSL, 2000). ACSL is Fortran-like in its statements, produces a very efficient executable code and is compatible with CSSL (Continuous System Simulation Language) software standards. ACSL programs, with their mathematically well-defined structure, are easily translated into other languages.

Integration is by Euler's algorithm (Thornley and France, 2005, Section 2.4.1). The integration step is mostly 1/128 day = 11.25 minutes, although for some conditions this must be decreased for stability. When using the deciduous option, a step length of 1/512 day is employed. We have found it helpful to use a step length that can be represented exactly in binary.

Applications of the Edinburgh Forest Model

Some applications of the EFM are described in order to show the range of investigations that lie within its scope. These applications extend those previously described by myself and colleagues, but use the latest version of the model. No specific parameter tuning has been applied. In a model of this type, there are many parameters whose value is not well established and which it is legitimate to change. Parameter adjustment can be applied to vary predictions of say yield class, allocation fractions, leaf area index and soil pools by a factor of two or more (yield class is the mean annual volume increment over a rotation).

Such parameter adjustment may sometimes be justified by objectives, but it can also be an unrewarding and time-consuming procedure with no clear end point. Meaningful parameter adjustment can be difficult in areas where the experimental data are quite variable and even conflicting. Moreover, it rarely leads to a better understanding, which can provide a more general applicability and may reveal important opportunities for intelligent intervention in system behaviour.

The initialization problem

A substantial and unavoidable difficulty associated with using any plant ecosystem simulator is how to initialize the model. This is sometimes referred to as the spin-up problem. The difficulty resides primarily with the soil sub-model.

Most soil sub-models, including that shown in Fig. 21.3, use discrete pools that are not amenable to experimental characterization (Smith *et al.*, 2002). A forest soil may take several hundred years to reach an equilibrium state, even given constant environment and management, which are unlikely to have occurred. Current soil state depends therefore on a long uncertain history. Any disturbance of the system, natural or human-made, causes increased respiration and mineralization. Realistically, soil pools can be expected to be below, possibly far below, their equilibrium values.

Here a pragmatic but objective recipe for initialization has been adopted for some of our simulations (in other simulations initial values are of little consequence and the constant but rather arbitrary initial values in the program are employed). The EFM was run over many rotations to equilibrium, using the mean monthly environment for southern Scotland (Eskdalemuir, 55°19′N, 3°12′W, 242 m; Meteorological Office, 1982), with atmospheric CO_2 concentration of 350 μmol/mol, mean annual air and soil temperatures of 7.35°C, annual rainfall of 1527 mm, mean daily photosynthetically active radiation receipt of 3.77 MJ/m^2, average fraction of bright sunshine hours is 0.26, average relative humidity of 0.808, mean wind speed of 4 m/s, constant N deposition of 10 kg N/ha/year, clay content of 0.3, zero fertilizer inputs, and a standard 60-year rotation and thinning regime typical of the area (eight thinnings at 20, 25, 30, 35, 40, 45, 50, 55 years, with fractions of the then standing stem numbers removed of 0.45, 0.4, 0.35, 0.3, 0.25, 0.2, 0.15 and 0.1), followed by end-of-rotation clear-felling at 60 years. The equilibrium (non-tree) state variables immediately after clear-felling and replanting, which are instantaneous and simultaneous events, were used to initialize the EFM. As can be seen in Fig. 21.8b, d, f, some of these initial values are highly dynamic.

Dynamics and death of an even-aged unthinned plantation

Figure 21.7 shows a simulation where the model is allowed to run until a numerical instability occurs, which is construed as tree death. Death occurs in the evening (21 h) of 9 June in the 294th year of the simulation. The instability cannot be avoided or significantly delayed by (for instance) decreasing the integration interval. It is of interest to examine the physiological events that give rise to the instability, which occur essentially over the last 20 years of the simulation (the timescale may not be realistic but is easily manipulated). Fine root mass (via which water is taken up) reaches an early maximum at 15 years (Fig. 21.7a), after which it declines continually on a year-to-year basis. Leaf area index (LAI) peaks at 40 years, again followed by a monotonic decline. Gross canopy photosynthesis reaches its maximum earlier than LAI because of the effect of decreasing stomatal conductance (Fig. 21.7b) on photosynthesis. Increasing stem mass (Fig. 21.7c) slowly increases maintenance respiration costs and decreases the ratio of net to gross photosynthesis (Fig. 21.7c). Under the conditions of the simulation, carbon substrate concentration in the fine roots (Fig. 21.7b) decreases at increasing speed during the last 50 or so years before death. Note that: (i) fine root turnover time is 200 days at 10°C; (ii) continual

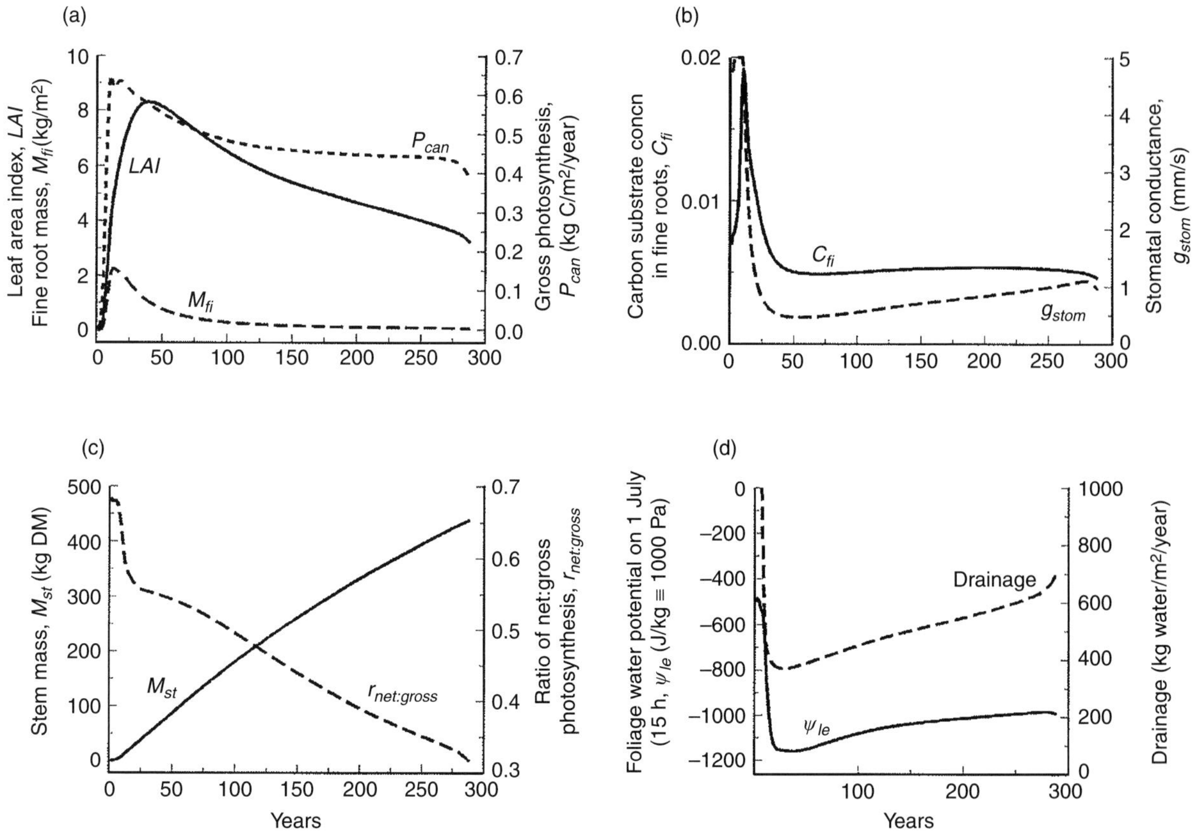

Fig. 21.7. Dynamics of death in an even-aged unthinned plantation. The EFM is run in evergreen mode until the simulation stops due to numerical instability, interpreted as tree death. Neither initial values nor parameters have been adjusted for any specific performance. The timescale may not be realistic but is easily adjusted.

fine root growth is required to maintain root function; and (iii) the fine roots are at the far end of the assumed carbon substrate delivery system (Fig. 21.2). At about 250 years, the tree enters its final stages, where increasing maintenance causes decreasing leaf area index and fine root mass at an increasing rate, decreasing water uptake, and eventually leading to increasing foliage water stress (Fig. 21.7d) and decreasing stomatal conductivity (Fig. 21.7b), further decreasing canopy photosynthesis (Fig. 21.7a), leading to further decrease in root mass. Small pool sizes in the fine root eventually cause numerical difficulties that bring down the simulation.

Dynamics of a thinned plantation over a 60-year rotation

This is illustrated in Fig. 21.8. Some data are listed in Table 21.2. The system is in equilibrium, as explained above. Leaf area index (Fig. 21.8a) is a sawtooth due to thinning, although tree mass (per stem) increases continually. The soil mineral N pool, comprising ammonium and nitrate N (Fig. 21.8b), is very high in the years immediately after planting, which occurs just after clear-felling. It decreases as litter decomposition decreases and tree N uptake gathers momentum (the small increase in the third year is thought to be a transient of no significance). Soil organic matter (SOM) C increases initially due to SOM production from the much increased litter pools following clear-felling. As the system is in equilibrium, SOM C (and N, not shown) ends the rotation with the same value as at the beginning. Tree respiration as a fraction of gross photosynthesis gradually increases (Fig. 21.8c), indicating that maintenance respiration is increasing relative to growth respiration (Cannell and Thornley, 2000). Soil N fixation (Fig. 21.8d) is inhibited in the early years of plantation growth by the high soil mineral N levels (Fig. 21.8b), but then increases as canopy closure occurs (Fig. 21.8a) and soil mineral N levels decrease. Soil biomass exhibits similar changes (Fig. 21.8d). Figure 21.8e shows the time course of the C and N substrate pools averaged over the five tree compartments (Fig. 21.2): in early growth N is in plentiful supply; canopy closure is accompanied by more limiting N; this is partly alleviated by thinning, which speeds up the recycling of N within the system. N losses from the system are high in the early years of plantation growth, decreasing as the trees become larger, transpiration increases and drainage decreases. In this simulation, volatilization is the principal loss flux of N over the rotation. After an initial transient, height:diameter ratio decreases from nearly 60 in the young tree towards about 30 as the end of the rotation approaches (Fig. 21.8g), although the growth ratio is more variable (this is the yearly height increment divided by the yearly diameter increment). Figure 21.8h shows soil and foliage relative water content (RWC) at 15 h on 1 July each year and similarly the stomatal conductance. Stomates are mostly fully open before canopy closure (about 15 years; Fig. 21.8h), after which there is considerable stomatal closure, alleviated by the thinning events that occur every 5 years from 20 years. Soil water on 1 July at 15 h is appreciably depleted after 10 years, leading to foliage water stress and stomatal closure. This is caused by the increasing LAI and transpiration.

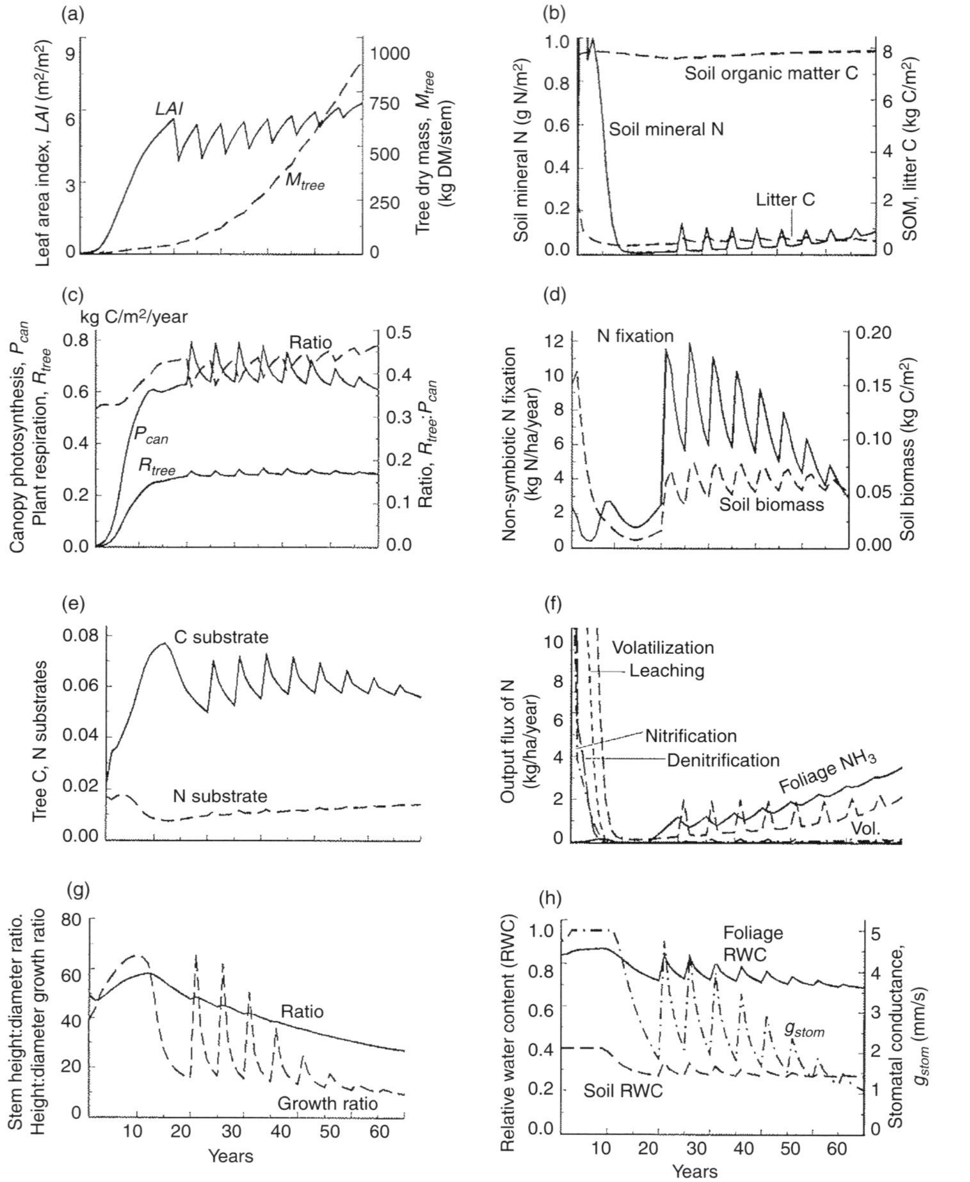

Fig. 21.8. Dynamics of a thinned conifer plantation over a 60-year rotation. The system is in equilibrium – that is, the state of the system is exactly the same every 60 years. Details of environment and management are given in 'The initialization problem' section. Labile quantities such as soil mineral N (b), tree substrate concentrations (e) and soil biomass (d) are annual averages. Fluxes are summed over each calendar year. In (h), values are for 15 h on 1 July in each year, when water stress is close to maximal.

Table 21.2. Performance of the Edinburgh Forest model, without site-specific tuning, simulating a generic thinned conifer plantation in the climate of Eskdalemuir, Scotland. Quantities are averages for a 60-year rotation, unless otherwise stated. Yield class is the mean annual volume increment over a rotation. Net photosynthesis is gross canopy photosynthesis minus whole-plant respiration. Quantities for beech (90-year rotation) given in {...}. Initial state is the conifer plantation equilibrium state; this is also used for the beech plantation.

Yield class: 11.5 {6.8} m^3/ha/year	
Carbon budget (kg C/m^2/year)	
Gross canopy photosynthesis	0.63 {0.50}
Net photosynthesis	0.34 {0.35}
Ratio (net/gross, dimensionless)	0.58 {0.70}
Local growth respiration	0.09 {0.08}
Residual maintenance respiration	0.13 {0.04}
Soil respiration	0.19 {0.21}
Leaching	0.0012 {0.005}
Products	0.16 {0.14}
Nitrogen budget (kg N/ha/year)	
Deposition	10 {10}
Fixation	5 {6}
NH_3 emission from foliage	2 {0.4}
Soil NH_3 volatilization	4 {6.1}
Nitrification	1 {0.6}
Denitrification	1 {0.5}
Soil gaseous N emission	5 {7.2}
System gaseous N emission	6 {7.5}
Leaching	1 {1.2}
Products	6 {7}
Water budget (m/year)	
Annual rainfall	1.53 {1.53}
Intercepted and evaporated rain	0.59 {0.26}
Plant evapotranspiration	0.40 {0.13}
Drainage	0.54 {1.13}
End-of-rotation variables	
Leaf area index {1 August 15 h}	6.6 {1.9}
Stem height	12 {21} m
Stem mass	732 {740} kg structural dry mass/stem
Stem height:diameter ratio	24 {68}
Some other variables	
Stomatal conductance, 15 h, 1 July	0.0026 {0.005} m/s (cf. 0.005 fully open)
Soil mineral N concentration	0.0002 {0.0003} kg N/m^2

Continued

Table 21.2. *Continued.*

Carbon sequestered (kg C/m^2)	
System	16.5 {14.4}
Soil (SOM, litter, biomass, C_{sol})	9.2 {9.3}
Tree	3.3 {3.2}
Products	2.6 {2.0}
C:N ratio of system	16.6 {15.8} kg C/(kg N)
C:N ratio of soil organic matter	10.7 {10.0} kg C/(kg N)
C:N ratio of decaying product pool	274 {250} kg C/(kg N)

Response of a thinned conifer plantation to N fertilizer

The simulated effects of a single application of 500 kg N/ha ammonium fertilizer on 1 May of the 16th year just as the canopy is becoming closed are illustrated in Fig. 21.9. The large spike in the soil mineral N pool (N_{min}) is quickly followed by increase in leaf area index, LAI (Fig. 21.9a). In this case, the increased N_{min} is sustained for some 30 years. With other parameter values, an increased N_{min} is only sustained for 5 to 10 years, after which N_{min} can take a value below the zero-fertilizer case; this is because an increased LAI (before and after thinning) with no stomatal closure is driving increased root N uptake. Soil organic matter N (Fig. 21.9b) increases slowly throughout the rest of the rotation, whereas N input to the litter pools (surface and soil, Fig. 21.3) moves back towards normal after *c.* 10 years. In this case, yield class (Fig. 21.9c) increases after fertilizer application, but this increase is not sustained (yield class is the mean annual volume increment over a rotation). With other parameters, yield class increases above normal for 10 years after fertilizer application, and this increase is sustained until the end of the rotation. The response of yield class to fertilizer application can be positive, zero or even negative.

Phenology in deciduous mode

When the EFM is run in deciduous mode, the phenology sub-model of Fig. 21.6 is used to time phenological events. The clock is updated every integration interval (1/512 day, Table 21.3, bottom line), but phenology changes are applied only at the beginning of a new day. Currently, the four principal processes, represented by the continuous arrows in Fig. 21.6, are affected only by instantaneous air temperature. The temperature dependences of the processes are drawn in Fig. 21.10 and are as follows.

Chilling

$$k_c = 0.5\,[k_{cz} + \text{abs}\,(k_{cz})],$$

$$k_{cz} = k_{c,opt}\,\frac{(T_{air} - T_{c,min})^{q_{c,min}}\,(T_{c,max} - T_{air})^{q_{c,max}}}{(T_{c,opt} - T_{c,min})^{q_{c,min}}\,(T_{c,max} - T_{c,opt})^{q_{c,max}}},$$

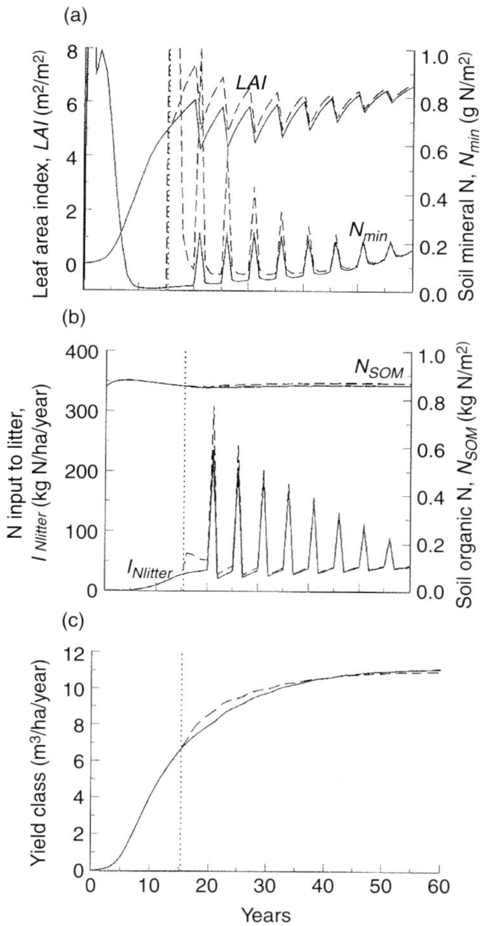

Fig. 21.9. Simulation of effects of a single application of ammonium fertilizer of 500 kg N/ha on 1 May of 16th year, indicated by the vertical small-dashed line. Continuous line: no fertilizer. Dashed line: fertilizer applied. See Fig. 21.8 legend.

$$
\begin{aligned}
T_{c,opt} &= \frac{q_{c,min}T_{c,max} + q_{c,max}T_{c,min}}{q_{c,min} + q_{c,max}}, \\
k_{c,opt} &= 2/\text{day}, \\
T_{c,min} &= -5,\ T_{c,max} = 15^\circ\text{C}, \\
q_{c,min} &= 2,\ q_{c,max} = 1,
\end{aligned} \tag{1}
$$

where abs(x) denotes absolute value of x. Chilling units are acquired between -5 and $+15$°C, with the maximum (optimum) rate at 8.33°C. The response is quadratic from -5°C and decreases linearly to zero at 15°C.

Forcing

Two options are provided. The first is to use the standard biological temperature function shown by the dashed line in Fig. 21.10b and given by equations (3.9a)

Table 21.3. Parameter changes when moving from coniferous to deciduous simulation. In conifer mode, phenology (Fig. 21.6) has no effect on bud growth, bud opening or leaf senescence, and dormancy is also without effect.

No increase in leaf senescence due to shading.
Lifetimes of branches, coarse and fine roots increased by factor of 10.
Increase fractions of recycled C and N from foliage and fine root structural litter fluxes from 0.5 to 0.8.
Switch off fine roots senescence when leaves not present.
Increase response of specific leaf area (SLA) to substrate C concentration by increasing $c_{SLA,C}$ of equation 3.6c (Thornley, 1998a) by a factor of 4.
Increase maximum SLA ($c_{SLA,max}$) of equation 3.6c (Thornley, 1998a) from 10 to 30 m^2 leaf/(kg structural DM).
Halve the synthetic rate constant for photosynthetic protein: parameter $k_{G,Nph}$ of equation 11 (Thornley, 1998b) from 0.03 to 0.015 kg photosynthetic protein/(kg structural DM)/day.
Increase proportionality constant, $c_{Pmax,Nph}$ (Thornley, 1998c, equation 10), between photosynthetic protein and light, CO_2-saturated, 20°C photosynthetic rate from 0.0008 to 0.002 kg CO_2/s/(kg photosynthetic N).
Increase canopy extinction coefficient (k_{can} of equation 3.2b of Thornley, 1998a) from 0.5 to 0.87.
Increase leaf transmittivity (X_{leaf} of equation 3.2b of Thornley, 1998a) from 0.03 to 0.1.
Decrease residual maintenance respiration by increasing parameter K_C of equation 17 of Thornley and Cannell (2000c) from 0.05 to 0.1 kg C substrate/(kg structural DM). This increases sugar levels and, through osmotic pressure, will cause stomates to be open for longer.
Increase root uptake efficiency. Parameter $\sigma_{N,20}$ of equation 3.4b of Thornley (1998a) is increased from 0.1 to 5 kg N/(kg root structural DM)/day.
Stem density increased from 350 to 500 kg structural DM/m^3.
Meristem activity constants (see parameters k_{iM20}, $i = l, b, s, c, f$ for foliage, branches, stem, coarse roots and fine roots of equations 3a, 7b, 9b, 11b and 13d in appendix 2 of Thornley, 1991) are changed from 200 (all) to 500 for foliage and buds, 100 (branches), 100 (stem), coarse roots (50) and 10 (fine roots)/(kg structural DM)^2 (kg substrate C) (kg substrate N)/day. The leaf-bud sub-model has not been written up. Details can be found in file efm.csl at www.nbu.ac.uk/efm (search file for 'kG').
Soil to fine root water transport parameters, $c_{W,so\text{-}rs}$ and $c_{W,rs\text{-}rt}$, of equation 6.2i (Thornley, 1998a) are changed from 100,000 to 20,000 m^2 and from 5 to 0.5 m/day, respectively (in efm.csl parameters are denoted by kso_fisurf (bulk soil to fine root surface) and kfisurf_fi (fine root surface into the fine root)).
Within-plant water transport resistivities are decreased by a factor of 10.
Planting density increased from 0.25 to 0.5/m^2.
Rotation length increased from 60 to 90 years.
Thinning regime changed (from that given in text above: see 'The initialization problem' section) to: 13 thinnings at 25, 30, 35, 40, 45, 50, 55, 60, 65, 70, 75, 80, 85 years, with fractions of the then standing stem numbers removed of 0.2, 0.32, 0.30, 0.26, 0.21, 0.18, 0.16, 0.15, 0.14, 0.13, 0,11, 0.1, 0.1.
Time step for the Euler integration is decreased from 1/128 day (11.25 min) to 1/512 day (2.8125 min).

DM, dry matter.

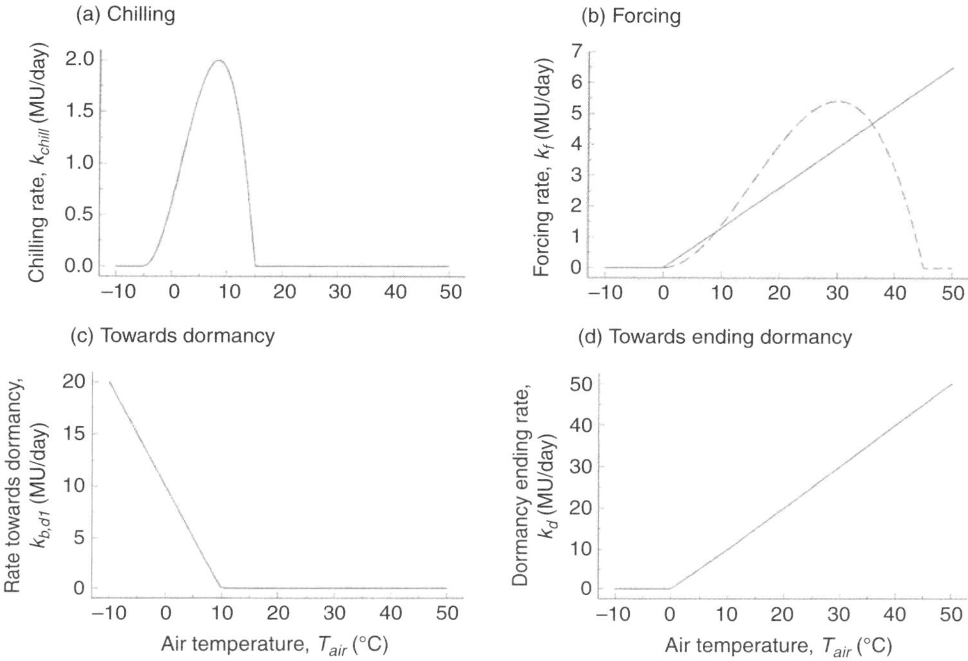

Fig. 21.10. Temperature dependencies of the developmental processes in the phenological sub-model (Fig. 21.6) are shown. Two options for the forcing function, k_f, are given in (b).

(multiply $f(T)$ by 4) and Fig. 3.6 (cubic) of Thornley (1998a). This is of the same type as equation (1) but with $T_{c,min} = 0$, $T_{c,max} = 45°C$. The second is to use a ramp function:

$$\begin{aligned} k_f &= 0.5\,[k_{fz} + \text{abs}\,(k_{fz})], \\ k_{fz} &= g_f\,(T_{air} - T_{f,min}), \\ g_f &= 0.13/(°C)/\text{day}, \;\; T_{f,min} = 0°C. \end{aligned} \tag{2}$$

Here the second option is employed.

Dormancy

Onset of dormancy and dormancy ending (Fig. 21.10c, d) are given by ramp functions similar to equation (2).

Phenology over 2 years is simulated in Fig. 21.11. In the first 2 years bud burst occurs on 7 June and 22 May (the difference is due to initialization on 1 January); otherwise, in both years, buds stop opening on 8 September; leaf fall and dormancy begin on 28 October; after 17 days, dormancy ends on 14 November and the tree can respond to forcing temperatures. In the second and subsequent years, buds open for 109 days and leaves are present for 158 days.

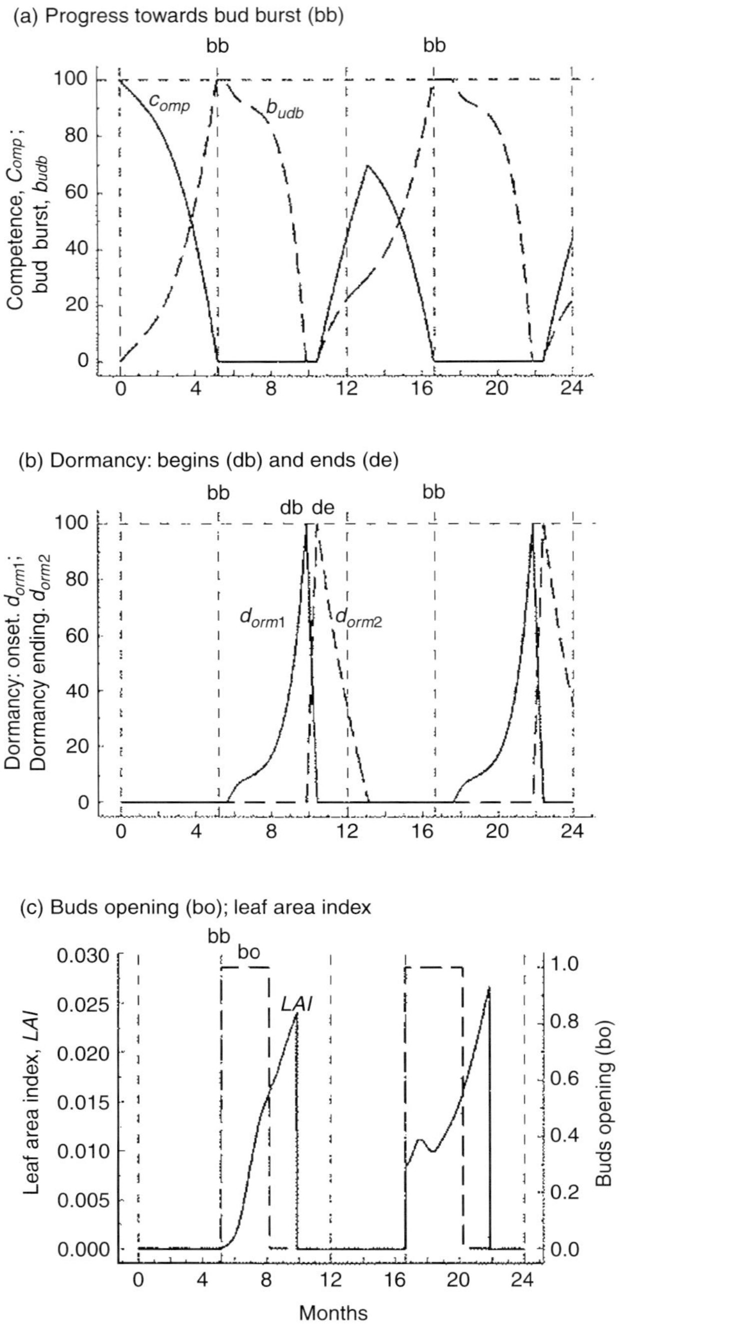

Fig. 21.11. Simulation of phenology (Fig. 21.6) in southern Scotland (Eskdalemuir) over 2 years.

Beech plantation dynamics

Running the model for a beech plantation (where some of the biological and management parameters have been altered as in Table 21.3) at Eskdalemuir (which would not be normal practice) gives results qualitatively similar to those in Fig. 21.8, but with decreased productivity. Table 21.2 lists beech data alongside the conifer equivalents.

Simulation of natural forest

The EFM is able to simulate natural forest, deciduous or evergreen. This is achieved by assuming empirical equations for tree mortality and regeneration but preserving the mean-tree concept of the model. While this will not help us understand the mechanisms of mortality and regeneration in forests, it may help us understand the role of the processes of mortality and regeneration in forest ecosystem responses.

A proportional stem mortality rate (m, per day) depends phenomenologically on the reciprocal of tree nutrient status (the product of carbon substrate and nitrogen substrate concentrations averaged over the tree, $<CN>_{\mathrm{tree}}$), on foliage water stress (via a shoot water function, $f_{\mathrm{W,sh}}$; see Thornley 1998a, p. 133; increasing stress gives increasing mortality) and positively on leaf area index, L_{AI} (as a surrogate for between-tree competition). In detail

$$m = \frac{5 \times 10^{-9} f_{\mathrm{T}}\, L_{\mathrm{AI}}}{(f_{\mathrm{Wsh}} + 0.1) <CN>_{\mathrm{tree}}}, \tag{3}$$

where f_{T} is the usual biological temperature function (dashed line in Fig. 21.10b). The 0.1 term in the denominator is to obviate numerical problems during the winter when shoot water potential is meaningless if using the deciduous option. For example, at 20°C, $f_{\mathrm{T}} = 1$, assume $L_{\mathrm{AI}} = 5$, $f_{\mathrm{Wsh}} = 0.9$ and $<CN>_{\mathrm{tree}} = 0.0001$ (with C and N substrate concentrations of 1% of dry matter), then $m = 5 \times 10^{-9} \times 5/0.0001 = 0.00025$/day = 9%/year. At 10°C, this decreases to about 3%/year.

Proportional stem regeneration rate (g, per day) is assumed to depend positively on leaf nutrient status ($<CN>_{\mathrm{leaf}}$; generates viable seeds), relative irradiance at ground level ($\exp(-0.5L_{\mathrm{AI}})$, promotes seed germination) and leaf area per stem (A_{leaf}, m^2/stem) (surrogate for a tree maturity and crown size), namely,

$$g = 0.2 f_{\mathrm{Tsoil}} f_{\mathrm{Wsoil}} [0.1 + \exp(-0.5 L_{\mathrm{AI}})] A_{\mathrm{leaf}} <CN>_{\mathrm{leaf}}, \tag{4}$$

where f_{Tsoil} is the soil temperature function (dashed line in Fig. 21.10b) and f_{Wsoil} is a soil water function (Thornley 1998, p. 133; decreasing soil water causes f_{Wsoil} to decrease below unity). With both these equal to unity, including the term in [. . .], and then assuming $A_{\mathrm{leaf}} = 10\ m^2$ and $<CN>_{\mathrm{leaf}} = 0.0001$ (with C and N substrate concentrations of 1% of dry matter), then $g = 0.2 \times 10 \times 0.0001 = 0.0002$/day = 7%/year.

A slow small instability was sometimes observed. This was not understood, although the water sub-model is involved. It was eliminated by assuming a constant height : diameter ratio of 60; in plantation mode the height : diameter growth ratio (Fig. 21.8g) is affected by nutrient status and water stress.

Applying this phenomenology causes the simulated forest to approach a steady state, where annual tree mortality and regeneration balance. The steady-state natural forest is often more convenient (than a plantation system) for examining the consequences of fire (Thornley and Cannell, 2004), pruning (Thornley and Cannell, 2000b) and climate change.

Seasonal dynamics of natural conifer and beech forests

The equilibrium state of natural forests has interesting seasonal dynamics. These are very different for conifer and deciduous forests. Figure 21.12 illustrates this for the two cases.

Figure 21.12a shows that the deciduous forest achieves higher rates of canopy photosynthesis during the summer, partly because the conifer undergoes more water stress and stomatal closure (Fig. 21.12d). However, over the year, conifer gross production is 0.49 kg C as opposed to a deciduous 0.43 kg C.

Figure 21.12b illustrates the considerable seasonal photosynthetic acclimatization predicted with the acclimatization sub-model (Thornley, 1998c). The photosynthetic N pool by which this is achieved is shown in Fig. 21.2a.

Soil mineral N and root N uptake are much affected by the presence or absence of leaves providing carbohydrates to drive N uptake, and the massive litter inputs caused by leaf fall (Fig. 21.12c). Internal N recycling can also be a factor in N uptake. Per year, beech roots take up only 43 kg N/ha compared with 181 kg N/ha for the conifer.

Annual non-symbiotic N fixation is 2.5 kg N/ha for the conifers and 8 kg N/ha for beech (Fig. 21.12d). For beech, there is very little summer stomatal closure, but this is considerable for the conifer forest (Fig. 21.12d), which is reflected in canopy photosynthesis (Fig. 21.12a).

Managing forests for wood yield and carbon storage

The EFM has been used to study theoretically which management regimes best achieve the dual objectives of high sustained timber yield and high carbon storage (Thornley and Cannell, 2000b). The conclusions are interesting. They relate to the model as formulated and parameterized at that time, but are likely to be of general validity.

More carbon was stored in the undisturbed forest than in any regime in which wood was harvested (35.2 kg C/m^2). Plantation management gave moderate carbon storage (14.3 kg C/m^2) and timber yield (15.6 m^3/ha/year). But, most notably, annual removal of 10 or 20% of woody biomass per year gave both a high timber yield (25 m^3/ha/year) and high carbon storage (20 to 24 kg C/m^2). The efficiency of the latter regimes could be attributed to high light interception and net primary productivity, with less evapotranspiration and

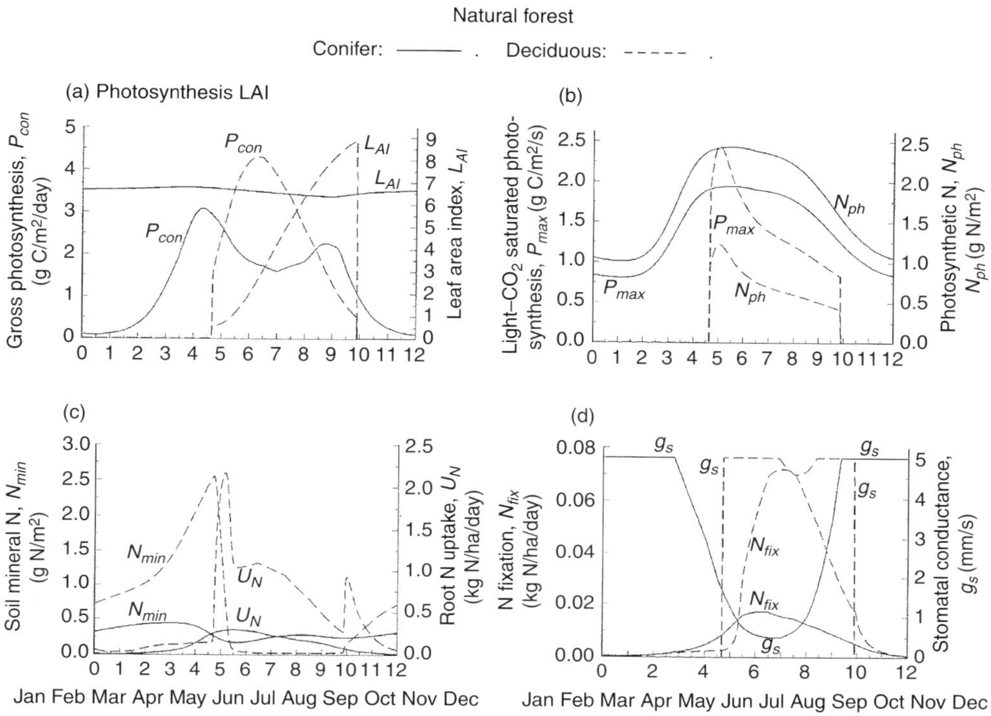

Fig. 21.12. Seasonal predictions for conifer and deciduous natural forest. Both simulated forests are in an equilibrium state. Forest self-thinning and regeneration are simulated using equations (3) and (4) for both forest types. (a) Gross photosynthesis is summed over each day: 50 g C/m²/day = 48 μmol C/m²/s [48 = 50 × 10⁶/(12 × 86400)]. (b) P_{max} is a light- and CO_2-saturated photosynthetic rate parameter referred to 20°C; see equation (3.2t) of Thornley (1998a) for a full explanation; the photosynthetic N pool is shown in Fig. 21.2a. (c) Soil mineral N comprises the ammonium and nitrate pools shown in Fig. 21.3; root N uptake is summed over each day. (d) Non-symbiotic N fixation in the soil is summed over each day; stomatal conductance is a daily 15 h value. Parameters changed for deciduous rather than conifer simulation are listed in Table 21.3; deciduous parameters are based approximately on beech data.

summer water stress than in the undisturbed forest, high litter input to the soil giving high soil carbon and N_2 fixation, low maintenance respiration and low N leaching owing to soil mineral pool depletion.

There is no simple inverse relationship between amount of timber harvested from a forest and amount of carbon stored. Management regimes that maintain a continuous canopy cover and mimic, to an extent, regular natural forest disturbance realize the best combination of high wood yield and carbon storage.

Aphid dynamics in natural conifer forest

The aphid model of Fig. 21.5 has been applied to the EFM running in evergreen natural forest mode in an Eskdalemuir environment without any specific parameter adjustment. The parameterization is mostly as described by Newman *et al.* (2003).

Figure 21.13 illustrates these simulations. Six treatments are applied: mean temperature is raised by +0, 2, 5°C, times CO_2 concentrations of 350, 700 μmol/mol. In each case the forest is in an equilibrium state when 5 aphids/stem are introduced at time zero.

At ambient Eskdalemuir temperatures (+0°C), the aphids quickly die out at both CO_2 concentrations (Fig. 21.13a, b). At +2°C, the aphid population settles down to a relatively low annually cyclic value; the cycle amplitude is greater at high CO_2 concentration (in other simulations less growth occurs at high CO_2 concentration – there can be profound differences between short- and long-term responses to elevated CO_2, particularly where N status is concerned: Thornley and Cannell, 2000a). At +5°C, aphid growth is altogether more volatile: at low CO_2 a stable annual cycle is achieved for many years; at high CO_2 the system appears to show more of the characteristics of chaos. Higher aphid numbers are accompanied by decreased phloem N concentrations (not shown) and decreased leaf area indices (Fig. 21.13c, d).

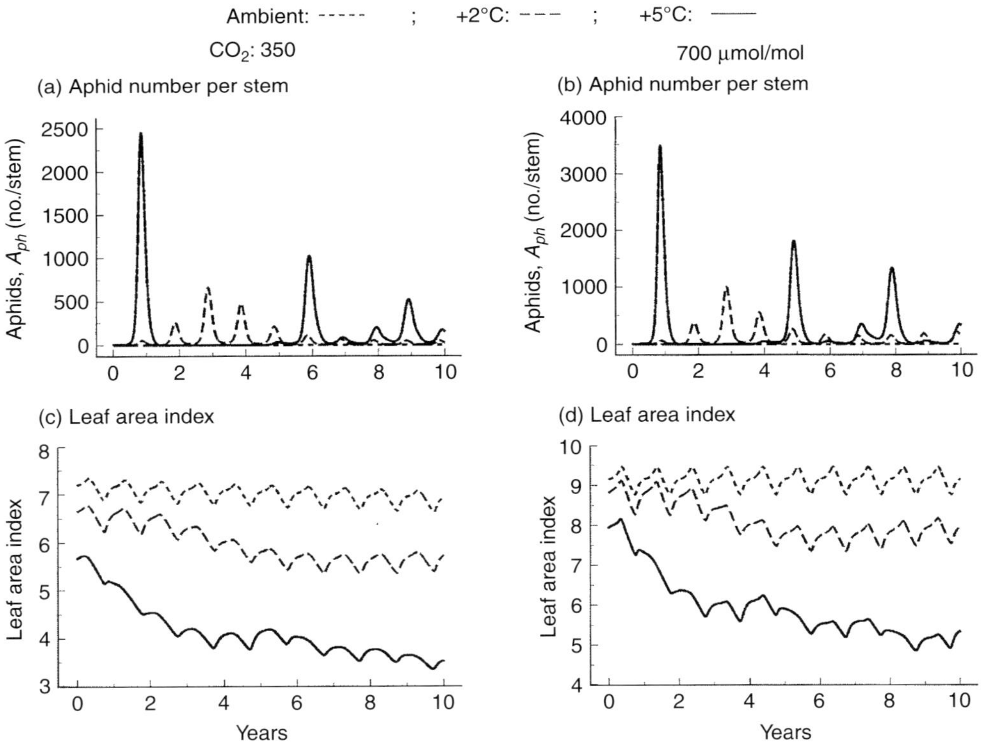

Fig. 21.13 Simulated effects of aphids on natural evergreen forest at three temperatures and two CO_2 concentrations. At time zero, the system is, in each case, in equilibrium when five aphids per stem are introduced.

These simulations suggest that temperature greatly affects both speed and volatility of aphid infestation, with CO_2 modulating the details of this response.

Aphid dynamics in a plantation

The simulated effects of aphid infestation on an evergreen plantation grown at Eskdalemuir at + 5°C are illustrated in Fig. 21.14. Note that most of the quantities plotted in Fig. 21.14 are annual averages, which can hide important seasonal changes. At these raised temperatures, aphids flourish (Fig. 21.13). Figure 21.14a, c shows that both leaf area index and yield class are depressed considerably by aphid infestation. (Note that raising the temperature by 5°C increases the no-aphid yield class from 11.1 (Table 21.2) to 12.8 m^3/ha/ year (Fig. 21.14c).) When aphids are introduced, the general effect is to decrease growth (Fig. 21.14a, c), although relatively complex dynamics arise from the interaction between the aphid life cycle and the changing nutrient status of an uninfected plantation (Fig. 21.8b, d, e). The mechanistic structure of the model allows these complications to be examined and their causes unravelled. In Fig. 21.14b, the initial aphid population outbreak is brought to an end by decreasing soil mineral N (Fig. 21.14d) and the increasing C-driven demand for N in the tree (Fig. 21.8e). Before the first thinning at 20 years, the aphid population goes through a second epidemic at about 17 years (the second spike from the left in the continuous line in Fig. 21.14b). The N-dependent component of mortality increases sharply and brings the population back down, contributing a peak to soil mineral N (Fig. 21.14d). Phloem N levels are much affected (Fig. 21.14e), and there are capricious changes in non-symbiotic N fixation.

Climate change

The last application presented here is concerned with climate change and the possible effects of climate change on forests. Recent reviews are by Curtis and Wang (1998), Saxe *et al.* (1998) and Pritchard *et al.* (1999). Climate change experiments are usually short term, perhaps of several years duration, and are often concerned with the responses to step changes in environmental variables. It is dangerous to extrapolate such results to the reality of climate change, which is slow, possibly spanning centuries, and may give ecosystems much time to respond. Indeed, these and other simulation results suggest that the results of short-term experiments can be misleading, even opposite to long-term effects. First, therefore, the simulated consequences of step changes to environmental factors are examined, before considering the simulated consequences of realistic climate change scenarios. The assumed climate change affects atmospheric CO_2, atmospheric N deposition and temperature only.

Figure 21.15 illustrates, for southern Scotland equilibrium conditions for a natural conifer forest, the effects over 20 years of step changes to ambient CO_2

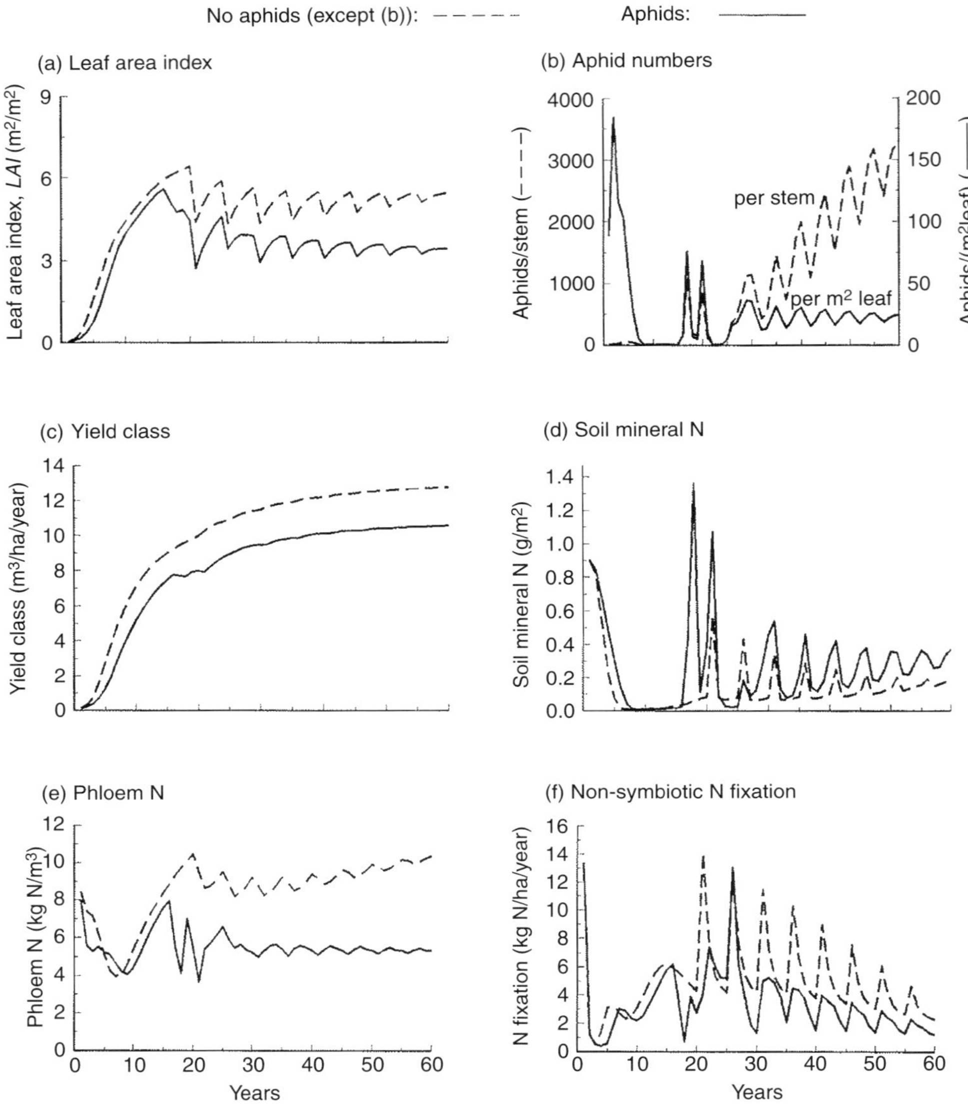

Fig. 21.14. Simulated effects of aphids on plantation conifer forest in an Eskdalemuir environment at +5°C. In (a)–(f), except for (b), the dashed lines represent the no-aphid control, and the continuous lines the situation where 6 aphids per m² aphid-accessible leaf surface are introduced at time zero. Aphid-accessible leaf surface is provided by age categories 2–4 of the four leaf age categories represented (Fig. 21.2). In (a) and (c) plotted values are instantaneous values at the end of each year; otherwise, annual averages or annual sums (of fluxes) are plotted. The initial state is the no-aphid equilibrium state for the Eskdalemuir environment at +5°C.

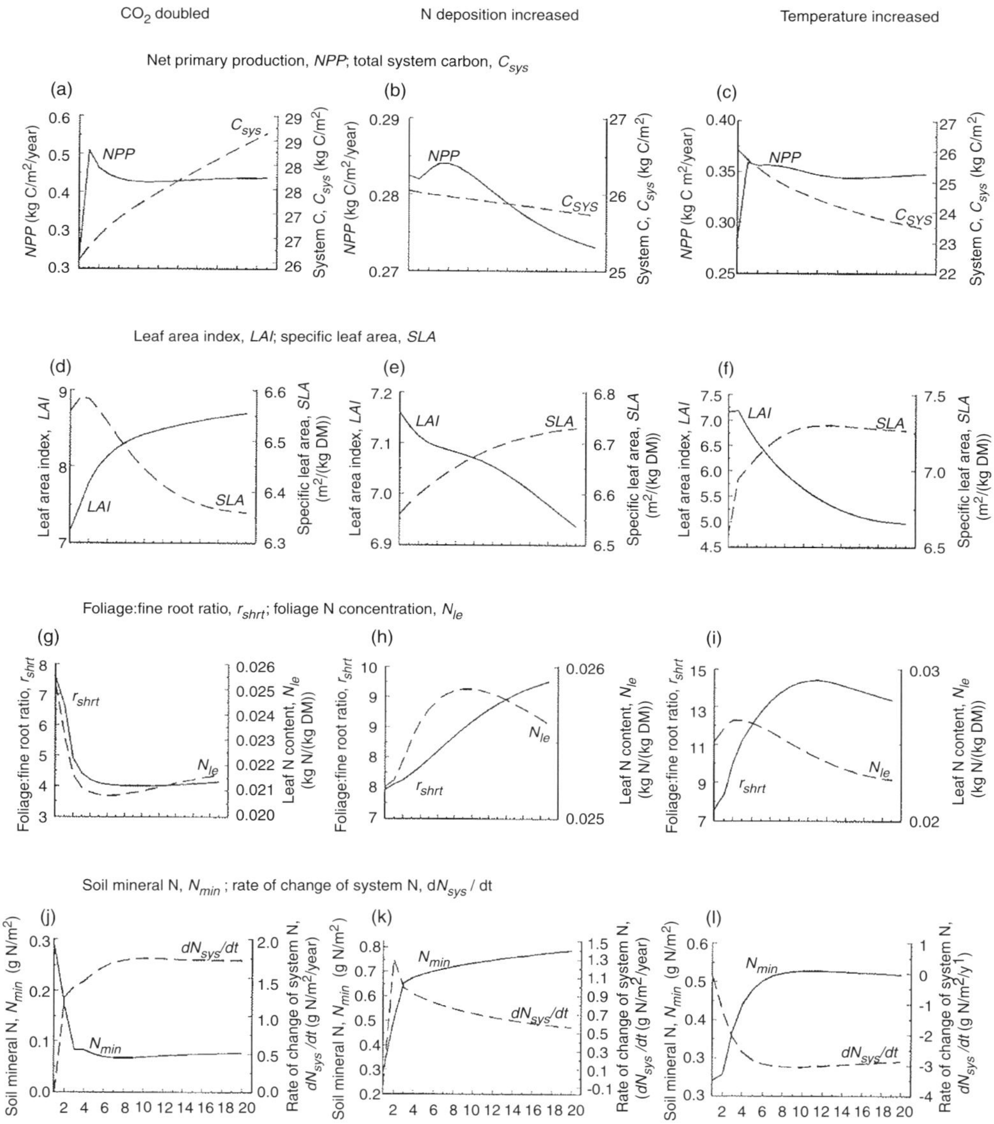

Fig. 21.15. Responses to step changes in ambient CO_2, N deposition and temperature. Ambient CO_2 is doubled, from 350 to 700 μmol/mol (left-hand column); N deposition is increased from 10 to 30 kg N/ha/year (centre column); all temperatures increased by 5°C (right-hand column). The environment is for Eskdalemuir (southern Scotland) using equilibrium initial values for a natural conifer forest (Fig. 21.12). The changes are imposed after 1 year in the equilibrium environment. Values of system C, leaf area index and specific leaf area are point values (i.e. at 0 h on 1 January each year). Otherwise, annual averages or sums (for fluxes) are plotted.

(+CO_2: doubled, from 350 to 700 μmol/mol), N deposition (+N: 10 to 30 kg N/ha/year) and temperature (+T: all temperatures increased by 5°C). The forest model is, using the language of dynamic systems, a classical 'stiff' system: the carbon dynamics are relatively fast, whereas the longer-term behaviour depends on nitrogen and its acquisition or loss by the ecosystem. This can be seen by comparing the ratio of carbon input, of the order of 0.5 kg C/m^2/year, to the carbon content, of the order of 25 kg C/m^2, giving a rate constant (by division) of 1/50 = 0.02/year, with the ratio of nitrogen input, of the order of 0.001 kg N/m^2 /year, to the nitrogen content, of the order of 1 kg N/m^2, giving a rate constant (by division) of 0.001/1 = 0.001/year. The latter is 20 times smaller, giving a correspondingly longer time constant. A more detailed explanation of ecosystem N dynamics for grassland is given by Thornley and Cannell (2000a).

The items +CO_2 and +T (first and third columns of Fig. 21.15) give some immediate responses that are partially or even completely reversed with the further passage of time: e.g. for net primary production, specific leaf area, shoot : root ratio, foliage N content and soil mineral N. Responses to N (centre column) are mostly simpler and smaller; note that equilibrium natural conifer forest is already an N-rich system. Managed systems, where products are regularly removed, are relatively N-poor and therefore can respond more strongly to increasing N. These results indicate plainly the anomalous and possibly misleading responses that can be obtained from short-term experimentation on the forest ecosystem.

The simulated effects of 250 years of climate change, from 1850 to 2100, are illustrated in Fig. 21.16. The climate scenario is given in Fig. 21.16a, where ambient CO_2 concentration (C), N deposition rate (N) and temperature (T) are drawn. The effects of these three assumed climate change components are shown separately and in combination. Figure 21.16b illustrates that high C sequestration is favoured by high CO_2 and low temperature; N deposition makes little difference to this already N-rich ecosystem. On the other hand, Fig. 21.16c indicates that net primary production is maximized by increasing CO_2 and temperature. Soil mineral N is decreased by +CO_2 (Fig. 21.16d), as are N losses from the system (not shown), but these are increased by +N and +T. Leaf area index is increased by +C (Fig. 21.16e), although other simulations can give a decrease (Fig. 8.16 of Thornley, 1998a) and experimental work gives a range of results (Fig. 4, Saxe *et al.*, 1998). Specific leaf area (Fig. 21.16f) responds according to expectation, as does the foliage : fine root ratio (Fig. 21.16g). Non-symbiotic N fixation (Fig. 21.16h) behaves almost oppositely to soil mineral N (Fig. 21.16d); the model assumes this process is inhibited by soil mineral N (equation 5.4c of Thornley, 1998a).

In an equilibrated natural forest, N supply is often good, even though external inputs are small. However, in a relatively N-poor system such as plantation forestry, where N is continually removed from the sytem, the combination of increased atmospheric CO_2 and increased N deposition can act powerfully, with very beneficial consequences for forest productivity (see Cannell and Thornley, 1998a). The impact of climate change on forests is therefore strongly dependent on forest type, local conditions and management.

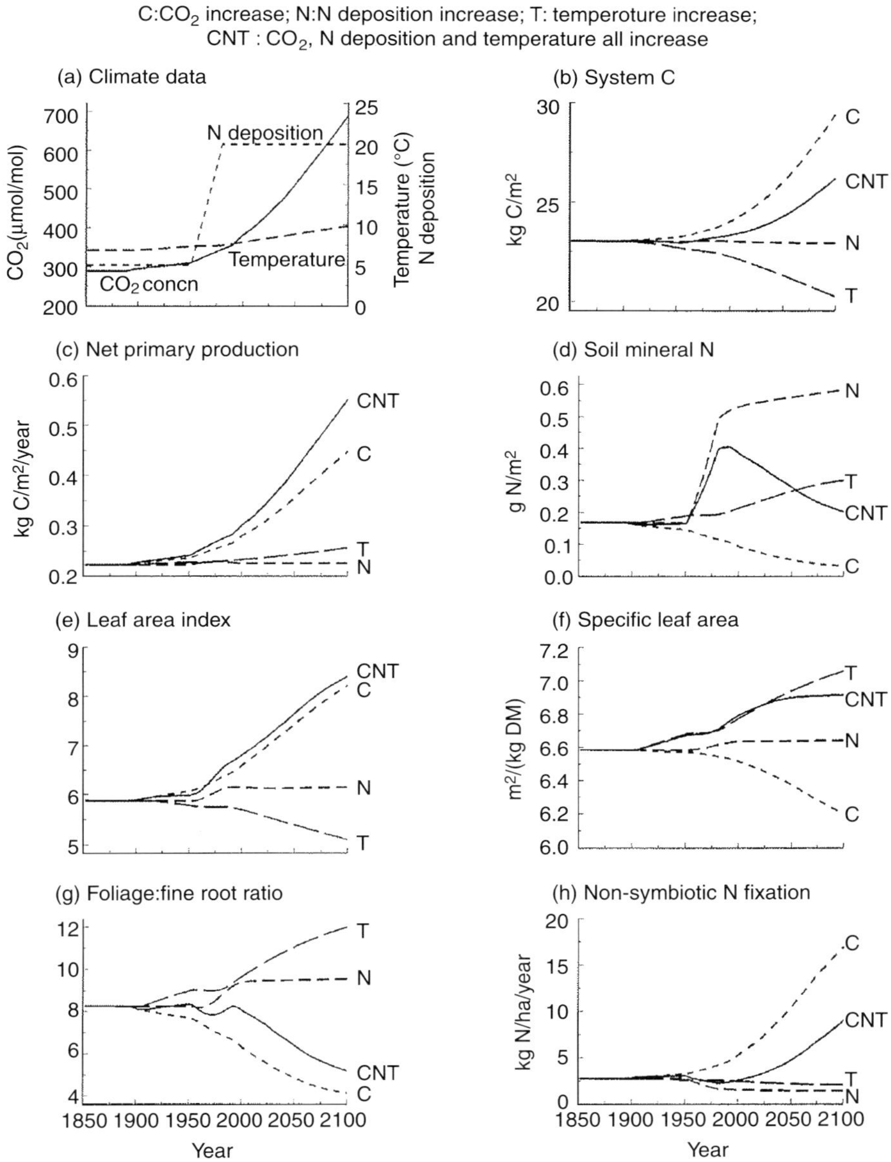

Fig. 21.16. Responses to climate change since 1850. Natural forest evergreen mode, without aphids. Base environment as for Eskdalemuir, but (a) shows the quantities assumed affected by climate change. Temperature is mean annual temperature. Nitrogen deposition has units of kg N/ha/year. (b) Total system carbon. (c) Net annual primary production. (d) Soil mineral nitrogen, comprising ammonium N and nitrate N pools. (b), (e) and (f) are beginning-of-year values (0 h, 1 January); (c), (d), (g) and (h) are annual sums or averages. The system is initialized to the 1850 equilibrium state.

Conclusions

The development of mechanistic models is part of a shift in research focus towards quantitative explanation, integration of complexity and prediction in the agricultural and ecological sciences. A forest research programme without a modelling component may be critically emasculated by this missing dimension. To do such work successfully requires appropriate organization and commitment. While models are, rightly, only a part of the research scene, they provide a framework for ideas that can be helpful to all involved in research and applications of that research.

References

ACSL (2000) Advanced Computer Simulation Language. Aegis Research Corporation, Huntsville, Alabama, USA. Email: acsl-sales@aegisrc.com. www.ACSL.sim.com

Cannell, M.G.R. and Thornley, J.H.M. (1998a) N-poor ecosystems may respond more to elevated [CO_2] than N-rich ones in the long term: a model analysis of grassland. *Global Change Biology* 4, 431–442.

Cannell, M.G.R. and Thornley, J.H.M. (1998b) Temperature and CO_2 responses of leaf and canopy photosynthesis: a clarification using the non-rectangular hyperbola model of photosynthesis. *Annals of Botany* 82, 883–892.

Cannell, M.G.R. and Thornley, J.H.M. (2000) Modelling the components of plant respiration: some guiding principles. *Annals of Botany* 85, 45–54.

Curtis, P.S. and Wang, X. (1998) A meta-analysis of elevated CO_2 effects on woody plant mass, form, and physiology. *Oecologia* 113, 299–313.

de Wit, C.T. (1970) Dynamic concepts in biology. In: Setlik, I. (ed.) *Prediction and Measurement of Photosynthetic Productivity*. Pudoc, Wageningen, Netherlands, pp. 17–23.

Meteorological Office (1982) *Tables of Temperature, Relative Humidity, Precipitation and Sunshine for the World. Part III Europe and the Azores*. Met.O.856c. UDC 551.582.2(4), Meteorological Office, Bracknell, Berkshire, UK.

Newman, J.A., Gibson, D.J., Parsons, A.J. and Thornley, J.H.M. (2003) How predictable are aphid population responses to elevated CO_2? *Journal of Animal Ecology* 72, 556–566.

Pritchard, S.C., Rogers, H.H., Prior, S.A. and Peterson, C.M. (1999) Elevated CO_2 and plant structure: a review. *Global Change Biology* 5, 807–837.

Roderick, M.L. (2001) On the use of thermodynamic methods to describe water relations in plants and soil. *Australian Journal of Plant Physiology* 28, 729–742.

Saxe, H., Ellsworth, D.A. and Heath, J. (1998) Tree and forest functioning in an enriched CO_2 atmosphere. *New Phytologist* 139, 395–436.

Smith, J.U., Smith, P., Monaghan, R. and Macdonald, A.J. (2002) When is a measured soil organic matter fraction equivalent to a model pool? *European Journal of Soil Science* 53, 405–416.

Thornley, J.H.M. (1991) A transport-resistance model of forest growth and partitioning. *Annals of Botany* 68, 211–226.

Thornley, J.H.M. (1998a) *Grassland Dynamics: An Ecosystem Simulation Model.* CAB International, Wallingford, UK.

Thornley, J.H.M. (1998b) Modelling shoot:root relations: the only way forward? *Annals of Botany* 81, 165–171.

Thornley, J.H.M. (1998c) Dynamic model of leaf photosynthesis with acclimation to light and nitrogen. *Annals of Botany* 81, 421–430.

Thornley, J.H.M. (2002) Instantaneous canopy photosynthesis: analytical expressions for sun and shade leaves based on exponential light decay down the canopy and an acclimated non-rectangular hyperbola for leaf photosynthesis. *Annals of Botany* 89, 451–458.

Thornley, J.H.M. and Cannell, M.G.R. (2000a) Dynamics of mineral N availability in grassland ecosystems under increased [CO_2]: hypotheses evaluated using the Hurley Pasture Model. *Plant and Soil* 224, 153–170.

Thornley, J.H.M. and Cannell, M.G.R. (2000b) Managing forests for wood yield and carbon storage: a theoretical study. *Tree Physiology* 20, 477–485.

Thornley, J.H.M. and Cannell, M.G.R. (2000c) Modelling the components of plant respiration: representation and realism. *Annals of Botany* 85, 55–67.

Thornley, J.H.M. and Cannell, M.G.R. (2004) Long-term effects of fire frequency on carbon storage and productivity of boreal forests: a modeling study. *Tree Physiology* 24, 765–773.

Thornley, J.H.M. and France, J. (2005) *Mathematical Models in Agriculture*, 2nd edn. CAB International, Wallingford, UK.

Thornley, J.H.M. and Johnson, I.R. (2000) *Plant and Crop Modelling. A Mathematical Approach to Plant and Crop Physiology*, reprint of 1990 edn. Blackburn Press, Caldwell, New Jersey.

22 Information and Knowledge Management for Sustainable Forestry

ALAN J. THOMSON,[1] H. MICHAEL RAUSCHER,[2] DANIEL L. SCHMOLDT[3] AND HARALD VACIK[4]

[1]Canadian Forest Service, Pacific Forestry Centre, Victoria, British Columbia, Canada; [2]USDA Forest Service, Southern Research Station, Asheville, North Carolina, USA; [3]USDA Cooperative State Research, Education and Extension Service, Washington, DC, USA; [4]Institute of Silviculture, Department of Forest and Soil Sciences, University of Natural Resources and Applied Life Sciences, Vienna, Austria.

Abstract

Institutional information and knowledge management often involves a range of systems and technologies to aid decisions and produce reports. Construction of a knowledge system organizing hierarchy facilitates exploration of the interrelationships among knowledge management, inventory and monitoring, statistics and modelling, and policy. Two case studies illustrate these interrelationships in institutional settings: (i) the FAO National Forest Assessment process; and (ii) knowledge management in supply chains. The development and adoption of knowledge management systems in institutions can be improved by considering the principles and studies generated by the social sciences, e.g. innovation diffusion, escalation of commitment and agency theory. Still, many of these principles and practices – as they relate to sustainability – have evolved primarily in the context of the developed world. Broader, more inclusive perspectives are needed as we mesh traditional Western thinking with the insights, cultures, practices and limitations of the developing world.

Introduction

Policies, established by governments and other organizations, both implicitly prioritize those problems that warrant societal (or organizational) attention and provide a broad agenda for the issues involved. This dynamic landscape constrains and focuses the ecological and environmental phenomena we measure and the analyses we perform with those data. Knowledge management (KM) activities, on the other hand, remain relatively invariant with respect to any specific policy direction, but provide the tools and techniques for creating, conserving and sharing knowledge, whatever that knowledge may be. The four thematic areas of this book – inventory and monitoring, statistics and modelling, knowledge management and policy – are interrelated in this very general way.

With ever-greater frequency, the keyword 'sustainability' enters into policy discussions, often as part of an objective or criterion, e.g. 'sustainable forest management' or 'sustainable development'. Because sustainable forest management can be evaluated at many scales and involves social, economic and environmental aspects, each of the four thematic areas comes into play in important ways. But how do they work together in a sustainable forestry context? In this chapter, we explore the interrelationships among these thematic areas by creating a knowledge system organizing hierarchy. Two case studies illustrate the range and interaction of such systems in operational settings: (i) the National Forest Assessment process of the Food and Agriculture Organization (FAO); and (ii) forestry supply chains. Several prevailing theories in the social sciences – innovation diffusion, escalation of commitment and agency theory – are used to illustrate knowledge management system development and adoption in institutions. This will provide guidance for managers who wish to use knowledge management tools in attaining sustainability to successfully integrate these tools into their operations. It will also guide knowledge management researchers in achieving successful integration of their products into existing processes. Vignettes illustrate the relationship of other chapters in this section of the book to particular points in the present chapter.

Interrelationships among Knowledge Management, Inventory and Monitoring, Statistics and Modelling, and Policy

People working in KM, inventory and monitoring, statistics and modelling, and policy tend to write for different journals, attend separate scientific conferences and regard themselves as belonging to different peer groups, as delineated by disciplinary boundaries. But it appears to us that all of these scientific disciplines have something in common. They each offer theory and tools to help identify, understand and solve problems. Therefore, it should be possible to reorganize these four thematic areas in order to highlight their interrelationships, using 'problem solving' as a common theme.

An organizing hierarchy

For the sake of this discussion, we shall assume that there are three broad classes of knowledge systems useful in problem solving: descriptive, predictive and prescriptive systems (Rauscher and Reynolds, 2003). Within each class are subclasses that represent different approaches to providing each class's tools – either descriptive, predictive or prescriptive. In some cases, more specific approaches (e.g. participatory decision making as one type of decision analysis method) could be nested further within these classes. Using these classes as an organizing framework, we might agree to the relational hierarchy below. With this cognitive map as a guide, we can more readily discuss how each theme supports the problem-solving process and how the themes might be interrelated and mutually supportive.

Descriptive tools
- Declarative knowledge management tools – know-what
- Inventory and monitoring
- Descriptive statistics

Predictive tools
- Procedural knowledge management tools – know-how
- Predictive statistics
- Expert-based heuristics
- Spatially aware and non-spatially aware modelling
    - Analytical models
    - Quantitative simulation models
    - Qualitative simulation models
    - Expert system models

Prescriptive tools
- Causal knowledge management tools – know-why
- Decision analysis methods
    - Single-criteria optimization
    - Multiple-criteria decision making (see Vignette 1)
    - Satisficing
    - Participatory (group) decision making
- Decision-support systems
    - Landscape scale
    - Forest scale
    - Project scale
- Policy science and forest management planning
    - Adaptive management
    - Options forestry

Descriptive tools

Descriptive tools focus on the management of declarative data, information and knowledge. The focus here is on what we know. The purpose is to create a shared, explicit and accessible understanding of concepts, ideas, relationships and categories that enables effective communication and understanding of a common societal knowledge base (Heinrichs *et al.*, 2003). It is important that all stakeholders of a particular issue be able to agree on a common descriptive set of knowledge. Such a common understanding of the descriptive, factual knowledge provides a sound basis for reasonable disagreement concerning interpretations, courses of action and values. Successful group decision making can only be based on an explicit identification and discussion of legitimate and factually based differences of opinion when they occur. The various methods and approaches used in KM, inventory and monitoring, and descriptive statistics should be viewed as complementary. We have typically concerned ourselves with intrinsic data quality, which deals with data bias, precision and accuracy. A KM focus on descriptive data also calls attention to how accessible the data is, how secure it is, how ethically it is treated (e.g. Thomson and Schmoldt, 2001), how understandable it is within a given context and how well it is presented to enhance its interpretation (Ribeiro *et al.*, 2004). In fact, improving the organization and accessibility of already existing data and information could achieve considerable technology transfer gains.

Predictive tools

Predictive tools focus on the management of procedural knowledge. The focus is on how activities occur, how things are changing in the real world, how specific problems are solved and how we predict the results of alternative courses of action (Heinrichs *et al.*, 2003). The organization and sharing of procedural knowledge, such as best management practices or how-to processes, creates better understanding and leads to more effective problem solving.

KM methods help us organize and share accepted 'nuggets' of procedural knowledge. This procedural knowledge can be associated with descriptive knowledge to improve understanding. This helps avoid the mindless application of how-to recipes in situations for which they are not appropriate. In fact, situational analysis and guidance should be required as an explicit component of every how-to, best management practice and predictive tool.

Predictive statistics are immensely useful to reduce the noise in information recorded about the natural world and to find the signal that can guide our current actions and help us predict future consequences. Expert-based heuristics, such as rules of thumb, are equally powerful guides when quantitative, predictive statistics are unavailable but when human expertise exists (Schmoldt and Rauscher, 1996; Gigerenzer and Todd, 1999). In the case of predictive statistics, uncertainty is inherent in this class of tool. For heuristic methods, there may be no estimate of reliability or uncertainty, in which case, modelling or decision science tools may help reduce uncertainty.

These tools can be combined into models that provide users with a structured, problem-solving environment. Some models focus on current conditions, such as the estimation of site index or the evaluation of habitat quality. Others, such as growth-and-yield models, use the past as a guide to predict the future. Thinking of this class of predictive tools as having a common purpose aids in organizing them, placing them into their correct context of use and making them more readily available and understandable to a broad variety of clients.

Prescriptive tools

Prescriptive tools deal with causality, judgement, values and choices. Causal knowledge and the prescriptive tools that manage it create the assumptions and theory and drive the choices and actions that directly affect the lives of individuals, organizations and nations (Heinrichs *et al.*, 2003). Although knowledge management researchers have a role to play in organizing causal knowledge and making it more accessible, they are not the prime players. Researchers in policy science should be thought of as organizing the assumptions, creating the theory and identifying the values, emotions and power positions of interested stakeholders. Decision-support systems (DSSs) are in many ways the ultimate integrating tools that bring together what we know in order to assist decision makers in making wise and supportable choices. Holsapple (2003) describes a DSS as 'a computer-based system composed of a language system, a presentation system, a knowledge system, and a problem-processing system whose collective purpose is the support of decision-making activities'. It is absolutely not the function of DSSs to serve up answers to managers (Holsapple, 2003). Although managers in many cases are attracted to apparent turnkey solutions to complex problems, DSSs are primarily communication and organizing tools: the computer model

> forces us to see the implications, true or false, wise or foolish, of the assumptions we have made. It is not so much that we want to believe everything the computer tells us, but that we want a tool to confront us with the implications of what we think we know.
>
> (Botkin, 1977, p. 217)

Many dimensions influence the decision process. The type of decision analysis, whether optimization, multi-criteria decision making or satisficing, represents just one dimension. The political and power dimension as well as the emotional and ethical dimension must also be considered (Rauscher, 1996, p. 265). But DSSs have primarily been constructed to support the technical dimension. It is unclear how the technical, power and value dimensions of the decision process interact with each other. Neither is it clear how to bring support tools for the power and value dimensions explicitly into a DSS framework. This is an important issue, because it is quite possible to have components of the power and value dimensions entirely dominate the technical/factual dimension. Policy science has the potential to teach us by helping to clarify these issues and suggesting ways to publicly clarify sometimes influential, but hidden, elements of the decision process.

Interrelationships

The function of a DSS is to organize the decision process and provide flexible, on-demand access to the full array of prescriptive, predictive and descriptive tools applicable to a particular problem situation. Ideally, a DSS should satisfy the user's need to know what society knows, to know how to use that knowledge and to know why different courses of action produce different expected outcomes. It should also help managers to understand and to explain that understanding to stakeholders.

There is an extremely important feedback loop from DSSs to predictive and descriptive tools. There is no easier way to dramatically demonstrate the limits of our descriptive knowledge base and our capability to use it for predictive purposes than to build and field-test a DSS. For example, in the southern Appalachian Mountain region in the USA, the ability to forecast established overstorey forest conditions over a 30–50 year time frame is quite good (Rauscher *et al.*, 2000). We are beginning to be able to understand and predict tree seedling regeneration following a stand-replacement disturbance (Kim *et al.*, 2000). However, we do not understand, nor can we predict, understorey tree and other woody species' dynamics in the presence of a significant overstorey canopy. What do we do? We assume a constant understorey over the life of a 50–100-year planning horizon. What else can we do? We know that our growth-and-yield predictions for overstorey growth start to seriously degrade after 30–50 years of simulation time from the present. What do we do? We use the models to predict further than 50 years because, once again, there is no other choice.

There are also logical relationships between inventory operations and predictive modelling capabilities. The tension between eco-physiological models and their need for non-standard inventory data is well known. A similar tension exists between DSSs and inventory data. For a goal to be operationally useful, it must have a measurement criterion that can be inventoried in a real forest somewhere. It is not unusual for a client to want goals with measurement criteria that are not available in the current inventory of the property. It may not even be possible to forecast the future value of those measurement criteria by using currently available prediction systems. Such examples are numerous and provide great opportunity to focus new research efforts to fill these major knowledge gaps.

In very general terms, determining exactly what data are inventoried, stored and made accessible is often driven by all four discipline areas. First, land-management policy questions often direct significant changes in data needs, including both variables measured and the scale of measurement. Secondly, the statistics needed by an organization and the models accepted for use, or being developed, also have an impact on data choices. Thirdly, internal organizational policies and cultures can have a significant bearing on knowledge management adoption, methods and successes. It is readily apparent, then, that, without some joint interaction among these disciplinary areas, it will be difficult to ensure that the proper information is available for problem solving (Reynolds *et al.*, 2005).

As with descriptive knowledge, land-management policies and organizational cultures can play a significant role in determining how and which predictive tools are used. Certain mathematical models may be acceptable for land-management planning purposes, but others are not. Conversely, well-grounded procedural knowledge and predictive statistics can inform – and change, in some cases – management policies and procedures. For example, predicted habitat loss for an endangered species can dramatically alter management guidelines for a broad geographical area, as illustrated by the Northwest Forest Plan (USDA Forest Service, 1994), where quantitative science (including inventory and analysis) drives policy.

Policy often determines what causal predictive knowledge we have available to us through the funding streams of science R & D investments. Furthermore, it partially defines the values that hold sway at any point in time, helps to establish the suite of choices that we are presented with and legitimizes certain trade-offs and compromises while effectively dismissing others. Many policy impacts are subtle and poorly understood and yet far-reaching and powerful.

Adaptive management is a way to explicitly acknowledge risk and uncertainty in the forest management process and deal with them logically (Walters and Holling, 1990). However, as Bormann and Kiester (2004) noted, many laws and policies governing natural resource management constrain efforts to properly implement an adaptive management approach. Furthermore, there is an unwillingness by individual managers, management organizations and the public to be wrong – which makes a no-action alternative unusually attractive. But no action has its own set of consequences. 'Nature, never having been constant, does not provide a simple answer as to what is right, proper, and best for our environment. There is no single condition that is best for all life' (Botkin, 1995). Policy scientists have an important role to play in helping scientists and managers deal with the socio-economic realities of sustainable forest management.

New tools face technological and institutional challenges for successful operational deployment. The two case studies, below, illustrate the complex information technology environments within which new tools must be deployed. The first case study illustrates the many issues to be considered in assembling a suite of systems to perform a complex analysis, i.e. a national forest resource assessment. The second case study illustrates issues in linking the system-related activities of multiple organizations, i.e. forestry enterprises interacting in supply chains. Theories that can guide the development and deployment process in institutions are then discussed.

Case Study 1: National Forest Assessment

FAO regularly monitors the world's forests through the Forest Resource Assessment (FRA) Programme, in which countries are required to complete 15 tables of information (Table 22.1) (FAO, 2005). Each item in Table 22.1 has its own policies, monitoring and assessment methods and information-processing approaches.

Countries have a wide range of methods of assessing their resources in order to complete these tables. FAO provides guidance for the process through an online knowledge reference.[1] The organization of the knowledge reference exhibits many similarities to the four themes of the present discussion: an introduction including a policy chapter, an inventory and data collection section, and analyses, outputs and cases sections. The analyses section includes chapters on information management and data registration and modelling for estimation and monitoring.

Information management and data registration

The chapter on information management and data registration (Thomson, 2004) is particularly relevant to the present discussion. The section headings from that chapter illustrate the range of information management considerations (Table 22.2).

The basic Forest Resource Assessment scenario (Table 22.2) is based on the case in which a single institution is conducting the assessment (see Vignette 1). However, real assessments almost always involve many institutions and many computers with complex distributed processing and data registration issues as well as infrastructure and institutional issues. Technical, semantic, political/human, inter-community, legal and international interoperability constraints in particular

Table 22.1. Information required by the FAO Forest Resource Assessment process.

Extent of forest and other wooded land
Ownership of forest and other wooded land
Designated functions of forest and other wooded land
Characteristics of forest and other wooded land
Growing stock
Biomass stock
Carbon stock
Disturbances affecting health and vitality
Diversity of tree species
Growing stock composition
Wood removal
Value of wood removal
Non-wood forest products removal
Value of non-wood forest products removal
Employment in forestry activities

Vignette 1. Indicators and multiple-criteria decision making (Vacik *et al.*, Chapter 23, this volume)

This chapter illustrates the manner in which the knowledge framework adopted by an agency can influence an analysis or assessment approach. For example, indicators have proved to be powerful tools for collecting and reporting information within a management system. Vacik *et al.* use the driving forces–pressure–state–impact–response (DPSIR) approach of the European Environmental Agency (EEA) for the evaluation of alternative management strategies at the forest management unit level. A set of indicators for sustainable forest management (SFM) is arranged according to the DPSIR framework to cover the causal chain of environmental and socio-economic drivers and pressures, to detect changes in the state of the system and to identify impacts on ecosystems and society. The study combines the strengths of tools that enhance system understanding and those of multi-criteria decision making for the purposes of SFM, while keeping the whole concept at least semi-quantitative by integrating ecosystem modelling results. This integration creates new perspectives on the communication of decision making, on the relationship between ecosystem modelling and decision modelling and on the applicability of established approaches per se.

Table 22.2. Information management topics from the FAO knowledge reference for Forest Resource Assessment.

1. Introduction
   - 1.1 What is information management?
   - 1.2 National and international requirements for forest resource assessments
   - 1.3 Current status of information management in national FRAs
2. A basic Forest Resource Assessment scenario
   - 2.1 Data
     - 2.1.1 Data models
     - 2.1.2 Data input
     - 2.1.3 Computer programs for data and information management
     - 2.1.4 Standards, metadata and data quality
       - Standards
       - Metadata and meta-information
       - Verification and validation
       - Backups and archiving
   - 2.2 Information
     - 2.2.1 Information demand and supply
     - 2.2.2 Information aggregation and integration
       - Information transformations
       - Expert opinion
   - 2.3 Information management and change assessment
     - 2.3.1 Data and information sources
     - 2.3.2 Monitoring
   - 2.4 Reporting and communication
     - 2.4.1 Reporting requirements and information management
     - 2.4.2 Maps, graphs and statistics
3. Extending the basic scenario: many institutions and many computers
   - 3.1 The Internet and other computer-related issues
     - 3.1.1 Distributed systems and interoperability
     - 3.1.2 Data registration
     - 3.1.3 Institutional and infrastructure issues
4. Putting a full national forest information system in place
   - 4.1 System design and development
     - 4.1.1 Requirements analysis
     - 4.1.2 System development
     - 4.1.3 Funding and financial mechanisms
5. Discussion

can be limiting factors (Miller, 2000; Thomson, 2005b). Interoperability can be viewed as operating on three levels: strategic (agreements, partnerships and objectives), tactical/operational (who does what?) and technological (information systems and standards – see Vignette 2), and examples of these can be found within the chapters of this book.

Standards, metadata and data registration procedures play a key role in interoperability. The use of standards in supply chains (case study 2 below) can be critical to success (Gopal and McMillan, 2005).

Vignette 2. Standards: Establishment Management Information System (Perks *et al.*, Chapter 24, this volume)

The term 'standards' generally refers to the role of information technology standards in the development of interoperable systems. However, Perks *et al.* illustrate another key role of standards, i.e. the idea that the decision support provided by a system must conform with a set of forest management standards. In addition to supporting a set of interoperable components, the Establishment Management Information System must generate information that is consistent with the UK Forestry Standard, the government's approach to sustainable forestry. The UK standard in turn conforms to the Helsinki Guidelines and Pan-European Criteria, and in the second edition (2004) deals with issues such as the devolution of forestry in Great Britain to England, Scotland and Wales as well as a range of legislation and policy changes.[a] It is critical that systems can easily be kept abreast of such changes.

[a]http://www.forestserviceni.gov.uk/press/2004/14th_may.htm (Accessed 3 August 2005.)

Case Study 2: Knowledge Management in Supply Chains

The following scenario provides one possible view of how wood resources may some day move from timber stand to wood processor.

> The most ambitious predictions indicate that pulp and paper buyers will simply dictate to a wrist-mounted computer (voice recognition enabled, of course) that they need x tons of grade y to be delivered in three days and an order confirmation and delivery time will come straight back at them. That is, of course, if they actually need to place an order at all. After all, with all the data processing technology that is becoming available, the computer will have already decided that it needed x tons of grade y and placed the request automatically.
>
> (Kenny, 1999)

Implicit in this scenario is the idea of a supply chain with a suite of systems operating in concert over a set of enterprises that may contain up to 20 companies (Thony, 2003). The term 'chain' implies a linear flow of products and information and this is reflected in most diagrams of supply chains, in which 'trees' at one end and 'end users' at the other are linked by boxes and arrows, with the directions of the arrows depending on whether the chain is a 'supply-push' or 'demand-pull' situation. The quotation above represents demand-pull, in which a requirement would have the end result of triggering harvest of a specific stand of trees.

In practice, each forest stand contains different product assortments suited for use in several different industries and supply chains, with specific markets requiring specific assortments, and with several forest companies operating in overlapping catchment areas (Forsberg and Rönnqvist, 2003). Key questions for forest management therefore include:

- How should forest inventories be conducted to optimize their use in supply chains?

- To what extent must current forest planning and harvest scheduling systems be modified to fit a supply chain setting?

An end user may have more than one supplier, while the initial supplier may have more than one customer, resulting in 'supply networks' (Fig. 22.1) rather than supply chains. Using this perspective:

> The conventional wisdom is that competition in the future will not be company vs. company but supply chain vs. supply chain. But the reality is that instances of head-to-head supply chain competition will be limited. The more likely scenario will find companies competing – and winning – based on the capabilities they can assemble across their supply networks.
>
> (Rice and Hoppe, 2001)

This leads to development of 'intelligent-webs' that use high-speed and real-time communications to link partners in a networked structure to satisfy consumer demand in a highly responsive manner (Hoppe, 2001). This will provide those well-connected companies with a competitive advantage by supplying products more responsive to customers' needs and time frames.

'Trust' (see Vignette 3) is emphasized as a key consideration for information sharing in supply chains. A key issue, therefore, is the manner in which trust for information sharing operates in a networked situation, not only among participating individuals and corporate entities, but also among software agents (Goel *et al.*, 2005) that negotiate in automated systems. In contrast with the developed world, supply chains in developing countries are tightly linked with long-standing social structures (Woods, 2004). Significant differences in trust development arise in that setting, and systems and processes designed for use in developed countries may not be appropriate for the developing world, especially where software agents are used.

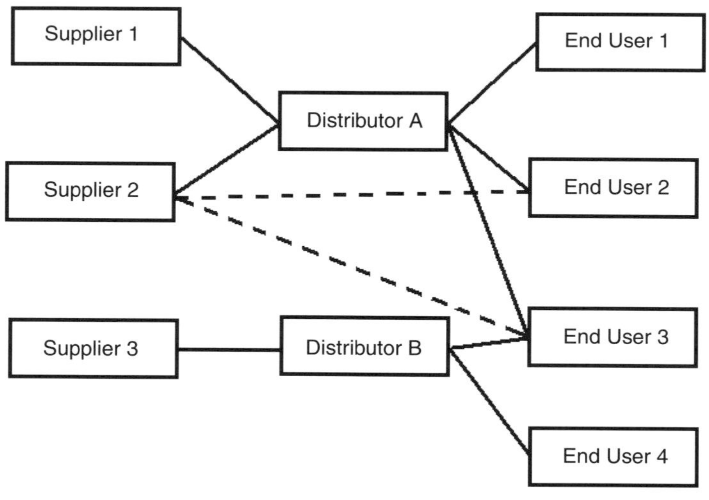

Fig. 22.1. A supply network. Connections may be either 'supply-push' or 'demand-pull' so directional arrow heads are omitted. The dashed lines indicate bypassing of steps in the chain or network.

Vignette 3. Trust: the virtual forester (Reinbolz and Hanewinkel, Chapter 25, this volume)

The ability of policymakers and the general public to understand concepts and issues and develop trust in research findings is key to sustainability. Trust has emotional and personalization aspects. The agent-based virtual forester in the innovative system described by Reinbolz and Hanewinkel is designed to provide personalized help on navigation through complex websites relating to sustainable forestry, and attempts to build an emotional connection with the system user.

Functions of multi-agent systems include (Frey *et al.*, 2003):

- Negotiations between enterprises
- Integrated process planning and scheduling
- Production planning and controlling (with focus on assembling industries)
- Production planning and controlling (with focus on batch production)
- Operational tracking of orders, including suborders, in supply chains
- Analysis of historical tracking information (tracing)

Software agents performing these functions must perform a range of activities including: negotiation of plans among supply-chain partners, monitoring of orders and related suborders, informing partners and internal planning systems when critical events are triggered, routinely forwarding information to trusted third-party supply-chain communication systems, performing internal rescheduling in reaction to a critical event, and renegotiating a plan of production between supply-chain partners due to a critical event. These activities require bridging not only technical differences between enterprises, but also cultural differences that involve work flows, processes, social expectations and established patterns of doing business with partner enterprises.

An Institutional Perspective on Tools for Sustainable Forestry

Institutional processes and cultures are complex and have generally evolved over many years. Not surprisingly, it is exceedingly difficult to introduce new software products for knowledge management or modelling, new sets of criteria or indicators to inventory and monitor, unfamiliar analysis methods that affect decision making, or novel management policies or fresh interpretations of existing ones. Consequently, institutional change – including transitioning towards sustainable forestry – is subject to great uncertainty, frequent missteps, voluminous debate and generally slow progress. A look at some recent developments in the social science literature can help us understand these problems and find ways to advance change.

Sustainable forest management can be aided by adoption of information technology applications. These include tools to track and enable compliance with regulations, reduce risks and increase ecological efficiency by assessing and reducing product and service life cycle costs (Waage *et al.*, 2003). Innovation diffusion theory (Rogers, 1995) can help guide adoption or explain observed patterns of adoption and abandonment of a particular system (Thomson *et al.*, 2004) or of an idea such as sustainable development (Innes *et al.*, 2005). Under Roger's theory, people exist within social systems and fall into five main categories with regard to adopting innovations. True innovators or pioneers comprise less than 3% of the population. The rest of the population is made up of 13% early adopters, 34% early majority, 34% late majority and the remaining 16% laggards. The adoption of innovations, therefore, follows a characteristic bell-shaped (cumulative S-shaped) curve over time. Adoption rate depends on five attributes of innovations: relative advantage, compatibility, complexity, trialability and observability. Pre-diffusion needs/problem awareness, basic and applied research and development and commercialization decisions can significantly affect the adoption process, leading to testable hypotheses in information systems research, such as in the use of Open Source software (Valier *et al.*, 2004).

Institutional and organizational behaviour often departs from such theoretical optimal paths of adoption and abandonment/replacement, and other concepts must always be considered when approaching the introduction of new tools or processes. One such concept, escalation of commitment, refers to a situation in which a decision maker commits additional resources to a failing course of action rather than adopting a new course or using resources for an alternative unrelated activity. The theory of escalation of commitment can explain the roles of different forms of agency commitment to failing approaches. For example, escalation in the IT sector can be related to an organizational reward structure in which a manager's performance is linked with the success of software process improvement activities (Abrahamsson, 2002). Salter and Sharp (2001) showed that the effect of an apparently small difference in national culture can explain differences in escalation of commitment to failing projects in two countries with significant cross-border investment (USA and Canada). Studies of de-escalation of commitment (Heng *et al.*, 2003; Pan *et al.*, 2004) can provide guidance for both researchers and managers to help avoid inappropriate escalation of commitment.

Organizational reward structures are also central to agency theory. Knowledge exists in both explicit and implicit (tacit) forms (Rauscher *et al.*, Chapter 26, this volume). 'The knowledge management literature specifically addresses the problem of converting the implicit to the explicit, while agency theory directs our attention to the costs of doing so' (Hall *et al.*, 2000). Knowledge management activities can require considerable time and effort for individuals in organizations (see Vignette 4). These are strategic activities for organizations, but can compete significantly with specific project-related activities for an individual's time and effort, and individual benefits can vary among team members. Furthermore, the institutional reward system may not adequately reflect KM contributions. These differences can have a significant impact on KM policies in project-oriented organizations, and can lead to conflict and project failure (Hall *et al.*, 2000). Balance between individual and organizational costs and benefits under the guidance of agency theory leads to different optimal strategies to enhance knowledge capture in different types of organizations (Hall *et al.*, 2000). Performing this balancing will

Vignette 4. Information and knowledge management (Rauscher *et al.*, Chapter 26, this volume)

Without attention to the key task of knowledge management, efforts in sustainable forest management may only have limited long-term success. Rauscher *et al.* make the case that proficient problem solving depends on an adequate foundation of relevant and readily applicable knowledge. Making good decisions can be extremely difficult when problems are not well structured and situations are complex, as they are when managing natural resources for multiple benefits and for users with differing values. It takes a well-coordinated, cooperative approach among people developing methodologies and techniques in the areas of knowledge management, decision-support systems and decision analysis methods to support sustainable forest management.

require some trade-offs between concern for the individual and for the organization. However, without some examination of those trade-offs, KM benefits to the organization may not be fully realized or, alternatively, individual commitment to KM may dissolve, owing to incorrectly perceived rewards.

Other useful theories, not dealt with specifically in this chapter but which can help understand the development and deployment of systems in organizations, include resource-based theory (Caldeira and Ward, 2003), the theory of computation,[2] complexity theory[3] and the theory of constraints.[4] Grounded theory (Orlikowski, 1993) is also commonly used in the information systems research literature as it specifically links data collection and analysis with theory development.

Discussion

The discipline of knowledge management covers not only specific tools, which we have explored through an organizational hierarchy, but also a process fundamental to the activities of individuals, institutions and organizations, which we have explored through theories such as innovation diffusion, escalation of commitment and agency theory. In the same manner that there is no single tool that fits all purposes and tools are made to interact to achieve particular aims, there is no single theory that fits all situations and theories also interact and overlap (Waters, 2004). Knowledge management, innovation and commitment theory and theories of organizational change can each provide valuable insights into the interpretation and application of principles from the others.

The manner in which information and knowledge management are used to meet institutional goals was illustrated in two case studies: (i) the creation of National Forest Assessments by FAO; and (ii) supply chains. Dealing with issues such as standards, metadata and interoperability contributes to successful outcomes. Interoperability includes not only computer system interoperability, but also political/human, inter-community, legal and international interoperability.

Vignette 5. Knowledge ecosystems (Thomson, Chapter 27, this volume)

Knowledge ecosystems can be defined as 'the complex and many-faceted system of people, institutions, organizations, technologies and processes by which knowledge is created, interpreted, distributed, absorbed and utilized.' Analogies with ecosystem processes can be used to guide activities such as design of forest planning processes. Under this concept, adaptive knowledge management (Thomson, 2005c) can be used to experiment with knowledge in the same manner that adaptive management, described elsewhere in this volume, is used to experiment with ecological management.

Vignette 6. Web services: habitat and rare species protection (Ray and Broome, Chapter 28, this volume)

A web service provides information for other applications that send messages to it over the Internet. Ray and Broome describe how one organization uses web services to deliver up-to-date advice on sustainable forest management in relation to habitat and rare species protection under complex and constantly changing biological and legislative constraints.

The concept of knowledge ecosystems (see Vignette 5) can also be helpful in guiding linkages among individuals, institutions and technology.

The conference on which this book is based highlighted many state-of-the-art applications that characterize sustainable forestry decision making in the highly interconnected setting of institutions in the developed world. However, forestry exists within the broader context of social, environmental and economic endeavours, and many of the drivers of forest-related decisions in less-developed parts of the world have their basis in these broader issues. The top three major social issues of global concern identified in the United Nations Millennium Declaration[5] are peace, security and disarmament; development and poverty eradication; and protecting our common environment. All three of these have consequences for forestry. Information and communication technology (ICT) will play a significant role in addressing these issues, with social and cultural drivers being paralleled by technological drivers (Thomson and Colfer, 2005). Agenda 21 has also been a driver of system development, particularly in relation to sustainability and enhanced participation (Thomson, 2005c). Service transformation (see Vignette 6), delivery of services in a way that meets the changing wants and needs of clients, is also a significant driver of system development, especially in government services (Thomson, 2005a).

Differences between developed and developing countries in their approach to information technology and knowledge management are often related to 'divides'. Differences in access to knowledge in the 'digital divide' concept are well recognized. However, other less-known ICT-related divides, such as democratic, gender, racial, knowledge (see Vignette 7), strategy and nanotechnology divides, may be more significant in the future, as may failures to address the

Vignette 7. The rational DSS model and the fact/value divide (Ekbia and Reynolds, Chapter 29, this volume)

Rational DSSs work well for situations in which there are well-defined, agreed-upon goals, all alternatives are known, preferences are clear and stable and there are no time and cost constraints. These conditions rarely apply in forestry, and Ekbia and Reynolds describe a range of alternative approaches for these types of more complex situations. These alternatives lie along a divide characterized by emphasis on facts, on one side, and on values, on the other.

issue of 'information literacy' (Thomson and Colfer, 2005). Institutions in developed and developing countries therefore face different challenges in creating and using systems to aid decision making and produce mandated reports.

Definitions of 'sustainability' have evolved with time and vary considerably (Innes *et al.*, 2005). However, many of the practical applications of the concept focus on certification processes for sustainability. Chain of custody is central to the process and is closely linked with supply-chain development:

> If the industry being certified has a fairly disintegrated supply chain, then certification can be used as a means to improve communication and information management, ultimately streamlining and integrating the supply chain. The result is a comprehensive plan that, prior to certification, did not exist and is one of 'the most valuable reasons for becoming certified' according to many interviewees.
>
> (CCIF, 2002)

Such examples demonstrate clear and tangible benefits from the interaction of policy setting (certification) and information and knowledge management.

Sustainable forestry is practised within a policy and planning hierarchy, ranging from national levels, for which Montreal process-type criteria and indicators[6] are developed, reported and compared, to local levels, at which a forest manager is trying to determine an appropriate silvicultural regime for a particular stand of trees. As described earlier in this chapter, there exists a hierarchy of descriptive, predictive and prescriptive tools to assist in these activities. However, adoption of a particular tool, integrating it with existing institutional processes and ensuring that its performance complies with current and changing policies must be carefully orchestrated. Fortunately, a range of organizational theories exists to avoid failure (Fortune and Peters, 2005) and increase the likelihood of success.

Notes

1. http://www.fao.org/forestry/site/7817/en
2. http://en.wikipedia.org/wiki/Theory_of_computation
3. http://en.wikipedia.org/wiki/Computational_complexity_theory
4. http://en.wikipedia.org/wiki/Theory_of_constraints
5. http://www.un.org/millennium/declaration/ares552e.htm (last accessed on 6 July 2005)
6. http://www.iisd.ca/forestry/mont.html

References

Abrahamsson, P. (2002) The role of commitment in software process improvement. PhD dissertation, Department of Information Processing Science, University of Oulu, Oulu. (Available at http://herkules.oulu.fi/isbn9514267303/isbn9514267303.pdf, accessed 18 August 2005.)

Bormann, B.T. and Kiester, A.R. (2004) Options forestry: acting on uncertainty. *Journal of Forestry* 102(4), 22–27.

Botkin, D.B. (1977) *Life and Death in a Forest: The Computer as an Aid to Understanding*. In: Hall, C. and Day, J. (eds) *Ecosystem Modeling in Theory and Practice. An Introduction with Case Studies*. John Wiley and Sons, New York.

Botkin, D.B. (1995) *Our Natural History*. G.P. Putnam and Sons, New York.

Caldeira, M.M. and Ward, J.M. (2003) Using resource based theory to interpret the successful adoption and use of information systems and technology in manufacturing small and medium sized enterprises. *European Journal of Information Systems* 12(2), 127–141.

Conservation and Community Investment Forum (CCIF) (2002) *Analysis of the Status of Current Certification Schemes in Promoting Conservation*. CCIF, San Francisco, California. Available online at http://www.cciforum.org/pdfs/Certification_Analysis.pdf (accessed 22 July 2005).

Food and Agriculture Organization (FAO) (2005) *State of the World's Forests 2005*. FAO, Rome.

Forsberg, M. and Rönnqvist, M. (2003) A flexible logistic system for forestry where several modes of transportation are integrated. In: *Proceedings of ODYSSEUS 2003: Second International Workshop on Freight Transportation and Logistics, Mondello (Palermo), Sicily, Italy, 27–30 May 2003* (abstract online at http://www.unipa.it/Odysseus/Odysseus2003_file/odysseus-main_file/pdf/ForsbergRonnqvist.pdf, accessed 20 July 2005).

Fortune, J. and Peters, G. (2005) *Information Systems: Achieving Success by Avoiding Failure*. John Wiley and Sons, New York.

Frey, D., Stockheim, T., Woelk, P.-O. and Zimmermann, R. (2003) Integrated Multi-agent-based Supply Chain Management. In: *Proceedings of the 1st International Workshop on Agent-based Computing for Enterprise Collaboration*. IEEE Computer Society Press, Los Alamitos, California.

Gigerenzer, G. and Todd, P.M. (1999) *Simple Heuristics that Make us Smart*. Oxford University Press, Oxford, UK.

Goel, A., Zobel, C.W. and Jones, E.C. (2005) A multi-agent system for supporting the electronic contracting of food grains. *Computers and Electronics in Agriculture* 48, 123–137.

Gopal, G. and McMillan, E. (2005) Innovations: synchronization, a cure for bad data. *Supply Chain Management Review* May, 58–62.

Hall, J., Sapsed, J. and Williams, K. (2000) Barriers and facilitators to knowledge capture and transfer in project-based firms. In: *Proceedings of a Conference on Knowledge Management: Concepts and Controversies, University of Warwick, 10–11 February 2000*, University of Warwick, Coventry, UK.

Heinrichs, J.H., Hudspeth, L.J. and Lim, J.S. (2003) Knowledge management. In: Bidgoli, H. (ed.) *Encyclopedia of Information Systems*, Vol. 3. Academic Press, Boston, Massachusetts, pp. 13–31.

Heng, C., Tan, B. and Wei, K. (2003) De-escalation of commitment in software matters? What matters? *Information and Management* 41(1), 99–110.

Holsapple, C.W. (2003) Decision support systems. In: Bidgoli, H. (ed.) *Encyclopedia of Information Systems*, Vol. 1. Academic Press, Boston, Massachusetts, pp. 551–565.

Hoppe, R.M. (2001) Outlining a future of supply chain management – coordinated supply networks. MSc thesis, Massachusetts Institute of Technology. Available online at http://web.mit.edu/supplychain/repository/network_hoppe.pdf, accessed 20 July 2005.

Innes, T., Green, C. and Thomson, A. (2005) Surprising Futures. In: Hetemaki, L. and Nilsson, S. (eds) *Information Technology and the Forest Sector*. IUFRO World Series Vol. 18, International Union of Forest Research Organizations, Vienna, Austria, pp. 24–48.

Kenny, J. (1999) SCM: are you ready for the future? *PPI* 41(11), 31–35.
Kim, G., Nute, D., Rauscher, H.M. and Loftis, D.L. (2000) AppBuilder for DSSTools: an application development environment for developing decision support systems in Prolog. *Computers and Electronics in Agriculture* 27, 107–125.
Miller, P. (2000) Interoperability. What is it and why should I want it? *Ariadne* 24. http://www.ariadne.ac.uk/issue24/interoperability/
Orlikowski, W.J. (1993) CASE tools as organizational change: investigating incremental and radical changes in systems development. *MIS Quarterly* 17(3), 309–340. Available online at http://misq.org/archivist/bestpaper/misq93.html (accessed 22 July 2005).
Pan, G.S.C., Pan, S.L.I., Newman, M. and Flynn, D.J. (2004) Unfreezing–changing–refreezing of actors' commitment: the transition from escalation to de-escalation of commitment to information technology projects. In: Leino T., Saarinen, T. and Klein, S. (eds) *Proceedings of the Twelfth European Conference on Information Systems*. Turku School of Economics and Business Administration, Turku, Finland. http://csrc.lse.ac.uk/asp/aspecis/default5.asp
Rauscher, H.M. (1996) Decision making methods for ecosystem management decision support. In: *Caring for the Forest: Research in a Changing World, Congress Report, IUFRO XX World Congress, 6–12 August 1995, Tampere, Finland*. Vol. II. IUFRO, Vienna, Austria, pp. 264–272.
Rauscher, H.M. and Reynolds, K.M. (2003) Natural resource management. In: Bidgoli, H. (ed.) *Encyclopedia of Information Systems*, Vol. 3. Academic Press, Boston, Massachusetts, pp. 239–265.
Rauscher, H.M., Lloyd, F.T., Loftis, D.L. and Twery, M.J. (2000) A practical decision-analysis process for forest ecosystem management. *Computers and Electronics in Agriculture* 27, 195–226.
Reynolds, K.M., Borges, J., Vacik, H. and Lexer, M.J. (2005) Information and communication technology in forest management and conservation. In: Hetemaki, L. and Nilsson, S. (eds) *Information Technology and the Forest Sector*. IUFRO World Series Volume 18, International Union of Forest Research Organizations, Vienna, Austria, pp. 150–171.
Ribeiro, R.P., Borges, J.G. and Oliveira, V. (2004) A framework for data quality for Mediterranean sustainable ecosystem management. *Annals of Forest Science* 61, 557–568.
Rice, J.B. and Hoppe, R.M. (2001) Supply chain vs. supply chain: the hype and the reality. *Supply Chain Management Review* September/October, 47–53.
Rogers, E.M. (1995) *Diffusion of Innovations*, 4th edn. Free Press, New York.
Salter, S.B. and Sharp, D.J. (2001) Agency effects and escalation of commitment: do small national culture differences matter? *International Journal of Accounting* 36(1), 33–45.
Schmoldt, D.L. and Rauscher, H.M. (1996) *Building Knowledge-based Systems for Natural Resource Management.* Chapman and Hall, New York.
Thomson, A. (2004) Information management and data registration for National Forest Assessments. In: *Knowledge Reference for National Forest Assessment*. FAO–IUFRO. http://www.fao.org/forestry/foris/webview/nfa-ref/index.jsp?siteId=2881&sitetreeId=7817&langId=1&geoId=0
Thomson, A.J. (2005a) Web-based decision support systems and information systems. Keynote presentation in *Proceedings of ForwardFORESTs Electronic Conference.* http://forward-forests.czu.cz/docs/key43.pdf
Thomson, A.J. (2005b) Editorial: Information interoperability and organization for national and global forest information systems. *Computers and Electronics in Agriculture* 47(3), 163–165.
Thomson, A.J. (2005c) Indicator-based knowledge management for participatory decision-making. *Computers and Electronics in Agriculture* 49, 206–218.
Thomson, A.J. (2006) *How Should We Manage Knowledge Ecosystems? Using Adaptive Knowledge Management*. In: Chapter 27, this volume, pp. 461–479.

Thomson, A. and Colfer, C. (2005) ICT and social issues. In: Hetemaki, L. and Nilsson, S. (eds) *Information Technology and the Forest Sector*. IUFRO World Series Volume 18, International Union of Forest Research Organizations, Vienna, Austria, pp. 172–196.

Thomson, A.J. and Schmoldt, D.L. (2001) Ethics in computer software design and development. *Computers and Electronics in Agriculture* 30(1–3), 85–102.

Thomson, A., Haggith, M. and Prabhu, R. (2004) Innovation diffusion: predicting success of system development. In: *Proceedings of 15th International Workshop on Database and Expert Systems, Saragossa, Spain, 30 August–3 September 2004*. IEEE Computer Society, pp. 627–631.

Thony, P. (2003) An integrated information and supply chain management system for the forest industry. In: Innes, T. (ed.) *Natural Resources Information Management Forum: Putting Knowledge to Work*. FORREX Series No. 8, FORREX – Forest Research Extension Partnership, Kamloops, British Columbia, pp. 55–61.

USDA Forest Service and USDI Bureau of Land Management (1994) *Record of Decision on Management of Habitat for Late-successional and Old-growth Related Species within the Range of the Northern Spotted Owl (Northwest Forest Plan)*. USDA Forest Service and USDI Bureau of Land Management, Portland, Oregon.

Valier, F.M., McCarthy, R.V., Aronson, J.E. and O'Neill, H. (2004) The province of diffusion of innovations: usable social theory for information systems research. In: *Proceedings of the Tenth Americas Conference on Information Systems, New York, August 2004*. Association for Information Systems, Atlanta, Georgia. http://aisel.isworld.org/subject-by-publication.asp?subject_ID=339

Waage, S., Shah, R. and Girshick, S. (2003) Information technology and sustainability: enabling the future in corporate environmental strategy. *International Journal of Corporate Sustainability* 10(4), 2-81–2-96.

Walters, C.J. and Holling, C.S. (1990) Large-scale management experiments and learning by doing. *Ecology* 71(6), 2060–2068.

Waters, J.F. (2004) Knowledge and commitment in innovation processes. PhD thesis, University of Western Sydney. Available on-line at http://library.uws.edu.au/adt-NUWS/public/adt-NUWS 20050615.153801/ (accessed 22 July 2005).

Woods, E. (2004) Supply-chain management: understanding the concept and its implications in developing countries. In: Johnson, G.I. and Hofman, P.J. (ed.) *Agriproduct Supply-chain Management in Developing Countries. Proceedings of a Workshop held in Bali, Indonesia, 19–22 August 2003.* ACIAR Proceedings No. 119, ACIAR, Canberra, Australia, pp. 18–26.

23 Integrating the DPSIR Approach and the Analytic Network Process for the Assessment of Forest Management Strategies

HARALD VACIK,* BERNHARD WOLFSLEHNER, RUPERT SEIDL AND MANFRED J. LEXER

Department of Forest and Soil Sciences, Institute of Silviculture, University of Natural Resources and Applied Life Sciences, Vienna, Austria.

Abstract

In this contribution, the driving force–pressure–state–impact–response (DPSIR) approach of the European Environmental Agency (EEA) is adopted for evaluating alternative management strategies at the level of forest management units. A set of indicators for sustainable forest management (SFM) is arranged according to the DPSIR framework in order to cover the causal chain of environmental and socio-economic drivers and pressures, to detect changes in the state of the system and to identify impacts on ecosystems and society. Responses may alter the pressure, but may also influence the socio-economic drivers. The approach allows enhancement of the systems-analysis view of the DPSIR framework by combining a set of criteria and indicators, multi-criteria decision-making techniques and a modelling approach. To account for the interconnectedness of indicators, the analytic network process is used to evaluate four forest management plans by modelling priorities of indicators and strategies resulting from their relationships within the network. Simulation outputs of the hybrid patch model PICUS v1.4 are used to outline the performance of the management plans under three climate scenarios for a planning period of 100 years of a case study in Austria. Challenges of combining the different methods and techniques in the proposed assessment tool are discussed in relation to the analysis of sustainable forest management.

Introduction

Concepts of sustainable forest management (SFM) have been widely adopted in society, but one will still not find a generic, universally valid definition of SFM or a uniform approach to monitoring and measuring. This observation is strongly rooted in the fact that there exists a diversity of historically developed and

* Corresponding author (harald.vacik@boku.ac.at)

ideologically driven terms since the concept of SFM was adopted by politicians, forest managers, ecologists and experts from ecological, economic and social sciences. This leads to heterogeneous views regarding the fundamentals and parameters for modelling, monitoring and evaluating sustainable forest management. Even within the scope of forest research there is a variety of approaches to SFM. The more traditional view, which comes from the original understanding of sustained (timber) yield, is implemented in modelling sustainable timber production systems and sustainable forest economic systems (see Pearce *et al.*, 2003). On the other hand, an eco-physiological understanding of SFM is now resulting in modelling of systemic states and dynamics as functions of ecosystem processes and substance flows (Peng, 2000; Sverdrup and Stjernquist, 2002).

When SFM came on to the political agenda and gained high public awareness in the late 1980s, it became evident that SFM includes a strong social component, resulting in potential conflicts among different cultures. Overall, an extended understanding of SFM includes people in decision making incorporating a pluralism of values in forestry concerns (Kant, 2003). In Europe, the Ministerial Conference on the Protection of Forests in Europe (MCPFE) has been guiding an SFM-developing process since 1990, developing *inter alia* criteria and indicators of SFM (MCPFE, 1998).

Defining objectives, gathering and amalgamating individual preferences and communicating decision-making methods and outcomes are major tasks of forest and ecosystem managers with regard to this new paradigm. In this context, methods serving communication demands, and also dealing with multiple criteria and objectives, have gained in importance in addressing areas for which the above modelling approaches do not operate convincingly. As a result, indicators have proved to be valuable tools for collecting and reporting information within a system, and are now the tools of choice for assessing or evaluating SFM (e.g. Franc *et al.*, 2001). Specific challenges arise when indicators are used to evaluate SFM in an operational, transparent and scientifically based way. Because there are no widespread consensual and operational definitions of a sustainable use of different resources, the monitoring of sustainable use of forest resources will not be undisputed. There is often a lack of identification of external pressures and constraints that might have significant influence on the overall performance of the evaluation object. Also, the prediction of multidimensional effects of sustainable management in a changing environment will not be captured by single indicators (see Prabhu *et al.*, 2001). Additionally, it is essential to integrate forest owners objectives to avoid top-down evaluations and decisions.

To address those requirements and improve their capabilities, indicators are often arranged in indicator systems, enriching problem perspectives and facilitating systemic understanding. The building of indicator systems goes beyond the boundaries of singular approaches, because there is a need for the analysis of linkages among indicators and checking for plausibility and for concepts about changes in the system (Kelly, 1998).

Based on the above evaluation, in this study the driving force–pressure–state–impact–response (DPSIR) approach of the European Environmental Agency (EEA) is adopted for the assessment of alternative management strategies at the level of forest management units. The DPSIR framework is capable of showing indicator information in an analytical manner when differentiating between causes and effects as well as between human measures and responses to reduce pressures and impacts (Hammond *et al.*, 1995). However, the DPSIR approach itself does not accommodate demands of multi-criteria analysis and decision making, because it suggests linearity in the relationship between human activities and environmental response and fails to capture and evaluate different management options.

Responding to these shortcomings, the analytic network process (ANP) is introduced to represent connections and interactions between the indicators more accurately and in a systemic view within the model. In this context, we demonstrate a case study of an indicator-based DPSIR framework. The ANP will be used to evaluate four different management strategies with regard to their present and future potential to fulfil the demands of sustainable forest management at the level of forest management units.

Materials and Methods

Indicators for sustainable forest management

Indicators are proven tools for assessing or evaluating certain aspects of SFM (Mendoza and Prabhu, 2000, 2003; Franc *et al.*, 2001; Raison *et al.*, 2001). They are custom instruments within political and certification initiatives (Rametsteiner and Simula, 2003). Indicators are powerful tools for collecting and reporting information within a system (e.g. SFM), which is usually characterized by a lack of knowledge, uncertainties and missing information about impacts, dependencies and feedbacks (Rametsteiner, 2001). On the other hand, there are shortcomings when indicators are not clearly defined, insufficiently reliable not targeted or too strongly simplifying (Brang *et al.*, 2002). Within SFM, indicators can serve various functions (Linser, 2001):

- As a reporting tool in terms of description and diagnosis of a situation.
- As a communication instrument to improve clarity on complex items.
- As forecasting tools for picturing future trends.
- As tools for the collection and processing of information and interests.
- As means of political control both as controlling and decision-making instruments.
- As instruments for checks of effectiveness of programmes and measures.

The DPSIR approach

Given a multitude of environmental indicators, there are different approaches to formatting them for public and stakeholder information. Most are examined one by one within a conceptual framework, while some of them are simply aggregated to an index or they are arranged according to a functional understanding. Regarding the latter context, the driving force–pressure–state–impact–response

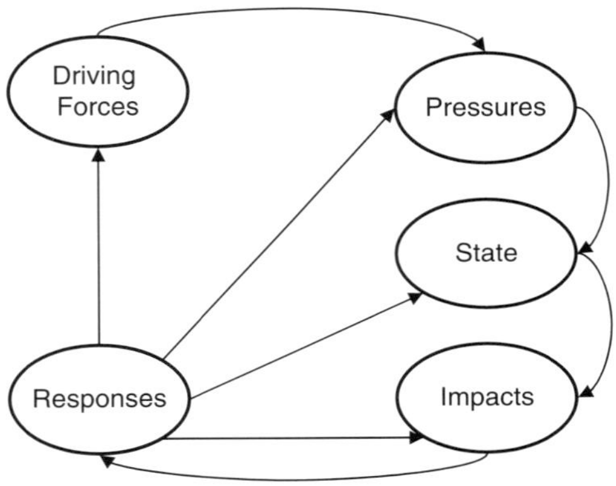

Fig. 23.1. The generic DPSIR approach as proposed by EEA (1999).

(DPSIR) framework is used by the European Environment Agency (EEA) for its reporting activities (EEA, 1999) (Fig. 23.1). In general, the DPSIR approach reflects a systems-analysis view of the relations between the environmental system and the human system. According to this view, social and economic developments are identified as driving forces (D) that exert pressures (P) on the environment. As a consequence, the state (S) of the environment changes, such as the conditions for health, availability and quality of resources or maintenance of biodiversity. Finally, these changes of the state lead to impacts (I) on human health, ecosystems and materials that may elicit a societal response (R) that feeds back on the driving forces, or on the state or impacts directly, through adaptation or curative action (OECD, 1993).

The links within the DPSIR chain are described by indicators, which have two main functions: (i) reducing the number of parameters measured; and (ii) simplifying the communication process by which information is provided for the user. The indicators for driving forces describe the social, demographic and economic developments in societies (e.g. industrial production), and the corresponding changes in lifestyles and production patterns (e.g. material intensity per service (MIPS) unit indicator). Pressure indicators describe developments in release of substances (e.g. CO_2 emissions) and physical and biological agents, the use of resources and the use of land. These indicators allow identification of the manner and extent of influence over the variety of natural processes. State indicators describe the status quo of the environment and the quality and quantity of resources that might be influenced by the pressures. Impact indicators are used to describe the impacts of social and economic functions on the environment (e.g. conditions for health, availability of resources, biodiversity) and their changes for a certain time step. Response indicators express the responses of society to environmental changes and concerns. This could be the number and kind of measures taken and the efforts of implementing, or the effectiveness of, those measures, and range from public (e.g. legislation, taxation, promotion) to private-sector responses (e.g. reduced consumption, recycling) (Linser, 2001). In practice, the assignment of indicators to one of the features of the framework is often uncertain and ambiguous (OECD, 1994).

The DPSIR framework has proved to be a logical, comprehensive tool to visualize environmental issues from an anthropocentric perspective. It is a powerful approach for communication and opinion-forming processes, but limited for scientific analysis. Current applications have a strong focus on marine development (Walmsley, 2002; EEA, 2004) and inland waters (La Jeunesse *et al.*, 2003) and their effects (Gobin *et al.*, 2004).

An adapted DPSIR framework

The DPSIR framework is capable of describing the relationships between the origins and consequences of environmental problems, but in order to understand their dynamics it is also useful to focus on the links among DPSIR elements. The DPSIR framework is considered for a superordinate level, which means that indicators should give information at national up to continental or global level. A few studies have described applications for a pressure–state–response (PSR) approach at regional level in natural and forest resource planning with the scope of supporting policy designs for a better understanding of monitoring data for both specialists and non-specialists (Crabtree and Bayfield, 1998; Kammerbauer *et al.*, 2001; Firbank *et al.*, 2003).

From the viewpoint of SFM, the original DPSIR framework faces some major limitations:

- It is generally designed for the national level.
- It deals only with ecological problems.
- It is not related to management concepts.
- It cannot cover value information, any weighing or preferences.

SFM indicators have been mainly introduced at the international level for national monitoring and reporting purposes. Yet there are some initiatives that address the regional or local level (Wedeles and Williams, 1999; Woodley *et al.*, 1999; Appanah and Kleine, 2001; Wolfslehner *et al.*, 2003; Mrosek *et al.*, 2006).

For the application of the DPSIR framework at the forest management unit (FMU) level, the original approach had to be modified rather rigidly. On the one hand, decisions about forest management should be addressed and, on the other hand, the objectives are transferred from environmental protection to SFM, including economic, ecological and social indicators. So the DPSIR framework is modified by: (i) considering an FMU as an object system; (ii) identifying driving forces that cause pressures at the FMU level; (iii) interlinking the indicators in the framework with a multi-criteria analysis method; and (iv) evaluating time series of management strategies to estimate their effects over a certain period.

The relationships among the indicators of the different functional groups (DPSIR) and the alternatives are indicated in Fig. 23.2 by arrows. Compared with the original approach, it is evident that there are feedback loops for the driving forces, pressures, impacts and responses indicating the relations among indicators also within a group. Management strategies are integrated bidirectionally, both influencing the performance of the indicators and being influenced by

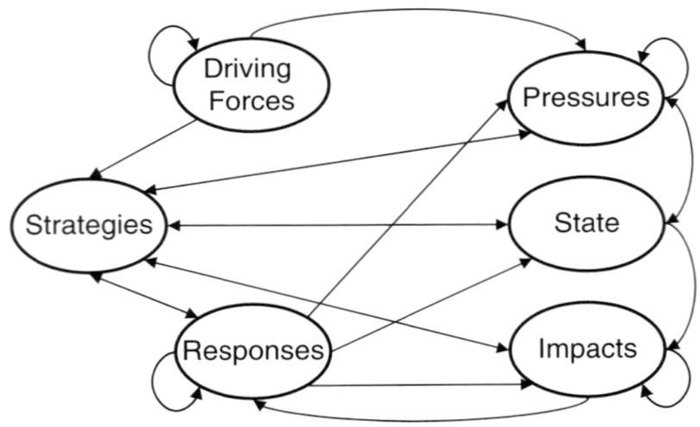

Fig. 23.2. An adapted DPSIR framework, including management strategies.

a preconditioned and changing environment. As an exception, there is no influence of strategies on driving forces, because they are assumed to be external.

Multi-criteria analysis by applying the analytic network process

Multi-criteria analysis (MCA), the method of choice to overcome the limitations of singular indicator approaches, provides an overview of indicators and their development over time. MCA methods are used to: (i) represent a higher degree of the complexity underlying the DPSIR framework; (ii) enhance analytical features to facilitate understanding of the indicator system; (iii) create aggregated priority or utility values for management alternatives; and (iv) analyse the trade-offs of indicator performances.

There is a multitude of approaches to deal with problems of this nature (see Kangas and Kangas, 2005). In this contribution, the ANP (Saaty, 2001) is proposed to evaluate management strategies within the adapted DPSIR approach. The ANP is a generalization of the analytical hierarchy process (AHP) that additionally allows inclusion of interdependencies among indicators into the decision model. Currently, there are only a few applications of ANP, mainly in logistic and environmentally friendly management of supply chains and manufacturing (e.g. Meade and Sarkis, 1998; Sarkis, 2003). Wolfslehner and Vacik (2004) and Wolfslehner *et al.* (2005) describe forestry applications. The building of our ANP model consists of several steps that illustrate the principles behind it:

1. Choice of a structure: a flat generic network or a control hierarchy defining a control criterion (goal).

2. Definition of criteria clusters (driving forces, pressures, states, impacts and responses).

3. Definition of indicators assigned to the criteria clusters.

4. Definition of a cluster of strategies.

5. Definition of interdependencies (influences) among indicators within a cluster or between clusters.

6. Interlinking the cluster of strategies with the criteria clusters to incorporate indicator performances.

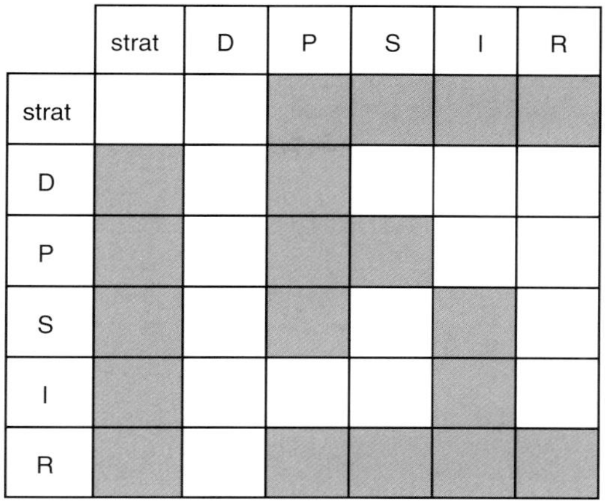

Fig. 23.3. Interconnections of indicators among and within strategy and criteria (DPSIR) clusters in the adapted DPSIR model.

7. Creating an unweighted super-matrix of ratio scales by pairwise comparison of the importance of elements with regard to a parent element.
8. Transforming performance data of the strategies on to a ratio scale.
9. Weighting the clusters according to their relative priority.
10. Calculating overall priorities for each strategy within a limit super-matrix.

Mathematically, an ANP model is implemented following a triple super-matrix calculation (Saaty, 2001). The unweighted super-matrix (*M*) is created by blocks of pairwise comparison matrices for both interconnections (step 7) and performances of the strategies (step 8), in which the relative weights are determined by equation (1).

$$Aw = \lambda_{max} w \quad (1)$$

in which A is a pairwise comparison matrix, λ_{max} is the largest eigenvalue of A and w is the priority vector.

To give a comprehensible picture of the model structure, Fig. 23.3 illustrates the interconnectedness of indicators among and within criteria and strategy clusters. Grey boxes indicate that there is at least one link between indicators of two different clusters or within a cluster. Linkages between indicators might be unidirectional or in both directions.

The unweighted super-matrix is transformed to the weighted super-matrix by multiplying their values with their associated cluster weights (step 9). By normalizing the weighted super-matrix, it is made column stochastic, i.e. the sum of the priorities under each column of the matrix is equal to 1. Finally, the limit super-matrix is processed by raising the entire super-matrix to powers until convergence in terms of a limit:

$$\lim_{k\to\infty} (1/N) \sum_{k=1}^{N} W^k \quad (2)$$

in which W is the weighted super-matrix, N indicates the number of iterations and k the exponent determined by iteration.

The limit priority values within this super-matrix indicate the flow of influence of an individual element towards the overall goal. Because the decision strategies build a particular cluster within the network, their limit priorities are synonymous with their contributions to the goal and are used for the ranking of strategies, being normalized within the cluster.

Predicting future effects of SFM

The intent of the study was to apply MCA to an indicator approach, but also to integrate eco-physiological model simulations with qualitative assessment under the umbrella of MCA. In this context, the application of a forest ecosystem model facilitates quantitative predictions of the effects of forest management decisions, and is furthermore able to take into account effects of changing environmental conditions.

The model applied in this study, PICUS v1.4, is a hybrid patch model, combining elements of a three-dimensional (3D) patch model and a process-based forest production model. The structural resolution is 10 m × 10 m, and spatial interactions among patches are modelled with a 3D canopy light regime and spatial seed dispersal. The concepts of inter- and intra-species competition for resources follow a modified patch-model concept (Lexer and Hönninger, 2001), whereas net primary productivity is derived from usable solar radiation and canopy quantum use efficiency, following the 3-PG approach (Landsberg and Waring, 1997). The coupling of both elements is accomplished, *inter alia, via* the leaf area of a stand, and is described in detail in Seidl *et al.* (2005). Below-ground processes are modelled by applying a biogeochemical process model of carbon and nitrogen cycling (Currie *et al.*, 1999). The PICUS modelling framework incorporates a flexible management module, allowing for different thinning and harvesting operations, as well as for planting operations, and is designed for decision support in complex alpine landscapes.

Application

Study area

Extensive areas supporting broadleaved species in warmer and drier lowlands of Austria have been transformed to conifer plantations dominated by Norway spruce (*Picea abies* L.) in the past. Scenarios of a possible climate change with higher temperatures and more frequently occurring drought periods increase the risk to such secondary spruce stands. To reduce economic and ecological risk in the context of sustainable forest management, the conversion of Norway spruce stands into mixed-species stands, which are better adapted to particular site conditions, is recommended (Spiecker *et al.*, 2004). We therefore draw our example from a 250 ha forest enterprise of almost pure Norway spruce stands located in Carinthia in the south of Austria. Elevation ranges from 540 to

640 m a.s.l., with average precipitation of 778 mm per year and annual average temperature of 8.3°C. According to Kilian *et al.* (1994), the potential natural vegetation (PNV) would mainly consist of forest associations of *Pino–Quercetum* and *Luzulo–Fagetum*. Forest management is influenced by gradations of *Lymantria dispar* (L.), *Pristiphora abietina* and bark beetle infestations as a follow-up of severe damage caused by several snow-breakage events.

Management strategies for multi-purpose forestry

The set of strategies for this study was designed to: (i) be realistic in terms of practical applicability; (ii) incorporate several alternative silvicultural concepts of sustainable forest management; and (iii) contain reference scenarios of forest development.

The first strategy (MS1) represents current management practice, which is a traditional even-aged forestry system with several thinnings from above and clear-cut after a rotation period of 90 years. Within this 'business as usual' (BAU) strategy, the stands are naturally regenerated in a shelterwood system. In contrast, regeneration in strategy 2 (MS2) takes place continuously, because the management goal is continuous-cover forestry (CCF) with target-diameter harvesting. Whereas in the first two strategies Norway spruce is dominant, strategy 3 (MS3) aims at a species change towards a higher share of deciduous species. This conversion is mainly accomplished by preplanting of beech below Norway spruce shelterwood or by afforestation of beech and oak after clear-cut. The prescribed management schedule in this strategy contains several thinnings from above and a final clear-felling. Strategy 4 (MS4) serves as a reference strategy for an undisturbed natural forest development without human interference, i.e., no management interventions are carried out.

For all strategies, a current inventory of the forest enterprise was used to classify the stands into 25 simulation entities, which serve as initial state for the computations of the 100-year simulation period.

Deriving priorities for management strategies

For the evaluation of the management strategies, 33 indicators have been related to the DPSIR clusters (Table 23.1). These indicators are based on several studies, transferring the principles of SFM from the European scale (MCPFE, 1998) to the FMU level (Wolfslehner *et al.*, 2003; Vacik and Wolfslehner, 2004). In practice, it is not an easy task to assign an indicator to one of the DPSIR clusters, because it is always a matter of perspective. Depending on the status in a causal chain, an Impact could equally well be a pressure or a state (Linser, 2001). By definition, an Impact results from changes in environmental quality (EEA, 1999). In order to delineate this problem properly and to broaden the scope towards the three dimensions of SFM, impacts within this adapted DPSIR framework are defined as the ratios of the current condition

Table 23.1. Data sources for the calculation of the DPSIR indicator performance values.

	Indicator	Data source
Driving force	Climate change	No input
	Groundwater recharge	No input
	Hunting	No input
	Recreation	No input
	Timber production	No input
Pressure	Browsing	Qualitative
	Droughts	Qualitative
	Increasing temperature	Qualitative
	Pests (bark beetles)	Qualitative
	Tourist frequency	Qualitative
	Low timber prices	Set equal
State	Biodiversity (Shannon–Weaver)	Simulation
	Growing stock	Simulation
	Contribution margin II	Simulation
	Damaged wood by bark beetles	Qualitative
	Employment within the enterprise	Qualitative
	Area with broadleaved regeneration	Simulation
	Naturalness of tree species composition	Simulation
	Water percolation quantity	Simulation
Impact	Ratio biodiversity (Shannon–Weaver)	Simulation
	Ratio growing stock	Simulation
	Ratio contribution margin II	Simulation
	Ratio damaged wood by bark beetles	Qualitative
	Ratio employment within the enterprise	Qualitative
	Ratio broadleaved regeneration area	Simulation
	Ratio naturalness of tree species composition	Simulation
	Ratio water percolation quantity	Simulation
Response	Introduction of broadleaved measures	Qualitative
	Habitat conservation actions	Qualitative
	Enhancing forest patchiness	Qualitative
	Intensity of silviculture/tending	Qualitative
	Controlled timber harvest	Qualitative
	Training activities for the staff	Qualitative

and the state of the next period in a prospective view for ecological, economic and socio-economic indicators (Fig. 23.4). Impacts are thus direct derivates of states over time. Input data for the ANP assessment were obtained from the output of the hybrid forest patch model, PICUS v1.4, and from qualitative assessments of experts (Table 23.1). For each management strategy, the performance of 33 indicators was calculated for a time period of 100 years at 20-year time steps, and for three different climate-change scenarios. For the

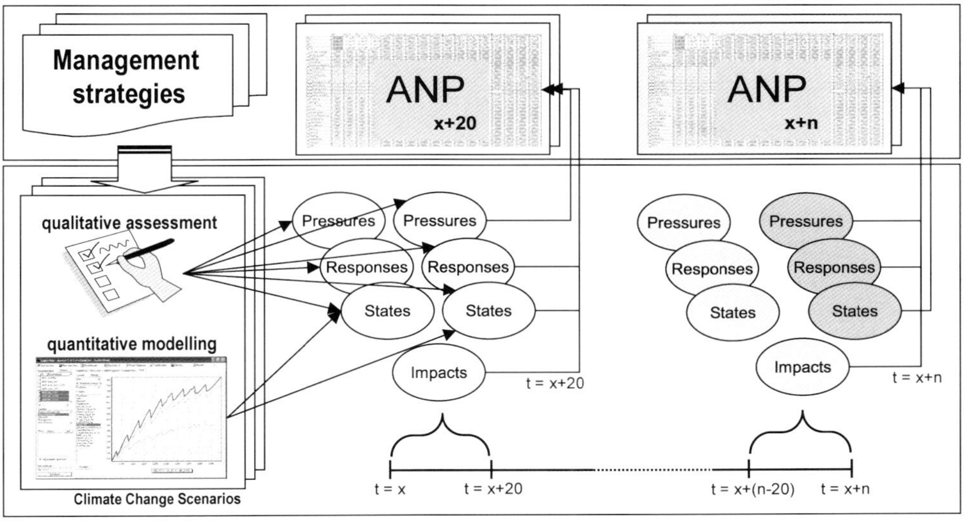

Fig. 23.4. Data and information flow in the adapted DPSIR framework.

climate-change scenarios, we used de-trended current climate for C1 (T = 7.6°C, P = 1013 mm), and two climate scenarios referring to the IS92a 'business as usual' scenario of the Intergovernmental Panel on Climate Change (IPCC, 1996). Climate scenarios C2 and C3 are described in detail in Kellomäki and Leinonen (2005), and are based on GCM simulations, with ECHAM4-OPYC3 for C2 (year 2100: ΔT = + 4.7°C, ΔP = + 90 mm) and HadCM2 for C3 (year 2100: ΔT = + 3.9°C, ΔP = – 203 mm).

Simulation outputs were transferred to indicator performances, using linear preference functions. The qualitative assessments of the management strategies are based on pairwise comparisons for each indicator recurring at different time steps. According to the derivation of impact indicators from state indicators, overall priorities of the management strategies are not calculated for time step t100.

The indicators were arranged in a flat generic network and the influences among the indicators defined within a cluster or between clusters. The cluster of strategies was linked with the criteria clusters to incorporate indicator performances of the management strategies. The unweighted super-matrix of ratio scales for the indicators was calculated by pairwise camparison of the importance of elements with regard to a parent element.

The indicator 'browsing' illustrates the manner in which individual indicators influence the performance of other indicators (Fig. 23.5). In this context, the indicator 'browsing' means the amount of regeneration that is damaged by deer in relation to undamaged regeneration. The indicator 'browsing' influences the performance of the indicators 'biodiversity', 'broadleaved', and 'naturalness'. Selective browsing of broadleaved regeneration will lead to a decrease in species richness and a reduction of broadleaved species, which will

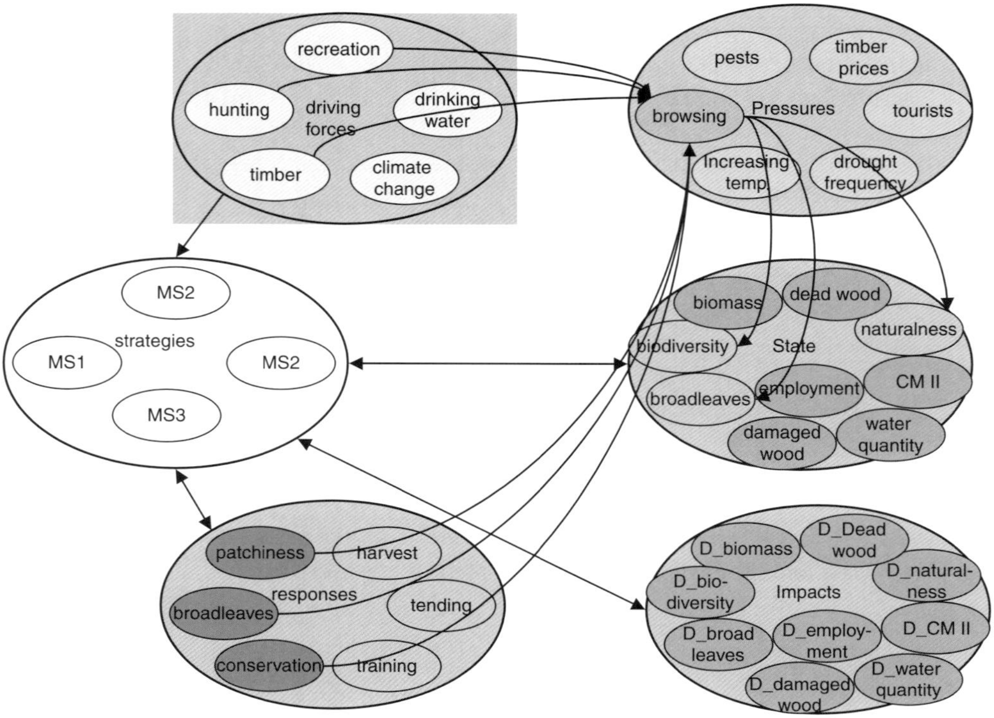

Fig. 23.5. Illustration of the direct effects of the indicator 'browsing' for the ANP implementation of DPSIR.

have an impact on the naturalness of the forests in the long run. The driving forces 'hunting', 'recreation' and 'timber production' influence the habitat environment. A high tourist frequency, low hunting rates and the type of forest management have an influence on the browsing activity of deer. On the other hand, responses of forest management, such as the quantity of artificial regeneration of broadleaved species, special conservation measures and the improvement of forest patchiness, will improve the habitat. In this context, pairwise comparisons have been done for all elements with regard to their related elements and the performance of the strategies. The weights for all clusters were set equal.

Results

Cumulative priorities for the four management strategies, with respect to four time steps and the overall goal of forest management to 'select the best management strategy with regard to sustainable forest management', are shown in Fig. 23.6. When conducting the first evaluation at year 20 (t20), strategies 1 (business as usual – BAU) and 2 (continuous-cover forestry – CCF) show

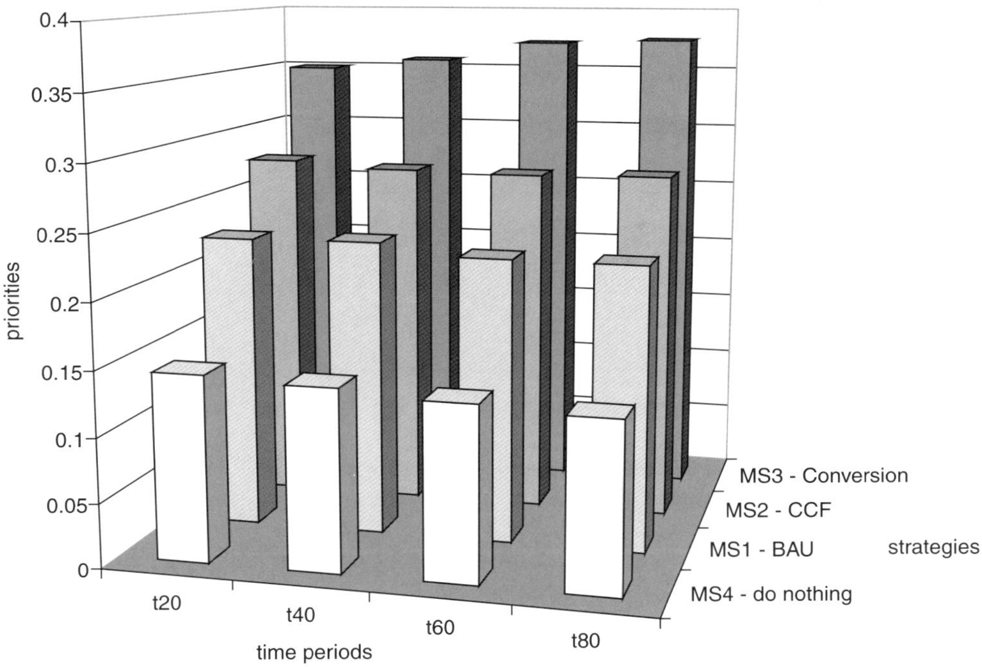

Fig. 23.6. Overall priorities for all four management strategies with respect to all four time steps.

decreasing priorities until t80, while strategy 3 (conversion) indicates increasing priorities. The continuous-cover regime can achieve stable priorities over the planning period at a relatively high level. With regard to the absolute priorities at the end of the simulation period, strategy 3 has the highest priorities (0.38), followed by strategy 2 (0.27), 1 (0.22) and 4 (0.13). With regard to different climate-change scenarios, no major deviation in the ranking and priorities of management strategies could be observed.

Apart from stable overall performance of the strategies, the priorities of single indicators vary widely from an equal-shared initial state at t0 (Fig. 23.7). The average contribution margin II, i.e. the profit minus harvesting and silvicultural costs, for strategy 3 indicates the lowest priorities at the end of the simulation period, strategy 2 the highest. The 'do nothing' scenario causes a contribution margin II of zero, because of abandoned management. The high costs for the introduction of broadleaved species lead to the decrease of priorities in strategy 3. Priorities related to the biodiversity indicator (species richness) indicate the supremacy of the conversion strategy 3 in relation to the other management strategies, which have no enrichment with other tree species. As there is only a slight change of species in the first 20 years, the priorities of strategy 3 are rather low, but they increase steadily with the duration of the simulation period.

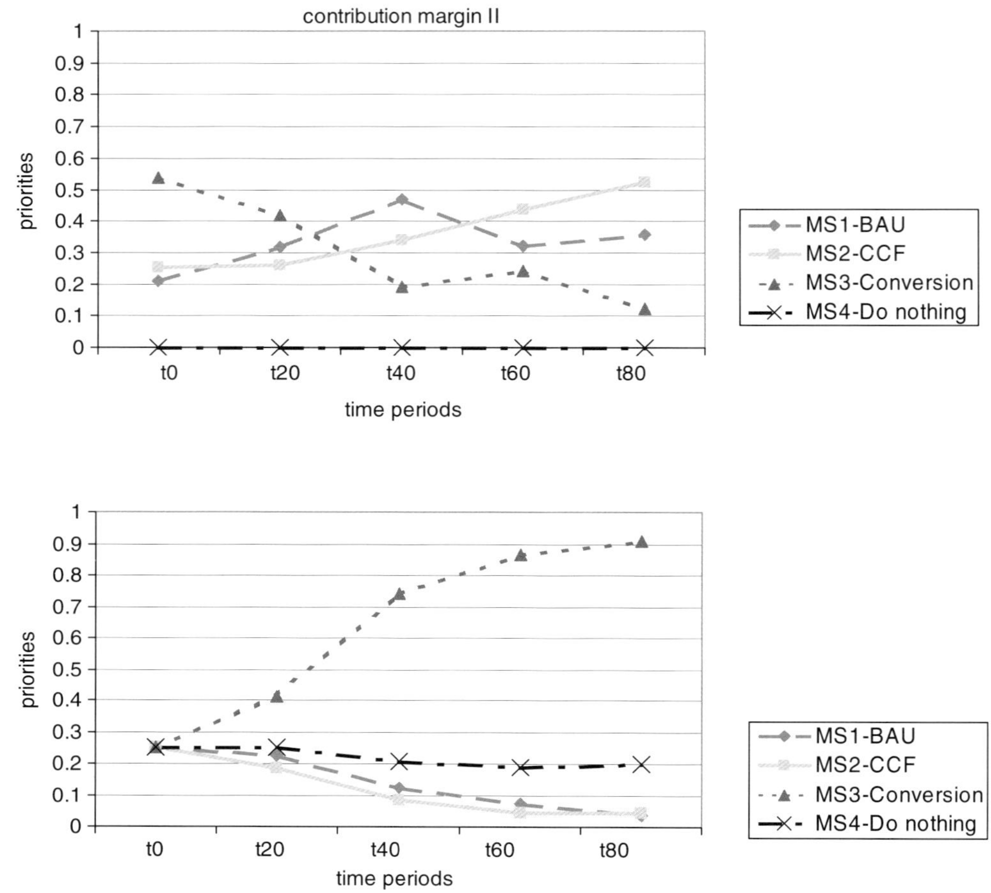

Fig. 23.7. Single indicator priorities for 'contribution margin II' and 'biodiversity' for all strategies with respect to four time steps, starting from the initial state t0.

Discussion

The present application of the DPSIR evaluation scheme provides a holistic view of the performance of strategies for forest management in response to pressures driven by external forces that influence the actual states of a forest and their impacts over time. In the example application, the overall results of the ANP evaluation show that conversion strategy 3 is optimal for all time steps and climate-change scenarios with regard to the principles of SFM. A transformation of the present conifer plantations dominated by Norway spruce to more close-to-nature forests by introducing and pre-planting deciduous tree species is promising in the long run. It will allow maintenance of key ecosystem processes, enhance resiliency of the forest ecosystems and restore a richer biodiversity, thus addressing major concerns of SFM (Spiecker *et al.*, 2004). These advantages even seem to compensate for economic fluctuations within our model. In comparison, maintaining a secondary spruce forest regime (strategy 1) may provide continuity, but will be subject to higher risks in terms of calamities, especially in light of climate-change scenarios (Lexer *et al.*, 2002). Strategy 2 may be seen as a compromise between persistence and conversion, which is also reflected by its multi-criteria evaluation preferences. The balancing of harvest and growth at the stand level allows achievement of relatively stable overall priorities at the FMU level. It is remarkable that strategy 4, the do-nothing approach, obviously would not correspond to the given definition of SFM. Consequently, a strategy without active management, leading to large-scale breakdowns of stands and large accumulation of biomass, would not correspond to the given definition of SFM, because at least the economic and social dimensions of SFM are not served.

The DPSIR framework seems highly capable of enriching the original simple state-based view of indicators to a figurative understanding of a higher-dimensional decision problem, while showing information in a causal way. The effects of pressures and drivers, as well as human responses to control the level of impacts, can be demonstrated to the public or practitioners. Most of the present sets of indicators are arranged without examining their interrelationships (Mendoza and Prabhu, 2003), not forgetting that many of those sets are politically negotiated and not scientifically based. Indeed, the original DPSIR approach has shortcomings in oversimplifying linkages and structures within the model and postulating linearity within them, which means reduced applicability for decision making and scenario analysis. Furthermore, the classical DPSIR approach is incapable of evaluating management strategies and dealing with stakeholders' preference information. To satisfy those demands in a more appropriate way, we have linked the DPSIR framework to an application of ANP. The ANP has been identified as a promising tool for integrating interdependencies within elements of SFM in an increasingly complex decision system (Wolfslehner *et al.*, 2005). Hence, ANP seems a reasonable method to process the adapted DPSIR approach, because of its generic power to cover a structure of interconnected clusters and to transfer them to a multi-criteria evaluation of different management strategies, while retaining the network features. Moreover, one could draw a high structural correspondence between DPSIR cause–effect chains and an ANP network model.

In addition, combined use of the multi-criteria evaluation technique of ANP and the simulation model PICUS v1.4 allowed extension of the original idea of the DPSIR approach as a reporting instrument. The set of tools was used to predict the future effects and consequences of forest management and to identify the most promising strategy in a changing environment. On the one hand, this approach should grant an at least semi-quantitative background to a decision model for 12 indicators (see Table 23.1) of the kind usually based only on experts' qualitative judgements, hence strengthening the reliability of the results. On the other hand, the use of a limited set of model outputs for evaluating the performance of different management strategies was enhanced by the use of qualitative assessments for 15 additional indicators (see Table 23.1).

However, there are some items that are handled imperfectly in the present approach. As in every model, there are trade-offs between different degrees of abstraction. The need to generalize input factors and to assign indicators to clusters causes a loss of detail in the DPSIR framework. This is especially true because the distinction between states and impacts is not fully clarified. We also found that there is strong compensation between indicators, making the model rather insensitive to changes of indicator performances with respect to the overall priorities of management strategies. We assume that the more complex the ANP structure appears and the more interconnections among clusters are incorporated, the more rigidly compensation dominates the process of prioritization. It must therefore be considered whether a more simplified structure could mitigate this phenomenon, while still being sufficiently representative. The identification of key indicators or key interconnections might be promising in this context (Wolfslehner *et al.*, 2005).

Operational drawbacks of the present approach are that there were no forest owners' preferences incorporated and that linear preference functions were used to calculate direct ratios from the simulation data. This approach carries the danger of covering priorities only implicitly, and thus potentially distorting the true preference pattern. Because this question was not the core of this study, the simplified approach sufficed for purposes of demonstrating model implementation.

Finally, the objective of integrating simulation results and qualitative assessments in the multi-criteria analysis is effective but complex, because the combination may cause redundant input values (with different aspects of a topic covered both by model outputs and MCA), and may hamper the distinction between direct and indirect, internal and external influences and effects. It is of importance that the experts charged with performing the qualitative assessments know about the characteristics and assumptions of the simulation model.

Conclusions

The adapted DPSIR approach responds to the perception that decision making often lacks a holistic prospective. In this context, models used to assess and monitor the effects of management may be seen as incomprehensible against the background of public communication. On the other hand, environmental communication instruments tend to be generalizing, remote from the operational level and often cut off from a scientific rooting or reference. It has been the rationale of this study to combine the strengths of the various methodologies, while keeping the whole concept at least semi-quantitative. The DPSIR approach enhances system understanding, while ecosystem modelling results have been integrated with multi-criteria decision-making techniques for the evaluation of SFM scenarios. This integration creates new perspectives on the communication of decision making, on the relationship between ecosystem and decision modelling and on the applicability of established approaches per se.

Adopting the original DPSIR approach and enriching it follow this paradigm at the price of raising new methodological questions, as discussed above.

Regarding the ANP application, it is questionable whether it is possible to keep up the attempts at complexity when the insensitivity of the evaluation model increases. Still, there are further potentials within the adapted DPSIR model concerning: (i) the formulation of specific preference functions for a proper transport of simulation results into the ANP; (ii) the design of relationships and changing preferences more time-specifically; and (iii) experimenting with the weighting of the DPSIR clusters.

Finally, appealing for a holistic and aggregated SFM evaluation should not prevent observance of the single indicator performance. Because it has been shown that single indicator values vary widely, while keeping the overall priority relatively unchanged, the degree to which results of MCA evaluation are inherent to structural assumptions and to actual performances need further examination.

References

Appanah, S. and Kleine, M. (2001) *Auditing of Sustainable Forest Management. A Practical Guide for Developing Local Auditing Systems based on ITTO's Criteria and Indicators*. FORSPA Publication 26/01, FAO, Bangkok.

Brang, P., Courbaud, B., Fischer, A., Kissling-Näf, I., Pettenella, D., Schönenberger, W., Spörk, J. and Grimm, V. (2002) Developing indicators for the sustainable management of mountain forests using a modelling approach. *Forest Policy and Economics* 4, 113–123.

Crabtree, B. and Bayfield, N. (1998) Developing sustainability indicators for mountain ecosystems: a study of the Cairngorms, Scotland. *Journal of Environmental Management* 52, 1–14.

Currie, W.S., Nadelhoffer, J.K. and Aber, J.D. (1999) Soil detrital processes controlling the movement of 15N tracers to forest vegetation. *Ecological Applications* 9, 87–102.

EEA (1999) *Environmental Indicators: Typology and Overview*. Technical Report No. 25, European Environment Agency, Copenhagen.

EEA (2004) *An Indicator-based Approach to Assessing the Environmental Performance of European Marine Fisheries and Aquaculture*. Technical Report No. 87, European Environment Agency, Copenhagen.

Firbank, L.G., Barr, C.J., Bunce, R.G.H., Furse, M.T., Haines-Young, R., Hornung, M., Howard, D.C., Sheail, J., Sier, A. and Smart, S. (2003) Assessing stock and change in land cover and biodiversity in GB: an introduction to Countryside Survey 2000. *Journal of Environmental Management* 67, 207–218.

Franc, A., Laroussinie, O. and Karjalainen, T. (eds) (2001) *Criteria and Indicators for Sustainable Forest Management at the Forest Management Unit Level*. EFI Proceedings 38, European Forest Institute, Joensuu, Finland.

Gobin, A., Jones, R., Kirkby, M., Campling, P., Govers, G., Kosmas, C. and Gentile, A.R. (2004) Indicators for pan-European assessment and monitoring of soil erosion by water. *Environmental Science and Policy* 7, 25–38.

Hammond, A., Adrianse, A., Rodenburg, E., Bryant, D. and Woodward, R. (1995) *Environmental Indicators: A Systemic Approach to Measuring and Reporting on Environmental Policy Performance in the Context of Sustainable Development*. World Resource Institute, Washington, DC.

IPCC (1996) *Second Assessment Report: Climate Change 1995*. Intergovernmental Panel on Climate Change, Cambridge University Press, Cambridge

Kammerbauer, J., Cordoba, B., Escolan, R., Flores, S., Ramirez, V. and Zeledon, J. (2001) Identification of development indicators in tropical mountainous regions and some implications for natural resource policy designs: an integrated community case study. *Ecological Economics* 36, 45–60.

Kangas, J. and Kangas, A. (2005) Multiple criteria decision support in forest management – the approach, methods applied, and experiences gained. *Forest Ecology and Management* 207, 133–143.

Kant, S. (2003) Extending the boundaries of forest economics. *Forest Policy and Economics* 5, 39–56.

Kellomäki, S. and Leinonen, S. (eds) (2005) *Management of European Forests under Changing Climatic Conditions*. Research Notes 163, Faculty of Forestry, University of Joensuu. Joensuu, Finland.

Kelly, K.L. (1998) A systems approach to identifying decisive information for sustainable development. *European Journal of Operational Research* 109, 452–464.

Kilian, W., Müller, F. and Starlinger, F. (1994) *Die forstlichen Wuchsgebiete Österreichs. Eine Naturraumgliederung nach waldökologischen Gesichtspunkten.* FBVA-Berichte 82, Forstliche Bundesversuchsanstalt, Vienna.

La Jeunesse, I., Rounsevell, M. and Vanclooster, M. (2003) Delivering a decision support system tool to a river contract: a way to implement the participatory principle at the catchment scale? *Physics and Chemistry of the Earth* 28, 547–554.

Landsberg, J.J. and Waring, R.H. (1997) A generalized model of forest productivity using simplified concepts of radiation-use efficiency, carbon balance and partitioning. *Forest Ecology and Management* 95, 209–228.

Lexer, M.J. and Hönninger, K. (2001) A modified 3d-patch model for spatially explicit simulation of vegetation composition in heterogeneous landscapes. *Forest Ecology and Management* 144, 43–65.

Lexer, M.J., Hönninger, K., Scheifinger, H., Matulla, C., Groll, N., Kromp-Kolb, H., Schadauer, K., Starlinger, F. and Englisch, M. (2002) The sensitivity of the Austrian forests to scenarios of climatic change: a large-scale risk assessment based on a modified gap model and forest inventory data. *Forest Ecology and Management* 162, 53–72.

Linser, S. (2001) *Critical Analysis of the Basics for the Assessment of Sustainable Development by Indicators.* Schriftenreihe Freiburger Forstliche Forschung Vol. 17, University of Freiburg, Freiburg, Germany.

MCPFE (1998) *Third Ministerial Conference on the Protection of the Forests in Europe. General Declaration and Resolutions Adopted*. Liaison Unit, Lisbon.

Meade, L. and Sarkis, J. (1998) Strategic analysis of logistics and supply chain management systems using the analytical network process. *Transportation Research Part E: Logistics and Transportation Review* 34, 201–215.

Mendoza, G.A. and Prabhu, R. (2000) Multiple criteria decision making approaches to assessing forest sustainability using criteria and indicators: a case study. *Forest Ecology and Management* 131, 107–126.

Mendoza, G.A. and Prabhu, R. (2003) Qualitative multi-criteria approaches to assessing indicators of sustainable forest resource management. *Forest Ecology and Management* 174, 329–343.

Mrosek, T., Balsillie, D. and Schleifenbaum, P. (2006) Field testing of a criteria and indicators system for sustainable forest management at the local level. Case study results concerning the sustainability of the private forest Haliburton Forest and Wild Life Reserve in Ontario, Canada. *Forest Policy and Economics* 8, 593–609.

OECD (1993) *Core Set of Indicators for Environmental Performance Reviews: A Synthesis Report by the Group on the State of the Environment*. Environment Monographs Vol. 83, Organisation for Economic Co-operation and Development, Paris.

OECD (1994) *Environmental Indicators.* Core Set, Organisation for Economic Co-operation and Development, Paris.

Pearce, D., Putz, F.E. and Vanclay, J.K. (2003) Sustainable forestry in the tropics: panacea or folly? *Forest Ecology and Management* 172, 229–247.

Peng, C. (2000) Understanding the role of forest simulation models in sustainable forest management. *Environmental Impact Assessment Review* 20, 481–501.

Prabhu, R., Ruitenbeek, H.J., Boyle, T.J.B. and Colfer, C.J.P. (2001) Between voodoo science and adaptive management: the role and research needs for indicators of sustainable forest management. In: Raison, R.J., Brown, A.G. and Flinn, D.W. (eds) *Criteria and Indicators for Sustainable Forest Management.* IUFRO Research Series 7, CABI, Wallingford, UK, pp. 39–66.

Raison, R.J., Brown, A.G. and Flinn, D.W. (eds) (2001) *Criteria and Indicators for Sustainable Forest Management.* IUFRO Research Series 7, CABI, Wallingford, UK.

Rametsteiner, E. (2001) SFM indicators as tools in political and economic context: actual and potential roles. In: Raison, R.J., Brown, A.G. and Flinn, D.W. (eds) *Criteria and Indicators for Sustainable Forest Management.* IUFRO Research Series 7, CABI, Wallingford, UK, pp. 107–130.

Rametsteiner, E. and Simula, M. (2003) Forest certification – an instrument to promote sustainable forest management? *Journal of Environmental Management* 67, 87–98.

Saaty, T.L. (2001) *Decision Making with Dependence and Feedback: The Analytic Network Process*. RWS Publishing, Pittsburgh, Pennsylvania.

Sarkis, J. (2003) A strategic decision framework for green supply chain management. *Journal of Cleaner Production* 11, 397–409.

Seidl, R., Lexer, M.J., Jäger, D. and Hönninger, K. (2005) Evaluating the accuracy and generality of a hybrid patch model. *Tree Physiology* 25, 939–951.

Spiecker, H., Hansen, J., Klimo, E., Skovsdard, J.P., Sterba, H. and von Teuffel, K. (2004) *Norway Spruce Conversion – Options and Consequences.* Report 18, EFI, Brill, Leiden, The Netherlands.

Sverdrup, H. and Stjernquist, I. (eds) (2002) *Developing Principles and Models for Sustainable Forestry in Sweden.* Managing Forest Ecosystems Vol. 5, Kluwer Academic, Dordrecht, The Netherlands.

Vacik, H. and Wolfslehner, B. (2004) Development of an indicator catalogue for the evaluation of sustainable forest management at the forestry management unit level. *Schweizerische Zeitschrift für Forstwesen* 155, 476–486.

Walmsley, J. (2002) Framework for measuring sustainable development in catchment systems. *Environmental Management* 29, 195–206.

Wedeles, C. and Williams, J. (1999) *Assessment of Indicators of Sustainable Forest Management. Lake Abitibi Model Forest. Summary Report.* ArborVitae Environmental Services, Cochrane, Ontario.

Wolfslehner, B. and Vacik, H. (2004) A pressure–state–response framework for multi-criteria analysis of sustainable forest management. In: *IUFRO Conference (WP 4.04.10), 25–27 August 2004, Tatras/Slovakia.* Department of Forest Management and Geodesy, Forestry Faculty, Technical University, Zvolen, Slovakia.

Wolfslehner, B., Vacik H., Lexer, M.J., Würz, A., Hochbichler, E., Klumpp, R. and Spörk, J. (2003) A system analysis approach for assessing sustainable forest management at FMU level. In: *XII World Forestry Congress – Forest's Source of Life, 21–28 September, Quebec.* FAO Forestry Department, http://www.fao.org/docrep/article/wfc/xii/0690-b4.htm

Wolfslehner, B., Vacik, H. and Lexer, M.J. (2005) Application of the analytic network process in multi-criteria analysis of sustainable forest management. *Forest Ecology and Management* 207, 157–170.

Woodley, S., Alward, G., Gutierrez, L.I., Hoekstra, T., Holt, B., Livingston, L., Loo, J., Skibicki A., Williams, C. and Wright, P. (1999) *North American Test of Criteria and Indicators of Sustainable Forestry.* Report No. 3, Inventory and Monitoring Institute, USDA Forest Service. Washington, DC.

24 Establishment Management Information System (EMIS): Delivering Good Practice Advice on Tree Establishment in the Uplands of Britain

MIKE P. PERKS, ALAN J. HARRISON AND STEPHEN J. BATHGATE

Forest Research, Northern Research Station, Roslin, Midlothian, UK.

Abstract

The Establishment Management Information System (EMIS) is a decision-support tool that integrates existing advice on tree establishment for upland restocking on a site-specific basis. It draws upon information from many technical and scientific publications to provide the user with 'walk through' systems delivering acceptable (site-constrained) tree establishment options for restock sites. Site information (user input) allows calculation of environmental variables that constrain species choice and identifies appropriate on-site management practices. EMIS currently provides advice encompassing the main commercial upland conifer species, plus birch. Whilst all potential system inputs and interactions have been investigated, primary drivers were identified to ensure that the system evolves, guided by operational requirement and existing knowledge. The web-based decision-support system integrates with another forest research tool (Ecological Site Classification, ESC) via shared components. Output of information from EMIS will be available as both html and pdf. EMIS will be a web-delivered tool, but the constituent models are also available as document-wrapped-style web services to allow integration with spatial data (GIS) systems. This will enable delivery of spatially explicit guidance on good practice in the future.

Introduction

In forest policy worldwide, the concept of sustainability has expanded to include sustainable forest management (SFM; Lane and McDonald, 2002). The application of silvicultural knowledge at establishment is therefore the first step towards sustainable forest management, on which all other decisions depend (Ray and Broome, 2003). In the UK, a number of measures have been introduced to formalize sustainable forestry, including the UK Forestry Standard (Anon., 2004) and the UK Woodland Assurance Scheme (UKWAS; Anon., 2000). Forests that have been UKWAS-certified are monitored to ensure that appropriate management practices are carried out (Anon., 2002).

Forestry can be defined as intervention in ecological processes to meet human needs or goals. Forestry practices in general, and silviculture in particular, are based on the premise that any activity in the forest is intended to assist the goals of the manager. Such considerations, applied to a restock setting, include condition of forest soils (e.g. the selective use of cultivation taking into account its potential impact), forest condition (e.g. minimizing chemical use according to site-specific needs) and timber production (e.g. identifying productive species well suited to the site). Indeed, identification of the landowner's objectives is the first step taught to silviculturists in forestry schools (Smith, 1986). In particular, strategic-level forest design planning has a wide range of competing goals that the forester has to appreciate and account for (Bell, 1998). Therefore, decision-support systems, intended to help forest landowners or managers determine appropriate actions, must focus on meeting management goals defined by the user (Mason, 1997). A method-sophisticated approach, in which interaction between ecological (and social) components must be known and considered, requires powerful decision-support systems (DSS; Rauscher *et al.*, 2000). DSS tools are computer-delivered programs that provide support for decision makers engaged in solving various semi- to ill-structured problems involving multiple attributes, objectives and goals (Turban and Aronson, 2000; Nemati *et al.*, 2002). A key feature of DSS tools is that the decision maker is as important a part of a DSS as any other component. People do not simply use DSS outputs; rather, they provide the system with judgements and values that are critical to, and often dominate, the decision-making process.

Successful tree establishment, whether reforestation or afforestation of upland sites, requires knowledge of site constraints to ensure suitable matching of species to site and appropriate site management techniques. In order to match species to a location, knowledge of site factors that influence tree establishment and growth is required. These include an understanding of the general site environment (e.g. soil type, lithology, soil moisture and soil nutrient status), an understanding of the local climatic environment (e.g. wind climate, oceanicity, elevation, temperature profile) and the interactions between these factors. The ability of the forester to assess site conditions and select well-suited tree species is therefore of fundamental importance, as is an understanding of the silvicultural options available to improve tree establishment and growth (Tabbush, 1988). Silvicultural considerations include choice of plant species and provenance, plant type, plant quality, plant storage, time and method of planting (e.g. Morgan, 1999), site cultivation (Sutton, 1993; Paterson and Mason, 1999), fertilization (Taylor, 1990a, b; Smith and McKay, 2002) and vegetation management (e.g. Willoughby and Dewar, 1995; Willoughby *et al.*, 2004).

In the present study, we describe the development of a prototype expert system (EMIS), developed to help with compliance to guidelines on sustainable forestry for establishment in British forests. The Establishment Management Information System (EMIS) attempts to model the complex interactions represented by the varied silvicultural options for the successful establishment and growth of trees on UK upland restock sites (i.e. sites previously used for the cultivation of trees), whilst taking into account the goals of sustainable forestry.

Guidance on good practice is via the web to the user, whilst the architecture provides decision support by recommending options for cultivation, fertilization, aspects of plant quality and species choice, matched to site constraints.

The System

The EMIS extensible-service-oriented framework (i.e. shared software code) integrates the Ecological Site Classification system (ESC; Ray *et al.*, 2001; Fig. 24.1). It will ultimately allow incorporation of additional simulations, knowledge bases and other forest research DSS components in the future (e.g. herbicide advisor, *Hylobius* management support system). The ESC DSS uses two models to assess suitability of tree species and native woodland community types, dependent upon climatic and edaphic site factors. EMIS integrates with the ESC tree species suitability model, and provides additional guidance on species-specific silviculture and plant quality. The complex EMIS schema, which is presented here as stand-alone constituent modules, was developed in the first instance using the Simile modelling environment.

For a given site, the suitability of individual species for timber production (on restock sites) is predicted on the basis of six Ecological Site Classification (ESC) factors as criteria for testing species suitability to a site (Pyatt and Suarez, 1997; Pyatt *et al.*, 2001):

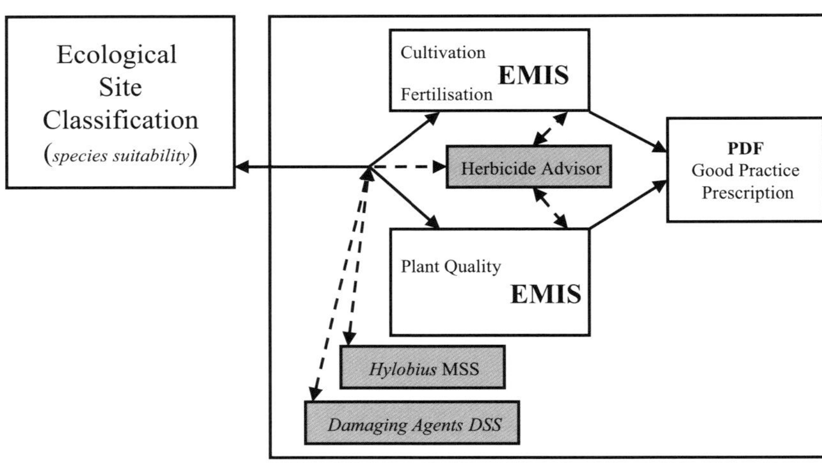

Fig. 24.1. Schematic of the major functional components of EMIS and interactions with other systems. White boxes and solid black arrows, indicating information flow, are currently functional. The elements of the EMIS framework relating to conifer restocking and site treatment are shown in the right hand pane. Integration with Herbicide Advisor and *Hylobius* MSS are currently in development (dashed arrows), whilst possible interactions with other systems codifying the impact of a range of damaging agents (e.g. deer, frost, pathogens, other insect pests) are included in the architecture, but are not yet functional. Arrows indicate directional flow of information.

- Four climatic factors: accumulated temperature, moisture deficit, windiness (by detailed aspect method scoring (DAMS); Quine and White, 1993) and continentality.
- Two soil quality factors: soil moisture regime (SMR) and soil nutrient regime (SNR).

ESC DSS (Ray, 2001) calculates climatic indices from user input of a grid reference (Fig. 24.2). A user is encouraged to check soil information via a soil survey rather than depending on the often inaccurate data (Kennedy, 2002). Soil quality (SMR, SNR) is estimated using the ESC soil type directly, as modification by site vegetation is not relevant for restock sites.

However, the added accuracy that an assessment of site vegetation offers has been retained, because an extension of EMIS to non-restock sites (e.g. farm woodlands) is envisaged in the future. Further modification (input) with respect to underlying lithology, the presence of heather and the depth of peat (for peat-based soil types) are required because these factors are known to affect site fertility (Taylor, 1990b). Soil quality is therefore the primary driver for advice on good practice relative to cultivation (Fig. 24.3A) and fertilization (Fig. 24.3B) options.

Parameters for plant quality are constrained in the first instance by recommended choices of species, although multiple scenarios, when several species are suitable for a site, can be considered. Plant quality parameters for consideration by the user include correct (constrained) choice of provenance in the first instance. Users are presented with all currently common plant-type specifications, by nursery production system (e.g. cell-grown vs. bare-root stock types), acceptable physiological limits such as routinely assessed by physiological plant quality tests (McKay, 1992), plant morphology (size classes denoted by acceptable height and root collar diameter ranges), and cell size for container-grown plants (Fig. 24.4).

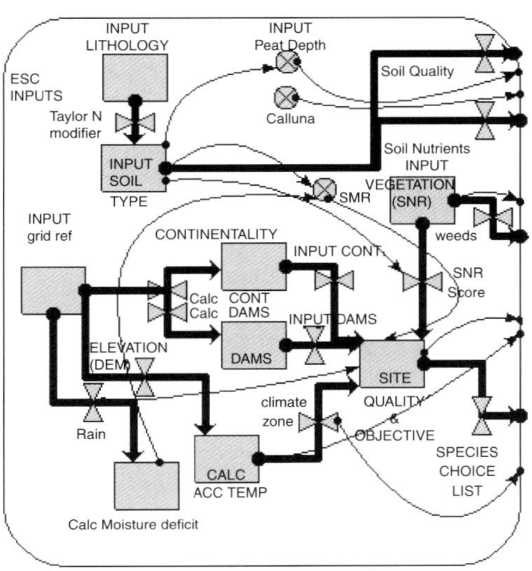

Key	Descriptor
Compartment	A compartment contains/requires either: (i) calculated data [for constraint of future options]; (ii) choice from a number of pre-determined options; (iii) data input by the user; and (iv) a combination of the above
Variable	A variable is modifier to the decision flow which influences the choices available to the manager. A variable can also act as a switch (e.g. peat depth alters nutrition)
Flow	A flow arrow shows the decision flow through the support system, ultimately leading to a series of good practice prescriptions.
Influence	An influence arrow shows that a factor [variable/compartment] has a modifying effect upon another factor and that this influence will modify the flow of information (i.e. decision process).

Fig. 24.2. Detailed schematic of the climatic and soil quality components within ESC, which are accessed by EMIS as 'species suitability service' after user input of location, soil type and lithology. The input of vegetation information to amend SNR is not required for restock sites.

Acceptable planting windows, dependent on plant specifications and climate zone (captured from accumulated temperature), are presented in tabulated format.

The complex interactions between and within modules were considered, through reference to the scientific literature (by the authors), and expert comment on the overall model schema was sought. Whilst all possible interactions between compartments, variables, flows and influences were investigated in an iterative process, the primary drivers (decision flows) and major influences have been identified and used to ensure that a system evolves that is guided by both operational requirements and existing knowledge. Conflicts have been avoided by delivery of all advice on good practice such that the user may identify acceptable suboptimal trade-offs (e.g. the user is aware of any conflict).

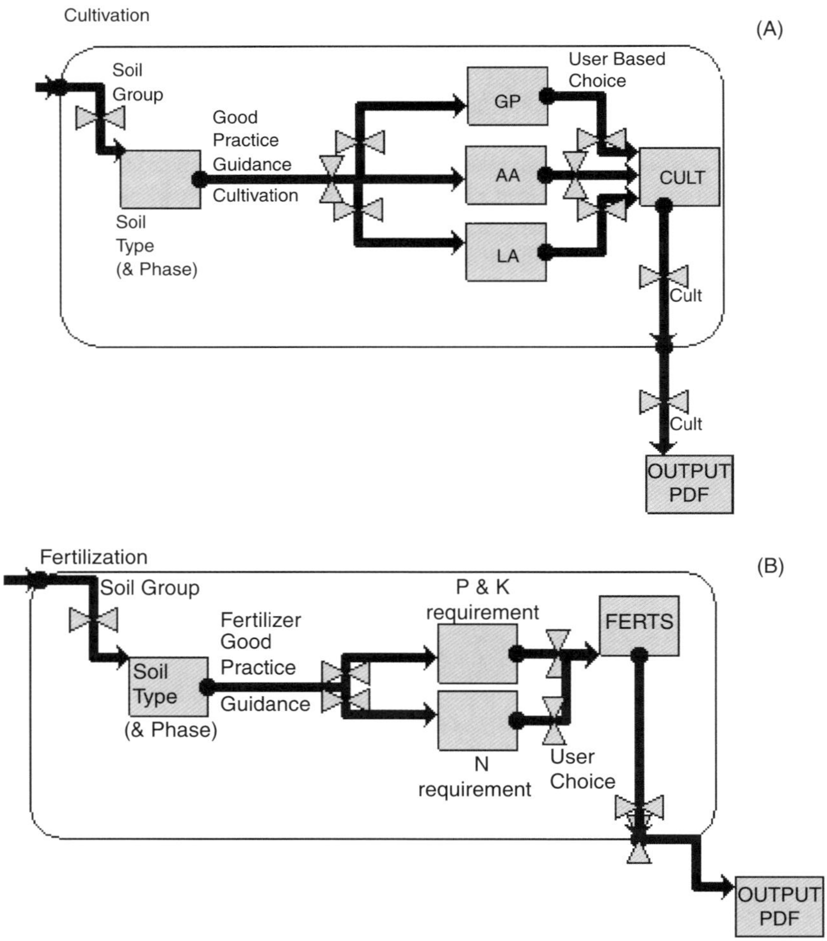

Fig. 24.3. Detailed schematic of the influence of soil quality components on output guidance. (A) Cultivation, for which options are delivered to the user, denoted good practice (GP), acceptable alternative (AA) and lesser alternative (LA: i.e. with some penalties), and (B) Fertilization, for which nitrogen, phosphorus and potassium specifications are given. Key – see Fig. 24.2.

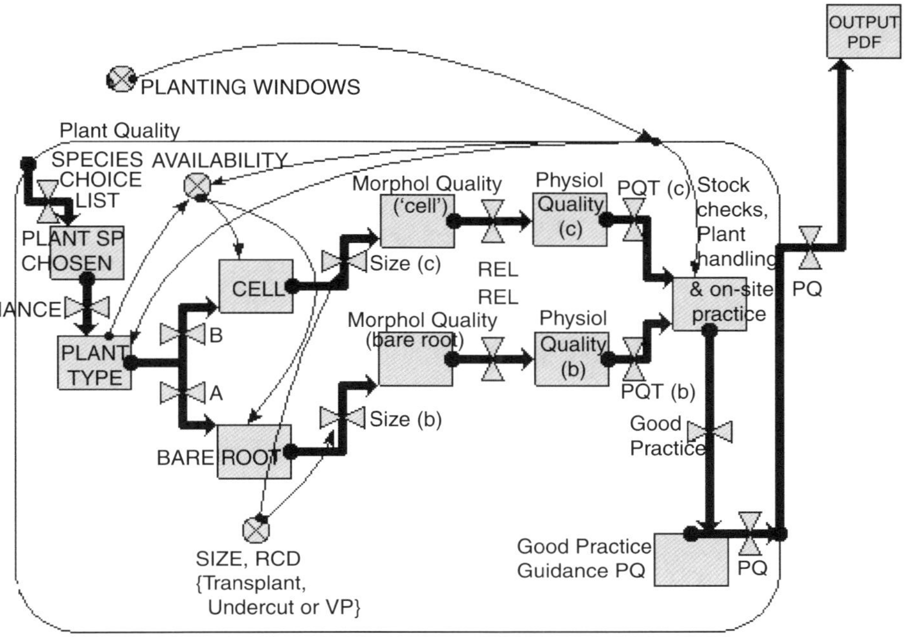

Fig. 24.4. Detailed schematic of the plant quality components considered in output guidance. The user is able to view all currently available plant specifications (plant type, morphology and acceptable physiological limits). Acceptable months for planting (windows) are also tabulated. Key – see Fig. 24.2.

The Web-based User Interface

Functional elements available in EMIS are accessed through a thin client web–browser interface (i.e. all business logic and data reside on an Internet server). EMIS is designed to take advantage of available graphical–user interface, point-and-click technology to create an intuitive environment. Further development of the EMIS interface will include testing to ensure usability and allow bespoke development of supplementary training materials. Adherence to Worldwide Web Consortium (W3C) standards, such as Extensible Hypertext Markup Language (XHTML) and Cascading Style Sheets (CSS), should ensure stable behaviour across modern browser implementations, as well as clear delineation between presentation style and content.

Development of an integrated treatment prescription

On the opening screen, the user selects the EMIS program, and is then instructed regarding the input parameter settings required to set up the necessary background drivers (grid reference, lithology, soil type, *Calluna*). The user is then asked to input the first of these parameter settings in a site-location screen by choice of the appropriate 100 km grid square and input of a six-figure grid reference.

The user is then required to input site-based assessment information regarding the dominant soil type, chosen from a drop-down list of 14 FC soil groups and their attendant soil types and phases (Kennedy, 2002). Underlying lithology is also chosen from a drop-down menu within EMIS: underlying solid lithology at 1 : 625,000 scale is considered acceptable, and can be obtained from British Geological Survey (BGS) maps or from the online BGS data portal. The user is able to input information on peat depth and heather presence (*Calluna/Erica* sp.) by means of tick boxes, when appropriate. Where more accurate site information has not been collected, within the boundaries of the national forest estate, the Forestry Commission's sub-compartment database may provide sufficient soils information for quick and approximate evaluation of site type. However, to meet the requirements of sustainable forestry, users will often require a more accurate site-based evaluation. The ESC models are interrogated and captured site values, for the six constraining factors, are then displayed within EMIS, and the opportunity to amend one or many of them is afforded the user. The ESC models are then interrogated for species yield, and an initial predicted (potential) yield estimated from accumulated temperature is displayed. This value is then modified by the limiting ESC site factor, and species suitability and predicted yield class are assessed against the continuous suitability functions that have been developed within ESC (Ray *et al.*, 1998) for the ten conifer species and two birch species considered by EMIS (Fig. 24.5). This process is entirely analogous to the ESC parameterization, and operates by calling the relevant ESC models behind the scenes.

The user chooses, from the ESC-derived suitability–yield table, species of particular interest for the site under consideration. EMIS then delivers relevant guidance on good practice (Fig. 24.6) from the cultivation, plant quality and fertilization modules via interrogation of the relevant models (Fig. 24.7).

Output

The EMIS web–browser interface delivers all the appropriate guidance whilst the user can also obtain the output in PDF for any of the scenarios that have been run.

Discussion

Implementation

During development, linkages between EMIS modules and among tools developed within the EMIS framework architecture were considered within the Simile schema. EMIS alone has been developed with reference to approximately 40 technical and scientific publications regarding site–species suitability and the attendant silvicultural management options.

EMIS implementation employed open standards and technology as well good software engineering practice such as utilization of object-oriented design, and automated testing to offer extensibility, portability and interoperability. By delivering EMIS as a web application, maintenance is reduced as the software and data

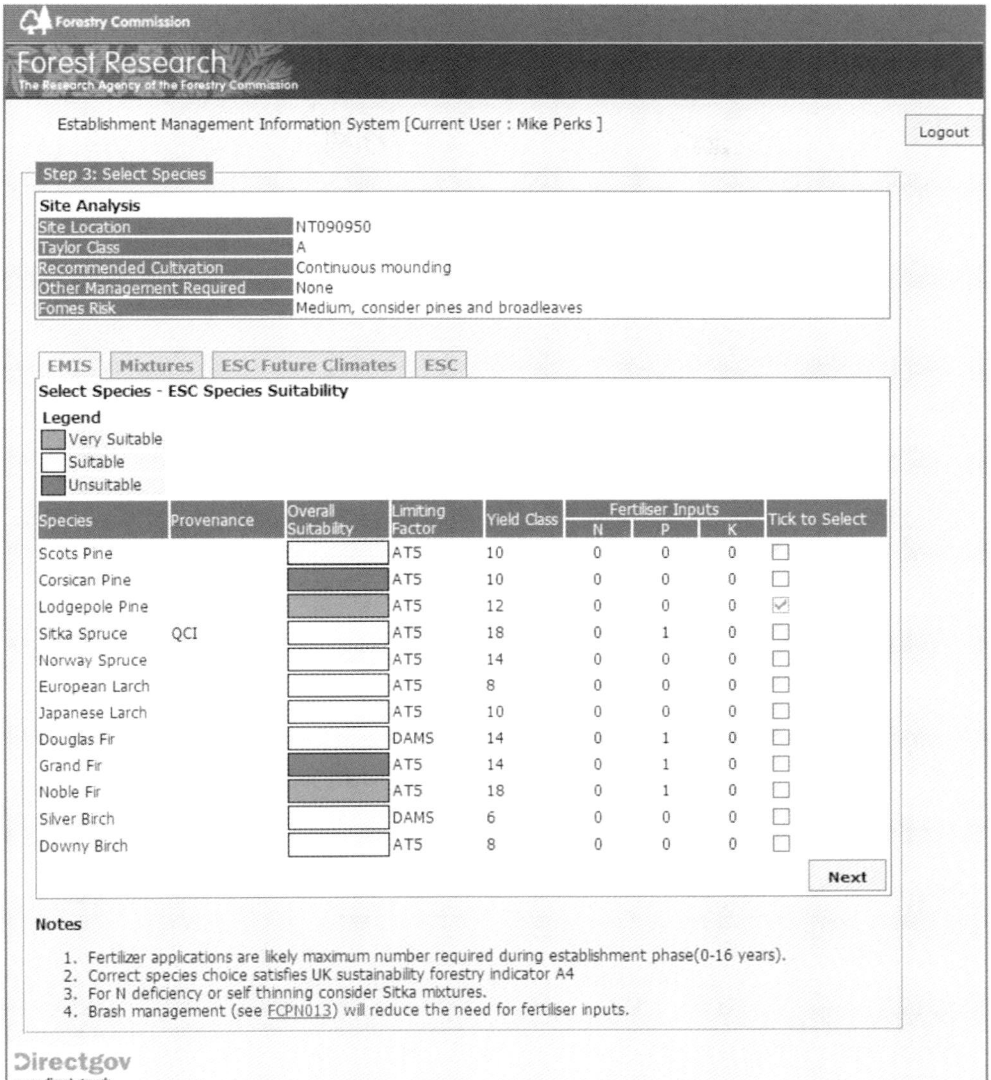

Fig. 24.5. Screenshot of ESC-derived EMIS output for species suitability and predicted yield (productivity) class. In this example, lodgepole pine has been selected for further analysis.

are held centrally; potential users simply require a web browser. Java was chosen as the implementation language, because of the availability of open-source products to support the development of EMIS and for compatibility with the FC Internet delivery platform, which is based on Java Enterprise Edition (J2EE) technologies. A variety of tools comprise the EMIS software; these include the Struts framework, DOM4J, Netbeans, iText and Axis, which together saved licensing costs and accelerated development time.

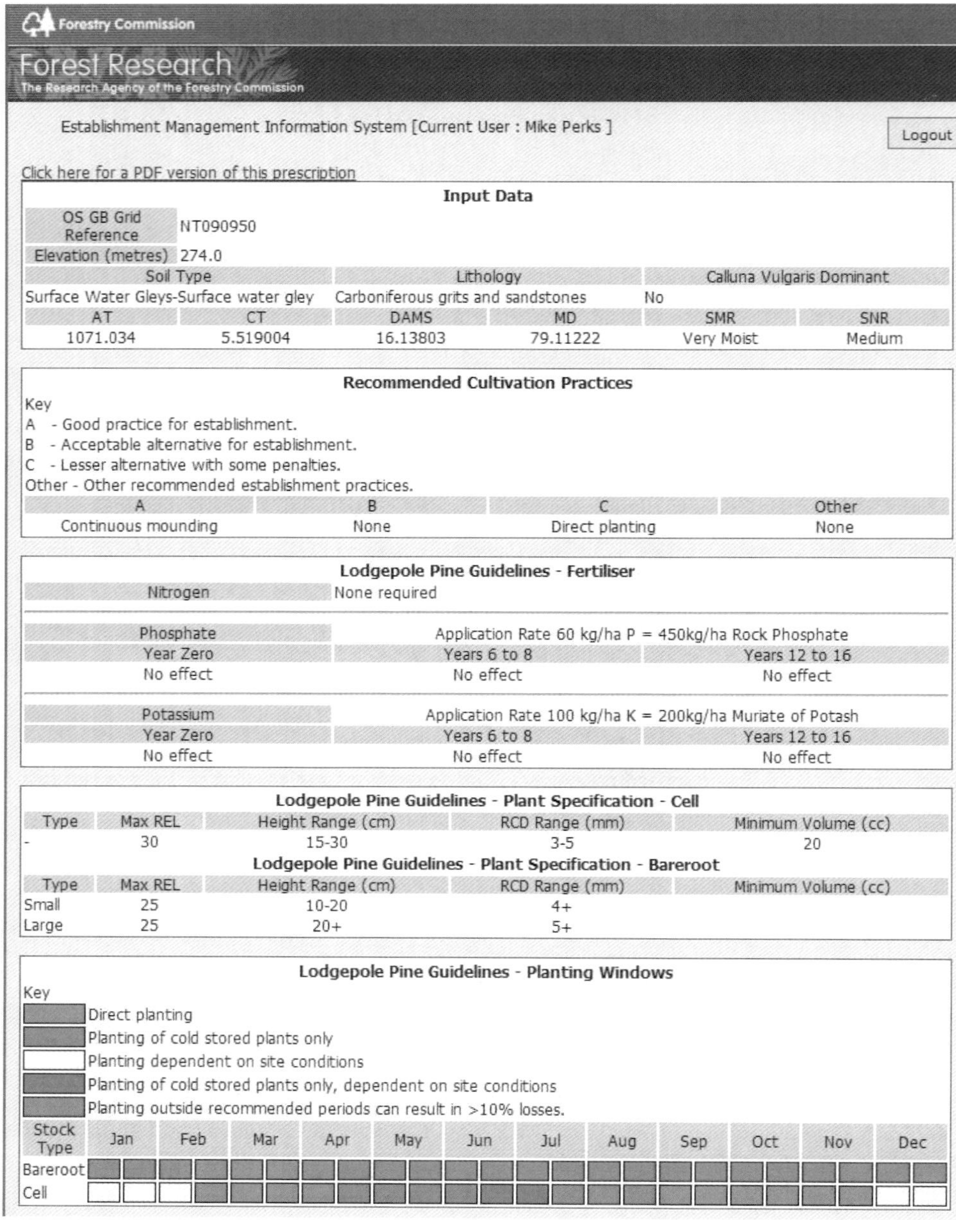

Forestry Commission

Forest Research

The Research Agency of the Forestry Commission

Establishment Management Information System [Current User : Mike Perks]

Logout

Click here for a PDF version of this prescription

Input Data

OS GB Grid Reference	NT090950
Elevation (metres)	274.0

Soil Type	Lithology	Calluna Vulgaris Dominant
Surface Water Gleys-Surface water gley	Carboniferous grits and sandstones	No

AT	CT	DAMS	MD	SMR	SNR
1071.034	5.519004	16.13803	79.11222	Very Moist	Medium

Recommended Cultivation Practices

Key

A - Good practice for establishment.

B - Acceptable alternative for establishment.

C - Lesser alternative with some penalties.

Other - Other recommended establishment practices.

A	B	C	Other
Continuous mounding	None	Direct planting	None

Lodgepole Pine Guidelines - Fertiliser

Nitrogen	None required	
Phosphate	Application Rate 60 kg/ha P = 450kg/ha Rock Phosphate	
Year Zero	Years 6 to 8	Years 12 to 16
No effect	No effect	No effect
Potassium	Application Rate 100 kg/ha K = 200kg/ha Muriate of Potash	
Year Zero	Years 6 to 8	Years 12 to 16
No effect	No effect	No effect

Lodgepole Pine Guidelines - Plant Specification - Cell

Type	Max REL	Height Range (cm)	RCD Range (mm)	Minimum Volume (cc)
-	30	15-30	3-5	20

Lodgepole Pine Guidelines - Plant Specification - Bareroot

Type	Max REL	Height Range (cm)	RCD Range (mm)	Minimum Volume (cc)
Small	25	10-20	4+	
Large	25	20+	5+	

Lodgepole Pine Guidelines - Planting Windows

Key

Direct planting

Planting of cold stored plants only

Planting dependent on site conditions

Planting of cold stored plants only, dependent on site conditions

Planting outside recommended periods can result in >10% losses.

Stock Type | Jan | Feb | Mar | Apr | May | Jun | Jul | Aug | Sep | Oct | Nov | Dec

Bareroot

Cell

Fig. 24.6. Screenshot of EMIS web output (HTML) after species selection by the user (the example shown is for lodgepole pine).

To provide GIS interoperability, which in the FC is based on Microsoft.NET technology, some functionality was exposed as document-literal wrapped web services (Butek, 2003). Inclusion of this technology will enable EMIS to deliver

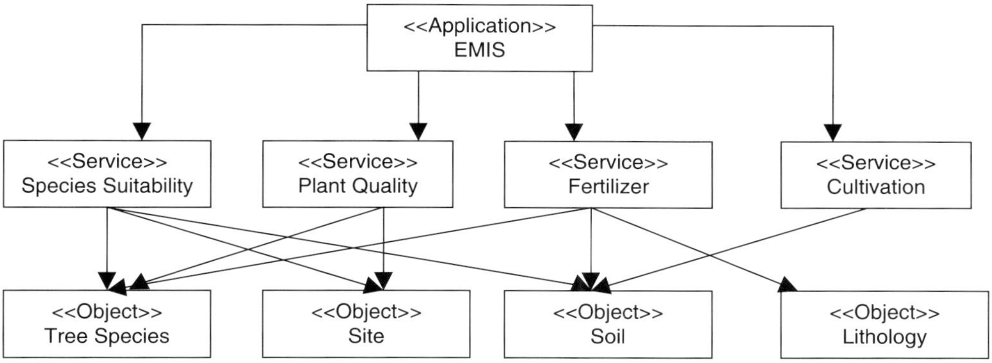

Fig. 24.7. Depiction of EMIS architecture, implemented in Java, which illustrates how various models (services) and objects (data) are composed and interact to form the EMIS DSS tool. The service layer is exposed as document-literal wrapped web services for interoperability with other applications. While EMIS uses an Oracle database, the EMIS software framework is based on open source products and standards.

decision support to both strategic and small-scale users within the British forestry sector, irrespective of their platform, because XML can be consumed by any common computer language. This architecture does mandate an Internet connection, however.

The climatic data (*c*. 34 million records) are stored at 100 m^2 resolution in an index-organized table (using an Oracle 10 g relational database), allowing query by UK Ordnance Survey 100 km tile code and six-figure grid reference. To connect to the relational database from Java components, Java Database Connectivity (JDBC) technology is used to pass Structured Query Language (SQL) queries to the database via the Oracle thin driver. The design of the EMIS software is shown in Fig. 24.7.

The object-level tier associates key parameters with suitable entities, for instance SMR with soil, although there is no decision-support functionality. The service tier (i.e. component models) is the intermediate level; it is this layer that orchestrates interactions between objects and accesses databases as necessary. The top tier is the application layer, in which we only have one application instance, namely EMIS. New DSS components can be introduced, and access any of the existing models, objects and data.

Interoperability

A key to effective decision support for ecosystem management is the interoperability of a variety of systems. Interoperability is the ability of two or more components to cooperate by exchanging services and data with one another (Twery *et al*., 2000). Interoperable systems promote communication between components and facilitate the integration of legacy and newly developed modules. EMIS displays interoperability at software and model level with the site classification DSS ESC. In the future, linkage with the *Hylobius* Management Support System (HylobiusMSS: Moore, 2004) and Herbicide Advisor tool (Thomson and Willoughby, 2004) are planned. An example of the type of linkage required is the influence of time of felling, prior species and distance to nearest clear-fell upon predicted *Hylobius abietis* (weevil) damage to newly established trees. Predicted incidence of damage may, therefore, affect selection of the appropriate seedling size class (Moore, 2004). As a consequence of this, a pop-up window regarding plant size will alert the user of the HylobiusMSS tool or preselect larger size classes if HylobiusMSS has already been run during the user session. Furthermore, strategies currently under development within forest research to avoid peaks in *Hylobius* populations may influence advice on good practice for both cultivation and herbicide use. These interactions are captured in the EMIS Simile schema, and can be implemented through linkage to a (new) *Hylobius* service.

Operational scale and use

EMIS has been designed initially for use at the stand scale. Within the British national forest estate a spatial (GIS) planning tool, Forester GIS, has been developed (Suarez *et al.*, 2003) as an extension to the ArcView GIS platform (ESRI, Redlands, California). The development of ESC as an extension to ArcView has been demonstrated, allowing the suitability of tree species to be analysed spatially using the same six site factors (Clare and Ray, 2001). This ESC–GIS model derives climate factors from a digital elevation model and calculates default values of soil quality (SMR and SNR) directly by data capture from digital soil maps. EMIS has therefore inherited this legacy, and, in recent trials, remote calls from the GIS system to EMIS modules have provided proof of concept of the interoperability of these tools, thereby enabling a spatial landscape-scale delivery of guidance on good practice to the forest planner in the future.

Whilst experienced UK foresters will have appropriate species, plant types and silviculture in mind when restocking harvest sites, EMIS may be consulted to provide a check, the added value being that any new research or guidance can be centrally updated. Forest planners may consider inappropriate species (for landscaping reasons), and EMIS would identify such instances. The guidance ensures that suitable silviculture, species and sustained yield are achieved, as highlighted by the UK Woodland Assurance Scheme (Anon., 2000), which is the UK sustainable forest management certification standard.

Future developments

The non-spatial EMIS decision-support tool described here is in an advanced stage of development and, following testing by research and field specialists, will be released in 2006. It is then intended that the silvicultural management options will be critiqued against a set of sustainability criteria, which will be developed and applied by an expert panel of stakeholders in order that the user may more clearly define and balance the ecological and production elements that underpin forest management in the 21st century. In addition, a visualization of the effects of such decisions upon establishment success and early growth is planned through integration with an initial growth model developed for the UK (McKay and Mason, 2001). Application of the system to upland afforestation sites is possible, whilst extension to other site types (e.g. ex-agricultural land) would require additional data input to existing fields (e.g. wider range of species) and expansion of the service layer. Furthermore, EMIS can utilize the ESC suitability models to provide species-specific predictions of suitability under future climate-change scenarios, which are outside the normal experiential knowledge of foresters.

References

Anon. (2000) *United Kingdom Woodland Assurance Scheme*. Forestry Commission, Edinburgh, UK.

Anon. (2002) *UK Indicators of Sustainable Forestry*. Forestry Commission, Edinburgh, UK.

Anon. (2004) *The UK Forestry Standard. The Government's Approach to Sustainable forestry*, 2nd edn. Forestry Commission and Department of Agriculture for Northern Ireland, Forestry Commission, Edinburgh, UK.

Bell, S. (1998) *Forest Design Planning: a Guide to Good Practice*. Forestry Commission Practice Guide 12, HMSO, London.

Butek, R. (2003) Which style of WSDL should I use? IBM developerWorks. www.ibm.com/developerworks/webservices/library/ws-whichwsdl/

Clare, J. and Ray, D. (2001) A spatial model of Ecological Site Classification for forest management in Britain, In: Konecny, M. (ed.) *Proceedings of the 4th AGILE Conference on Geographic Information Science, Brno, Czech Republic, 19–21 April*. AGILE, Brno, Czech Republic.

Kennedy, F. (2002) *The Identification of Soils for Forest Management*. Forestry Commission Field Guide, HMSO, London.

Lane, M.B. and McDonald, G. (2002) Towards a general model of forest management through time: evidence from Australia, USA, and Canada. *Land Use Policy* 19, 193–206.

McKay, H.M. (1992) Electrolyte leakage from fine roots of conifer seedlings: a rapid index of plant vitality following cold storage. *Canadian Journal of Forest Research* 22, 1371–1377.

McKay, H.M. and Mason, E.G. (2001) Modelling the effects of nursery and site management on the early performance of *Picea sitchensis* (Bong.) Carr. *New Forests* 22, 111–131.

Mason, E.G. (1997) Planning forest establishment operations with a computerized decision-support system: a case study analysis of decision-making over a full rotation. *New Zealand Journal of Forest Science* 26, 222–234.

Moore, R. (2004) *Managing the Threat to Restocking Posed by the Large Pine Weevil*, Hylobius abietis: *the Importance of Time of Felling of Spruce Stands*. Forestry Commission Information Note 61, HMSO, London.

Morgan, J.L. (1999) *Forest Tree* Seedlings: *Best Practice in Supply, Treatment and Planting*. Forestry Commission Bulletin 121, HMSO, London.

Nemati, H.R., Steiger, D.M., Iyer, L.S. and Herschel, R.T. (2002) Knowledge warehouse: an architectural integration of knowledge management, decision support, artificial intelligence and data warehousing. *Decision Support Systems* 33, 143–161.

Paterson, D.B. and Mason, W.L. (1999) *Cultivation of Soils for Forestry*. Forestry Commission Bulletin 119, HMSO, London.

Pyatt, D.G. and Suárez, J.C. (1997) *An Ecological Site Classification for Forestry in Great Britain: with Special Reference to Grampian Scotland*. Forestry Commission Technical Paper 20, HMSO London.

Pyatt, D.G., Ray, D. and Fletcher, J. (2001) *An Ecological Site Classification for Forestry in Great Britain*. Forestry Commission Bulletin 124, HMSO, London.

Quine, C.P. and White, I.M.S. (1993) *Revised Windiness Scores for the Windthrow Hazard Classification: the Revised Scoring Method.* Research Information Note 230, Forestry Commission, Edinburgh, UK.

Rauscher, H.M., Lloyd, F.T., Loftis, D.L. and Twery, M.J. (2000) A practical decision-analysis process for forest ecosystem management. *Computers and Electronics in Agriculture* 27, 195–226.

Ray, D. (2001) *Ecological Site Classification Decision Support System (v.1.2)*. Forestry Commission, Edinburgh, UK.

Ray, D. and Broome, A. (2003) Ecological Site Classification: supporting decisions from the stand to the landscape scale. In: *Forest Research Annual Report 2001–2002*. HMSO, London. pp. 40–49.

Ray, D., Reynolds, K., Slade, J. and Hodge, S.J. (1998) A spatial solution to Ecological Site Classification for British forestry using Ecosystem Management Decision Support. In: Abrahart, R.J. (ed.) *Proceedings of the 3rd International Geocomputation Conference, Bristol University, September 1998*. SIRC, Otago, New Zealand. 18 pp.

Smith, D.M. (1986) *The Practice of Silviculture*, 8th edn. Wiley, New York.

Smith, S.A. and McKay, H. (2002) *Nutrition of Sitka Spruce on Upland Restock Sites in Northern Britain*. Forestry Commission Information Note 47, Forestry Commision, Edinburgh, UK.

Suárez, J., Smith, S. and Ditchburn, B. (2003) The use of GIS in forestry. *Quarterly Journal of Forestry* 97, 265–270.

Sutton, R.F. (1993) Mounding site preparation: a review of European and North American experience. *New Forests* 7, 151–192.

Tabbush, P.M. (1988) *Silvicultural Principles for Upland Restocking*. FC Bulletin 76, Forestry Commission, HMSO, London.

Taylor, C.M.A. (1990a) *The Nutrition of Sitka Spruce on Upland Restock Sites*. Forestry Commission Research Information Note 164, HMSO, London.

Taylor, C.M.A. (1990b) *Forest Fertilisation in Britain*. Forestry Commission Bulletin 95, HMSO, London.

Thomson, A.J. and Willoughby, I. (2004) A web-based expert system for advising on herbicide use in Great Britain. *Computers and Electronics in Agriculture* 42, 43–49.

Turban, E. and Aronson, J. (2000) *Decision Support Systems and Intelligent Systems,* 6th edn. Prentice-Hall, New Jersey.

Twery, M.J., Rauscher, H.M., Bennett, D.J., Thomasma, S.A., Stout, S.L., Palmer, J.F., Hoffman, R.E., DeCalesta, D.S., Gustafson, E., Cleveland, H., Grove, J.M., Nute, D., Kim, G. and Kollasch, R.P. (2000) NED-1: integrated analyses for forest stewardship decisions. *Computers and Electronics In Agriculture* 27, 167–193.

Willoughby, I. and Dewar, J. (1995) *The Use of Herbicides in the Forest*. Forestry Commission Field Book 8, HMSO, London.

Willoughby, I., Evans, H., Gibbs, J., Pepper, H., Gregory, S., Dewar, J., Nisbet, T., Pratt, J., McKay, H.M., Siddons, R., Mayle, B., Heritage, S., Ferris, R. and Trout, R. (2004) *Reducing Pesticide Use in Forestry*. Forestry Commission Practice Guide 15, HMSO, London.

25 On Chatbots and Avatars – the Virtual Forester as a Guide to Knowledge about Sustainable Forest Management

A. REINBOLZ AND M. HANEWINKEL

Baden-Württemberg Forest Research Institute, Freiburg, Germany.

Abstract

A chatbot (the term is a combination of 'chat' and 'robot') is an engine that answers questions asked via an electronic medium such as the Internet in natural language. A chatbot is personalized by an avatar, a virtual animated character, and is able to use emotion as a way to transport information. A chatbot can be used to navigate through complex websites, to interact with users by answering questions, to link to large databases and to guide to a desired source of knowledge. With a chatbot, different media such as flash-animation, audio and video can be integrated on a website. This chapter presents a chatbot representing a virtual forester. It is used as a guide through a website that contains knowledge issued from a large research programme in five project regions of Germany dealing with sustainable forest management, especially forest transformation. Target groups of this website range from the general public, secondary school students (aged 13–14 years) to scientists and the press. The design and structure of the website, including virtual excursions on three different knowledge levels, a forest game for students and a scientific database, as well as teaching material for secondary schools, are outlined. The abilities of the chatbot and the usability of the website are demonstrated. Further objectives of the research, which aim at linking the chatbot to a content-management system and to an information marketplace where forest experts give advice in order to create a comprehensive knowledge management system, are presented.

Introduction

The necessity of effective knowledge transfer – lessons learned from the forest-decline disaster in Germany

In the recent past, Germany has seen many large programmes in forest research. In the 1980s, the perception of forest decline led to a huge research programme in forest sciences and related disciplines. Close to €250 million was spent on research initiatives to determine the causes of forest deterioration, which was occurring predominantly in Germany (Anders and Uekötter, 2004).

Because the majority of project finances were spent on basic research and little was assigned to professional knowledge transfer or knowledge management, the perception of results in the broader public was dominated not by reliable scientists but by the mass media. Assumptions were mixed with half-truths, and many unreliable hypotheses linking forest decline to a number of absurd reasons were publicized (Anders and Uekötter, 2004). Public perception of the scientific discourse on the topic and the communicative skills of scientists was disastrous (Zierhofer, 1998). In the end, the whole phenomenon was viewed as a cliché (Holzberger, 1995), and in many cases the research initiative was either completely neglected or judged as an overreaction (Keil, 2004). Severe doubts arose as to how scientific results can be used to advise politicians on the acceptance of scientific results in general (Pregernig, 1999; Roll-Hansen, 2002).

The result was a distinct loss of reputation for forest researchers (Keil, 2004) and a decline in research funding assigned to forest ecosystem studies. As a consequence, research budgets in the field of forest sciences were redirected during the 1990s from basic research to more applied research. The main lessons learnt from this experience were that inputs from stakeholders must be incorporated from the beginning of the project and that the methods used to transfer knowledge to different user groups are of crucial importance. One of the research initiatives that followed this new policy was the Future-oriented Forestry programme.

The Future-oriented Forestry programme

The Future-oriented Forestry programme was launched in 1999 by the German Federal Ministry for Education and Research (BMBF), and was initiated as an applied trans-disciplinary research effort that included input from stakeholders and practitioners right from the application phase of the project. From 1999 to 2003, more than 100 research projects were conducted involving around 200 researchers in five project areas distributed across Germany, reaching from the south-west (South Black Forest) to the north-east (north-eastern lowlands) of the country.

The main focus of the project was analysis of the effects of transformation and conversion of monospecific even-aged stands dominated by conifers into mixed uneven-aged forests in order to enhance sustainable forest management in Germany. The topic was investigated within various forestry disciplines and related research fields ranging from forest ecology, soil sciences, silviculture, growth and yield to socio-economics, politics and technology (von Teuffel *et al.*, 2005). The overall financial volume of the programme was around €22 million. Based on the experiences of former research programmes, where knowledge transfer was largely neglected (see 'The necessity of effective knowledge transfer'), the Federal Ministry decided to allocate a further €700,000 for the synthesis and dissemination of project results once scientific research was complete.

Objectives and Problem Definition

Objectives

How can the scientific results of a €22 million research project dealing with sustainable forest management be effectively transferred to a general audience? This study looks at how a virtual forester has been used to guide users through a website, and to explain the scientific results of the research programme 'Zukunftsorientierte Waldwirtschaft' (Future-oriented Forestry), which was carried out between 1999 and 2003. The sheer mass of results, comprising many thousands of pages, has made it necessary to break away from traditional web communication methods. Instead, techniques based on pedagogic reduction, personalization and emotionalization have been developed from the lessons learned from previous approaches, and are used to present the results.

How to present the results, and to whom

The first question posed in the synthesis and dissemination phase of the project was: To whom should the results be transferred? Aside from the traditional groups targeted by scientific projects, such as the wider scientific community and forestry practitioners, it was decided to make the results of the Future-oriented Forestry project available to decision makers in politics and society. As the Federal Ministry is also responsible for education, teachers and pupils were also identified as core target groups to be reached by Future-oriented Forestry. Additionally, it was politically decided that both the press, as an important distributor of information, and the general public should also be addressed. When considering this very broad and heterogeneous target group, it became obvious that the question of how to transfer the scientific results was of crucial importance.

Traditional information channels involving publications in peer-reviewed journals were selected as the preferred method for disseminating results to the scientific community. For investigations conducted in the Southern Black Forest project area, for example, the results were published in English in two issues of the *European Journal of Forest Research* (e.g. Bieling, 2004; Ludemann *et al.*, 2004; Heisner *et al.*, 2004; Vossbrink and Horn, 2004).

Results from the Southern Black Forest project area were additionally published by a well-known publishing company in a scientific book for forest practitioners and interested scientists (von Teuffel *et al.*, 2005). The language of this publication is German, as many forestry practitioners are not sufficiently familiar with English.

Fritz (2006) presents a summary of the major findings of the entire research programme in clear, easy to understand language, together with a glossary explaining technical terms, for decision makers in politics and administration, and information disseminators such as the press and other publishers. This publication is produced in the German language.

For the other target groups (pupils and teachers, the interested public), which are numerically the most important of the target audiences, it was clear that the

Fig. 25.1. Screenshot of start page at www.zukunftswald.de. The chatbot Horst Förster is located on the left-hand side, offering his assistance. Teaser boxes of different sizes offer highlights from the content of the page.

results could only be effectively presented through a generally accessible and widely used medium, such as the Internet. As a result, the website www.zukunftswald.de (www.futureforest.de) was created (Fig. 25.1).

As a project of knowledge transfer, the communication of the scientific results had to cope with different types of knowledge. In general terms, it is not possible to store knowledge in a database, but various forms of information (e.g. Haun, 2002). This information, used in the different media, is used to create knowledge as the user reads hypertext, interacts with the chatbot or plays the forestry game offered. Additionally, within the pedagogical concept, the desired user knowledge had been defined. In the context of this chapter, the term knowledge is used in a broad way to describe the information content of the site.

The www.zukunftswald.de website

The www.zukunftswald.de website will not be presented in detail in the present chapter. However, a brief overview of the various sections developed for each target group is outlined in this section.

A database containing information on all research projects undertaken within the Future-oriented Forestry programme has been compiled for forestry scientists. Practitioners will find summarized information on the major findings of the research project relevant to forest management. Specially designed teaching material has been developed for teachers by the University of Education in Freiburg (Stollenwerk and Rieß, 2005), and is available for download in pdf format. Students have access to web pages that have been adapted to the teaching material, and that provide background information on forest ecosystems and forestry practices. A forest management game, which allows Internet users to interactively learn interrelations between ecology, economics and risk in forest ecosystems, is also available on the site, and is mainly aimed at elementary school pupils and secondary school students. Pupils and the broader public should also be attracted by three virtual excursions offering information on forests, forestry practices and forest research. The excursions have been designed using Flash animation technology, and contain interactive graphical displays.

This short description of the complex website reveals the multitude of approaches to knowledge transfer required to reach so many different target groups. It also highlights the need for new ways of navigating websites beyond traditional tools such as search engines or site maps. Without this, good usability and easy orientation for users would have been hard to accomplish. Unfortunately, this is a common problem that many nature and science websites have had to struggle with (Plieninger and Reinbolz, 2003).

More importantly, the subject matter must be made as interesting as possible to reach the broad public and not just a small group of forestry specialists. The material must stand out from other categories found on the Internet, such as health, sport or entertainment, to attract the widest possible audience. The core question is how to interest people in issues concerning forests and forest management on the Internet.

Getting people interested in forestry issues

There are many ways to generate interest in scientific issues. The most efficient methods are those that convince people that the subject is of great importance to society and to them as an individual. Medical issues, for example, are relatively easy, because subjects dealing with health are usually of high importance to the public. Generating interest in forestry and forest sciences, however, is much more difficult. Despite the almost mythical relationship between Germans and their forests (Pogue Harrison, 1992), forestry is at present of little economic importance and does not really attract public attention to a large extent. This was certainly different during the 'forest-decline' campaign, which seized the German public's attention for a couple of years at the beginning of the 1980s. The forest-decline

Table 25.1. Examples of Horst Förster's dialogue on www.zukunftswald.de (translated from German) with associated visualization.

Bot image	Dialogue	Explanation
	Good afternoon! My name is Horst Förster and I will try to answer all your questions on forest conversion. Just type your question into the form underneath my picture. What is this all about?	Opening sentence – adjusted according to the time of the day.
	The results of the research programme Future-oriented Forestry are presented on this website. The main topic is forest conversion and the ways in which sustainable forest use protects the environment and soil. More on this topic is displayed on the right-hand side. Would you like to go on a virtual field trip with me? I would love to!	The bot recognizes the relatively broad question, and gives some general information. It also proactively offers an excursion, which is a good starting point to learn more about forests.
	If you have any further questions, just type it into the form and I'll be back. Is Germany going to win the world championship?	The bot recognizes the follow-up and disappears, before reappearing in the adjacent flash window as the excursion opens.
	Sorry, I don't know anything about this. But may I tell you something about biodiversity? A lot of very interesting research has been done on this topic. OK.	The bot does not recognize this question and, instead of answering, actively offers a topic it prefers to talk about.
	Biodiversity describes the variety of living species. While three levels are usually defined, they all overlap: genetic diversity, species diversity and ecosystem diversity. In this research project, we discovered that forests which are close to nature can have an optimal, but not always the maximal biodiversity. [...] More about this?	Again, the follow-up is recognized and the bot talks about biodiversity. For further information, the bot opens an appropriate page on biodiversity on the website.

phenomenon, however, was partly presented with horror scenarios, based on largely exaggerated and false predictions of the complete extinction of European forests within a short period of time (Holzberger, 1995). As a result, the whole affair, and thus forestry topics in general, was linked to negative feelings. This first led to a suppression of the topic, and later to the almost complete disappearance of the subject at the beginning of the 1990s, with the exception of some groups of environmental activists.

An effective way of making scientific issues interesting to the public is to link them to positive feelings (Reinbolz and Pretzell, 2003). Once a website is able to provoke passion or enthusiasm for a topic, the number of people visiting the site will increase and public attention will also rise. Emotions are therefore a key to the user's attention. In order to link knowledge about forests and forestry to positive feelings and emotions, it is necessary to present the information in a personalized manner. The virtual person presenting the information has to be visually attractive, and must be able to make contact with the user of the website. Furthermore, if scientific knowledge is to be communicated, the 'person' presenting the information must to be credible. A virtual person, known as an avatar, can be integrated into a website to achieve such a goal. The interaction between the web user and the avatar is realized with the help of chatbot technology.

The Approach – Chatbots and Avatars

From avatars to chatbots

The term avatar has a religious background. The word is of Sanskrit origin, where Avatara means: 'the descent of God',[1] or simply 'incarnation'. In English, the word has come to mean an embodiment, a bodily manifestation of the Divine. A newer interpretation of the term sees an avatar as an artificial person, or a graphical representation of a real person in a virtual world, for example in a computer game. Avatars are depicted as pictures or icons or as 3D figures of human, or any other, beings. The usage of the term avatar in this context was made popular in 1992 by Neal Stephenson in his science fiction novel *Snow Crash*.[2] For the present task of transferring knowledge on forests and forestry, an avatar in the form of a virtual forester was determined to be the most appropriate approach.

Chatbot technology allows an avatar to communicate with the website user by answering questions or by assisting navigation through the website. The term chatbot is a hybrid of the words 'chat' and 'robot'. The chatbot unifies the necessary personalization and visual attractiveness of the avatar in order to provoke emotions when transferring information, and provides the avatar with the capability to reliably answer questions and to interact directly with the web user.

How chatbot technology works

The birth of chatbot technology is often cited as the ELIZA programme by Joseph Weizenbaum (1976), who set up a virtual psychotherapist at MIT between 1964 and 1966. In a similar way to current systems, this chatbot used a pattern-matching input analysis and a series of templates for responses. Chatbot technology at that time was seen as an opportunity to study and develop artificial intelligence applications. Competitions based on the famous Turing Test (Turing, 1950) and the Loebner prize[3] evaluate the grade of intelligent behaviour and award prizes to the most humanoid chatbots.

This has often led to misunderstandings within the scope of artificial intelligence, because chatbots still perform relatively poorly and can seldom simulate real human behaviour (Storp, 2002). Nevertheless, chatbots are successfully used in many other areas where advanced interactivity is more important than real intelligence. On websites, chatbots provide a functionality far beyond other means of navigation and interaction, and are used as virtual customer-service agents. To understand the difference between these approaches, the composition of a common chatbot should be considered.

A modern chatbot consists mainly of three elements: front-end, software and database back-end, which is often called the knowledge base. The front-end is the face of a chatbot. This is where the user communicates with the bot, where responses are displayed and where emotional interaction takes place. The front-end must therefore reflect the intended personality of the virtual contact person. The front-end usually comprises a form field for user input, a textbox for responses and a visual representation of the avatar, implemented as a flash movie or as static pictures. Visual representations can be altered in accordance with corresponding responses to add an emotional feedback that written text alone could not provide.

Once the user poses a question, the content of the form field is submitted to the chatbot software. The software is usually implemented as a program on the web server to access a database with answers and recognitions (the knowledge base). The term recognition in this context means a programmed expression for the recognition of pattern. Using these patterns, the chatbot can analyse the user input.

Depending on the software, different steps are performed to provide an adequate reaction to user input. The functionality of the Lingubot[4] engine by Kiwilogic, which has been used in this project, is briefly described in the following section (Kiwilogic, 2003). Other software packages may use slightly different procedures. However, the main principles are comparable.

In the so-called preprocessing phase, the given input is initially partitioned into sentences and then single words. The software then attempts to adjust capitalization, spelling and umlauts, before commencing the process of pattern matching. Every predefined answer has a set of recognitions stored with it in the database. A recognition could be as simple as 'includes the words "forest" and "deciduous"' but can also include more complex expressions concerning the total number of words, the distance between terms and logical operations. All answers have a predefined rank, which allows the software to return the match with the highest rank in cases where more than one recognition matches the user input. A powerful feature of the system is the ability to store recognitions in so-called macros. A macro could, for example, cover a predefined list of words used for forest topics such as forest, wood, bark and grove. Very complex recognitions can be defined by nesting these macros, as illustrated in the following example: 'Is it useful to regenerate my spruce/pine forest with beech?' (German: 'Ist es gut, meinen Kiefernwald/Fichtenwald mit Buchen zu verjüngen?'). The recognition for this question looks like the following:

> %ZW_BUCHEN&(%ZW_KIEFER/%ZW_FICHTE/%ZW_WALDUMBAU/gut/besser)&(%ZW_ANPFLANZEN/%ZW_NATÜRLICHE_VERJÜNGUNG)

The main elements in this recognition are macros covering predefined expressions, identifiable by the prefix %. The prefix ZW_ was added for clear separation of the different macro libraries of this project from more general libraries. In this case, the given question must contain something about beech trees (%ZW_BUCHEN). The input must contain something about pine trees (%ZW_KIEFER, %ZW_FICHTE), or some form of positive connotation (gut, besser – good, better). The third element required for a positive recognition is the context regeneration, which is represented by the macros %ZW_ANPFLANZEN (planting) and %ZW_NATÜRLICHE_VERJÜNGUNG (natural regeneration).

While the macros used in the database are generally just collections of words, they can also integrate a series of macros or more sophisticated macro combinations. The macro %ZW_BUCHEN used above reads as follows (other macros are made up similarly):

> Buche/Buchen/Buchenart/Buchenarten/Buchenbestand/Buchenbestandes/Buchenbestände/Buchenbeständen/Buchenforst/Buchenforste/Buchenforsten/Buchenkultur/Buchenkulturen/Buchenwald/Buchenwaldes/Buchenwälder/Buchenwäldern/((buchendominiert/buchendominierte/buchendominiertes/buchendominierter/buchendominiertem/buchendominierten/buchenreich/buchenreiche/buchenreicher/buchenreichen)&(%ZW_WALDBESTAND/%ZW_MISCHBESTAND))/(fagus+sylvatica)/rotbuche/rotbuchen

The system of ranking predefined answers allows the bot to return very specialized responses when the recognition is very specific, and to use broader answers as a fallback in cases where the question is more vague in nature. If the question is not recognized at all, so-called catch recognitions are used to bring the user back to a topic with which the chatbot is familiar or to simply continue the dialogue.

To simplify the creation of a new chatbot, nearly all products come with a predefined database of recognitions for small talk and common dialogue situations. Answers can either be static or be connected to expected follow-ups, trigger further actions or link dynamic data from a database. Once the recognition process has been completed, a response is returned to the user.

A chatbot is usually set up using an authoring tool that allows the developer to keep track of all recognitions, answers and actions. While open-source chatbots offer a high level of functionality, the tools available for authoring commercial chatbots are far more sophisticated.

Capabilities and limits

Modern chatbots offer a high level of interoperability. All the commands that can normally be incorporated into a web link can also be used by the chatbot to communicate with other applications or media. A chatbot can, for example, control flash animations, websites or media to be displayed. Interfaces with database servers are also very common. An integrated scripting language allows the designer to influence the process at virtually any point during preprocessing, recognition and output generation. Outputs can cover not only predefined answers, but also calculations and even real-time data, such as 'How cold is it?' By predefining short dialogues in a hierarchical manner, the chatbot can also ask for more information about the user and their expectations. All information gathered by the chatbot can be stored internally for reuse in dialogues, scripts or virtually any other manner.

As another feature, the chatbot can open web pages associated with the answer. These pages can be static, for example a corresponding article, but can also be dynamic. A possible application would be a dynamic search page controlled by the chatbot and using information collected during the chat. With these techniques, logically sophisticated but intuitively usable navigation can be realized.

The knowledge base may experience some difficulties in certain circumstances because the chatbot cannot distinguish between questions it does not understand and questions without answers, due to the pattern-matching process used for recognition. As a consequence, the chatbot does not know what it does not know, which can be problematic when a chatbot attempts to find an answer in the database.

Example – the Horst Förster Chatbot as a Virtual Forester

Design and implementation

Within the scope of the present research project, the chatbot on the Future-oriented Forestry website was designed both as a site guide and as a virtual expert on forest conversion. As a result, a number of special needs had to be met while designing the avatar and its integration into the website.[5]

During the design process, the avatar was adapted to the expectations of the target group. As many studies have previously shown (e.g. Institut für Forstpolitik, 1999), a forester is widely associated with competence in matters concerning forests and forestry. As a result, the avatar was designed as a virtual forester. The model was selected according to additional attributes, such as being down to earth, friendly and candid, in order to emphasize a positive image. It was agreed that the model should be about 30 years of age to assure a high level of knowledge without losing contact with younger members of the target audience. The model is dressed with a shirt and a green scarf evocative of a forester's uniform, and also reminiscent of a scout uniform, which prevents the model's image from being too formal. The figure was named Horst Förster (forester) to imply the avatar's profession. Twenty-seven movie clips capturing a range of different moods were recorded in a film studio with a digital video camera and converted into Flash clips.[6]

Documents concerning the research project were collected and fractionized into FAQs (frequently asked questions) for the knowledge base. This collection of about 500 questions and answers was used as a base for the scientific knowledge. The collection was increased by a further 200 questions and answers to cover questions expected on more general aspects of forestry. The basic communication capabilities of the bot were provided by a library of default recognitions and responses included with the software, which had to be largely modified to meet the avatar's personality. A framework of recognitions was developed to cover the complete collection of questions and answers. The rest of the website was designed at the same time, using a simple content-management system (CMS). The site's content was then input into the chatbot's knowledge base, allowing it to help users navigate through the site.

Networking the chatbot and content

One of the main features of the site is a virtual excursion application, which is implemented as interactive Flash animations using real pictures and videos, and allows the user to visit three different forests. Horst Förster is used as a guide in these excursions in order to emphasize the competence of the avatar. While the figure is acting in this role, the chatbot window is empty, except for a sign displaying 'I'm next door' (German: 'Bin nebenan'). During this time, the chatbot functionality is still active and the bot can answer questions during the excursion. The bot can also actively propose to commence an excursion and can even show certain sections of the excursion if they help to answer the user's question. Horst Förster also presents the forest simulation game incorporated in the website.

Testing and optimization

An extensive optimization process is crucial to achieving high recognition rates for chatbots. A two-staged optimization process was used in this project, once the chatbot was judged to be ready for regular use. The first stage involved a group of ten target-group users, who were instructed to test the bot in multiple dialogues using ten questions. The dialogues were analysed from log files to eliminate severe mismatching and address knowledge gaps. In the second stage, the site was tested again, with further mismatching and knowledge gaps being identified and revised. After setting the site online, the chatbot logs were analysed continuously. Three types of result could be achieved from these complete conversations, and have been used for optimization:

1. Error correction. Due to the complexity of hundreds of recognitions, interferences and mismatching could not be eliminated completely during development. By manually checking user questions and the given chatbot answers, the majority of these errors could be eliminated.

2. User feedback. Analysing the dialogues provides information about user behaviours, types of user questions and topics with limited chatbot knowledge. Based on this information, about 200 new questions and answers have been added, providing more information and an improved chat experience.

3. Quality statistics. Chatbot server logs provide detailed information on recognition quality. According to these, about 75% of all user inputs are answered correctly by the bot.

Discussion and Outlook

Opportunities for technology

Chatbot technology was used in this project for emotionalization and user guidance. Although this intuitive way of navigating the website provided many advantages, many possibilities have not been thoroughly explored. A chatbot could help users find relevant data, such as documents in large databases. This is particularly helpful for web users without any knowledge about the structure of the data provided, as is often the case with knowledge repositories. In this case, the chatbot could be connected to server-based agents that use interactively gained information about the user's needs to assemble a tailor-made package of information. Such a need-based access system could be used to provide improved support for forestry practice and to connect forestry professionals with experts.

Emotionalizing knowledge

Publicity will become an increasingly essential aspect of research activities in the future, because it determines the impact of research on society, plays an important role in the success of fund-raising initiatives and is instrumental in winning respect within the scientific community. These factors are particularly important in the present context of declining public funding. To maintain sustainability in forestry, public awareness is crucial because it is a key factor for the embedding of forestry in society. Scientists from different fields compete not only with each other for attention in the media but also with a tremendous flood of news, advertising and entertainment. Especially with regard to forest sciences, there is often something more exciting (space flights, aeronautics), of higher personal impact (finances, medicine) or simply more spectacular. However, none of these fields is closer to the daily experience of people. In order to take advantage of this fact, a positive emotional context can strengthen this personal contact and enhance communication between scientists, forestry professionals and the general public. As initial experiments with students indicate, Horst Förster is widely accepted as a competent and friendly dialogue partner. He provides orientation on the website and expands the user's navigation experience with an emotional dimension. More efforts will be necessary to take full advantage of personalization in all forms of communication with the public. As this new approach has proved to be effective, the forestry community will have to adapt and become – despite this being sometimes difficult for scientists – more emotional.

Notes

1. http://www.avatara.org/essay.html
2. http://de.wikipedia.org/wiki/Avatar
3. http://www.loebner.net/Prizef/loebner-prize.html
4. 'Lingubot' is a registered trademark of the Kiwilogic AG Hamburg.
5. The concept and design of the 'Horst Förster' avatar were developed by the authors in association with Pixelwings Berlin/Vienna. Programming and implementation were carried out by Pixelwings (www.pixelwings.com).
6. Macromedia Flash is a quasi-standard for displaying animations on the web.

References

Anders, K. and Uekötter, F. (2004) Viel Lärm ums stille Sterben: Die Debatte über das Waldsterben in Deutschland. In: Uekötter, F. and Hohensee, J. (eds) *Wird Kassandra heiser? Die Geschichte falscher Ökoalarme.* Steiner, Stuttgart, Germany, pp. 112–138.

Bieling, C. (2004) Non-industrial private-forest owners: possibilities for increasing adoption of close-to-nature forest management. *European Journal of Forest Research* 123, 293–303.

Fritz, P. (2006) *Ökologischer Waldumbau in Deutschland–Fragen, Antwosten, Perspektiven.* Oelsom Verlag, München, Germany.

Haun, M. (2002) *Handbuch Wissensmanagement – Grundlagen, Umsetzung und Praxis beispiele.* Springer, Berlin and Heidelberg.

Heisner, U., Raber, B. and Hildebrand, E.E. (2004) The importance of the soil skeleton for plant-available nutrients in sites of the Southern Black Forest, Germany. *European Journal of Forest Research* 123, 249–257.

Holzberger, R. (1995) *Das sogenannte Waldsterben: Zur Karriere eines Klischees. Das Thema Wald im journalistischen Diskurs.* Eppe, Bergatreute, Germany.

Institut für Forstpolitik, University of Freiburg (ed.) (1999) *Repräsentative Bevölkerungsumfrage in der BRD zu den gesellschaftlichen und politischen Bedingungen für die Entwicklung von Laubwälder.* University of Freiburg, Freiburg, Germany.

Keil, G. (2004) Chronik einer Panik. *Die Zeit* 51 (9 December), 66–67.

Kiwilogic (2003) *Lingubot-Schnittstellen und ihre Einsatzmöglichkeiten.* Training document. Kiwilogic, Hamburg, Germany.

Ludemann, T., Michiels, H.-G., Nölken, W. (2004) Spatial patterns of past wood exploitation, natural wood supply and growth conditions: indications of natural tree species distribution by anthracological studies of charcoal-burning remains. *European Journal of Forest Research* 123: 283–292.

Plieninger, T. and Reinbolz, A. (2003) Deutsche Nationalparks im Internet - Eine Evaluation von Benutzerfreundlichkeit und Informationsgehalt. *Naturschutz und Landschaftsplanung* 35: 235–241.

Pogue Harrison, R. (1992) *Wälder: Ursprung und Spiegel der Kultur.* Hanser, Munich and Vienna.

Pregernig, M. (1999) *Die Akzeptanz wissenschaftlicher Erkenntnisse: Determinanten der Umsetzung wissenschaftlichen Wissens am Beispiel der österreichischen 'Forschungsinitiative gegen das Waldsterben'.* Europäische Hochschulschriften 22 (339), Soziologie 339, Lang, Frankfurt am Main, Berlin, Berne, Brussels, New York and Vienna.

Reinbolz, A. and Pretzell, D. (2003) Strategische Öffentlichkeitsarbeit: Mehr Aufmerksamkeit für den Naturschutz. *Jahrbuch für Naturschutz und Landschaftspflege* 53, 259–272.

Roll-Hansen, N. (2002) *Ideological obstacles to scientific advice in politics? The case of 'forest death' and 'acid rain' – Makt- og demokratiutredningen 1998–2003*. Rapportserien Nr. 48, November, Oslo University, Oslo.

Stollenwerk, N. and Rieß, W. (2005) Unterrichtsmaterialien für Lehrer zu Wald und Forstwirtschaft, University of Education, Freiburg. Available online at http://www.zukunftswald.de

Storp, M. (2002) Chatbots. Möglichkeiten und Grenzen der maschinellen Verarbeitung natürlicher Sprache. *Networx* 25 (online: http://www.websprache.net/networks/docs/networx-25.pdf). Access date: 1 June 2005.

Turing, A.M. (1950) Computing machinery and intelligence. *Mind* 59, 433–460.

von Teuffel, K., Baumgarten, M., Hanewinkel, M., Konold, W., Spiecker, H., Sauter, H.-U. and von Wilpert, K. (eds) (2005) *Waldumbau für eine zukunftsorientierte Waldwirtschaft*. Springer, Berlin and Heidelberg.

Vossbrink, J. and Horn, R. (2004) Modern forestry vehicles and their impact on soil physical properties. *European Journal of Forest Research* 123, 259–267.

Weizenbaum, J. (1976) *Computer Power and Human Reason. From Judgement to Calculation*. Freeman and Company, San Francisco, California.

Zierhofer, W. (1998) *Umweltforschung und Öffentlichkeit. Das Waldsterben und die kommunikativen Leistungen von Wissenschaft und Massenmedien*. Westdeutscher Verlag, Opladen, Germany.

26 Information and Knowledge Management in Support of Sustainable Forestry: a Review

H. MICHAEL RAUSCHER,[1] DANIEL L. SCHMOLDT[2] AND HARALD VACIK[3]

[1]USDA Forest Service, Southern Research Station, Asheville, North Carolina USA; [2]USDA Cooperative State Research, Education and Extension Service, Washington, DC USA; [3]Institute of Silviculture, Department of Forest and Soil Sciences, University of Natural Resources and Applied Life Sciences, Vienna, Austria.

Abstract

For individuals, organizations and nations, success and even survival depend upon making good decisions. Doing so can be extremely difficult when problems are not well structured and situations are complex, as they are for natural resource management. Recent advances in computer technology coupled with the increase in accessibility brought about by the Internet have increased our ability to solve complex problems in natural resources. Scientific disciplines that have evolved to exploit this new computer-based technology include knowledge management and decision science. Impressive computer-based systems have been developed, but their use in natural resource management has been limited. Widespread adoption will require close cooperation among people working in research as well as management, across disciplines, up and down the administrative structure in state and federal agencies and in the private sector.

The objectives of this chapter are: (i) to briefly review the history and recent advances in natural resource information and knowledge management, including decision-support systems and multiple-criteria decision making; (ii) to discuss some of the interrelationships among inventory and monitoring, statistics and modelling, information and knowledge management and policy science; and (iii) to offer some ideas on how to best support sustainability as a forest management paradigm, given the new capabilities afforded by progress in information and knowledge management.

Knowledge Management

Overview

Until fairly recently, many people did not think in terms of 'managing knowledge'. They felt that knowledge was a personal asset accumulated from experiences,

 Sustainable Forestry: from Monitoring and Modelling to Knowledge Management and Policy Science (eds K.M. Reynolds, A.J. Thomson, M. Köhl, M.A. Shannon, D. Ray and K. Rennolls)

education and trusted colleagues (Plunkett, 2001). As computer technology improved and became cheaper in the early 1990s, researchers began to explore the gains that could be made by organizing knowledge, codifying it and sharing it more widely. Innovators demonstrated that improving the management of knowledge could: (i) help scientists improve communication of research results to users (Rauscher, 1987); (ii) help government cope with downsized budgets and increased work (Plunkett, 2001); and (iii) help private industry to gain competitive advantages (Heinrichs *et al.*, 2003). The advancement of information management technologies presented new opportunities for business and governmental organizations. In some cases, the implementation of information technologies represented major changes for organizations, as knowledge was now viewed by some as a resource, much like facilities, finances, equipment or workers (Nesbitt *et al.*, 1996; Evans and Wurster, 2000). Some existing applications of information technologies for managing – primarily natural resources – knowledge appear in Table 26.1.

Knowledge exists in either explicit or tacit states. Explicit knowledge is that which has been codified in some way, such as in scientific journal articles, operating procedures, best management practices and simulation models. Tacit knowledge is that which people carry in their minds. It consists of facts, opinions, intuition, feelings and judgements. People seldom fully understand their own knowledge stores. As Polyani (1958) said, 'We know more than we know how to say.'

Knowledge management (KM) can be defined as the systematic strategy of creating, conserving and sharing knowledge to increase performance (Plunkett, 2001; Heinrichs *et al.*, 2003). KM provides methods for managing both explicit and tacit knowledge. Some methods help people to exchange knowledge. Others make existing explicit knowledge more readily accessible (Hansen *et al.*, 1999). But KM also concentrates on methods that help to codify tacit knowledge so that it can be converted to explicit knowledge for general use (Heinrichs *et al.*, 2003). Nonaka and Takeuchi (1995) describe four processes for conversion of knowledge from one form to another:

- Socialization. Tacit knowledge is shared through shared experiences.
- Externalization. Tacit knowledge is articulated into explicit knowledge.
- Combination. Explicit knowledge is organized, systematized and refined.
- Internalization. Explicit knowledge is converted into tacit knowledge.

Knowledge about natural resource management is multifaceted and spans the biological, physical and social sciences (Simard, 2000; Innes, 2003). Such knowledge includes facts, propositions, laws and theories that provide general knowledge about the behaviour and functioning of ecosystems and their interactions with social systems. It also includes knowledge about places, events at specific times and implications for management.

KM uses information technology to identify, create, structure and share knowledge, with the goal of improving decision making (Tyndale, 2002). A number of technologies commonly associated with the term 'knowledge management' have been evaluated for their potential to support management processes (Ruggles, 1997; Plunkett, 2001; Tyndale, 2002). Table 26.1 provides a

Table 26.1. A classification of types of knowledge management tools.

Class of KM tool	Description	Links
Knowledge maps	Establish a classification scheme called a taxonomy of knowledge, provide a frame of reference for many knowledge management products, and serve as a critical first step for identifying available knowledge.	forest.cse.ogi.edu/portal cmap.ihmc.us
Electronic yellow-page directories	Aid in finding hard-to-access tacit knowledge resources by providing access to experts. They also organize existing websites and serve up a variety of explicit knowledge assets in understandable ways.	sref.info/www.forestryguide.de/ www.srs.fs.usda.gov/
Apprenticeship programs	Are typically one-to-one type relationships where an expert coaches a less experienced person in various forms.	www.treeguide.com/forum/
Communities of practice	Support groups of individuals with similar work responsibilities but who are not part of a formally designated work team. Many communities of practice communicate through a web-based system.	groups.yahoo.com/group/ dead_wood/dss.boku.ac.at/
Best practices and lessons learned	Typically present the situation, the options, choices taken and the results for a typical decision problem. They are widely used in natural resource management and can be extensively found on the Internet.	www.kyphilom.com/ www/wood/bmp.html www.forestrybmp.net
Lectures and storytelling	Allow people to gain more understanding and have greater recall then they do from written reports. Stories can be used to capture lectures on a particular topic, to capture after-action reports, to record difficult-to-codify tacit knowledge, and for many other purposes. Web-based software systems exist that support this knowledge management tool.	www.fsl.orst.edu/geowater/ morphology/
Frequently asked questions	In the course of performing a job, people naturally identify questions that their co-workers or their clients ask repeatedly. It is worthwhile to document and develop useful and standardized answers for these types of repetitive questions. Web-based systems also exist that specialize in the management of these questions.	www.answerlink.info

Continued

Table 26.1. *Continued.*

Class of KM tool	Description	Links
Web-based learning	Allows translation of a typical classroom experience to an online media to offer students the opportunity to learn codified knowledge in a structured way at their own pace.	www.forestandrange.org waldbau.boku.ac.at/lehre/
Scientific content management sites	Collects knowledge in some kind of web-based content management system. First, the knowledge has to be found, organized, synthesized, reviewed for quality and uploaded for availability. Secondly, the knowledge content has to be updated and maintained so that it keeps its currency. Software systems exist that support both of these functions.	forestencyclopedia.net www.waterontheweb.org www.cabi.org.compendia.asp
Simulation models	Mathematical models are a popular way to organize specific problem-solving knowledge and provide precise, quantitative answers to guide natural resource managers. Most such models have not yet been converted to be executed over the Internet; however, many simulation models can be downloaded from the Internet and then executed on a stand-alone computer.	www.fs.fed.us/fmsc/fvs/ www.cnr.usu.edu/online/simulation/
Free-content information collaboratories	Create and distribute free information content, e.g. encyclopedia. Articles are edited by volunteers and are subject to change by nearly anyone. They cover a wide range of topics, but lack the authority of traditional materials and lack the opportunity for quality control regarding the content.	wikipedia.org
Time maps	A visual-matrix index of the events, research topics, people and publications, organized by time, for a specific area. An electronic zoomable canvas allows the embedding of a large amount of information in a single plane.	www.fsl.orst.edu/geowater/timemaps/lter/

Databases	A common way to organize original source material is a database structure. It is irrelevant whether the data are numerical or graphic or computer files. Web-based methods have been developed to manage databases online. Spatial databases, as in geographic information systems, are included here.	www.archives.gov/aad/
Metadata	Databases make a great deal of information and knowledge readily available, but they don't contain any knowledge about those data, i.e. who, when, how, etc., that indicates their appropriate use.	www.nbii.gov/datainfo/metadata/
Library services	Managing and making accessible published books and scientific journal articles have long been the province of science libraries. These services are also available on the Internet either free of charge or fee-based.	www.srs.fs.usda.gov/pubs/index.htm www.waldwissen.net
Online scientific journals	More and more scientific journals have placed all or part of the content of their original research articles online. Search engines allow the user to find relevant articles and the number of citations referring to them.	www.fbmis.info/ www.scirus.com
Web portals	Provide links to many other sites that can either be accessed directly or be found by following an organized sequence of related categories. The provider of a web portal is responsible for structuring and filtering of web addresses relating to a special theme.	frames.nbii.gov

partial list of some web-based methods that are currently being used for KM in the natural resource field; this list is not intended to be exhaustive.

One example of the application of knowledge management tools has occurred in the area of scientific content management. Other KM tools in Table 26.1 have similar histories but, for brevity, are not reviewed here. Rauscher (1987) introduced the concept of modern knowledge management to the natural resource field in the same year that the first hypertext software programs became available for IBM[1] PC and Apple Macintosh[1] computers. Rauscher (1991) then provided the first electronic hypertext encyclopedia in forestry – 'The encyclopedia of AI applications to forest science'. The purpose of this encyclopedia was to demonstrate the functional difference between electronic hypertext and print-based methods by taking the same content published in the scientific journal *AI Applications* and providing it on a disk as an insert for that issue. Other hypertext products for non-networked personal computers followed in rapid succession: 'Managing the global climate change scientific knowledge base' (Rauscher *et al.*, 1993); 'Computer-assisted diagnosis using expert system-guided hypermedia' (Thomson *et al.*, 1993); 'A hypermedia reference system to the forest ecosystem management assessment team report' (Reynolds *et al.*, 1995); and 'Oak regeneration: a knowledge synthesis' (Rauscher *et al.*, 1997) among others.

As the Internet became more popular, it was obvious to some that KM systems using web-based hypertext had an enormous competitive advantage over stand-alone systems. Saarikko (1994) authored an early comprehensive summary of forestry information resources available on the Internet. He concluded in 1994 that Internet activity had been growing exponentially and that such growth would continue. In a pioneering effort, Thomson *et al.* (1998) combined knowledge-based systems processing and a hypertext user interface (HTML) to provide forest tree disease diagnosis over the Internet. A primary benefit of this approach was that anyone with a web browser could access the diagnostic software from any Internet-connected computer. Universally available access and inexpensive updating appear to be the critical elements for making scientific content management in natural resource management an attractive alternative to traditional, paper-based methods. Examples of scientific content management for natural resources on the Internet can be found at the Forest Encyclopedia Network, which contains a growing number of scientific encyclopedias (Kennard *et al.*, 2005). More and more knowledge management services of different types are appearing at a dizzying rate.

Future directions

Plunkett (2001) observed that 'Knowledge management consists of three fundamental components: people, processes, and the supporting technology.' As extensive web-based KM sites illustrate, we have progressed quite dramatically in our technical capabilities. Within the context of natural resources, it is much less clear that we have advanced equally far in changing our institutional processes and educating our people to use KM efficiently. It is challenging for organizations to properly value and use both tacit and explicit knowledge in the management of environmental problems while promoting a climate of learning that encourages the recognition and sharing of employees' experiences (Boiral, 2002). While we know of no study to objectively document the low level of use of KM tools in the natural resource field, anecdotal evidence suggests that many workers are very reluctant to use them. This may be a more serious problem for the public sector than the private sector, where organizational policies can be enforced more rigorously. In any case, a thorough evaluation of how natural resource management and science institutions support KM processes and how workers in this field use the available KM tools does not currently exist.

An effective KM strategy for an organization might have the following goals:

- All new employees are introduced to the KM strategy and their roles as consumers and contributors of knowledge.
- All employees are part of an explicit KM system for their entire professional life and are rewarded appropriately for how well they share their tacit as well as explicit knowledge.
- Processes for KM exist and are strongly supported at all levels of an organization. Among these processes are periodic stocktaking assessments, after-action reporting, mentoring and apprenticeship programmes, storytelling opportunities, effective and active communities of practice, periodic assessments of both internal and external client groups, and visible and vocal leadership endorsement of all effective KM processes.
- KM support technologies (see Table 26.1) are available and effective. They are used routinely by all members of an organization.
- Mission-critical knowledge has been identified, structured and codified. It is readily available. A system exists to monitor the availability and quality of this mission-critical knowledge, either tacit or explicit.
- Security of confidential and proprietary information is high where necessary. Knowledge is routinely re-evaluated to make sure that those people who should have access do have access.

KM is a young discipline, so a generally accepted framework for it has not yet been established. Instead a variety of approaches to KM have been implemented across a variety of organizations (Rubenstein-Montano *et al.*, 2001). Knowledge gains economic value when it is used to solve problems, explore opportunities and make decisions that improve performance. Hence, the problem-solving process is the vehicle for connecting knowledge and performance (Gray, 2001). Ways must be sought to enhance and promote KM in organizations (Girad and Hubert, 1999), because KM increases efficiency and generates value. These are important whether the organization is a commercial enterprise trying to reduce costs or a government agency competing for shared funding resources.

> We must remember that knowledge is only in the past tense. Learning is only in the present tense, and prediction is only in the future tense. To manage sustainable forests, we need to be able to know, to learn, and to predict.
>
> Chris Maser, 1994

Decision-support Systems

Overview

For the purposes of this discussion, we adopt the definition of decision-support systems (DSSs) from Holsapple (2003, p. 551): 'A computer-based system composed of a language system, a presentation system, a knowledge system, and a problem-processing system whose collective purpose is the support of decision-making activities.' Two key attributes in Holsapple's definition are a problem-processing system and purposeful support of a decision-making process. A decision-making process is a method that guides an individual or group through a series of tasks from problem identification and analysis to design of alternatives and selection of an alternative (Mintzberg *et al.*, 1976).

> Systems that generally fulfil the Mintzberg and Holsapple definitions include multi-criteria analysis techniques like the analytic hierarchy process, Electre, and Promethee; knowledge-based systems that provide a framework for applying procedural or reasoning knowledge to decision problems; and, perhaps somewhat more arguably, optimization systems including linear or goal programming. While geographic information, spreadsheet, and database systems may be critical components of a DSS, it stretches the definition of a DSS beyond usefulness to classify these types of applications as DSSs.
>
> (Reynolds, 2005)

Numerous simulation systems have been developed to support many aspects of forest planning (Schuster *et al.*, 1993; Mowrer *et al.*, 1997), but most should be considered as potential tools in a DSS framework as opposed to DSS *per se* (Reynolds, 2005). In most cases, these systems aid in the analysis phase of the decision process, but fail to directly support the other decision steps.

The adaptive management process provides the theoretical framework for most modern DSSs (Rauscher, 1999; Reynolds, 2005). Adaptive management consists of four components (Walters, 2002):

- Planning. The planning and decision-making process of Mintzberg *et al.* (1976) is a useful supplement to the adaptive management model, the end point of which is a set of goals and constraints that guide the action phase of the process. This phase prescribes how, when and where to implement activities to achieve the identified goals within the constraints. Identifying suitable hypotheses concerning expected outcomes of actions is critical at this stage to support the experimental nature of management actions and promote efficient learning. KM tools may support the identification of possible treatments and likely outcomes.
- Implementation. This phase puts the plan into action. Each activity implemented on the landscape must be adequately documented so that it is clear what the conditions prior to action were and how the management activity changed those prior conditions.
- Monitoring. This phase requires periodically examining the implemented management activities and recording current conditions over time. Adequate documentation must be developed because different people may need to

use this information over many years in order to learn whether goals were achieved by a treatment or not. KM tools may be used here to convert tacit knowledge to explicit knowledge.

- Evaluation. Hypotheses are tested in the evaluation phase by comparing actual and expected outcomes. The learning part of adaptive management comes from the differences found between the actual and expected outcomes and what changes these differences require in the next cycle. Results from this phase provide a starting point for the next iteration of planning (e.g. lessons learned, storytelling).

Future directions

Reynolds (2005) concludes that recent versions of DSSs designed to support ecosystems and sustainable forest management represent great improvements. He writes:

> The task posed to the DSS development community, to deliver effective, integrated decision support for sustainable forest management, was too large and complex to be achieved in a single development cycle. It has required, instead, an incremental and adaptive approach to system design.

In retrospect, it is difficult to understand that we did not anticipate this result. Not surprisingly, much work still needs to be done especially in the following areas.

Improve DSS niche identification

Even analysing only three out of possibly a dozen candidate DSSs, Reynolds (2005) found that each one occupies a different niche based on scale (landscape, forest level, operational level), support for all or only some parts of the four components of adaptive management listed above, access to different decision-analysis methods and an ability to make the entire planning process understandable to the user. There is a great need for all DSSs to be accessible to users and well documented and supported. In addition, there is a need to classify many popular software products such as spreadsheets, GIS and growth-and-yield models to clarify their supporting role within larger DSSs (Rauscher, 1999). Right now, these products reside in a poorly organized 'grab bag' of tools. Many, if not most, supporting resources used by DSSs were not designed to be a part of a DSS. As DSSs proliferate, the problem of clarifying the role of various supporting resources is compounded and logistically, technologically and administratively challenging.

Improve communications between DSSs operating at different scales

Multiple DSSs have been designed to function at the landscape, forest and project scales. Most DSSs embrace a hierarchical approach to planning (e.g. Martell *et al.*, 1998). Only a few definitive examples of integrated, multi-scale forest-resource planning have been described (Rose *et al.*, 1992; Reynolds and Peets, 2001; Vacik and Lexer, 2001). DSSs are rarely able to provide information either up the scale to the next higher level or down the scale to the next lower level (Mowrer *et al.*, 1997; Reynolds, 2005). This issue has been identified, and in some cases explicitly addressed (Camenson, 1998), but these examples are the exception rather than the rule.

Uniformly good support for biophysical, social and economic goals

Most DSSs still do a better job of accounting for biophysical goals than social and economic goals (Mowrer *et al.*, 1997). European approaches to decision support appear to focus primarily on the forest stand or forest enterprise level, with heavy emphasis on timber management support (Rauscher *et al.*, 2005). Due to the heterogeneity of European ecosystems, the landscape level has not been a frequent focus of DSSs so far. The concept of forest sustainability has a long tradition in Europe, and the paradigm is shifting from sustained yield and constant forest cover towards sustainability of an increasing diversity of values, goods and services. Researchers and forest managers in the USA and Canada have had a longer tradition in the development and application of DSSs. So current North American approaches have focused more heavily on non-timber forest products, such as clean water, wildlife and aesthetics, than their European counterparts (Rauscher *et al.*, 2005). More emphasis needs to be placed on improving support for non-timber goals in most DSSs.

Improve group consensus tools within DSSs

Mowrer *et al.* (1997) reported that tools for building group consensus were entirely missing from all DSSs examined. This deficiency is no longer as pronounced as it once was, as we review several recent developments in the section on multiple-criteria decision making below. Wide-scale acceptance and use of DSSs by the forest management community will depend to a great extent on how well these applications permit group participation in decision making. Successful group consensus tools will require designing the software with a human behaviour focus rather than a technical focus.

Improve use of DSSs by the management community

Capabilities of DSSs are expanding, and the systems are being used more frequently. Nevertheless, DSSs have not yet been widely adopted as standard tools for forest management in most areas of the world (Lexer *et al.*, 2005). We sense, however, that we are approaching a turning point – existing DSSs may soon be mature enough for forest managers to routinely use for complex decisions. More evaluations and case studies need to be conducted with management participation in order to gain widespread acceptance.

Improve support for implementation and monitoring

Evaluation and planning are typically performed over short time spans and in close temporal proximity to each other. However, implementation and monitoring are recurring activities that may be spread over many years in a typical adaptive management cycle. Speculating a little, one might imagine a collection of new system components, such as task-scheduler and task-management agents, helping to assure that implementation plans are staying on track and that data are updated and summarized. They could even be helping to spot evidence that might trigger initiation of a new management cycle. It is difficult to overstate the significance of decision-support implementations for these phases of the process. When we have learned how to provide practical support for the implementation and monitoring phases of the decision process, forest managers will be able to provide convincing demonstrations of successful adaptive management (Reynolds, 2005). Currently such efforts are extremely limited by time, money and support tools.

Provide greater flexibility for choosing alternatives

Simon (2003) observed that:

> In decision theory, the discussion of choice criteria has largely focused on two polar opposite alternatives: optimizing, that is, selecting the best alternative according to some criterion; and satisficing, that is, selecting an alternative (not necessarily unique) that meets some specified standard of adequacy.

Optimization methods have been criticized primarily because: it is difficult to incorporate real-world complexities into the mathematical approaches; solution mechanics have made it extremely difficult to explain to users why the solution came out as it did; and it takes a great deal of formal education and mathematical expertise to properly formulate a problem (Gigerenzer and Todd, 1999; Simon, 2003). The satisficing methods have been criticized primarily because solutions are not optimal. Despite a few dissenting voices (Howard, 1991), forest scientists are found overwhelmingly in the optimization camp. Many modern DSSs designed to address ecosystem and sustainable forest management problems use the satisficing method (Andersson *et al.*, 2005; Lexer *et al.*, 2005; Twery *et al.*, 2005). We believe that both satisficing and optimizing methods are legitimate strategies for choosing among alternatives. DSSs should routinely offer a wider range of alternative selection tools than they typically provide today.

Multi-criteria Decision Making and Satisficing

Overview

A traditional decision analysis has three elements: (i) a decision maker; (ii) a set of feasible alternatives; and (iii) a well-defined criterion (or objective) by which each alternative can be mathematically evaluated (Romero and Rehman, 2003). This objective function is commonly augmented with a set of mathematically defined constraints. One of many optimization algorithms can then be applied (depending on the form of the problem) to find the optimal value for the objective function within the feasible space.

The early successes of operations research methods resulted primarily from their focus on well-constrained problems in tactical planning. These methods were developed to address needs in industrial and business operations, where inputs, outputs, resources, actors, flows and other problem components could be described with completeness and certainty. Gradually these operations research methods were applied to planning in forest and natural resources management. In an era of forest management that was focused primarily on timber harvesting – with its real similarities to industrial operations – direct application of these optimization methods seemed to work well.

Traditional optimization-based decision analysis depends, however, upon some critical fundamental assumptions. One of these assumptions is that the decision maker is seeking to optimize a well-defined single objective (Klein and Methlie, 1990). A second assumption is that not only is there one best 'right answer' but it can be discovered through an objective technical process (Smith, 1997, p. 420). A third assumption is that the natural world is both predictable and knowable in a mechanistic and deterministic way (Smith, 1997, p. 420). All that is needed is to collect enough data. Finally, it is assumed that the values provided for mathematical expression are known quantities with no uncertainties. In situations where any one of these assumptions cannot be reasonably justified, any selected solutions, while possibly good ones, are no longer optimal. Given these constraining preconditions, then, some would ask why we use such exacting methods. In fact, optimization analyses have been quite successful in many situations. They have excelled in addressing mathematically well-defined problems and in addressing the quantitative components of larger problems.

However, Romero and Rehman (2003) posit that it is equally true that there exists considerable evidence that undermines the validity of the assumptions behind optimization. The primary argument is that, while traditional optimization processes are mathematically and logically sound, such formulations simply do not reflect real-life problems faithfully enough (Mendoza and Martins, 2005). As Levy *et al.* (2000) noted, ill-defined and messy problems create a decision environment containing imprecision and conflicting performance indicators. The previously applied optimization methods were difficult to fit reliably and effectively into this framework. In many situations, problem complexity prevents formulation of an adequately representative utility function (Gigerenzer and Todd, 1999; Romero and Rehman, 2003; Simon, 2003). Natural systems 'can be surprisingly simple and still be too complex for us to predict exact conditions only moderately far into the future' (Smith, 1997, p. 421). Furthermore, considerable uncertainty often exists in the data used to formulate the problem. Even multiple-objective programming (Mendoza *et al.*, 1987) and multiple-objective fuzzy linear programming (Gupta *et al.*, 2000) are essentially extensions of the linear programming model to several well-defined objectives. To address this new class of decision problems, multiple-criteria decision-making (MCDM) techniques were developed, beginning in the 1970s. Stewart (1992) reviewed those approaches, and Mendoza and Prabhu (2000) identified several important features of MCDM, including:

- Accommodation of multiple criteria (or attributes) in the analysis.
- Allowing direct involvement of multiple experts, interest groups or stakeholders.
- Not data-intensive: expert opinion can be used in place of data.
- Effective for both quantitative and qualitative data.
- Transparent analyses so that participants could understand what was happening and why.

In reality, the decision maker is frequently looking for an optimal compromise among several objectives. That situation may call for MCDM or for satisficing a set of goals. We first discuss MCDM in more detail, followed by a very brief overview of satisficing.

Multiple-criteria decision-making theory

One of the earliest MCDM methods is multi-attribute utility theory (MAUT) or value theory (Keeney and Raiffa, 1976; von Winterfeldt and Edwards, 1986). In its simplest form, MAUT provides a set of utility functions (one function for each attribute, or performance indicator), and then scores each decision alternative on each attribute. Scores across all attributes are combined for each alternative (often using an additive model), with individual attribute scores being appropriately scaled for comparability and weighted according to importance. The decision alternative with the highest aggregate utility (or value) score is then preferred. This general MAUT framework has spawned a large number of variants. Each variant modifies one or more aspects of the traditional implementation: to assign weighting values, to scale attribute scores, to combine scores across attributes (e.g. non-additive models), to elicit utility functions, etc. As a result, MAUT has been modified in a wide variety of ways. Furthermore, MAUT has been augmented by other multiple-criteria decision methods and other decision aids to make it more complete in some cases and to improve its applicability in other cases.

Another MCDM approach that was developed about the same time as MAUT and has all its components is the analytical hierarchy process (AHP). The AHP provides for: decomposition of the decision problem into a multi-level hierarchy of criteria, direct pairwise comparisons of the decision alternatives (or, alternatively, rating them individually) and rigorous mathematics to generate a preference structure for the alternatives. Schmoldt *et al.* (2001) describe many applications of AHP to environmental and natural resources decision making. While many have used the AHP as an MCDM technique by itself, others have used it in combination with other MCDM methods (e.g. Prato, 1999; Lexer, 2000; Hill *et al.*, 2005). In other MCDM developments, Leskinen *et al.* (2003) have used statistical methods to estimate ecological values and to account for interactions among the decision variables. Drechsler (2004) demonstrated the integration of quantitative population models with MCDM to evaluate decision conflicts. Recently, the network structure of the analytical network process has been used to model the complexity of forest decision problems in evaluating sustainable forest management strategies by using a criterion and indicator approach (Wolfslehner *et al.*, 2005). These several examples, and many others not cited, attest to the versatility and extensibility of MCDM methods in general.

Future directions for MCDM

Research in MCDM methodologies continues at a rapid pace to improve their problem-solving power. The following R & D areas represent several of the most active.

Integrating geographic information systems

Given the geo-spatial context of most forestry decision problems, it is not surprising that considerable recent effort has gone into integrating geographic information systems (GISs) and MCDM (Jankowski and Nyerges, 2001; Jankowski *et al.*, 2001; Feick and Hall, 2004; Sakamoto and Fukui, 2004; Hill *et al.*, 2005). Map-based displays enable better elicitation of utility functions and present decision results in ways that encourage understanding and iterative analysis.

Support for group decision making

Because single individuals rarely make natural resource management decisions, group decision making has become a priority topic in MCDM. Empirically, we have come to understand that group decisions are typically better than average individual performance, but rarely as good as the best individual. Hence, while there are many other reasons to engage groups in decision processes, making the best decision choices is not one of them (Schmoldt and Peterson, 2000). These include broad ownership in the decisions and their implications, support for implementation of decisions, more complete coverage of all pertinent issues and viewpoints during the decision process, and the appearance of an open and inclusive decision process. Examples from this rapidly expanding MCDM focus area include Mendoza and Prabhu (2000), Schmoldt and Peterson (2000), Jankowski and Nyerges (2001), Jankowski *et al.* (2001), Feick and Hall (2004) and Thomson (2005). Given the growing importance of inclusive and open decision processes in forestry, it is reasonable to assume that participatory decision making will continue to generate new developments in and variants of MCDM. Some efforts have already gone into adding MCDM to virtual meeting environments, which currently include such things as group whiteboards, shared documents, audio/video connectivity, voting and Delphi processes (Kangas and Store, 2003). As with many other technologies, MCDM will need to become network-accessible so that decision-making processes and/or results can be shared among geographically dispersed participants and stakeholders.

Expanding the suite of tools compatible with MCDM methods

Opportunities exist for expanding the repertoire of techniques available for MCDM. For example, the above applications of the AHP and GIS, in combination with other MCDM methods, have proved to be very effective. Traditional optimization or simulation techniques can also be used to generate biologically possible or economically feasible alternatives for evaluation within MCDM approaches (Kangas *et al.*, 2000; Kangas and Store, 2002). There exists a large toolkit of quantitative methods that can be very useful as part of a broader MCDM framework.

Incorporating risk and uncertainty into MCDM methods

Because all decisions occur in an uncertain environment and decision results can have uncertain consequences, greater effort is needed in measuring the uncertainty and risk associated with alternatives. While a decision alternative may be chosen precisely because it is expected to lead to a desired future condition, there may also be considerable risk associated with that alternative. The likelihood of attaining the future condition may be low or the potential for undesirable side effects may be high. Currently much of this type of risk assessment needs to be considered by the decision maker outside the formal MCDM process, where it is exogenous, rather than integral, to the decision analysis process.

Satisficing solutions

As a decision-making method, satisficing differs markedly from optimization or MCDM. The originator of the satisficing concept, Herbert Simon, argues that a large body of evidence shows that people rarely actually engage in optimization (Simon, 2003). Even knowledgeable experts in the fields of optimization and decision analysis readily admit that they rarely apply formal decision methods in their personal lives (von Winterfeldt and Edwards, 1986), even when a high-stakes decision is involved. Why? Because obtaining detailed data and trying to predict future behaviour of complex natural systems is a time-consuming and expensive proposition (Smith, 1997). People are inclined naturally towards the satisficing approach. They limit their input data to the most important and available information. They limit their alternative courses of action to a small set of possibilities. They choose the most acceptable option from that limited set rather than attempting to seek an optimal solution (Smith, 1997; Simon, 2003). In addition, wise satisficers make sure to monitor results of decisions to detect failure as soon as possible and then decide on corrective actions, which may lead to choosing another alternative.

Satisficing implemented as a decision process in DSSs depends upon the existence of goals, also called objectives or targets. Each goal must have one or more measurement criteria, which can be determined by measuring the current state of affairs or estimated for some simulated future state (Nute *et al.*, 2000; Rauscher *et al.*, 2000). Once these goals and their measurement criteria exist, decision makers: (i) generate possible courses of action in numerous ways; (ii) measure or predict the goal criteria; and (iii) evaluate how well the goals as a set, and individually, have been achieved. This process of alternative generation and simulation is repeated until: (i) the goal criteria are satisfied; (ii) the satisficing criteria for unachievable goals are adjusted downward so that they can be achieved (reality adjusts perceived values); or (iii) satisficing criteria for easily achievable goals are adjusted upward to obtain and maintain goal achievement at a higher level than originally thought possible.

Satisficing has numerous attractive characteristics (Simon, 2003):

- The decision process is relatively cheap because all possible alternatives need not be addressed.
- Since measurement criteria for goals are commonly defined as some threshold value, e.g. grassy openings must occupy at least 5% of the area of the property, the requirement for precision in simulating the future is significantly reduced.
- There is no requirement to create and agree upon a single utility function from multidimensional goals.

- Because the computational task is relatively small and the mathematical formulation of the problem is very simple, a high educational level is not required to implement the process.
- The causal relationships among the goals, the measurement criteria and the simulated state of forest ecosystems are relatively easy to explain to decision makers and stakeholders.

Future directions for satisficing

The satisficing method of alternative selection and goal satisfaction is currently underutilized in the natural resource field. More research is needed to compare and contrast satisficing with the various MCDM methods. It may be entirely possible to combine the two approaches in a very effective way.

Alternative selection and goal satisfaction methods differ significantly in effort and cost. Research to better match the risk of arriving at a poor solution with the total effort expended can be expected to yield effective guidance to managers. Most problem situations require forecasting the consequences of actions taken today on the forest landscape at some point in the future. The further out into the future the forecast, the more unknowable and uncertain the predictions are likely to be, until a point is reached where even pretending to predict the future is absurd. We need to scientifically examine when and whether it makes sense to invest large amounts of effort and money on choosing alternatives based on forecasting changes in the forest landscape unrealistically far into the future. The proponents of satisficing would argue that it is better to spend less effort and money on the front-end of the decision problem and more on inventory and monitoring to detect failure as soon as possible. It is this failure detection that is at the heart of the adaptive management idea. Coupled with the rather low-cost satisficing decision approach, increased inventory and monitoring would allow early detection of problems based on reality rather than overextended theory and potentially untrustworthy forecasts.

Conclusions: How Can We Best Support Sustainability?

The scientific community is slowly coming to grips with the concept of sustainability. An inherent difficulty with this concept has been that, unlike traditional scientific investigations that seek to explain how things currently function or how previous events led to current phenomena, sustainability research is forward-looking, with the goal of understanding how both current and future societal needs can be met (Schmoldt, 2004). This goal is further complicated by an imperative for maintaining ecological and environmental integrity as well as staying within the parameters set by what is socially acceptable and economically feasible. The function of a decision-support system is to organize the decision process and provide flexible, on-demand access to the full array of methods and tools applicable to a particular problem situation (Rauscher and Reynolds, 2003). In this context a 'good' decision is one that is made based on a thorough understanding and analysis of the problem (Holloway, 1979). There is no guarantee that a good decision will always achieve a good outcome in terms of sustainability. The consequences of a 'good' outcome are favourable with respect to the preferences of the decision maker. A decision resulting in a bad outcome could still be considered a good decision as long as the decision-making process indicated the possibility of a bad outcome. The presented methods and tools provide a proper documentation of the decision-making process. Thus, rationales and information used in arriving at a decision can be compared with the achieved outcome, which enables better decisions in the future.

As the new sustainability sciences, such as ecological economics, industrial ecology, environmental geology and ecological engineering, organize themselves and advance, we can cooperate with scientists in these new fields to help them reorganize what we currently know, using our full KM toolkit. Much progress can be made relatively rapidly by drawing new boundaries of knowledge or by connecting existing data, information and knowledge in new ways (e.g. Vacik *et al.*, Chapter 23, this volume). As new data, information and knowledge are created, we can use our KM tools to rapidly integrate that new content into the existing body of knowledge. We can also share that content cheaply, widely and immediately. There is no doubt that such integration is technologically feasible; what is missing now is a shared vision and the political will to implement it. The KM sites of Table 26.1 provide excellent support tools for this objective.

We need to be concerned with how to make available both tacit and explicit knowledge. There is a temptation to concentrate primarily on explicit knowledge because it is tangible; we can readily organize it, check it and build taxonomies for it. Tacit knowledge, on the other hand, is messy. It moves from person to person and group to group. It grows and changes as it moves. KM professionals will have to continue to facilitate this movement of tacit knowledge (Dixon, 2000). The inexplicit nature of our understanding regarding sustainability concepts (their definitions) and practice implies that our knowledge management skills and tools will be severely challenged. This problem is further exacerbated by varied and inexplicit value systems (national, local and organizational), which must enter into any sustainability discussion. The struggle to implement the concept of sustainable forest management can be supported by many of the more tacit knowledge-oriented tools of Table 26.1: communities of practice, free-content information collaborations and best practices and lessons learned sites.

We can use our existing DSSs to define new goals and measurement criteria that better address forest sustainability. The use of MCDM and satisficing techniques and their various combinations will allow us to model the trade-offs between conflicting resource goals and will help us to identify compromise solutions. From there, we can identify in very precise terms what data, information and knowledge are needed to make the kinds of decisions we are looking for. On the other hand, a tension exists between the application of DSSs and the present use of inventory data. It may not even be possible to inventory a measurement criterion in a real forest or forecast the future value of that measurement criterion by using currently available prediction systems. Such examples are numerous and provide great opportunity to focus new research efforts to fill these major knowledge gaps.

Some people have advocated models designed to signal ecological problems before they occur – *anticipatory* research. Most ecological models are not constructed with this intent in mind, and few researchers place sufficient confidence in their models to extrapolate to unrecognized problems. However, with proper attention to ecological scale combined with a system science approach to investigation, we may have the tools in place to conduct anticipatory research. Given a sufficiently sound understanding of a system's dynamics and modelling those dynamics at the appropriate scale(s), we may be positioned to make informed predictions about system behaviour within an ecologically meaningful context. Armed with this knowledge, we will be positioned to direct policy discussions – if not policy itself – towards sustainable practices. As our environmental problems become more complex and far-reaching, the ability to anticipate impacts (or understand them shortly after they are uncovered) will greatly improve our reaction time and inform our alternatives.

Policy science must play an important role in the effort to define sustainable forest management and design implementation strategies. There is great opportunity to better understand the nature of the environmentally based conflicts that have energized a broad spectrum of the world's public. Policy scientists must take the forefront in analysing and communicating results from such studies. Another important focus area for policy science is the area of group decision-making dynamics and methods. There is an urgent need to analyse and summarize the experience of the last 20 years and communicate the results as lessons learned and best practices for group decision making. Finally, policy scientists should increase their focus on examining the environmentally based conflicts from a global perspective with a view to crystallizing the descriptive, factual knowledge. Who produces what kind of forest product and who consumes it in what quantities? Which forest regions in the world are best suited to sustainably produce what kind of product and in what quantity? What are the social, economic and ethical consequences for a country to reduce its forest product production while at the same time increasing the quantity of forest products it imports from other regions of the world? Such questions and their answers need to be widely known and commonly agreed upon before a truly sustainable regional and national forest management policy can be successfully implemented. The many powerful tools of KM can help us achieve these goals.

Note

1. Trade names are used for informational purposes only. No endorsement by the US Department of Agriculture is implied.

References

Andersson, M., Dahlin, B. and Mossberg, M. (2005) The Forest Time Machine – a multi-purpose forest management decision support system. *Computers and Electronics in Agriculture* 49(1), 114–128. http://dx.doi.org/10.1016/j.compag.2005.02.006

Boiral, O. (2002) Tacit knowledge and environmental management. *Long Range Planning* 35, 291–317.

Camenson, D.M. (1998) Hierarchical decision support tools for ecologically-based planning and management. In: Nabuurs, G.J., Nuutinen, T., Bartelink H. and Korhonen, M. (eds) *Forest Scenario Modeling for Ecosystem Management at the Landscape Level. Proceedings of the International Seminar and Summer School, Wageningen, Netherlands, 26 June–3 July 1997*. Proceedings No. 19, European Forestry Institute Wageningen, Netherlands, pp. 35–43.

Dixon, N.M. (2000) *Common Knowledge: How Companies Thrive by Sharing What they Know*. Harvard Business School Press, Cambridge, Massachusetts.

Drechsler, M. (2004) Model-based conservation decision aiding in the presence of goal conflicts and uncertainty. *Biodiversity and Conservation* 13(1), 141–64.

Evans, P. and Wurster, T.S. (2000) *Blown to Bits: How the New Economics of Information Transforms Strategy*. Harvard Business School Press, Cambridge, Massachusetts.

Feick, R.D. and Hall, B.G. (2004) A method for examining the spatial dimension of multi-criteria weight sensitivity. *Journal of Geographical Information Science* 18(8), 815–840.

Gigerenzer, G. and Todd, P.M. (1999) *Simple Heuristics that Make us Smart*. Oxford University Press, New York, 416 pp.

Girad, N. and Hubert, B. (1999) Modelling expert knowledge with knowledge-based systems to design decision aids. The example of a knowledge-based model on grazing management. *Agricultural Systems* 59, 123–144.

Gray, P.H. (2001) A problem-solving perspective on knowledge management practices. *Decision Support Systems* 31, 87–102.

Gupta, A.P., Harboe, R. and Tabucanon, M.T. (2000) Fuzzy multiple-criteria decision making for crop area planning in Narmanda River basin. *Agricultural Systems* 63(1), 1–18.

Hansen M.T., Nohria, H. and Tiernex, T. (1999) What's your strategy for managing knowledge? *Harvard Business Review* 77 (2), 106–116.

Heinrichs, J.H., Hudspeth, L.J. and Lim, J.S. (2003) Knowledge management. In: Bidgoli, H. (ed.) *Encyclopedia of Information Systems*, Vol. 3. Academic Press, Boston, Massachusetts, pp. 13–31.

Hill, M.J., Braaten, R., Veitch, S.M., Lees, B.G. and Sharma, S. (2005) Multi-criteria decision analysis in spatial decision support: the ASSESS analytic hierarchy process and the role of quantitative methods and spatially explicit analysis. *Environmental Modelling and Software* 20(7), 955–976.

Holloway, C.A. (1979) *Decision Making under Uncertainty: Models and Choices*. Prentice Hall, New York.

Holsapple, C.W. (2003) Decision support systems. In: Bidgoli, H. (ed.) *Encyclopedia of Information Systems*, Vol. 1. Academic Press, Boston, Massachusetts, pp. 551–565.

Howard, A.F. (1991) A critical look at multiple criteria decision making techniques with reference to forestry applications. *Canadian Journal of Forest Research* 21, 1649–1659.

Innes, T. (ed.) (2003) *Natural Resources Information Management Forum: Putting knowledge to work*. FORREX Series No. 8, FORREX – Forest Research Extension Partnership, Kamloops, British Columbia.

Jankowski, P. and Nyerges, T. (2001) GIS-supported collaborative decision making: results of an experiment. *Annals of the Association. of American Geographers* 91(1), 48–70.

Jankowski, P., Andrienko, N. and Andrienko, G. (2001) Map-centered exploratory approach to multiple criteria spatial decision making. *Journal of Geographical Information Science* 15(2), 101–127.

Kangas, J. and Store, R. (2002) Socioecological landscape planning: an approach to multi-functional forest management. *Silva Fennica* 36(4), 867–871.

Kangas, J. and Store, R. (2003) Internet and teledemocracy in participatory planning of natural resources management. *Landscape and Urban Planning* 62(2), 89–101.

Kangas, J., Store, R., Leskinen, P. and Mehtätalo, L. (2000) Improving the quality of landscape ecological forest planning by utilising advanced decision-support tools. *Forest Ecology and Management* 132(2/3), 157–171.

Keeney, R.L. and Raiffa, H. (1976) *Decisions with Multiple Objectives.* Wiley, New York.

Kennard, D.K., Rauscher, H.M., Flebbe, P.A., Schmoldt, D.L., Hubbard, W.G., Jordin, J.B. and Milnor, W. (2005) Using hyperdocuments to manage scientific knowledge: the prototype Encyclopedia of Southern Appalachian Forest Ecosystems. *Forest Ecology and Management* 207, 201–213.

Klein, M. and Methlie, L.B. (1990) *Expert systems: a Decision Support Approach*. Addison-Wesley, Reading, Massachusetts.

Leskinen, P., Kangas, J. and Pasanen, A. (2003) Assessing ecological values with dependent explanatory variables in multi-criteria forest ecosystem management. *Ecological Modeling* 170(1), 1–12.

Levy, J.K., Hipel, K.W. and Kilgour, D.M. (2000) Using environmental indicators to quantify the robustness of policy alternatives to uncertainty. *Ecological Modeling* 130(1–3), 79–86.

Lexer, M.J. (2000) A multi-attribute utility model for silvicultural decision-making in secondary Norway spruce forests. *Forstwissenschaftliches Centralblatt* 119(6), 377–394.

Lexer, M.J, Vacik, H., Palmetzhofer, D. and Oitzinger, G. (2005) A decision support tool to improve forestry extension services for small private landowners in southern Austria. *Computers and Electronics in Agriculture* 49, 81–102.

Martell, D., Gunn, E. and Weintraub, A. (1998) Forest management challenges for operational researchers. *European Journal of Operations Research* 104, 1–17.

Maser, C. (1994) *Sustainable Forestry: Philosophy, Science and Economics*. St Lucie Press, Boca Raton, Florida, 373 pp.

Mendoza, G.A. and Martins, H. (2005) Multi-criteria decision analysis in natural resource management: a critical review of methods and new modelling paradigms. In: Reynolds, K.M. (ed.) *Proceedings of Sustainable Forest Management in Theory and Practice, 5–8 April 2005, Edinburgh, UK.*

Mendoza, G.A. and Prabhu, R. (2000) Multiple criteria decision making approaches to assessing forest sustainability using criteria and indicators: a case study. *Forest Ecology and Management* 131(1–3), 107–126.

Mendoza, G.A., Bare, B.B. and Campbell, G.E. (1987) Multiobjective programming for generating alternatives: a multiple use planning example. *Forest Science* 33(2), 458–468.

Mintzberg, H., Raisinghani, D. and Theoret, A. (1976) The structure of unstructured decision processes. *Administrative Science Quarterly* 21, 246–275.

Mowrer, H.T., Barber, K., Campbell, J., Crookston, N., Dahms, C., Day, J., Laacke, J., Merzenich, J., Mighton, S., Rauscher, H.M., Reynolds, K., Thompson, J., Trenchi, P. and Twery, M. (1997) *Decision Support Systems for Ecosystem Management: an Evaluation of Existing Systems.* RM-GTR-296, US Department of Agriculture, Forest Service, Interregional Ecosystem Management Coordinating Group, Decision Support Systems Task Team, Rocky Mountain Forest and Range Experiment Station, Fort Collins, Colorado.

Nesbitt, A., Kemeny, C.R., Broshy, E. and Wurster, T.S. (1996) Managing for a wired health care industry. *IN VIVO: The Business and Medicine Report,* 1 July, pp. 1–8.

Nonaka I., and Takeuchi, H. (1995) *The Knowledge-creating Company.* Oxford University Press, Oxford, UK.

Nute, D., Rosenberg, G., Nath, S., Verma, B., Rauscher, H.M., Twery, M.J. and Grove, M. (2000) Goals and goal orientation in decision support systems for ecosystem management. *Computers and Electronics in Agriculture* 27, 355–375.

Plunkett, P. (2001) *Managing Knowledge @ Work: an Overview of Knowledge Management.* Knowledge Management Working Group of the Federal Chief Information Officers Council, US General Services Administration, Office of Information Technology, pearl.butler@gsa.gov.

Polanyi, M. (1958) *Personal Knowledge.* University of Chicago Press, Chicago, Illinois.
Prato, T. (1999) Multiple attribute decision analysis for ecosystem management. *Ecological Economics* 30(2), 207–222.
Rauscher, H.M. (1987) Increasing scientific productivity through better knowledge management. *AI Applications* 1(2), 21–31.
Rauscher, H.M. (1991) The encyclopedia of AI applications to forest science. *AI Applications* 5(2): insert (592 KB, 235 chunks, 449 links (electronic)).
Rauscher, H.M. (1999) Ecosystem management decision support for federal forests in the United States: a review. *Forest Ecology and Management* 114, 173–197.
Rauscher, H.M. and Reynolds, K.M. (2003) Natural resource management. In: Bidgoli, H. (ed.) *Encyclopedia of Information Systems*, Vol. 3. Academic Press, Boston, Massachusetts, pp. 239–265.
Rauscher, H.M., Alban, D.H. and Johnson, D.W. (1993) Managing the global climate change scientific knowledge base. *AI Applications* 7(4), 17–40.
Rauscher, H.M., Loftis, D.L., McGee, C.E. and Worth, C.V. (1997) Oak regeneration: a knowledge synthesis. *FORS Compiler* 15(1), insert (electronic).
Rauscher, H.M., Lloyd, F.T., Loftis, D.L. and Twery, M.J. (2000) A practical decision-analysis process for forest ecosystem management. *Computers and Electronics in Agriculture* 27, 195–226.
Rauscher, H.M., Reynolds, K. and Vacik, H. (2005) Editorial: Decision support systems for forest management. *Computers and Electronics in Agriculture* 49, 1–5.
Reynolds, K.M. (2005) Integrated decision support for sustainable forest management in the United States: fact or fiction? *Computers and Electronics in Agriculture* 49, 6–23.
Reynolds, K.M. and Peets, S. (2001) Integrated assessment and priorities for protection and restoration of watersheds. In: *Proceedings of the IUFRO 4.11 Conference on Forest Biometry, Modeling and Information Science, 26–29 June 2001, Greenwich, UK.* Published online at http://cms1.gre.ac.uk/conferences/iufro/proceedings/ReynoldsNew.pdf
Reynolds, K.M., Rauscher, H.M. and Worth, C.V. (1995) *A Hypermedia Reference System to the Forest Ecosystem Management Assessment Team Report and Some Related Publications.* General Technical Report PNW-357 + electronic files, USDA Forest Service, Pacific Northwest Research Station, St Paul, Minnesota.
Romero, C. and Rehman, T. (2003) *Multiple Criteria Analysis for Agricultural Decisions.* 2nd edn. Elsevier, Amsterdam.
Rose, D.W., McDill, M. and Hoganson, H.M. (1992) Development of an environmental impact statement of statewide forestry programs: a Minnesota case study. *Compiler* 10(4), 18–27.
Rubenstein-Montano, B., Liebowitz, J., Buchwalter, J., McCaw, D., Newman, B., Rebeck, K., and the Knowledge Management Methodology Team (2001) A system thinking framework for knowledge management. *Decision Support Systems* 31, 5–16.
Ruggles, R. (1997) *Knowledge Management Tools.* Butterworth-Heinemann, Oxford, UK.
Saarikko, J. (1994) Forestry information resources on the Internet. http://www.funet.fi/pub/sci/forest/doc/guides/Saarikko-1994b.html
Sakamoto, A. and Fukui, H. (2004) Development and application of livable environment evaluation support system using Web GIS. *Journal of Geographical Systems* 6(2), 175–195.
Schmoldt, D.L. (2004) Turning small snapshots into a bigger picture for sustainability. In: Gao, W. and Shaw, D.R. (eds) *Remote Sensing and Modeling of Ecosystems for Sustainability.* SPIE, Bellingham, Washington, 5544, pp. 1–11.
Schmoldt, D.L., and Peterson, D.L. (2000) Analytical group decision making in natural resources: methodology and application. *Forest Science* 46(1): 62–75.
Schmoldt, D.L., Kangas, J., Mendoza, G.A. and Pesonen, M. (eds) (2001) *The Analytic Hierarchy Process in Natural Resources and Environmental Decision Making.* Kluwer Academic Publishers, Dordrecht, The Netherlands.

Schuster, E.G., Leefers, L.A. and Thompson, J.E. (1993) *A Guide to Computer-based Analytical Tools for Implementing National Forest Plans*. General Technical Report INT-296, US Department of Agriculture, Forest Service, Intermountain Forest and Range Experiment Station, Moscow, Idaho.

Simard, A.J. (2000) *Managing Knowledge at the Canadian Forest Service*. Science Branch, Canadian Forest Service, Ottawa.

Simon, H.A. (2003) Decision theory. In: Bidgoli, H. (ed.) *Encyclopedia of Information Systems*, Vol. 1. Academic Press, Boston, Massachusetts, pp. 567–581.

Smith, G.R. (1997) Making decisions in a complex and dynamic world. In: Kohm, K.A. and Franklin, J.F. (eds) *Creating a Forestry for the 21st Century*, Island Press, Washington, DC, pp. 419–435.

Stewart, T.J. (1992) A critical survey on the status of multiple criteria decision making theory and practice. *Omega* 20, 569–586

Thomson, A.J. (2005) Indicator-based knowledge management for participatory decision-making. *Computers and Electronics in Agriculture* 49, 206–218.

Thomson, A.J., Sutherland, J.R. and Carpenter, C. (1993) Computer-assisted diagnosis using expert system-guided hypermedia. *AI Applications* 7(1), 17–27.

Thomson, A.J., Allen, E. and Morrison, D. (1998) Forest tree diagnosis over the World Wide Web. *Computers and Electronics in Agriculture* 21(1), 19–32.

Twery, M.J., Knopp, P.D., Thomasma, S.A., Rauscher, H.M., Nute, D.E., Potter, W.D., Maier, F., Wang, J., Dass, M., Uchiyama, H., Glende, A. and Hoffman, R.E. (2005) Ned-2: a decision support system for integrated forest management. *Computers and Electronics in Agriculture* 49, 24–43.

Tyndale, P. (2002) A taxonomy of knowledge management software tools: origins and applications, *Evaluation and Programm Planning* 25, 183–90.

Vacik, H. and Lexer, M.J. (2001) Application of a spatial decision support system in managing the protection forests of Vienna for sustained yield of water resources. *Forest Ecology and Management* 143(1–3), 65–76.

von Winterfeldt, D. and Edwards, W. (1986) *Decision Analysis and Behavioral Research*. Cambridge University Press, Cambridge.

Walters, C.J. (2002) *Adaptive Management of Renewable Resources*. Blackburn Press, Caldwell, New Jersey, 374 pp.

Wolfslehner, B., Vacik, H. and Lexer, M.J. (2005) Application of the Analytic Network Process in multicriteria analysis of sustainable forest management. *Forest Ecology and Management* 207(1–2), 157–170.

27 How Should We Manage Knowledge Ecosystems? Using Adaptive Knowledge Management

ALAN J. THOMSON

Canadian Forest Service, Pacific Forestry Centre, Victoria, British Columbia, Canada.

Abstract

The knowledge ecosystem concept and the related concept of information ecology are key foundations of participatory processes relating to sustainable development. Knowledge ecosystems can be defined as the complex and many-faceted system of people, institutions, organizations, technologies and processes by which knowledge is created, interpreted, distributed, absorbed and utilized. The analogy of knowledge ecosystems to natural ecosystems is explored in relation to forest planning processes.

Introduction

Many jurisdictions mandate stakeholder participation in forest planning processes. Criteria and indicators used in monitoring of forest practices and in certification of their sustainability include measures of the success of such consultation and communication. For example, the African Timber Organization's Indicator IV.2.2 is 'There is efficient communication between stakeholders' (van Buren and Blom, 1997). This is an example of the general aim of encouraging public participation in environmental decisions in Agenda 21[1] of the Rio Declaration on Environment and Development, adopted in 1992. Issues of participation, information, indicators, communication and decision making are key components of Chapters 8, 26, 31 and 40 of Agenda 21. Paragraph 31.4(h) of Agenda 21 specifically promotes information technologies to enhance dissemination of information for sustainable development. Paragraph 40.23 addresses establishment of information access, dissemination formats and communication interfaces, while 40.25 deals with transfer of information to and from non-electronic approaches. Other processes, such as forest certification in sustainable forest management (SFM), have also increased participation.

As the interaction of people and technology increases, we need to understand how individuals, organizations, cultures, forms of knowledge and knowledge transformations relate in order to develop appropriate systems and processes. This need has led to the related concepts of information ecology and knowledge ecosystems (Davenport and Prusak, 1997; Nardi and O'Day, 1999; Por, 2000).

> In the context of an evolving information society, the term information ecology was coined by various persons in the 1980s and 1990s. It marks a connection between ecological ideas with the dynamics and properties of the increasingly dense, complex and important digital informational environment and has been gaining progressively wider acceptance in a growing number of disciplines. 'Information ecology' often is used as metaphor, viewing the informational space as an ecosystem.[2]

For the present study, a knowledge ecosystem is defined as: the complex and many-faceted system of people, institutions, organizations, technologies and processes by which knowledge is created, interpreted, distributed, absorbed and utilized. That is, information ecology is the approach, based on analogy with the science of ecology, whereas a knowledge ecosystem is a specific community of interacting individuals in a particular environment or habitat. Using this analogy, the place where a participatory meeting is held becomes the environment.

In this study, I extend the current view of knowledge ecosystems by developing new analogies between knowledge processes and ecosystem concepts, and then illustrate these analogies in a situation in which the meeting space is virtual, using a computer system designed to mimic a particular type of meeting, an adaptive management workshop held as part of a forest planning process. Results are generalized to other meeting formats in the discussion. I then explore the question, 'If there are knowledge ecosystems, how should we manage them?' The management approach proposed is to use adaptive knowledge management (AKM), a new system design philosophy specifically geared to participatory processes (Thomson, 2005b).

AKM was defined (Thomson, 2005b) as:

> a formal process for continually improving communication policies and practices, by learning from the outcomes of operational programs. Its most effective form – 'active' adaptive knowledge management – is characterized by communication programs that are designed to experimentally compare selected policies or practices, by testing alternative hypotheses about the communication process.

In AKM, hypotheses can be developed regarding the relative importance of costs (both time and money), roles, expertise, responsibilities, flow and transformation of information, products and indicators of success. Human–computer interactions can be explored. Samples of potential products can be developed and surveys conducted to determine end-user satisfaction with different hypothetical report formats. Groups can experiment with communication methodologies in small-scale trials before committing to involvement in large-scale endeavours. In this process, construction of prototype systems such as that described in the present study can greatly facilitate decision making by shedding light on potential costs and benefits. Discussion and resolution of communication trade-offs in advance may reduce dissatisfaction with the subsequent process. The present study extends the concept of AKM, developed in relation to customized reporting, to cover the more general issue of stakeholder meetings and their electronic analogues and the role of knowledge management systems in the process.

The system (Thomson, 2005b) on which the AKM concept was developed focused on knowledge management for communication, particularly reporting, and was not concerned with particular meetings used to generate knowledge. On the other hand, a virtual adaptive management workshop environment was developed by Thomson (2000a) to mimic a particular type of meeting, the workshop process described by Holling (1978). Whereas the system described by Thomson (2005b) is text-oriented, the workshop system is graphics-oriented and illustrates many technological aspects of knowledge management related to knowledge ecosystems.

Knowledge Ecosystems

There are many approaches to forest planning, but all have features in common, which are listed here for reference as a prelude to exploring a knowledge-ecosystems approach to forest planning. A forest plan describes the process and identifies participants for that plan; it identifies the temporal and spatial scope of the plan along with needs, risks and information requirements; it defines land zones and sets objectives and targets; and it develops strategies for managing resources as well as monitoring, communication and opportunities for plan revision. Planning provides a future vision for land and resources, the steps to follow to achieve that vision and the processes to determine the vision and steps. In forest planning, there is a hierarchy of levels, from operational to strategic plans. Each level involves a different set of sectors, institutions and stakeholders. Stakeholders have a range of education, technological expertise and world views, with the world view often being related to the institution or organization the stakeholder represents. Outputs of a planning process are, by definition, plans, i.e. a recommendation for a specific course of action, with plans at one level often being constrained by requirements of higher-level plans, as well as constraining lower levels. Planning processes often mandate stakeholder involvement. Higher-level planning in particular is often a lengthy process involving many meetings.

Trophic levels and food webs

The analogy of trophic levels and food webs reflects the hierarchy of meetings in the planning process. Trophic levels represent the feeding position in a food chain, such as primary producers (first trophic level), herbivore (second trophic level) and carnivore (third trophic level and higher). Ecological pyramids show decreases in numbers, biomass or energy towards the top of the pyramid. On average, about 10% of the energy available in one trophic level will be passed on

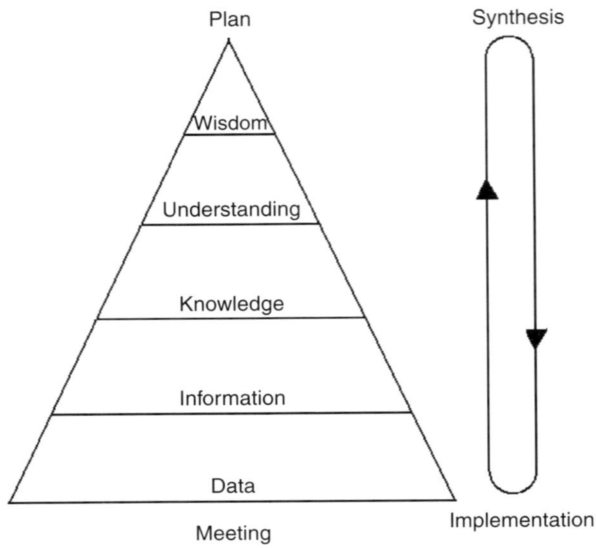

Fig. 27.1. Stages of synthesis between initial meetings and the actual plan are illustrated in a knowledge pyramid.

to the next. Death of organisms returns material to the base of the pyramid, giving flow in the opposite direction. In the knowledge analogy for forest planning, a synthesis takes place (Fig. 27.1), with meetings and debate occurring at each transition from one level to another. Knowledge filtering during synthesis will result in only key elements being passed from one level to another of the synthesis process. A flow can also occur in the opposite direction during the implementation of a plan, ending with meetings to indicate what data must be collected in a particular situation. In practice, plans go through cycles of creation and revision.

The ecosystem approach applied to the knowledge pyramid focuses attention on issues for participants as their knowledge amalgamates with other sources and is filtered during the synthesis process. Issues include: Who decides on the language and concepts included in a summary? Who draws the inferences from the material presented? Who prepares the summary? Is the meeting recorded? Are transcripts verbatim? Do participants have an opportunity to see how their information is used and provide feedback before the summary is transmitted to the next level? Where does the summary go – to what individual or group – and what are their roles and responsibilities? What response is expected from the next level, and what topics will be addressed in the response? How transparent is the process, i.e. is it open to the public, or is only the report public, in whole or in part?

Organisms in food chains interact through processes such as herbivory, competition or predation, with interactions often being illustrated by food webs. The knowledge equivalent is the relationship web (Fig. 27.2), which shows the interactions of individuals and organizations. Such webs play a key role in decisions to adopt innovative ideas (Haggith *et al.*, 2003). The ecosystem approach applied to relationship webs focuses attention on interactions within the meeting, as well as between participants and non-participants. In particular, competition is a key process occurring in the knowledge ecosystem, and may occur among individuals, organizations or forms of knowledge representation. The ecological concept of species in a niche is a good analogy for competition in the knowledge ecosystem. While in many cases competition among individuals and institutions, such as between representatives of environmental groups and forest industries, can be balanced by cooperation among those with similar views (e.g. all representatives of environmental groups), similar-appearing groups may not always have the same goals. There will be many kinds of roles (species of individual): translators, facilitators, scribes, mediators, knowledge creators and knowledge holders. Analogies for some of these roles in other studies include identification of librarians as 'keystone species' (Nardi and O'Day, 1999) and 'hunter-gatherers of the knowledge economy' (Berreby, 1999). There may also be competing formats (species of knowledge) such as spoken words (including stories and anecdotes), graphs, equations, documents, pictures, maps, graphs and equations, with each person or group having a preferred format.

If a meeting is a transient manifestation of a relationship web (Fig. 27.2), issues include: How can people not physically in attendance have their knowledge incorporated: by informal pre-meeting discussions or by a formal process, such as allowing submission of position statements? Competition can be a significant feature of many meetings: is the process and meeting format equally supportive of each person's mode of knowledge expression and way of working? Are there any gender issues? Do time and space constraints favour one group

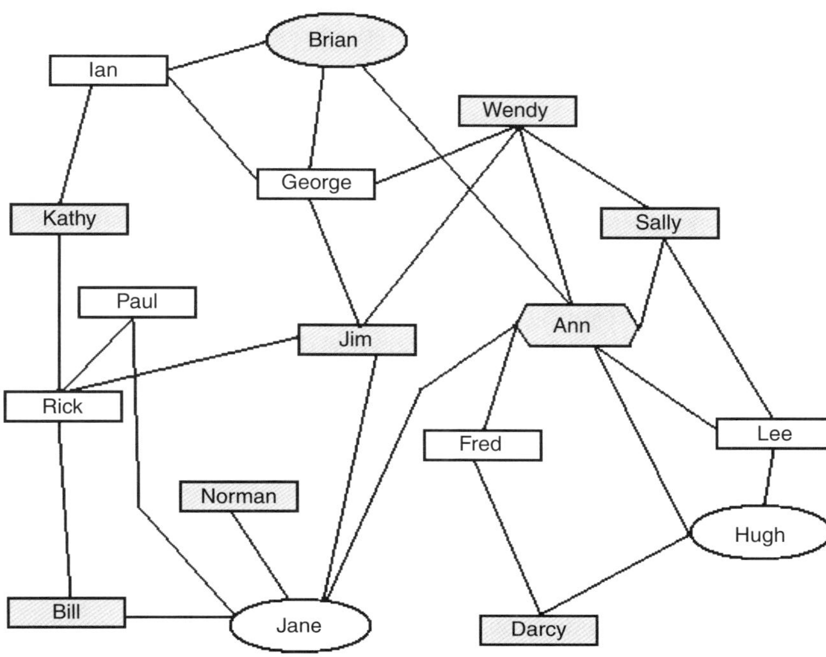

Fig. 27.2. A relationship web showing how only certain individuals (shaded) may actually be present at a particular meeting. The different symbols could represent competing groups or formats.

over another? Do knowledge formats favour one group over another? Can a group increase the competitiveness of their knowledge in the process? Are there any conflict-management processes in place? By mapping information flow and collaboration patterns among the people attending a meeting, one can pinpoint individual bottlenecks, essential participants and those who have been pushed to the periphery or whose expertise is underutilized (Cross and Parker, 2004).

Meta-populations and succession

At a particular position in the planning hierarchy, there may be many meetings at different locations involving different groups of people and organizations; this is captured by the analogy of meta-populations. A meta-population is a group of subpopulations, each isolated in a patch of habitat. The different populations are able to exchange individuals by migration and to recolonize sites where the species has recently become extinct. As succession proceeds, some subpopulations disappear, although migration can occur, permitting survival in a new setting. The theory is important to conservation biologists attempting to understand the process of regional extinctions. The equivalent in the planning process is the many local meetings that feed into the higher-level processes (Fig. 27.3). Some individuals from the lower-level meetings may continue to participate at higher levels, although possibly with a different role, while new individuals become involved. Some groups have no representation at the higher level, so their input to the process must be through some medium such as a meeting summary.

The subgroups of Fig. 27.3 may not simply get together on a single occasion. Rather, they may operate periodically over a considerable length of time as some participatory processes continue for many years, and during this time some groups may disband or have a high turnover of members. The individual meetings in Fig. 27.3 therefore go through their own distinctive evolution in addition to the succession to higher-level groups. In the same manner that populations go through cycles of increase and decrease and individuals go through a cycle of birth, growth, reproduction and death, meetings also go through cycles of activities. There may be development of an initial concept for a meeting, followed by pre-meeting arrangements and decisions, the meeting itself and finally post-meeting activities. Preserving knowledge during those cycles in the evolution of human institutions is a critical issue for knowledge management.

The meta-population concept highlights issues related to multiple groups in a hierarchical process. How uniform are the conditions under which each group operates? How are the outputs of the different groups synthesized? How is representation at higher levels determined, and to what extent is the representative bound by group decisions? Do all meetings at one level have the same agenda and, if not, what are the consequences? How do individuals' roles vary with venue and stage of the process? How are extraneous sources of information included? At what level do most extraneous sources occur? How much loss of knowledge and communication can result if a particular group disbands?

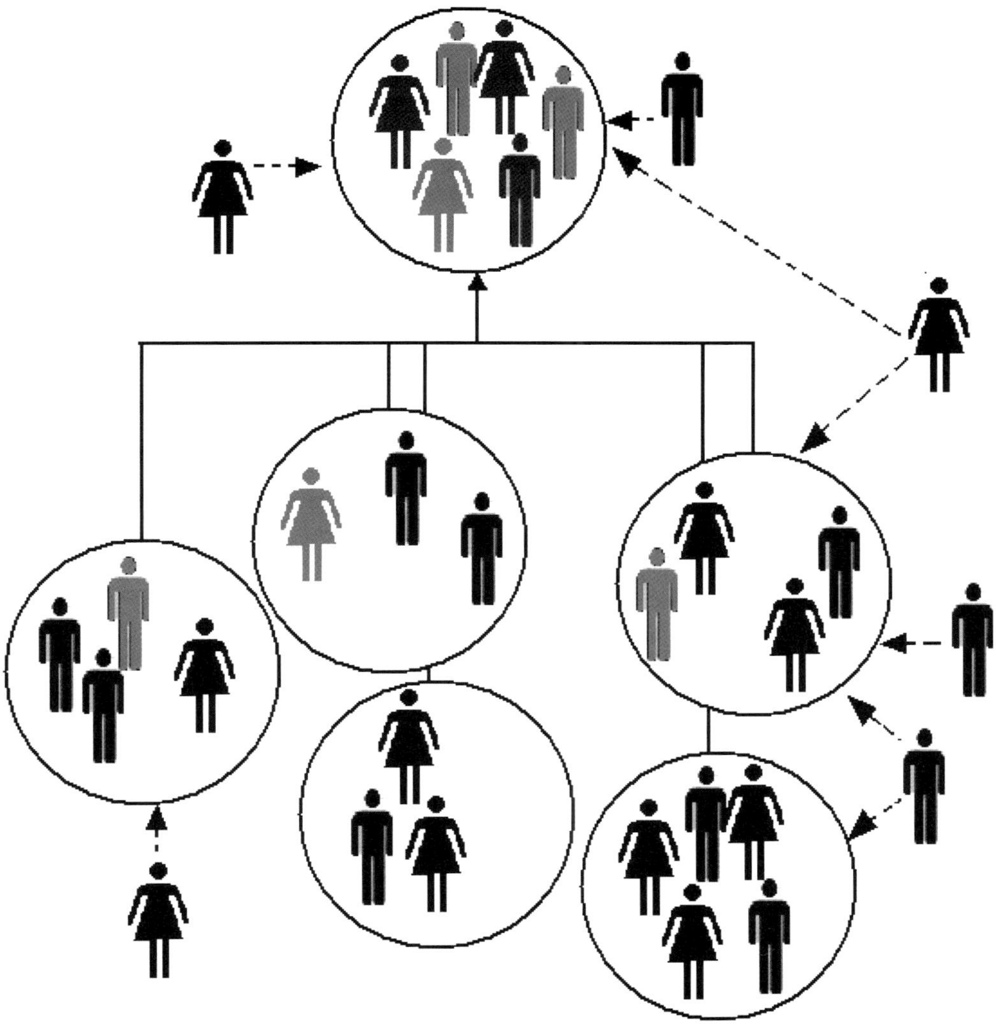

Fig. 27.3. Some individuals (grey) from a meta-population of meetings participate in higher-level meetings along with individuals (dark grey) who did not participate at the lower level. Dashed lines indicate possible extraneous influences.

The spheres

The examples above focus on knowledge, individuals and processes in a hierarchy. This section addresses the knowledge at the level of a particular meeting in relation to spheres of knowledge. The geochemical model of the earth is based on four spheres: atmosphere, lithosphere, hydrosphere and ecosphere or biosphere. As suggested by the World Future Society,[3] knowledge can be partitioned into six spheres: biosphere, sociosphere, technosphere, econosphere, politisphere and futuresphere. Star diagrams based on this concept can be used to characterize the nature of a meeting. For example, a system-design meeting

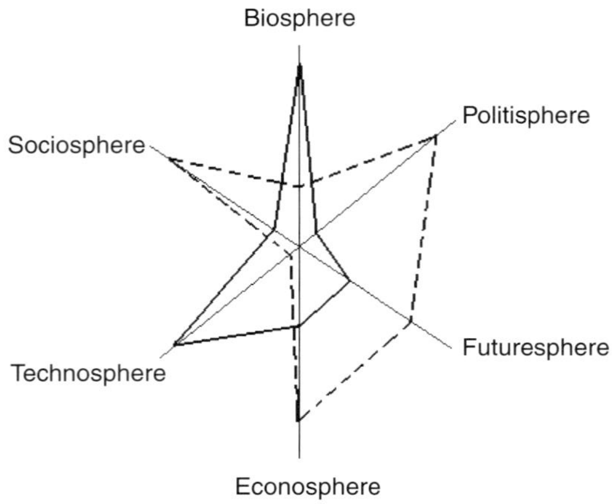

Fig. 27.4. Star diagrams can be used to characterize meetings. A science meeting (heavy solid line) and a forest employment policy design meeting (dashed line) are illustrated.

might be heavily weighted to the technosphere and sociosphere poles, whereas a meeting to define forest policy would have little weighting towards the technosphere pole (Fig. 27.4). Distance along an axis may reflect the range of expertise included for that domain. This approach can be used to categorize meeting types to provide a framework to organize knowledge as this approach is pursued, as well as providing a guide to the range of participants.

Meeting size influences group dynamics, so how should the needs of breadth of knowledge (Fig. 27.4) be balanced against optimal numbers? As there are often different schools of thought within disciplines, how can one ensure that a single expert will bring the required breadth to the deliberations? To what extent can the other meetings at the same level be relied on to provide breadth? Star diagrams (Fig. 27.4) can also be used to ensure an appropriate mix over a set of meetings or to evaluate the extent to which actual participation met a desired level.

Other ecosystem concepts

Other concepts contributing to an expanded view of knowledge ecosystems include sustainability, resilience, ecosystem restoration, information pollution and semantic drift. Sustainability is an important concept for ecosystem management, and there are many definitions and measures, each providing useful analogies for knowledge ecosystems with regard to the effort required for successful long-term processes. For example, 'the energy needed to regenerate waste into useful material for other processes in the system' (Maine, 2003) suggests the idea of difficulty in recovering when a process goes wrong.

The concept of resilience also pertains to recovery:

> ecological resilience can be defined in two ways. The first is a measure of the magnitude of disturbance that can be absorbed before the (eco)system changes behavior. The second, a more traditional meaning, is as a measure of resistance to disturbance and the speed of return to the equilibrium state of an ecosystem.[4]

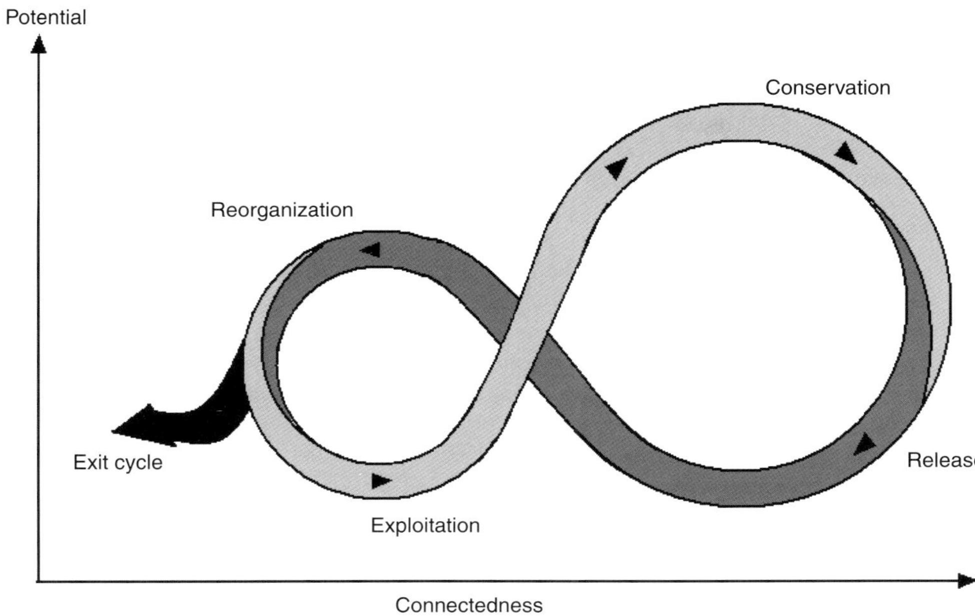

Fig. 27.5. The adaptive cycle of panarchy. For full description see text.

The related concept of panarchy (Gunderson and Holling, 2002) involves cycles of destruction and reorganization, providing opportunities for reorganization of the structure and processes of a system (Fig. 27.5). The individuals participating in meetings during the early stages may be different from those maintaining the process once it is well established. Disturbance of the status quo may provide opportunities for growth and redevelopment, but, if the disturbance is extreme, the process may fail. A hierarchy of cycles within a process can lead to a disruption at one level cascading through other levels.

Ecosystem restoration, the process of re-establishing the structure, function and composition of ecosystems, implies returning to an original format rather than taking advantage of new opportunities. A key question is, therefore: Should change be encouraged (panarchy) or discouraged (ecosystem restoration)? Other questions include: Does the process provide any mechanism to increase resilience? If a group falls apart, are there any contingency plans for its restoration? Is restoration possible (or desirable)? Will dissatisfied participants attempt to cause disruption in hopes of exploiting opportunity? When a participatory process ends in break-up of the group, some people abandon the process. However, in analogy to immigration or emigration, others join different groups at the same or different parts of the process. While this can result in a significant transfer of ideas, it can also spread problems when individual personalities may have been at the root of the original group break-up.

Information pollution includes wrong (or outdated) data, ambiguity, incompatibility of systems and languages, underuse of hardware, hacking, viruses, addressing systems to the wrong 'epistemic who' and lack of responsibility of software suppliers (Capurro, 1990). As knowledge flows through a hierarchy of meetings (Fig. 27.1), it is filtered and transformed; however, information pollution can result in inaccuracies in that process. Genetic drift refers to the constant tendency of genes to evolve even in the absence of selective forces. Similarly, semantic drift (Davenport and Cronin, 2000) suggests that, in an extended process, the meaning of terms can change. Semantic drift may contribute to information pollution.

Whether it will be possible for a process to evolve with experience is a key decision made at the start of the process. The concept of fitness landscapes[5] is key to process evolution. A fitness landscape can be considered as representing how likely a set of conditions is to lead to a successful process. Experience may indicate the optimal set of conditions. However, the existing process, although not the best possible, may be working satisfactorily. The fitness landscape can then be used to indicate whether an attempt to switch to the optimal process might first require passing through transition stages with low likelihood of success, indicating maintenance of the status quo.

Knowledge Management for the Meeting Space Environment

The knowledge ecosystem concepts and the questions raised above can be applied to any participatory process, without any technological components. However Agenda 21 promotes the use of information and communication technologies in participatory processes. This issue will be explored by considering the situation in which one of the meetings is virtual, i.e. a virtual adaptive management workshop.

In ecosystems, organisms interact through processes such as predation and herbivory; in knowledge ecosystems, individuals interact through processes such as debate. The system developed by Thomson (2000a) brought together a diverse set of knowledge management tools developed as Java applets. However, experience with these tools indicated that they did not easily track required attributes of debate. Since that time, the system has therefore been redesigned using a consistent approach based on xml. While the opening page of the new system (Fig. 27.6) is similar in appearance to the former applet-based system, the underlying structure and construction method is very different, and subsequent pages for the range of tools have been redesigned in a manner that facilitates tracking debate.[6]

Plenary and breakout sessions

The first step in the workshop process, similar to the meetings described above, is the identification of context, objectives and players (including experts, facilitators and stakeholders). An individual with a management role, who has particular goals and objectives in mind, normally initiates this. It involves establishing a steering committee to organize the meeting, arrange the meeting space and issue

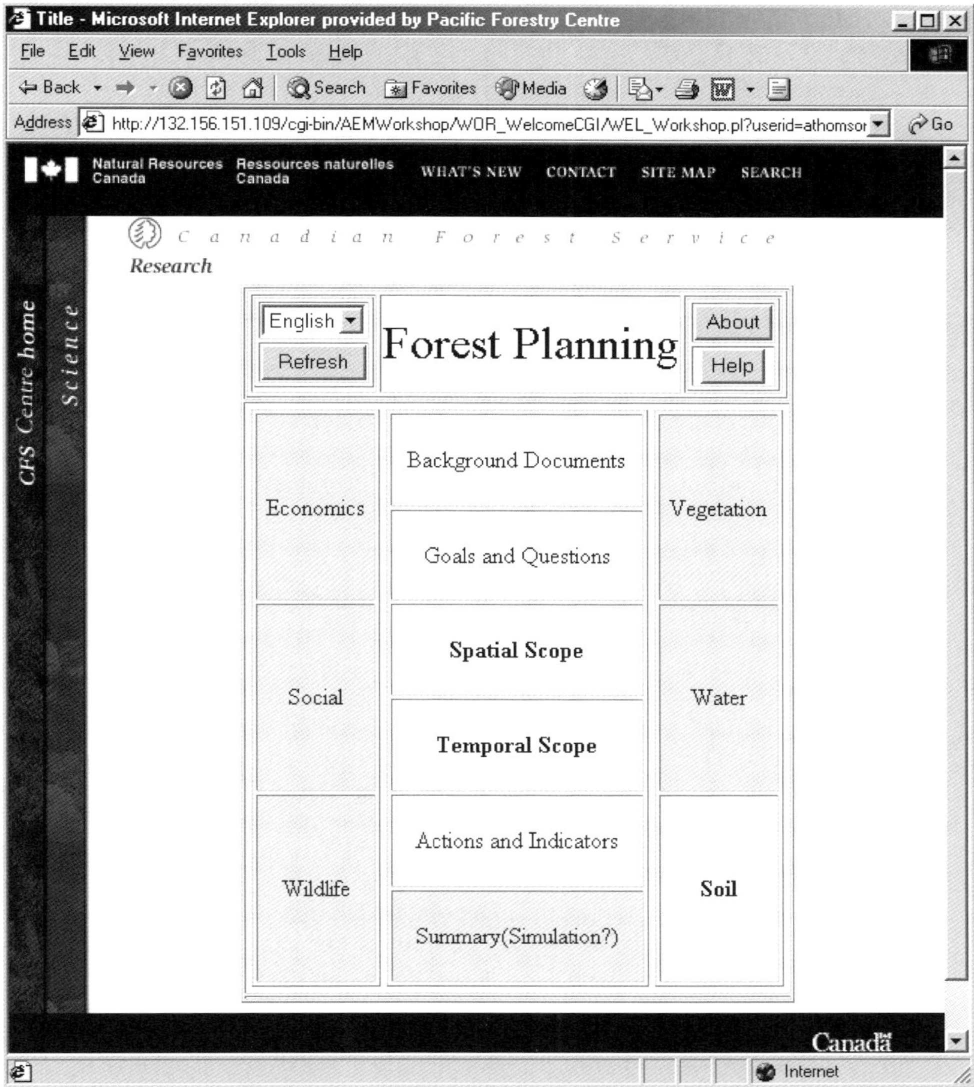

Fig. 27.6. The virtual workshop for a hypothetical evaluation of grizzly bear population management, showing the functions of the plenary session and their status, as well as the available subgroups. The shading and bolding pattern represents a workshop in progress: in light grey cells, bold text represents open and normal text represents closed; darker grey shading represents a pending activity not yet open for interaction.

invitations (Holling, 1978). Adaptive management (AM) workshops have a goal of developing a model, and are structured with both plenary sessions and subgroup or breakout sessions.

In both the real and the virtual workshop environment, spatial and temporal scoping of the system and identification of management actions and indicators occur in the plenary session, the primary consensus-building arena. During the plenary session, participants identify major subsystems and their inputs and outputs, and each subsystem becomes the focus of a separate subgroup session. Subgroup sessions develop algorithms to relate outputs to inputs, involving issues of data collation and data analysis. Competing hypotheses are identified and appropriate test data sets defined. Major knowledge gaps are identified and decisions for future activities resolved. Simulation modelling may link the subsystems to permit exploration of scenarios.

Tracking debate in a knowledge management environment

The general approach to tracking debate is illustrated for the spatial scoping tool (Fig. 27.7). Spatial scoping is map-based, with debate occurring over specified themes. In the example (Fig. 27.7), two themes are illustrated: threatened populations and elk habitat. The term 'grizzly bear' is implicit in the first theme from the context of the workshop. Indentation shows the structure of the debate. The names furthest left in bold text indicate separate views expressed about the winter range. Thelma Hyack and James Level give two different perspectives (equal indentation status) on Bruce Jones's point of view. Two different perspectives on Jean Sidney's point of view (Richard Martin and Norman Olsen) have also been given, but Richard Martin's view was then debated by Greg Jameson, whose view was in turn debated by John Femur. Timings of entry of a point of view are indicated, as well as the initial text of accompanying comments. In the virtual environment, all three of the initial points of view can be debated concurrently.

At this stage, debate is not aimed at determining right and wrong, but at defining alternative hypotheses or views. For example, when the debate on spatial scope closes, a consensus view may be reached, or there may be two or more alternative hypotheses. Outcome of the debate is represented by the workshop facilitator assigning a number to each point of debate to indicate to which generalized view (hypothesis) the point contributes. For example the '(1)' after the name Bruce Jones indicates that his contribution is captured in the generalized view number 1.

The virtual workshop is structured for multilingual situations; i.e. the data entered are language-neutral. 'English comment' is the default value of the comment field when the English language is selected for display of the scoping tool. There is actually only a single version of each tool, with the interface being constructed using symbolic terms, replaced at run time with the term of the appropriate language (selection at the top left of Fig. 27.7). Comments are entered in the language of choice and translated subsequently by human translation services, as machine translation is not yet of sufficient quality.

The nature of debate

The nature of debate varies for each type of tool, and is illustrated by a description of the spatial scoping tool. Thelma Hyack and James Level debated Bruce

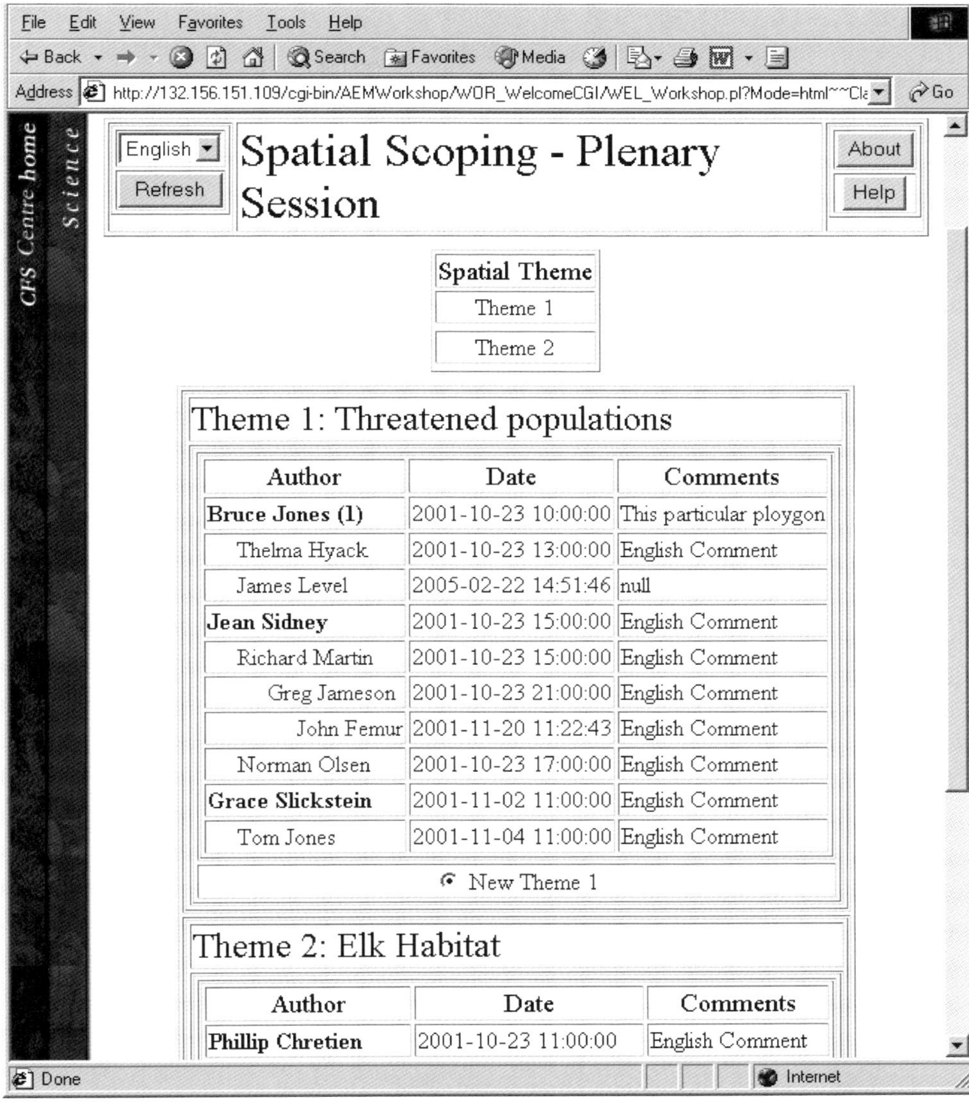

Fig. 27.7. Tracking debate in the spatial scoping tool. For full description, see text.

Jones's point of view: by clicking on Bruce Jones's name (Fig. 27.7), his point of view, expressed as a map and accompanying comment, is displayed in a web page that permits the system user (Thelma or James in this case) to provide their perspective. Note that the title indicates the original authorship. They cannot change the data entered by Bruce, but they can propose additions (through the 'Start new polygon' button) or deletions ('Start new opening') to Bruce's entry. The use of the 'Composite history' checkbox combines the history of additions and deletions in a single view. The comment box allows provision of an explanation for the change. The comment can also include a locator for a file of data or a reference to an article supporting the viewpoint.

A particularly important function of the comment box from the knowledge ecosystem perspective relates to transformations. Ecosystem processes transform energy and matter; in knowledge ecosystems, knowledge is transformed. For example a holder of traditional knowledge who is not computer-literate could work with another individual who would enter their knowledge into the system. Verbally transmitted knowledge or knowledge captured in a sketch map would then be transformed into digital form. The comment box might then contain text such as 'Entered for . . . by . . .'

The spatial scoping tool actually provides very basic GIS functionalities (Fig. 27.8), but is not intended to supplant the sophisticated GIS capabilities offered by formal GIS systems. These latter systems may play major roles in the final stages of the planning process; however, the spatial scoping tool has a quite different role. First, it is integrally linked with the debate-tracking hierarchy (Fig. 27.7). Secondly, it puts all contributors on an equal footing, as will be discussed later. Thirdly, it avoids use of proprietary GIS software, so the system can be freely distributed. It is also possible to run the system without an Internet connection in a remote location. This allows offline meetings and even working with

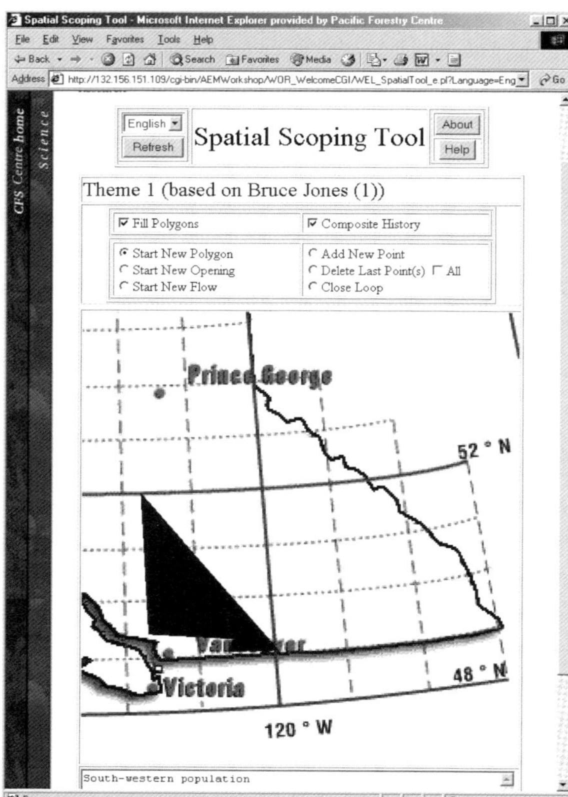

Fig. 27.8. The spatial scoping tool ready for a new user to provide their perspective on an earlier entry.

an individual, giving time for consideration and discussion, facilitating offline consensus before entering a point of view.

A similar hierarchical debate structure to that in Fig. 27.7 is also used with the temporal scoping tool. Three temporal formats can be selected: hours within a day, months within a year or years within a span of years (for example, to record drought cycles or insect outbreak patterns). Other tools have different debate structures. For example, an indicator interaction tool (Fig. 27.9) retains a hierarchical structure, but in this case a thumbnail image of the general graph form proposed is illustrated, and, within the hierarchy for a single graph form, people debate the relative positions of inflection points.

Virtual meetings can mimic their real-world counterparts in many respects, and provide additional features to enhance a number of aspects of participatory processes. In particular, they address the issue of competition, both of individuals

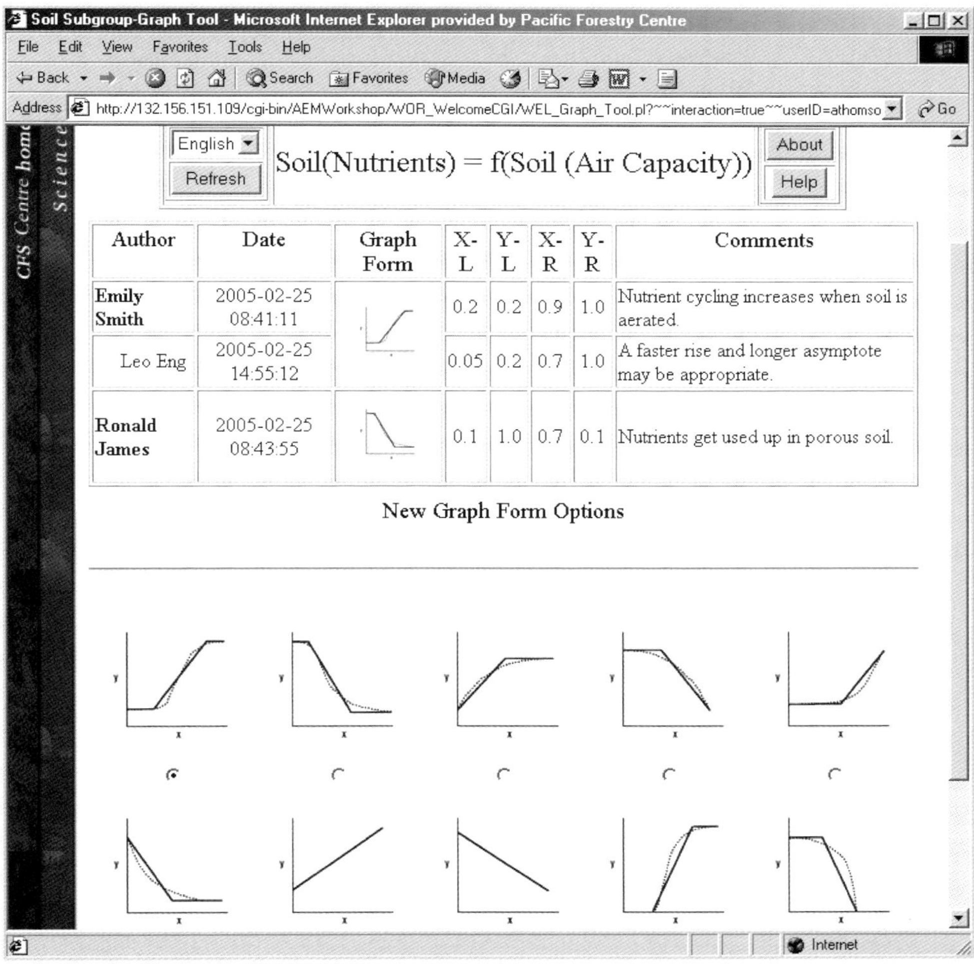

Fig. 27.9. A tool for debating the form of relationships between indicators.

and of knowledge formats. A key feature is the ability to extend the process over time and space, providing more opportunities for individuals or groups to work in their preferred style. In addition, a common set of tools is used, removing format competition. A login process is used that can provide a level of anonymity, although this raises several ethical issues.

Adaptive Knowledge Management

Adaptive knowledge management develops hypotheses about participatory processes and experimentation to develop optimal approaches. The knowledge ecosystem concept provides guidance on prioritizing elements of a particular process that might be detrimental to the desired outcome. These elements would therefore be the primary focus of debate and experimentation. For example, how important is the ability to track contributions through the synthesis process (Fig. 27.1), compared with the ability to have input to a meeting for which there is no representative (Figs 27.2 and 27.3)? Does the meeting format favour some groups over others? Are the technological options (Figs 27.6–27.9) worth the cost and effort? How can other approaches to knowledge be integrated?

Software products in knowledge ecosystems

The tools described above mimic activities in a workshop setting. However, many other knowledge management tools can augment participation, and these can also be incorporated within the system. For example, a traditional knowledge study (Thomson, 2000b) includes questions, answers, inferences, indicators, pictures and stories in an online database.[7] Questions centred on topics of hunting, gathering, community and learning; direct links can be provided to these subsets, showing tracking of inferences drawn from statements, and the use of the inferences in the development of indicators. Similar links can be made to subsets of principles, criteria and indicators in a customized reporting system[8] (Thomson, 2005b). In contrast to the traditional knowledge system and the virtual adaptive management system, which have a role in the synthesis process (Fig. 27.1), the customized reporting system works during implementation. Use of a common interface, linking to different pre-existing knowledge management approaches, facilitates experimentation with the costs and benefits of software development.

Interlinking of systems, people and institutions is the main tenet of interoperability (Thomson, 2005a). '[Interoperability ensures] that systems, procedures, and culture of an organization are managed in such a way as to maximize opportunities for exchange and re-use of information whether internally or externally' (Miller, 2000). Miller (2000) recognizes six types of interoperability: technical, semantic, political/human, inter-community, legal and international. Human and organizational issues are therefore as important as computer-related issues in ensuring interoperability.

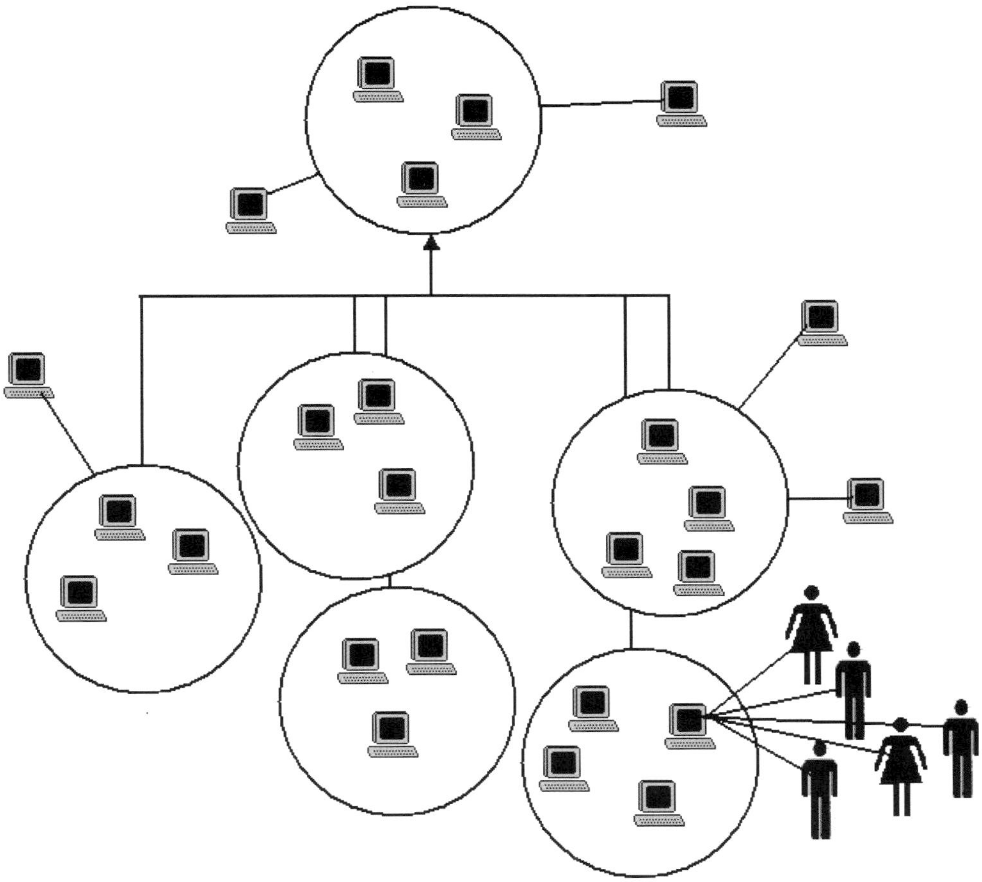

Fig. 27.10. A process based on virtual meetings, with subgroups linked to a plenary session (upper circle). Individuals can have full privileges (symbols inside circles), or only browse access (symbols outside circles). A group of individuals (bottom right) can come together at a common location, such as at a facilitated workshop, to collaborate on their interaction.

A process based entirely on virtual meetings (Fig. 27.10) represents an extreme situation that could be evaluated. Benefits over real-time meetings (Fig. 27.3), such as the reduction of competition, have already been indicated. Other benefits include the ability to automate generation of reports, the transparency of the whole process and the lack of a necessity to integrate the synthesis over many separate meetings.

Discussion

Real-world ecosystems provide a powerful analogy for knowledge ecosystems, explored in relation to forest planning processes. The exploration extended the existing view of the concept by introducing analogies: trophic levels to information synthesis in hierarchical processes; competition in food webs to competition in relationship webs; meta-populations, succession and cycles to the life cycle of parallel meetings at one level of a hierarchical process. These analogies permit the knowledge ecosystem approach to provide a number of concepts to guide creation of successful future participatory processes and also to guide participants in existing processes.

Virtual meetings can be included in a process to increase opportunities for participation, reduce competition based on a meeting format favouring a particular group and facilitate tracking knowledge through the transformations of synthesis. A registration policy may be adopted to permit active (knowledge entry), as opposed to passive (browsing), participation (Thomson, 2000a). Both plenary and breakout sessions can be emulated. In an electronic meeting environment, with both 'drop-in' and 'concurrent' users, the role of the core group in managing the process, and who perform modelling and analysis, may change from that described in Holling (1978). A major role for such meetings is to allow competing hypotheses to be identified, major knowledge gaps to be identified, appropriate test data sets to be defined and decisions for future activities to be resolved.

By helping to identify and prioritize process issues, the knowledge–ecosystem analogy enhances the role of adaptive knowledge management in designing and conducting participatory processes that can be addressed by debate and experimentation. Other knowledge-related analogies could be pursued to provide different insights into participation. These include the social life of information (Brown and Duguid, 2000), geographies of knowledge and politics of space (Kratoska *et al.*, 2005), navigating social-ecological systems (Berkes *et al.*, 2003) and information metabolism (Powell, 1995). Powell raises the useful concept of a 'diagnostician' to monitor the process and to keep it running smoothly – a new role to include in planning. Analogies guide development of new processes and tools, but ethical issues often become more limiting than the technological capability to develop the tools (Thomson and Schmoldt, 2001). Finally, once the technical and ethical issues are resolved, there remains the final barrier of adoption of the process and tools by the parties involved (Thomson *et al.*, 2004).

Notes

1. http://www.un.org/esa/sustdev/documents/agenda21/english/agenda21toc.htm (accessed 12 October 2005).
2. http://en.wikipedia.org/wiki/Information_ecology (accessed 12 October 2005).
3. http://www.wfs.org/8ga.htm (accessed 12 October 2005).
4. http://biodiversity-chm.eea.eu.int/CHMIndexTerms/Glossary/E/ecological_or_ecosystem_resilience
5. http://en.wikipedia.org/wiki/Fitness_landscape (accessed 12 October 2005).
6. For technical details and usage arrangements, contact the author.
7. Available at http://www.pfc.cfs.nrcan.gc.ca/programs/tek/
8. Available at http://www.pfc.cfs.nrcan.gc.ca/management/cfmtools/

References

Berkes, F., Folke, C. and Colding, J. (eds) (2003) *Navigating Social-ecological Systems*. Cambridge University Press, Cambridge, UK.

Berreby, D. (1999) The hunter-gatherers of the knowledge economy. *Strategy and Business*, Third Quarter 16, 52–64.

Brown, J.S. and Duguid, P. (2000) *The Social Life of Information*. Harvard Business School Press, Cambridge, Massachusetts.

Capurro, R. (1990) Towards an information ecology. In: Wormell, I. (ed.) *Information Quality. Definitions and Dimensions*. Taylor Graham, London, pp. 122–139.

Cross, R. and Parker, A. (2004) *The Hidden Power of Social Networks: Understanding How Work Really Gets Done in Organizations*. HBS Press, Boston, Massachusetts.

Davenport, E. and Cronin, B. (2000) Knowledge management. Semantic drift or conceptual shift? *Journal of Education for Library and Information Science* 41, 294–306.

Davenport, T.H. and Prusak, L. (1997) *Information Ecology, Mastering the Information and Knowledge Environment*. Oxford University Press, Oxford, UK.

Gunderson, L.H. and Holling, C.S. (eds) (2002) *Panarchy: Understanding Transformations in Human and Natural Systems*. Island Press, Washington, DC.

Haggith, M., Prabhu, R., Colfer, C.J.P., Ritchie, B., Thomson, A. and Mudavanhu, H. (2003) Infectious ideas: modelling the diffusion of ideas across social networks. *Small-scale Forest Economics, Management and Policy* 2, 225–239.

Holling, C.S. (1978) *Adaptive Environmental Assessment and Management*. Wiley, London.

Kratoska, P., Nordholt, H.S. and Raben, R. (eds) (2005) *Locating Southeast Asia: Geographies of Knowledge and Politics of Space*. NUS Publishing, Singapore.

Maine, T. (2003) Towards a metric of sustainability. In: *In Search of Sustainability, Proceedings of an Internet Conference held February–November 2003*. http://www.isosconference.org.au/papers/Maine.pdf (accessed 3 March 2005).

Miller, P. (2000) Interoperability. What is it and why should I want it? *Ariadne* 24 (21 June). http://www.ariadne.ac.uk/issue24/interoperability/

Nardi, B.A. and O'Day, V.L. (1999) *Information Ecologies: Using Technology with Heart*. MIT Press, Cambridge, Massachusetts.

Por, G. (2000) Nurturing systemic wisdom through knowledge ecology. *The Systems Thinker* 11, 1–5.

Powell, T. (1995) The information metabolism. *Competitive Intelligence Review* 6, 41–45.

Thomson, A.J. (2000a) Knowledge elicitation tools for use in a virtual adaptive environmental management workshop. *Computers and Electronics in Agriculture* 27, 57–70.

Thomson, A.J. (2000b) Elicitation and representation of traditional ecological knowledge, for use in forest management. *Computers and Electronics in Agriculture* 27, 155–165.

Thomson, A.J. (2005a) Editorial: Information interoperability and organization for national and global forest information systems. *Computers and Electronics in Agriculture* 47, 163–165.

Thomson, A.J. (2005b) Indicator-based knowledge management for participatory decision-making. *Computers and Electronics in Agriculture* 49, 206–218.

Thomson, A.J. and Schmoldt, D. (2001) Ethics in computer software system design and development. *Computers and Electronics in Agriculture* 30, 85–102.

Thomson, A., Haggith, M. and Prabhu, R. (2004) Innovation diffusion: predicting success of system development. In: *Proceedings of the 15th International Workshop on Database and Expert Systems, Saragossa, Spain, 30 August–3 September 2004*. IEEE Computer Society, Los Alamitos, California, pp. 627–631.

van Buren, E.M.L. and Blom, E.M. (1997) *Hierarchical Framework for the Formulation of Sustainable Forest Management Standards*. Tropenbos Foundation, Backhuys Publishers, Netherlands.

28 An Information Retrieval System to Support Management of Habitats and Rare Priority and Protected Species (HaRPPS) in Britain

D. RAY AND A.C. BROOME

Ecology Division, Forest Research, Roslin, Midlothian, UK.

Abstract

Recent legislation in the UK levies heavy penalties for failing to carefully maintain protected and rare species and habitats. This has precipitated an immediate requirement for information on management procedures to reduce the risk of damage and disturbance to species and habitats. The people of the UK consider the provision of wildlife habitat as a key social function of forests, and it brings with it economic benefits. Competent management of the wildlife resource, and especially rare and threatened species and habitats, is of key importance. To elevate the standards of sustainable forestry, managers need relevant, clearly presented, highly accessible information recommending the best management practices for managing woodlands and associated open habitats.

There is an increasing volume of information describing species autecology, species distribution, guidance on species and habitat management and the legal requirements of land management. This information is available in a variety of forms from a range of different sources. Forest managers are faced with the task of sifting, selecting and integrating information that is relevant to their requirements. Habitats and Rare Priority and Protected Species (HaRPPS) is a new decision-support tool that provides managers with easy access to this information. The system's foundation is a knowledge database, which is populated using a systematic review process to assess available knowledge and its quality. For each plant and animal species, information can be recovered for a standard set of topics, including appropriate operations to use in forest management. This assists in decision making, as the user is presented with a consistent arrangement and weighting of information that is relevant to their particular subset of requirements. A citation and data-quality tag enables users to assess the authority and potential impact of information. HaRPPS has been developed as a web service, providing access from and to other information systems used by foresters, private woodland owners and land managers. The interoperability of the system with spatial GIS and modelling applications allows consideration of alternative scenarios for rare species and their habitats.

Introduction

Over recent decades it has become more widely recognized that the continued spiral of economic development at the expense of future generations is unsustainable for the planet (Anon., 1992a). The Brundtland Report defined sustainable development as development that 'seeks to meet the needs and aspirations of the present without compromising the ability to meet those of the future' (Brundtland, 1987). In the early 1990s, international forest policy embraced the concept of sustainable development, with, for example, the Helsinki Ministerial Conference on the Protection of Forests in Europe (Anon., 1993), which defined sustainable forest management (SFM) as:

> the stewardship and use of forests and forest lands in a way, and at a rate, that maintains their biodiversity, productivity, regeneration capacity, vitality, and their potential to fulfil, now and in the future, relevant ecological, economic and social functions, at local, national and global levels, and that does not cause damage to other ecosystems.

The UK government responded by publishing several strategic planning documents; two of these – *Sustainable Forestry: the UK Programme* (Anon., 1994a) and *Biodiversity: the UK Action Plan* (Anon., 1994b) – laid the foundation for the forestry agenda and the development of SFM guidelines, and pioneered the use of targets in nature conservation. *The UK Forestry Standard* (Anon., 1988, revised 2004) identified key issues, with targets and indicators for forest management, underpinned by SFM principles. The standard sets out the government's approach to sustainable forestry, for which the Forestry Commission has statutory responsibility within Britain; the Northern Ireland Forest Service assumes this role in Northern Ireland. More recently the standard has been adopted as Britain's implementation approach to the 'Pan-European Criteria' (PEC), which are applied throughout Europe in all forests. Since the introduction of the UK Woodland Assurance Scheme, in 2000, which is a certification standard of SFM, 40% of the UK's forests have been certified, accounting for 60% of the total timber production. This has led to overall improvements in the standard of management, including biodiversity, conservation and monitoring of these areas (Garforth and Thornber, 2003).

In the UK, as in most countries (Marcot *et al.*, 1994; Wisdom *et al.*, 2000; Huff *et al.*, 2001), forest is viewed as a resource that should be managed for multiple benefits (timber quality and its economic value, recreational and aesthetic quality of forests for people and environmental benefits such as forest biodiversity). As a result of forest policy changes, forest managers have adopted the principles of SFM; management of state-owned forest follows government policy, and such adherence is a prerequisite for privately owned woodlands receiving forestry grants. Along with these changes, the level of species protection in EU member countries continues to increase (Anon., 1992b). In the UK, recent changes in legislation, and regulations currently under review, e.g. Conservation (Natural Habitats & c.) Regulations 1994, have increased the protection for many additional species and habitats not covered by the EU Habitats Directive (Anon., 1992b). Forest managers are now more accountable for the quality of management for biodiversity, and are legally responsible for species and habitats within their care.

Research in support of SFM guidelines, and for conservation of biodiversity, is providing a growing information resource. Within the sphere of research on forest biodiversity, there is an increasing knowledge base on the broad principles of forest management and its impact on biodiversity (e.g. Humphrey *et al.*, 2003), as well as data on the distribution and status of species and habitat (e.g. Hill *et al.*, 2005). Studies of habitat and species autecology are feeding into management guidance (e.g. Hill *et al.*, 1997; Anon., 2001; Vanhinsbergh *et al.*, 2002); most government bodies concerned with land management, along with the wildlife non-government organizations (NGOs) and charities, produce information on their priority species or habitats.

Forest managers are faced with a plethora of information, from a variety of sources, and in addition have to deal with the complex legislation covering the protection status of species and habitats, e.g. EU Species and Habitats Directive (92/43/EEC), the Wildlife and Countryside Act 1981 (as amended), and the Nature Conservation (Scotland) Act 2004. Managers who wish to follow policy, law and good practice need clearer advice. For those who are reluctant to manage for fear of legal responsibilities, encouragement is needed through well-presented advice and support. A single, open-access gateway is needed, with information and guidance presented in an organized way that is relevant to forest practitioners and their operations. Our view has been that this could be most effectively offered by a combined knowledge and information decision-support system. Such a proposal has been offered and accepted by key professionals in the forest industry.

This chapter describes such a system, the Habitats and Rare Priority and Protected Species (HaRPPS) decision-support system. It is designed to aid forest managers in maintaining high standards of SFM, while minimizing disturbance to protected species and reducing damage to their habitats. The HaRPPS application structures, simplifies and delivers complex information on the protection of existing species and their habitats, and in addition it encourages careful habitat management for potential colonizing species. Its aims are to: (i) make clear the threats posed and opportunities offered by management operations; (ii) provide the information to help resolve conflicts associated with managing multiple species or planning multiple operations; and (iii) assist decision making by presenting relevant information in a logical framework.

HaRPPS Design

The decision-support system is composed of several components that allow easy linkage from internal (HaRPPS) and external sources (other web applications). The model of information flow is a schematic representation of the relationships between data tables within the database, and is a useful way of demonstrating the linkages and dependencies in understanding the kind of data queries that can be constructed.

System architecture

HaRPPS has been developed as a web application to provide open access with exposed web services, and to allow interoperability and rapid linkage by third-party systems. Figure 28.1 shows a schematic of the architecture.

The HaRPPS application has five main elements: two database components, a database interface, various business–logic components and a presentation layer. The HaRPPS database holds information on the species, habitat, structure, operations, location and protection status, while the second is a 'user and site detail' database, holding information input by the user relating to site analyses. Business–logic is split into two components that hold the business rules and logic of the HaRPPS application, providing access between the presentation and data sources for both the HaRPPS database and external web services. The first business–logic component handles queries from the HaRPPS application, while the second contains the system application that handles how the HaRPPS component is accessed by the user, maintaining registrations and other user information held in the 'user and site detail' database. This separation ensures that the application maintains potential for access from different interfaces such as web services.

In order to allow the dynamic configuration of queries by users on the client side using the HaRPPS application interface on a web browser, and at the same time to provide access to third-party applications to query the database, the client-presentation component sends pertinent web pages as JSPs (Java server pages) as an xml message. The information is then translated into different formats

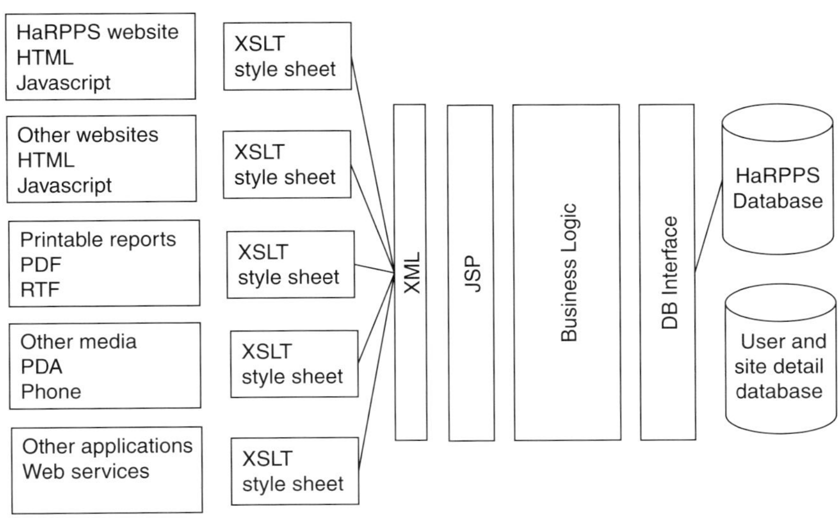

Fig. 28.1. Schematic system architecture of the HaRPPS system. DB – represents the database interface; business logic – handles access permissions and organizes user queries into database requests; JSP – handles data within web pages in XML format; XSLT – creates the web page format and style for different end-user applications.

using style sheets. For example, the style sheet will convert the web page information to html and Java script on older web browsers. More recent browsers will handle xml style sheets and, for reports, the style sheets convert the information to Adobe®.pdf and Microsoft® Word®.rtf format for printing.

Information model

The HaRPPS system is underpinned by an Oracle relational database, simplified schematically in Fig. 28.2. In reality, there are five similar models of information flow that account for differences required in the ecology tables between the main taxonomic groups. For example birds, invertebrates and herpetofauna have tables that hold information on eggs and nests, mammals have tables for information on young and higher plants and lower plants contain tables holding information on seed and spore production.

Each of the five variants of the information-flow model share the features shown in Fig. 28.2; however, the number of tables contained within the autecology factor changes by variant. Figure 28.2 shows linkages between species and associated factors: habitat type, forest operations, autecology, location and protection status (holding information about legislation, regulation and policy).

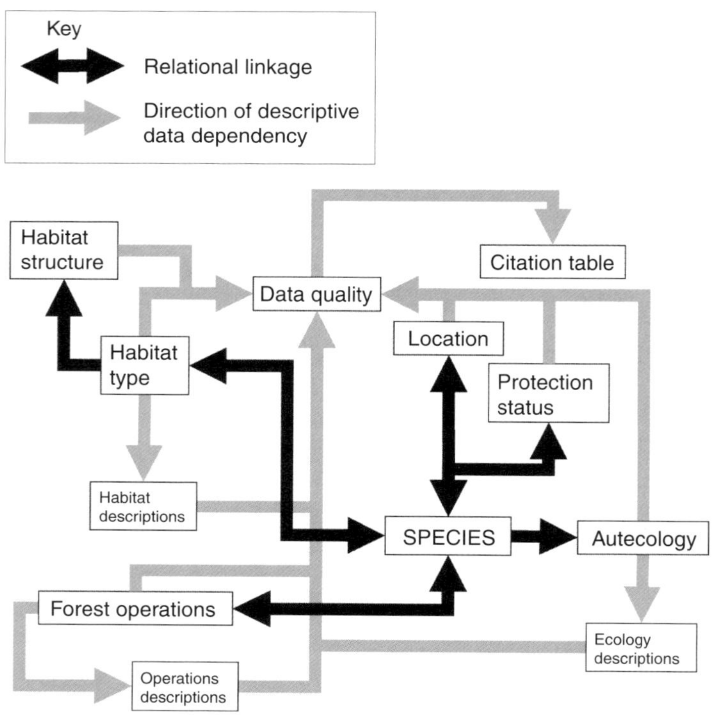

Fig. 28.2. Relational linkages and the association of descriptive data to forest operations, autecology and habitat within the HaRPPS database.

Each factor contains several attribute tables. For example, invertebrate autecology contains tables that hold the species information about: minimum area requirements, home range and dispersal, adult food, young food, breeding, predation, hibernation, activity period, fecundity, larvae/pupae characteristics and signs. Similarly, the habitat-type factor contains known relationships between habitat types classified according to the National Vegetation Classification (NVC) (Rodwell, 1991), habitat action plans and broad habitat types (Anon., 1994b).

An important feature of the information flow is the development of two-way relational linkages between species and habitat type, species and forest operations, species and location and species and protection status, allowing queries to be made on species or any linked factor. In addition, two one-way dependencies between ecological data and species and between habitat type and habitat structure were created, allowing queries on species to provide a range of detailed autecological information and allowing very detailed habitat queries to be defined. Dependencies were created between habitat, forest operations and ecology tables to associate descriptive information with a species, and the numerical and descriptive information in all tables was linked to a table for data-quality classification and a citation index.

Knowledge acquisition

Linkages between the attributes in different feature groups have been formed from a systematic review of the research literature and from the experiences and expert opinion of practitioners within forest management. Figure 28.2 identifies how the data quality and a citation table operate as core features within the system, and these tables offer the transparency required when mixing evidence and experience-based information presented to a user or reviewer of the database. The data-quality table provides a means of classifying the type and quality of information used in the database. Information is tagged in the review process to identify the source and its quality. Information gained from the results of scientific research and published in peer-reviewed papers and books is given a classification value of 5 (Table 28.1). Information may also be derived from un-replicated trials (value 3), and observations from rangers and managers (value 1), allowing its

Table 28.1. The classification of data quality in HaRPPS to allow the user or reviewer of the system to assess the reliability of the information provided by the system.

Quality class	Description
5	Peer-reviewed papers and books
4	Websites (known quality review process)
3	Unpublished internal reports
2	Websites (unknown review process)
1	Anecdotal experience

reliability, and hence some of the uncertainty associated with the information, to be considered.

The citation table holds the reference(s) from which the information was obtained. Every piece of information in the database has at least one citation, and every citation has a data-quality class associated with it.

Temporal and spatial resolution of data and HaRPPS architecture

The current version of HaRPPS (version 1, June 2005) is not linked to a geographic information system (GIS) and provides no spatial context for queried information. However, we know that species or their location, land-cover types (habitat) and forest operation types, as well as controlling biophysical variables, all vary through space and time. Species distributions provide a good example; records from the two decades 1985–2005 from Butterfly Conservation, accessed through the National Biodiversity Network (NBN, 2004), show a contraction in the range of the pearl-bordered fritillary (*Boloria euphrosyne* L.) in the south of England and a range expansion in Scotland. The HaRPPS version 1 database holds data on species location at the resolution of a vice-county (a vice-county is an administrative unit area for botanical recording), and vice-counties vary in size but on average cover areas of about 50,000 ha. This crude spatial resolution potentially limits the use of HaRPPS in site-based queries, but we plan to use information on species location from other sources (e.g. web services) in future versions. This solution has the advantage of eventually making fast-changing information (e.g. species location) independent of HaRPPS. The likely sources of data on species distribution are the NBN and the Forestry Commission GIS systems (*Forester*).

HaRPPS Operation

Simple queries

The HaRPPS data-browser interface contains two sections: an upper section for building a query, and a lower section in which the results are returned (Fig. 28.3). Simple species queries are performed by selecting the species from the list shown on the pane under the 'Species' tab. Figure 28.4 provides an example of ecological information on the pearl-bordered fritillary (*Boloria euphrosyne* L.).

Complex queries

Complex queries can also be set up in HaRPPS. For example, it is possible to extract rare priority and protected species that occur in management units, such as forest districts, conservancies, vice-county or within the vice-county containing the specified grid reference and/or are associated with a habitat type and/or

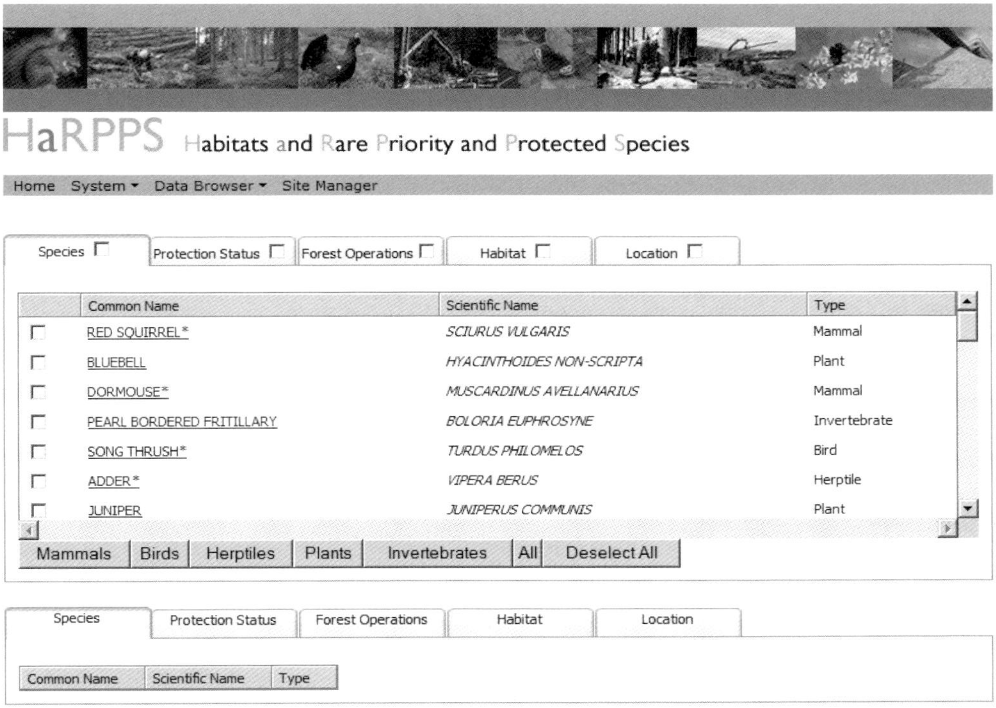

Fig. 28.3. The HaRPPS Version 1 query builder interface (upper section) and lower section where results are displayed. Radio buttons on tabs allow the factors to be switched on and off quickly without disturbing detailed selections in sub-tabs (not shown).

which will be affected by forest operations and/or have a particular protection status. In Fig. 28.3, selecting the 'Location' tab provides access to defining an area for the query; selection of the 'Habitat' tab provides the user with an interface to describe the habitat type and structure, and so on for forest operations and protection status. When the query has been constructed, it is sent to the database on the server, and the results are returned in the lower section of the window. Information on the quality of the data is displayed interactively, when the cursor is held over the item (Fig. 28.5).

Discussion

SFM is demanding more of forest managers, and the knowledge supporting SFM is proliferating. As a result, there is growing interest in decision-support systems (Turban and Aronson, 2000) that synthesize and deliver the knowledge to managers (Rauscher *et al.*, 2000); examples in the UK include ESC (Pyatt *et al.*, 2001; Ray, 2001), ForestGALES (Dunham *et al.*, 2000) and EMIS (Perks *et al.*, Chapter 24, this volume). Some of these decision-support tools have been

Fig. 28.4. Ecological data on the pearl-bordered fritillary (*Boloria euphrosyne* L.) presented in HaRPPS Version 1.

designed to deliver complex models in a simple manner – for example, matching biophysical variables to tree physiological preferences (ESC). HaRPPS has been designed to deliver complex information in a structured and accessible way. Other examples in the UK include Tree Doctor (Bourgery *et al.*, 2002), and Herbicide Advisor (Thomson and Willoughby, 2004). Another Windows®-based computer decision-support system with similar objectives (NEWILD-, within the NED family of decision-support systems) (Thomasma *et al.*, 1998; Twery *et al.*, 2000), developed in the USA, has been designed in a broadly similar way with links between species, habitat type and structure. NEWILD has codified expert knowledge on 338 species, centred mainly in the eastern USA. Although it does not contain explicit links between forest operations, impacts and status of legal protection, it has been developed to deliver SFM within the framework of the Endangered Species Act (1973).

Fig. 28.5. Data quality information on one of the food sources for the larval stage of the pearl-bordered fritillary (*Boloria euphrosyne* L.). Quality 5 indicates that the information has been extracted from peer-reviewed sources; the citation is also shown.

Organizing and delivering information in HaRPPS

A key concept of HaRPPS is the interplay of priority organisms (for which the Forestry Commission in England, Scotland and Wales has either a legal obligation or a duty of care to protect), their habitats, distribution and sensitivity to forest operations, and the implication of this to their continued viability. The database is structured to allow queries to be constructed from knowledge of any one or more of these factors, and to provide information on any of the others. Existing information on the autecology of species and habitats is dispersed throughout the literature in scientific papers, books, magazines and specialist websites. The review process in HaRPPS aims to search the literature and select information that is relevant and reliable. The key data requirements of the review cascade naturally from the predetermined database structure, and the review was conducted by an experienced ecologist. The relevance of information was largely based on the reviewer's judgement, and the process sources information to supply the same suite of facts for each species. The literature search inevitably identifies information of varying quality: information that is relevant and specific to the data requirements is extracted directly, classified and cited. In the absence of specific information of any quality, the reviewer applies a judgement or inference, sometimes from combined sources, to populate the database. In such cases, a low score for data quality (value 1) and appropriate citation reflect the nature of the data and any judgement applied to the data value. This results in the user being presented with the same arrangement and consistency of information on which to base decisions, even though the data quality is acknowledged as variable from species to species.

Reliability is assumed from the data quality. The peer-review process ensures that information is screened for accuracy, even when this is not explicitly stated. Failing that, reliance is placed on the 'grey literature' (internal and unpublished reports, literature that has not been peer-reviewed and web page content) as this information has been assimilated and placed in the context of other work during the writing process, and is therefore assumed to be better than experience-based and anecdotal evidence for which results are observed in isolation. For many species, little is known about aspects of their ecology or their response to interactions with habitat or operations. In this case, we are dependent upon the experience of wildlife rangers, land managers and ecologists for information, again with the proviso that all information is classified in terms of quality and contains the source reference. The approach used in HaRPPS, to grade and reference all information, allows multi-sourced data to be held on species and presented in a transparent form to users. This contrasts with a purely evidence-based approach extracted by a systematic review process (Pullin and Knight, 2001, 2003; Stewart *et al.*, 2005), initially developed within the field of medicine. Evidence-based knowledge is rigorously tested in review, and so provides a robust knowledge base. However, the acquisition process would take time and considerable resources to complete for the 45 species and their interaction with forest operations included within HaRPPS. It is important that the systematic review continues, but in the meantime HaRPPS provides qualified best evidence, and a system of identifying where both the research and review effort might be focused.

A future development in HaRPPS version 2 will facilitate feedback from users on the efficacy of the operations that they practise in the field, which may or may not have been recommended by HaRPPS. This information will be considered within the regular review programme of the HaRPPS database, and may be used to refine advice. One aspect of the provision of this information that requires further thought is recording the efficacy of operations in relation to the variation in the application of operating standards. For example, the response to bracken (*Pteridium aquilinum* (L) Kuhn) control using a herbicide such as glyphosate to improve conditions for the pearl-bordered fritillary (*Boloria euphrosyne* L.) might vary with antecedent weather conditions. It would be advantageous to attempt to record such information to avoid the implicit variation in operation standards adding to uncertainty.

In support of species and habitat protection

HaRPPS principally addresses the nature conservation aspects of SFM, for which there is a stated requirement that 'biodiversity in and around woods and forests is conserved and enhanced' and that, in particular, 'species and habitats subject to EU Directives or priorities of the UK Biodiversity Action Plan are conserved or enhanced' (Anon., 2004). In Britain, there has been a long preoccupation with conservation in the form of preservation of habitats for rare and threatened species (Ratcliffe, 1977). Recent work, though, has shown that preservation alone is not an effective means of preventing decline in biodiversity (Margules and Pressey, 2000). Indeed, evidence suggests that the maintenance and enhancement of biodiversity are achieved only when a more holistic approach is adopted, perhaps by management planned at the landscape scale (Forman and Godron, 1986). While this view may be true for biodiversity generally, it does not always follow that rare and threatened species will respond solely to management at the landscape scale. Many very rare species are sedentary, have very specialized habitat requirements or require a long continuity of the habitat to remain viable.

Foresters are perhaps unique among land managers in their approach to species and habitat management. The UK forest industry is engaged in the active sustainable management of land, with a history of improving woodland for production, biodiversity and other environmental benefits, as well as public enjoyment. As a result, in forestry, the problem is not so much overcoming the reluctance to intervene, but providing evidence- and experienced-based knowledge to make the method and timing of interventions successful in all aspects of SFM. Clearly, readily accessible information may be a key to encouraging management for rare and threatened species.

HaRPPS lists both the positive and negative impacts of forest operations on species, and provides current recommended practices to avoid disturbance of species or their habitats. It also indicates the species that have potential to be present in a management area, because they occur locally and their habitat is present. Managers have the option to plan management and create habitat for colonization opportunities. With the help of HaRPPS, management can be focused to mitigate the impact of forest operations, enhancing species conservation and encouraging biodiversity.

Recommending management practice to the forest industry

Amendments in the legislation for species protection make it critically important that forest managers are aware of whether operations could lead to the damage of protected species or their habitats. HaRPPS reveals potentially damaging operations, describes how they are likely to cause damage or disturbance if badly timed or implemented, and suggests alternative methods to avoid such problems. It also provides context decision support, in providing a potential list of affected species from location, habitat type and all operations planned. This important feature flags possible impacts on species that may be present on a site or that may not have been considered. The manager may then use the information to look for signs or other evidence of species presence.

Within the Forestry Commission, HaRPPS has been acknowledged as a principal delivery mechanism for the dissemination of good forestry practice to maintain high SFM standards and reduce the likelihood of statutory infringements from forest operations damaging habitats and species. HaRPPS maintenance and regular review cycles, which incorporate peer review, are thus essential elements of the system, in order to ensure that information and recommendations remain current.

Resolving management conflicts

Management operations have the potential to seriously impact different species in a variety of complex and often conflicting ways. A good understanding of the habitat requirement of each species of interest, coupled with knowledge of the larger landscape context and of the forest and habitat types that comprise the landscape, is helpful for resolving these conflicts (Liu *et al.*, 1995). Managing conflicts is best approached from a position of knowledge, where decisions are made after assessing various model scenarios that attempt to minimize conflicts and after discussion with interested parties. HaRPPS provides the evidence or experience base from which a forest manager may: (i) quickly reassess the pertinent knowledge; (ii) be in an informed position when discussing objectives with regulatory authorities; (iii) show willingness and skill in offering alternative solutions; and (iv) set priorities based on the designation and local importance of the different species involved. Solutions will involve developing a risk assessment of alternate approaches, an understanding of habitat and niche requirements of species (particularly sedentary species) and assessment of the distribution of species in question. The benefits of this approach are that forest managers become more knowledgeable about the ecological implications of forest management they are considering and, when discussing operational conflicts with regulatory authorities, are able to offer solutions to problems and negotiate clear management prescriptions.

Future modelling needs

Potentially, there are two interesting links between the HaRPPS type of decision-support system and other forest models. For example, growth models such as 3PG (Landsberg and Waring, 1997) can add predictive elements to HaRPPS by informing the rate and direction of woodland habitat creation, giving a time frame for the development of newly established (or regenerated) woodland components and structures that provide the niche requirements for colonization.

It is acknowledged that stand development trajectories are highly stochastic; even so, integration with growth models might be achieved by incorporating model outputs from scenario predictions into the HaRPPS database, and coding the data quality accordingly (as a model prediction with a highly uncertain outcome, and a citation). Secondly, data in HaRPPS can be and have been used as input for other predictive models. For example, the spatial model BEETLE (Ray *et al.*, 2004; Watts *et al.*, 2005), which analyses the functional connectivity of habitat patches within landscapes, requires information on the home range and dispersal distance of focal species; HaRPPS provides this information.

Networking with inventories

Interoperability is an essential aspect of many decision-support tools, allowing the exchange of data and services between components of different systems (Twery *et al.*, 2000). HaRPPS has been developed specifically with this functionality. The XML components allow linkage with national digital data sets (MAGIC, 2002). For example, HaRPPS version 2 is being designed to query the NBN to access distribution data for species. HaRPPS will also provide autecological and recommended management guidance to the Forester GIS™ system (Ditchburn *et al.*, 2005), which holds management information on Britain's state forests. Interestingly, following the agreement of forest management and expansion plans, HaRPPS could predict the distributions of rare species to populate inventory scenarios of the future.

The social rationale for rare species management

Rare and threatened species and habitats are important in the public perception of forest management and its success. The *UK Public Opinion of Forestry 2003 Survey* (Anon., 2003a) reported that '48% of all adults had seen or read about forests, woodlands or trees . . . on the television, radio or in the newspapers. As in previous years, the topics most widely seen were "birds or other animals in the woodlands".' Selecting from a list of 14 issues, 72% of respondents believed the main public benefit of forests was in providing wildlife habitat, and this response had the highest ranking.

It has been estimated that approximately 570 million visits were made to UK woodlands in 2002/2003. The nature conservation function of forests, including the sensitive management of forests for biodiversity and the management of rare species and habitats, is therefore of major social consequence. HaRPPS provides a tool for managers to improve management of rare and threatened species and deliver this important social benefit.

Setting aside the social rationale for protecting and enhancing woodland biodiversity and rare species, there are also good economic advantages of doing so. The rare woodland species, particularly birds and butterflies, draw the public to the forests (e.g. osprey – *Pandion haliaetus* L., red squirrel – *Sciurus vulgaris* L., chequered skipper – *Carterocephalus palaemon* Pallas), and feature viewpoints and exhibition/viewing centres have been developed to raise the profile of rare-species management. This has stimulated a 'wildlife tourism' industry (Bryden, 2003; Martin, 2003) and, as a result, more people are choosing to visit forests as part of a recreational–learning event, and this is likely to benefit rural economies in terms of services, products and employment (Anon., 2003b).

The *UK Public Opinion of Forestry 2003 Survey* (Anon., 2003a) reported that nearly 70% of respondents would like more woodland in their part of the country. It is therefore clear that foresters must continue to disseminate the success and benefits of managing rare species, and of SFM in general. This important fact shows that there is a major role for open-access web applications supporting information systems, and decision-support tools to inform and support forest practitioners and managers in the direction of improved SFM and, in Britain, to help support continued forest expansion.

Acknowledgements

The concept of HaRPPS developed out of discussions with Gordon Patterson and Andy Brunt. Chris Vials and Una McEvoy developed the database framework, Tracy Brown undertook the literature review for HaRRPS version 1, and Andrew Mason developed the user interface. We would also like to thank the HaRPPS Steering Group, in particular Sallie Bailey, Patrick Green, Michelle Bromley and Fred Currie. We also wish to thank Chris Quine, Helen McKay, Paul Hessburg and Alan Thomson for comments on an earlier draft.

References

Anon. (1992a) Earth Summit. In: *United Nations Conference on Environment and Development.* Regency Press, London and Rio de Janeiro.

Anon. (1992b) Council of the European Communities European Union Habitats Directive, COUNCIL DIRECTIVE 92/43/EEC of 21 May 1992 on the conservation of natural habitats and of wild fauna and flora. *Official Journal of the European Communities* L 206, 22/07/1992 p. 0007–0050. http://eur-lex.europa.eu/LexUriServ/LexUriServ.do? uri=CELEX:319921.0043:EN:HTML

Anon. (1993) Resolution H1: General guidelines for the sustainable management of forests in Europe. Second Ministerial Conference on the Protection of Forests in Europe, Helsinki, June 1993. Ministry of Agriculture and Forestry, Helsinki, Finland. http://www.mcpfe.org/resolutions/helsinki

Anon. (1994a) *Sustainable Forestry: the UK Programme*. HMSO, London.

Anon. (1994b) *Biodiversity: the UK Action Plan*. HMSO, London.

Anon. (2001) *Great Crested Newt Mitigation Guidelines:* Version 2001. English Nature, Peterborough, UK.

Anon. (2003a) *UK Public Opinion of Forestry 2003 Survey*. Forestry Commission Report, Forestry Commission, Edinburgh, UK.

Anon. (2003b) *Forests for Recreation and Nature Tourism (FORREC)*. Technical Annex to Draft Membership of Understanding for the Implementation of a Concerted European Research Action designated as COST Action E33. Council of the European Union, Brussels. http://www.cost.esf.org/index.php?id=153

Anon. (2004) *The UK Forestry Standard,* 2nd edn. Forestry Commission and Forest Service Northern Ireland, Forestry Commission, Edinburgh, UK.

Bourgery, C., Grange, C., Rose, D.R., Tourret, V., Kopinga, J., Moraal, L., Tagliaferro, F., Aversenq, P., Chauvel, G., Nageleissen, L.M. and Edelin, C. (2002) *Tree Doctor*. Institut pour le Développement Forestier (IDF), Paris.

Brundtland, G. (ed.) (1987) *Our Common future: The World Commission on Environment and Development*. Oxford University Press, Oxford, UK.

Bryden, D. (2003) *Quality of Wildlife Tourism Survey*. Highlands and Islands Enterprise Report, Inverness, UK.

Ditchburn, B., Halsall, L., Kennelly, C. and Lennon, B. (2005) Using the Forester Geographic Information System to help deliver sustainable forest management in Britain's national forests. In: *Sustainable Forestry in Theory and Practice, IUFRO Edinburgh, April 2005*. IUFRO Edinburgh, UK.

Dunham, R., Gardiner, B.A., Quine, C.P. and Suarez, J. (2000) *ForestGALES: A PC-based Wind Risk Model for British Forests – User's Guide*. Forestry Commission Edinburgh, UK.

Forman, R.T.T. and Godron, M. (1986). *Landscape Ecology*. Wiley and Sons, New York.

Garforth, M. and Thornber, K. (2003) *The Impacts of Certification on UK Forests*. Report by the UKWAS Support Unit LTS International, Penicuik, UK. www.forestry.gov.uk/pdf/Finalreport27Nov.pdf/$FILE/Finalreport27Nov.pdf.

Hill, D., Hockin, D., Price, D., Tucker, G., Morris, R. and Treweek, J. (1997) Bird disturbance: improving the quality and utility of disturbance research. *Journal of Applied Ecology* 34, 275–288.

Hill, M.O., Arnold, H.R., Broad, G.R., Burton, V.J., James, T.J., McLean, I.F.G., Preston, C.D., Rowland, F. and Roy, D.B. (2005) *Biological Records Centre Report 1999–2004*. JNCC Report No. 370, Joint Nature Conservation Council, Peterborough, UK.

Huff, M., Mellen, K. and Hagestedt, R. (2001) A model to assess potential vertebrate habitat at landscape scales: HABSCAPES. In: Johnson, D.H and O'Neil, T.A. (eds) *Wildlife–Habitat Relationships in Oregon and Washington*. Oregon State University Press, Corvallis, Oregon.

Humphrey, J.W., Ferris, R. and Quine, C.P. (2003) *Biodiversity in Britain's Planted Forests: Results from the Forestry Commission's Biodiversity Assessment Project*. Forestry Commission, Edinburgh, UK.

Landsberg, J.J. and Waring, R.H. (1997) A generalized model of forest productivity using simplified concepts of radiation use efficiency, carbon balance and partitioning. *Forest Ecology and Management* 95, 209–228.

Liu, J., Dunning, J.B. and Pulliam, H.R. (1995) Potential effects of a forest management plan on Bechman's sparrows (*Aimophila aetivalis*): linking a spatially explicit model with GIS. *Conservation Biology* 9(1), 62–75.

MAGIC (2002) Multi-Agency Geographical Information for the Countryside website. http://www.magic.gov.uk/. MAGIC Steering Group.

Marcot, B.G., Wisdom, M.J., Li, H.W. and Castillo, G.C. (1994) *Managing for Featured, Threatened, Endangered, and Sensitive Species and Unique Habitats for Ecosystem Sustainability – Eastside Forest Ecosystem Health Assessment*. USDA Forest Service General Technical Report PNW-GTR-329, Pacific Northwest Research Station, Portland, Oregon.

Margules, C.R. and Pressey, R.L. (2000) Systematic conservation planning. *Nature* 405, 243–253.

Martin, S. (2003) Wildlife tourism and forests in Scotland. Forest Research, Roslin, unpublished internal report to Forestry Commission Scotland.

NBN (2004) National Biodiversity Network website. http://www.searchnbn.net/. National Biodiversity Network Trust, Newark, New Jersey.

Pullin, A.S. and Knight, T.M. (2001) Effectiveness in conservation practice: pointers from medicine and public health. *Conservation Biology* 15, 50–54.

Pullin, A.S. and Knight, T.M. (2003) Support for decision making in conservation practice: an evidence-based approach. *Journal for Nature Conservation* 11, 83–90.

Pyatt, D.G., Ray, D. and Fletcher, J. (2001) *An Ecological Site Classification for Forestry in Great Britain.* Bulletin 124, Forestry Commission, Edinburgh.

Ratcliffe, D.A. (1977) *A Nature Conservation Review,* Vols I and II. Cambridge University Press, London.

Rauscher, H.M., Lloyd, F.T., Loftis, D.L. and Twery, M.J. (2000) A practical decision-analysis process for forest ecosystem management. *Computers and Electronics in Agriculture* 27, 195–226.

Ray, D. (2001) *Ecological Site Classification Decision Support System V1.7.* Forestry Commission, Edinburgh, UK.

Ray, D., Watts, K., Hope, J. and Humphrey, J. (2004) Developing forest habitat networks in southern Scotland. In: Smithers, R. (ed.) *Landscape Ecology of Trees and Forests.* Royal Agricultural College, Cirencester, Gloucestershire, UK, International Association of Landscape Ecology, IALE (UK) and the Woodland Trust, Grantham, UK.

Rodwell, J.S. (1991) *British Plant Communities, 1: Woodlands and Scrub.* Cambridge University Press, Cambridge.

Stewart, G.B., Coles, C.F. and Pullin, A.S. (2005) Applying evidence-based practice in conservation management: lessons from the first systematic review and dissemination projects. *Biological Conservation* 125, 270–278.

Thomasma, S.A., Thomasma, L.E. and Twery, M.J. (1998). *NEWILD (version 1.0) User's Manual [Computer program].* General Technical Report NE-242, US Department of Agriculture, Forest Service, Northeastern Research Station, Radnor, Pennsylvania.

Thomson, A.J. and Willoughby, I. (2004) A web-based expert system for advising on herbicide use in Great Britain. *Computers and Electronics in Agriculture* 42, 43–49.

Turban, E. and Aronson, J. (2000) *Decision Support Systems and Intelligent Systems.* 6th edn. Prentice Hall, Upper Saddle River, New Jersey.

Twery, M.J., Rauscher, H.M., Bennett, D.J., Thomasma, S.A., Stout, S.L., Palmer, J.F., Hoffman, R.E., DeCalesta, D.S., Gustafson, E., Cleveland, H., Grove, J.M., Nute, D., Kim, G. and Kollasch, R.P. (2000) NED-1: integrated analyses for forest stewardship decision. *Computers and Electronics in Agriculture* 27, 167–193.

Vanhinsbergh, D., Fuller, R.J. and Noble, D. (2002) *A Review of Possible Causes of Recent Changes in Populations of Woodland Birds in Britain.* BTO Report No. 245, British Trust for Ornithology, Thetford, UK.

Watts, K., Humphrey, J.W., Griffiths, M., Quine, C.P. and Ray, D. (2005) *Evaluating Biodiversity in Fragmented Landscapes: Principles.* Forestry Commission Information Note, Forestry Commission, Edinburgh, UK.

Wisdom, M.J., Holthausen, R.S., Wales, B.C., Hargis, C.D., Saab, V.A., Lee, D.C., Hann, W.J., Rich, T.D., Rowland, M.M., Murphy, W.J. and Eames, M.R. (2000) *Source Habitats for Terrestrial Vertebrates of Focus in the Interior Columbia Basin: Broad-scale Trends and Management Implications,* 3 vols. General Technical Report PNW-GTR-485, US Department of Agriculture, Forest Service, Pacific Northwest Research Station, Portland, Oregon.

29 Decision Support for Sustainable Forestry: Enhancing the Basic Rational Model

H.R. EKBIA[1] AND K.M. REYNOLDS[2]

[1]*University of Redlands, Department of Computer Science, Redlands, California, USA;* [2]*USDA Forest Service, Pacific Northwest Research Station, Corvallis, Oregon, USA.*

Abstract

Decision-support systems (DSS) have been extensively used in the management of natural resources for nearly two decades. However, practical difficulties with the application of DSS in real-world situations have become increasingly apparent. Complexities of decision making, encountered in the context of ecosystem management, are equally present in sustainable forestry. Various writers have criticized the classical rational/technical solutions commonly employed in DSS and have proposed new approaches based on consensus building, collaborative learning and social DSS, for example. We propose that rational models provide necessary, but not sufficient, tools for effective decision support of sustainable forestry. On this basis, we hypothesize that rational models are not intrinsically limiting, as some have suggested. Rather, perceived limitations of rational models in contemporary DSS might be more a consequence of our methods or of how we have chosen to employ them. Using the Ecosystem Management Decision Support system as an example, we describe how solutions based on rational methods can be integrated with new approaches, such as collaborative learning, to better cope with the practical difficulties of decision support for sustainable forestry. The chapter also discusses eco-phenomenology as an outlook that offers a more integrative view of humans and nature and that avoids some of the basic issues facing the rational model.

Introduction

Decision-support systems (DSS) have evolved significantly during the last four decades. However, their capabilities are still very limited. Elgarah *et al.* (2002), for instance, writing on a project to develop a DSS for decision making on urban infrastructure for the city of Houston, report that they 'know of no DSS design methodology suitable for use in such a complex, conflict-filled situation as this'. This sort of observation is somewhat disturbing in the face of decades of research and practice on DSS. This chapter proposes some remedies for this situation.

The major reason for the current shortcomings of DSS, we argue, has to do with the techno-centric nature of the development of these systems. The history of DSS suggests that development has been largely driven by changes and innovations in computer technologies such as data and knowledge bases, expert systems, software agents and, more recently, web-based tools, and reveals the dominance of a techno-centric view in DSS development, which has sought purely technical solutions to problems of various natures. We argue that most DSS fall into the category of rational models. By rational models, we mean any DSS system or system component that uses a framework or procedure to implement a form of formal reasoning such as decision or logic modelling. We call these basic rational models because they provide the decision maker with basic necessary tools for dealing with situations such as those in sustainable forestry, but they may not be sufficient to account for the full complexity of such situations. Similarly, ecosystem management, since its inception more than a decade ago, has been dominated by a view that characterizes ecosystem management as a 'wicked problem'. Examining these views might provide fresh insights and may be a first step towards alternative ways of thinking about decision making in ecosystem management in general and forest sustainability in particular.

The chapter continues in the next section with a brief overview of ecosystem management and its relation to sustainable forestry, and introduces possible alternative formulations of ecosystem management, such as the socio-ecological approach of the Resilience Alliance. The third section provides an overview of DSS development in the last few decades, and the fourth reviews current proposals for improving decision making in ecosystem and natural resource management that might be used to enhance rational models. The fifth briefly introduces the Ecosystem Management Decision Support (EMDS) system, a relatively successful DSS model in ecosystem management, and discusses specific examples of how enhancements can be implemented. The final section discusses some implications of the alternative approaches for models such as EMDS, and draws some conclusions in terms of future directions of DSS development, especially from the perspective of eco-phenomenology.

Ecosystem Management and Sustainable Forestry

Ecosystem management is a relatively new and popular conceptual framework for land and forest management (Grumbine, 1994). Although a shared understanding of this framework is far from attainable, there is a consensus that the goal of ecosystem management is to create a balance between the long-term protection of the natural environment and the current use of its resources for improving the quality of life of a growing human population (Rauscher, 1999). This formulation of the goal also reveals an inherent tension within the notion of ecosystem management, which manifests itself both in the formulation of the 'ecosystem management problem' and in the characterization of the relation between its two major components – namely, 'ecosystem' and 'management'.

Rauscher (1999, p. 175), for instance, characterizes the ecosystem management problem (particularly in the US context) as one that is riddled with ambiguity and conflict in terms of social goals and values, complex and unclear in terms of legal mandates, variable in terms of stakeholder participation, permeated with missing and uncertain data, subject to political pressure and lacking in agreed-upon processes of decision making. He concludes that 'The ecosystem management debate is a competitive, conflict-laden social process that determines how power flows in resource management' (Rauscher, 1999, p. 175; see also Grumbine, 1994). This might be a fair assessment of the ecosystem management problem in many respects – for instance, in pointing out the interrelationship between power and knowledge (Foucault, 1980) – but it fails to provide a positive characterization of the problem – that is, one that would tell us the essential constitution of ecosystem management. On the basis of the above description, for example, the ecosystem management problem is usually thought of as an instance of unstructured or wicked problems, which are believed to have (Rauscher 1999, p. 175):

- No definite correct formulation – anyone can define it in their own terms
- No stopping rule to identify when they are solved
- No true or false solutions, rather good and bad ones
- No enumerable or exhaustively describable set of solutions
- No (or limited) susceptibility to rational-based methods

This characterization largely stems from the complexity of ecosystem management as an integration of natural and human systems at different levels. The third point about the evaluation criteria (good or bad rather than true or false), for example, derives from the fact that ecosystem management problems are far too complex, blurry, and contestable to warrant abstract logical distinctions based on truth and falsity. The above list, however, provides a negative characterization of wicked problems and, for that matter, of the ecosystem management problem itself. A careful consideration of these characteristics would reveal that, by this definition, most interesting real-life problems are, in fact, 'wicked'. This suggests that the negative characterization of the ecosystem problem (and the negative label 'wicked' attached to it) might derive from a particular world view. A more positive perspective on ecosystem management emerges if it is considered in the context of adaptive management and sustainable forestry. In this view, ecosystem management provides principles that help define the basic goals of management, adaptive management provides a general methodology to support the goals (Walters, 1986) and sustainable forestry provides specific objectives with which to evaluate progress towards goals (Maser, 1994). Yet another alternative world view might understand the ecosystem management situation differently and welcome, rather than lament, decisions based, for instance, on 'local, pragmatic inventions of necessity' or on 'missing and uncertain data'. Such is the world view presented by the phenomenological school of thought (Thomson, 2004). It is beyond the scope of this chapter to go into a discussion of the tenets of phenomenology in all its variations, but we discuss some insights derived from this view in the concluding section.

Traditionally, approaches to ecosystem management fall into one of two categories (Bousquet and Le Page, 2004):

- Ecological: an ecological system subject to anthropogenic disturbance.
- Economic: a social system subject to natural constraints.

In the first approach, scientists focus on the dynamics of the resource, while managers focus on sustained exploitation; in the latter, scientists focus on maximization of resource usage, assuming the same decision model for all actors. Recently, there has been a trend towards an integrated and interactive approach, which substitutes interactions among all actors for the traditional dynamics under constraints (Bousquet and Le Page, 2004). This trend is best captured in the notion of resilience, which integrates the social and natural environments. The Resilience Alliance, a research organization of scientists and practitioners from various disciplines who have adopted resilience as a central concept in ecosystem management, defines resilience as follows (http://resalliance.org):

> Ecosystem resilience is the capacity of an ecosystem to tolerate disturbance without collapsing into a qualitatively different state that is controlled by a different set of processes. A resilient ecosystem can withstand shocks and rebuild itself when necessary . . . Resilience in social systems has the added capacity of humans to anticipate and plan for the future. Resilience is a property of these linked social-ecological systems (SES).

According to the above characterization, resilience as applied to ecosystems, or to integrated systems of people and the natural environment, provides these systems with stability as well as a capability for self-organization, learning and adaptation. We consider these to be crucial features of sustainable forestry as well, and suggest that DSS for sustainable forestry should be of such a character as to support and enhance resilience. Current DSS used in forestry often apply rational models of decision making, such as multi-criteria evaluation (Carver, 1991; Malczewski, 1999). As mentioned earlier, basic rational models may not be sufficient to account for the full complexity of situations faced in forestry. To be sufficient, they need certain enhancement. Various alternatives have been proposed to the rational models, and we shalll draw upon some of these alternatives to suggest enhancements to the basic model. We specifically discuss those that support an integrated approach to the social and ecological aspects of ecosystem management, which is also the key insight behind the notion of resilience.

Decision-support Systems

Decision making, as an area of study, originates in organization science (Simon, 1960). According to Simon, organizational issues that cannot be solved by the optimization models of classical operations research can be successfully handled by modern computer technologies. DSS are, therefore, broadly understood as computer technologies used to support complex decision making in organizations (Keen and Scott Morton, 1978). This characterization applies to various types of DSS – data-driven, model-driven, group support system, intelligent DSS, and so on.

There has been a close parallel between the evolution of the concept of DSS and the development of computer technologies and tools (Shim *et al.*, 2002). In the era of data processing and management information systems, for example, the emphasis in DSS was on databases and data models. Later, with the advent of expert systems and executive information systems, the scope of DSS extended

Table 29.1. The development of DSS in relation to technology.

Stage	Approximate period	Dominant concept of DSS	Technologies
I	1960s–1970s	Data modelling	Databases, Management Information Systems
II	1980s	Collaborative and group decision support (GSS)	Knowledge bases, expert systems, Enterprise Information Systems
III	1990s	Knowledge management	Online analytical processing, data warehouse, data mining
IV	2000s	Web-based and active DSS	Internet, client-server tools, software agents

to group and corporate levels. Then the growing interest in intelligent systems and knowledge bases brought about the notion of organizational knowledge management. Most recently, the expansion of the world-wide web and wireless technologies is giving rise to web-based DSS and to new concepts of decision making from multiple perspectives (Courtney, 2001).

Table 29.1 summarizes DSS development as it relates to the supporting technologies (Shim *et al.*, 2002). As Table 29.1 illustrates, development has been dominantly bottom-up and technology-driven, with the available technical tools supporting the traditional concept of decision making as a basically rational process (Simon, 1960). We believe that many of the shortcomings of current DSS, as pointed out, for instance, by Elgarah *et al.* (2002), are due to this techno-centric view. This view considers technology as the panacea for all issues that emerge in human contexts (organizations, communities, societies, etc.), and, as such, underplays the integration of human and technical aspects that is proved to be crucial in socio-technical systems (Kling, 1980). The emphasis in DSS history on emerging new technologies is a clear manifestation of this view.

Approaches to Enhancing the Basic Rational Model

Various new approaches to decision support have recently emerged within and outside ecosystem management. This section reviews three approaches – namely, agent-based, learning and dialectic models. For each, we briefly describe the underlying principles that motivate the approach, give an overview of how each is implemented and in the section 'Application to the EMDS System' provide brief examples of how these approaches could be employed in the EMDS context to better address some of the difficulties surrounding decision support for sustainable forestry noted earlier.

Multi-agent-based simulation

Multi-agent-based simulation is an approach that is gaining a lot of momentum in ecosystem management, especially in Europe. As its name implies, it involves two major components: multi-agent systems and simulations. The notion of agent used in this approach originates in artificial intelligence (AI) and computer science, where an agent is viewed as anything that can perceive its environment through sensors and act upon the environment through actuators (Bousquet and Le Page, 2004). Thus, a human being, a robot and a software process are each examples of agents. Multi-agent systems typically involve several to many heterogeneous agents that share a common environment, resources and knowledge, communicate with each other and are controlled in a distributed manner. This means that there is no central process that collects information from all agents and then decides what action to take. Rather, each agent is an autonomous decision maker that is brought in contact with other agents through a coordination mechanism. Decision making is, therefore, a key component of agent-based modelling, which links it to DSS as well.

Multi-agent systems have found applications in social sciences (e.g. in Agent-based Computational Economics (ACE, 2005), ecosystem management (Bousquet and Le Page, 2004), common resource management (Bousquet *et al.*, 1998)) and elsewhere. In each area, various empirical methods of modelling have been employed. In resource management, for instance, in which the challenge is to model the interaction between groups of agents and resource dynamics, there are two methods. In one method, the emphasis is on the relationship between people who affect resources, and the model represents exchanges of information, services, contracts and agreements among agents. In another method, the emphasis is on representations that determine how agents and resources interact. Each agent develops and acts on its own representation of a resource, and in doing so transforms the resource for other agents. The model is used to study these representations of actions and attempts to generate coordination among agents through the environment.

The idea of simulation is also used in diverse areas such as computer science, artificial life, experimental economics and, more recently, ecosystem management (Becu *et al.*, 2003b). Ecosystems are typically complex systems characterized by: (i) slow dynamics (relative to other similar complex systems such as physical or even socio-economic systems); and (ii) involving people and livelihoods. It is therefore very difficult to conduct experiments on real ecosystems. This makes simulation an attractive tool for modelling ecosystems. Simulations are useful not only because they provide the opportunity for controlled, simulated experimentation – i.e. playing around with and observing the impact of parameters without affecting the real world – but they are also useful because they 'enable the observation and recording of the background of planning, decision-making, and evaluation processes that are usually hidden' (Dörner, 1996, pp. 9–10). In short, simulations can be conceived as social laboratories that provide an effective way of learning, which is different from the known alternative of learning by doing (Bousquet *et al.*, 1999).

Computer simulations can take different forms. In the context of multi-agent systems, agents are computer processes with distinct roles, capabilities and goals. Bousquet *et al.* (1999) introduce an approach called companion modelling, which uses multi-agent systems and simulations, not as prediction tools, but as learning environments. They also emphasize the importance of scenarios in the following way:

> Because the very long term is beyond the scope of prediction, and we wish to take it into account in the analysis of environmental problems, we must give ourselves very long-term reference points or objectives to guide the possible or impossible pathways of development. The long-term approach must inevitably be based on a scenario.
>
> (Bousquet *et al.*, 1999, p. 119)

Becu *et al.* (2003a) have proposed a methodology for eliciting and modelling stakeholders' representations with agent-based modelling. In this method, each agent is defined with specific abilities, goals and strategies – i.e. with a specific representation of the environment. Unlike traditional views that take individual representations as permanent and stable, this approach considers them 'temporary constructs elaborated through social interactions and communication' (Becu *et al.*, 2003a, p. 132). As such, 'any elicited representations should be used as a basis for discussion rather than decision' (p. 133). Such discussion may take different forms. If there are fundamental conflicts of opinion because people refer to the same process with different labels or indicators – e.g. if one person assesses soil moisture according to its colour while another looks at the aspect of the crop being cultivated – in that case differences are kept and the different indicators will be modelled in the same model using specific viewpoints. If, on the other hand, there is minor disagreement about the value of a threshold or a quantitative result, further discussion is encouraged for resolving the differences. Finally, if individual representations are found to be explicitly and unambiguously contradictory, each view is modelled separately and the overall model is split into two separate models (Becu *et al.*, 2003a).

Learning models

Another class of models of decision making adopts a framework centred on human skills and capabilities such as observation, learning, negotiation, and so on. These models come in different flavours and have distinct labels – e.g. consensus building, collaborative learning, social DSS, participatory DSS, etc. – but they are based on similar insights and premises.

Consensus building

Consensus-building models try to provide a structured environment to explore the intensity and source of conflict and to generate compromise alternatives among multiple decision makers. Like basic rational models, consensus building is based upon the evaluation of a set of alternatives according to a number of relevant criteria. In distinction from those models, however, consensus building recommends a flexible, interactive and transparent mode of decision making, and supports methods that expose the preferences and objectives of multiple decision makers. One group of such methods, which belongs to multi-criteria analysis (Vincke, 1992), is discrete compensatory methods that focus on problems with a finite number of alternatives and allow trade-offs to be made among criteria. Therefore, an alternative with a low score on some criteria can compensate by high scores on other criteria, subject to the priorities of a decision maker (Feick and Hall, 1999, p. 18).

Feick and Hall (1999) describe a spatial DSS, TourPlan, which implements a consensus-building model through a tight coupling of geographic information systems and multiple-criteria analysis. This model uses the concept of developing scenarios to present decision makers with alternative views of the future. The spatial component of the model makes it possible for each participant to develop their preferred solutions within a scenario, and for other participants to visualize and evaluate those solutions according to their own criteria (Feick and Hall, 1999, p. 22).

Collaborative learning

Another approach within this group, called collaborative learning (Daniels and Walker, 1996), tries to respond to both scientific and political realities by putting emphasis on mutual learning. Because 'no single party, agency, organization or discipline holds the key to understanding a particular resource management situation', these authors argue, it is crucial that the various participants learn from one another. That is why negotiation, as 'joint decision-making among parties with interdependent yet incompatible interests', is a central component of this approach. To foster collaborative learning, this approach 'emphasizes activities that encourage systems thinking, joint learning, open communication, and focuses on appropriate change' (Daniels and Walker, 1996). In so doing, it combines features of soft systems approach – e.g. the creation of 'a temporarily shared culture in which conflicts can be accommodated so that action can be taken' (p. 81) – with techniques of alternative dispute resolution, such as bargaining strategies. In other words, collaborative learning tries to seamlessly integrate the human and technical components in the decision-making process. In short, collaborative learning adopts a pragmatic approach, and emphasizes improving the situation rather than problem solving, communication and negotiation over concerns and interests rather than bargaining over positions, making progress towards desirable and feasible change rather than achieving a particular set of future conditions.

Social DSS

A third approach, called social DSS, tries to 'facilitate the integration of diverse views into a growing knowledge base' (Turoff *et al.*, 2002). This approach is mostly suited for situations that involve numerous participants – for example, citizens in a public discussion of the costs and benefits of EZ-Pass systems, an electronic toll collection system that allows drivers to pay toll-road fees in the form of a subscription fee. There are a number of features of social DSS that might be useful. First is the dynamic character of the voting process and the relevant knowledge base in the sense that there is no predetermination of what can be represented in the process. Secondly, in the system design, there is an emphasis on ease of learning over time rather than immediate ease of use. Finally, it introduces a new kind of uncertainty or measure of confidence that has to do with the percentage of individuals who have not voted.

Some social DSS use a scaling method called Thurstone's law of comparative judgement, which provides a formal technique for dealing with this kind of uncertainty (Li *et al.*, 2001). Scaling methods provide techniques of measuring human subjective judgements, and might therefore be as important as data collection techniques. In a voting system, for instance, people vote on a set of alternatives based on their preferences (for example, by rank-ordering them), and the system produces a group result on the basis of individual rankings. The question is how to best represent this result to the group so as to facilitate their understanding and decision making. The Thurstone method creates this facility by transforming individual preference data into a composite interval scale that allows individuals to more clearly observe differences among alternative objectives (Li *et al.*, 2001). A group of researchers at New Jersey Institute of Technology have created a web-based voting system that uses this method (Li *et al.*, 2001).

The dialectical approach

The dialectical approach attempts to accommodate conflicting views into the design process itself (Elgarah *et al.*, 2002). This is done through a dialectical process, in which a thesis that is initially formed on the basis of some mental model of the situation is later negated to form an antithesis. The synthesis then emerges as a result of debate and dialogue between the two opposing views. Ideally, the synthesis resolves the conflict, but practically the dialectical process continues on an iterative basis.

Elgarah *et al.* (2002) have proposed a methodology called the multiple perspective and dialectical approach, which incorporates different perspectives (personal, organizational, technical, ethical, aesthetic and enlightenment) into a dialectical process. The multiple perspective and dialectical approach is a conflict-driven approach that promotes heterogeneous views of decision making, and involves seven major steps: stakeholder identification, multiple perspective identification, decision factor determination, world-view formulation, conflict identification, resolution generation and resolution evaluation. It encourages decision makers to examine their own assumptions and to justify their choices of decision factors. To achieve this, Elgarah *et al.* have developed a model with multiple perspectives such as aesthetic, ethical, etc. The main advantage of the multiple perspective and dialectical approach is that, by producing multiple designs in an iterative process, it increases the likelihood of arriving at the most effective design. Its disadvantage is its inefficiency in dealing with well-structured problems. 'For clear-cut problems, conflict may be a time-consuming nuisance' (Elgarah *et al.*, 2002).

Application to the EMDS System

EMDS is a decision-support system for integrated landscape evaluation and planning (Reynolds *et al.*, 2003). It has been used in various applications for ecosystem management, including decision support for sustainable forestry

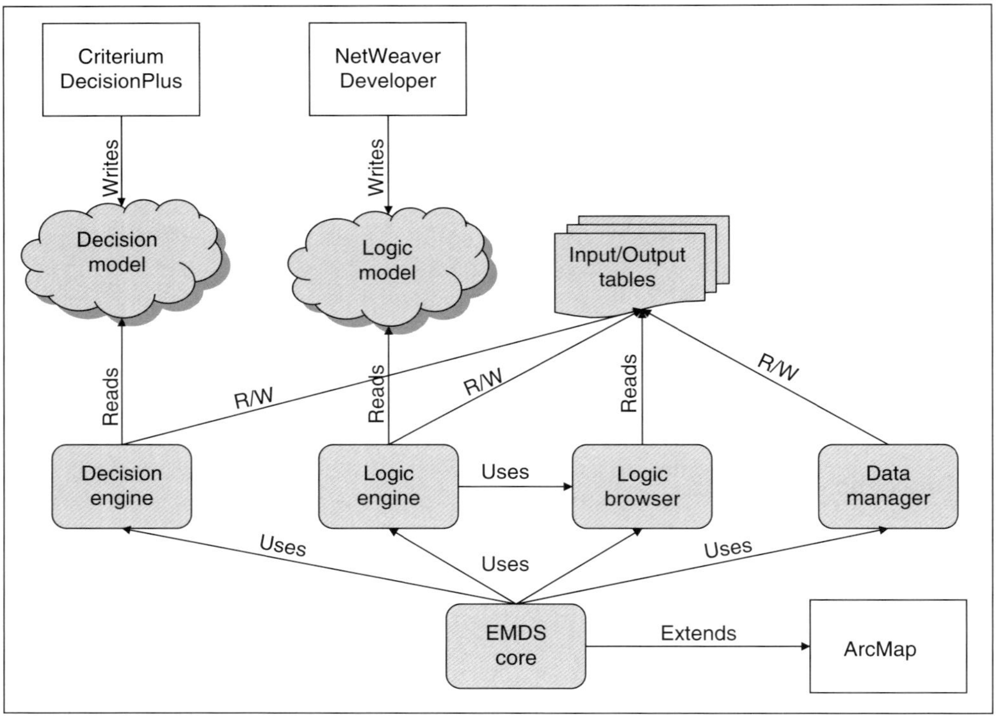

Fig. 29.1. Component diagram for the EMDS extension to ArcMap. Shaded components are internal to the EMDS extension. Relations between components are described by directed arrows. For example, the EMDS core extends ArcMap, and the logic engine reads a logic model written with NetWeaver Developer. The relation R/W indicates reads/writes.

(Reynolds, 2001, 2005). The system (Fig. 29.1) provides decision support for landscape-level analyses through logic and decision engines integrated with ArcMap® (a component of the ArcGIS geographic information system, Environmental Systems Research Institute, Redlands, California.[1] The logic engine evaluates landscape data with a logic model written with the NetWeaver Developer® system (Rules of Thumb, Inc., North East, Pennsylvania) to interpret ecosystem conditions such as sustainability. The decision engine evaluates NetWeaver outcomes and data related to additional logistic considerations, such as feasibility and efficacy of land management actions, with a decision model, written with Criterium DecisionPlus® (InfoHarvest, Seattle, Washington State), which prioritizes landscape elements for management. Criterium DecisionPlus models implement the analytical hierarchy process (Saaty, 1992), the simple multi-attribute rating technique (Kamenetzky, 1982) or a combination of the analytical hierarchy process and simple multi-attribute rating technique methods. Detailed descriptions of how knowledge is represented in the logic and decision models used by EMDS are not pertinent to the present discussion, but can be found in Reynolds *et al.* (2003). For our immediate purposes, it is sufficient to note that knowledge representation in both of these EMDS components clearly conforms to the concept of a rational model. For subsequent discussion, it is also useful to note that both the NetWeaver and Criterium DecisionPlus development systems use graphical design for model construction and that model design can be performed by individuals or groups.

In what follows, we discuss possible ways to enhance the modelling components of EMDS, based on what we presented in the section 'Approaches to Enhancing the Basic Rational Model', concerning alternative approaches to the basic rational model.

Working with logic models

Each alternative discussed in 'Approaches to Enhancing the Basic Rational Model' includes aspects that could be used to enhance the use of logic models as they are implemented by EMDS. In collaborative learning, for instance, people come together on the basis of common objectives and interests. Collaborative learning as such is not a confrontational context; rather, it is more like a planning environment, in which the core participants are motivated by a need to cooperate to accomplish their mission. There is no reason why the methods and principles of collaborative problem solving could not be applied to the design of logic models. Indeed, use of logic models can actually facilitate collaboration by providing a common framework within which people can more easily develop a shared representation of a complex problem.

Similarly, there is no reason why methods and principles of the dialectical approach could not be implemented. For example, a single logic model can contain multiple representations that would support a dialectical approach to decision making. In developing the logic model, a group of individuals might generate two or more variants of model components with a common core. Evaluating such variants side by side within a single logic model, participants, acting as agents in the sense of multi-agent-based systems, can go through a reasoned decision-making process that might eventually lead to a shared logic model. However, even in contexts in which views and opinions are so divergent that realizing a shared representation may be beyond reasonable expectations, logic models can still serve a useful role by helping participants gain a better understanding of the basis for contrasting perspectives, thus helping to clarify the points of agreement and disagreement and better focus discussions.

Finally, in terms of social DSS, logic models can be implemented as Internet-based applications, designed for broad public access and represented in a format that is easily understandable for the public. The issue, in the latter case, would be one of interface design for ease of communication and comprehension of concepts, not one of basic model structure. Table 29.2 lays out the different approaches and indicates how logic models can be designed using the ideas in each of them.

Working with decision models

Comments in the previous section pertaining to use of alternative decision methodologies with logic models are equally applicable in the context of

Table 29.2. Different approaches to enhancing logic models.

Approach	Main idea	Contribution to logic model
Collaborative learning	Joint learning and open communication	Provide a common framework for creating a shared representation
Dialectical approach	Accommodate conflicting views into the design process	Provide a reasoned decision process among variants or a better understanding of differences
Social DSS	Dynamic integration of diverse views	Provide a suitably understandable interface for public participation

decision-modelling components such as the decision engine in EMDS. In fact, if anything, the case for enhancing the utility of rational decision models with the alternative methodologies can be made more convincingly because there is a substantial body of literature documenting the methods for and experiences with collaboration and application of social DSS (Pereira and Corral Quintana, 2002) in the context of decision models.

Working with scenarios

Multi-agent-based systems have sometimes been implemented with people performing the role of agents, as discussed in 'Approaches to Enhancing the Basic Rational Model'. We have already noted one possible application of the multi-agent-based systems concept in the context of alternative logic representations. Similarly, people can also operate as agents in the context of exploring alternative logic scenarios in which either the logic or the data are modified within a running EMDS application to test the outcomes of alternative scenarios on a much more ad hoc basis.

Discussion

As mentioned earlier, DSS evolution has been largely bottom-up and technology driven, in the sense that DSS have evolved in such a way as to be able to accommodate the latest computer and information technologies of the time. Although this evolution has brought about beneficial changes in the way DSS are used and understood, the technical orientation has also tended to preclude the possibility of a deep reconceptualization of decision making as originally formulated by Simon (1960). Simon's view has had a lasting impact on the development of DSS in at least two ways. One is through the idea of bounded rationality, which basically portrays decision making as a weighing of alternatives according to a set of predetermined criteria. Indeed, it can be safely asserted that many subsequent models of decision making – e.g. optimization-based DSS (which involve three stages – formulation, solution and analysis), multi-criteria decision analysis or the analytical hierarchy process – are variants of the original Simon model, although they have become increasingly sophisticated in terms of the number and classification of criteria, in terms of mathematical formalisms, etc.

The second major impact of Simon's work is manifested through the idea of 'cognition as problem solving', which is mostly elaborated in his joint work with Allen Newell in artificial intelligence (AI; Newell and Simon, 1972), but its influence goes far beyond AI to areas such as organization and management science. A key tenet of the problem-solving paradigm is its emphasis on mental representations of external situations. According to this view, people deal with external situations by building more or less faithful models of the problem in their heads. Therefore, decision making consists mainly of the manipulation of these internal models and symbols. In other words, problems are in our heads, as are solutions to problems. A close look at the list of properties of so-called wicked problems in the section 'Ecosystem Management and Sustainable Forestry' precisely illustrates this point. There is almost nothing in that list that is external to people's minds – everything is in the decision maker's head.

The above view of cognition, thinking and decision making has dominated DSS as well as ecosystem management until recently. However, alternative views are on the rise, some of which we examined in 'Application to the EMDS System'. A common tenet of most of these approaches is that rationality is not the modus operandi of decision making. The point is not that rationality is ill conceived, but that the conditions under which it is sufficient by itself are relatively rare (Weick, 2001, p. 34). Rationality prevails if the environment changes slowly, if there are few social groups and if the situation is reasonably well controlled by agents with central authority (Kling, 1980, pp. 90, 100) – all of which are rare in real situations. Another common tenet of the new approaches is the idea, put simply, that problems are not so much in the head as they are in external situations. In other words, what we often have to deal with are problematic situations, not problems as mental models of those situations. This means that problems do not present themselves as given; rather, 'they must be constructed from the materials of problematic situations, which are puzzling, troubling, and uncertain' (Weick 1995, p. 9).

These shifting views have important implications for DSS design and implementation. Subsequently, we explore the possibility of accommodating these views in a system such as EMDS. In what follows, therefore, we shall examine some major ideas in these proposals, and possible methods of enhancing EMDS in particular and DSS in general in the light of those ideas.

Sense making

Recent models focus attention on the process of sense making rather than the product of decision making. A key feature of human sense making has been described as retrospective justification (Weick, 1995). Studies of decision making in juries have indicated that they are largely outcome-driven – 'The outcome comes before the decision' (Garfinkel, 1967, p. 114). That is, jurors do not seem to first evaluate the harm, then allocate blame and finally choose a remedy. Rather, they first decide a remedy and then decide the 'facts' from among alternatives that justify the remedy. In short, they retrospectively justify a decision that is being made on grounds other than (or beyond) facts. Garfinkel concludes from this study that:

> decision making in daily life would thereby have, as a critical feature, the decision maker's task of justifying a course of action. The rules of decision making in daily life . . . may be much more preoccupied with the problem of assigning outcomes to their legitimate history than with questions of deciding before the actual occasion of choice the conditions under which one, among a set of alternative possible courses of action, will be elected.
>
> (Garfinkel, 1967, p. 114)

The inattention of decision makers to this phenomenon might have an impact on their decisions in an undesired manner, and methods should be devised to counter that possibility. A system such as EMDS, for instance, should allow the participants to iteratively relive and review the process by, for instance, explaining it to others. This would reduce the possibility of retrospective justification biasing the evidence and jeopardizing the decision-making process.

Enacting situations

Retrospection also highlights the interleaving of thoughts and actions. People do not face a situation as a given; rather, they enact and produce the situations of which they are a part. As Garfinkel describes them:

> in the course of a career of actions, [people] discover the nature of the situations in which they are acting . . . [T]he actor's own actions are first order determinants of the sense that situations have, in which, literally speaking, actors find themselves.
>
> (Garfinkel, 1967, p. 115)

The active role of participants in enacting and constructing situations also highlights the difference between sense making and interpretation, as it is commonly understood (Weick, 1995, p. 13). Here, the emphasis shifts from the question of what people know to the issue of how people go about knowing what they know. In other words, it shifts towards people enacting situations, rather than taking alternatives as given.

The use of scenarios in EMDS is a good simulacrum for enactment, and the inclusion of role playing, where possible, can enhance this capability. This would allow the participants to observe the consequences of their decisions in a simulated environment, to take note of critical parameters and hence to lower costs and risks to some extent.

Transparency

Finally, the emerging views also suggest that the governing norms of decision-making processes should shift from accuracy and certainty of data towards transparency and plausibility of process. Traditionally, the emphasis in DSS has been on capturing, encoding and providing as much knowledge as possible for decision makers in order to support informed, documentable and responsible decisions (Pereira and Corral Quintana, 2002, p. 97). Responsible in this context has often meant the use of best (expert) scientific knowledge in decision making, not necessarily socially responsible, 'because the social context would not explicitly be taken into account' (Pereira and Corral Quintana, 2002, p. 97). Similarly, documenting the decision has been considered a preamble for the legitimizing and quality assurance of the decision process. While these are important criteria, it is suggested that the emphasis be shifted towards the plausibility, reasonableness and coherence of decisions as well as the transparency of the decision-making process. This shift would highlight the importance of meaningful (as opposed to informed) decisions and of reassuring and legitimate (rather than ensured and authoritative) processes.

Weick (1995, pp. 55–61) provides some reasons why accuracy is nice but not necessary in the sense-making process – e.g. that people need to filter signal from noise in order not to be overwhelmed with data, that there is often a tradeoff between speed and accuracy, that there is always a subjective, interpersonal component present in any decision-making situation, that accuracy is pragmatic and project-specific, that it is impossible to guarantee accuracy prior to action, and so on. Weick concludes from this that what people need is not more information, but 'values, priorities, and clarity about preferences to help them be clear about which projects matter' (p. 27). In a similar vein, those involved in management and policymaking have increasingly emphasized that, in matters of public or group concern, the quality and transparency of the process is no less important than the certainty and validity of the outcome (Pereira and Corral Quintana, 2002).

From its inception, design criteria for the current EMDS system placed strong emphasis on the reasonableness and transparency of results generated by the system. Both the logic and the decision engines in EMDS were selected with these considerations in mind first and foremost. However, among the array of DSS used in contemporary natural resource management, not all systems are equal in this respect, and there remain substantial opportunities to enhance contemporary systems along these lines.

Further on

> Full of merit, yet poetically, man Dwells on this earth.
>
> (Hölderlin, cited in Heidegger, 1975)

> we tend to reduce every entity we encounter to the status of an intrinsically meaningless 'resource' (Bestand) merely to be optimized as efficiently as possible, leveling all attempts to say what matters to us down to empty optimization imperatives (such as the ubiquitous: 'Get the most out of your potential!'). Thus we come to treat even ourselves (modernity's vaunted 'subject') in the terms underlying our technological refashioning of the world, as just another resource to be optimized, ordered, and enhanced with maximal efficiency.
>
> (Thomson, 2004, p. 397)

In this chapter, we have explored the possibility of enhancing a system such as EMDS for the purposes of sustainable forestry by drawing upon recent findings and developments in DSS and ecosystem management. Having examined some of the major views in these areas, we illustrated some of the ways in which these enhancements could be realized in EMDS and other systems. We showed how the logic and decision models of EMDS can be enhanced by incorporating ideas and methods borrowed from multi-agent simulation, as well as learning and dialectical approaches. Furthermore, we discussed a number of shifts based on recent findings and ideas from the sense-making framework – most importantly, a shift of focus towards iterative decision-making processes, enacted scenarios and transparency. We believe that these shifts can significantly contribute to the applicability and improved utility of DSS in ecosystem management and forestry.

But there is always room for improvement, and, as one possible direction, we would like to conclude this chapter by exploring the phenomenological approach, originally based on the views of Martin Heidegger. Heidegger is not unknown to ecologists, environmentalists and conservationists, many of whom have looked upon his writings as sources of insight and inspiration, or have alternatively sought in those writings a philosophical articulation of their own views. Heidegger's ideas have, therefore, been linked to 'deep ecology' (Zimmerman, 1983), environmental ethics (Beckman, 2000; Thomson, 2004) and sustainability (Peters and Irwin, 2002), but they have remained largely foreign to ecosystem management. This strangeness, we believe, has partly to do with the rather incongruous premises of the two approaches. Whereas ecosystem management commonly views the natural environment as a source of resources or services to society, Heideggerian phenomenology questions this attitude and offers a more integrative view of humans and nature. Consequently, ecosystem management seeks to optimize the utilization of natural resources by human beings, while phenomenology recommends a notion of dwelling (*wohenen*) that invites us to be at home with the environment and with 'being as such' (Thomson, 2004, p. 400). Through this shift of perspective, entities acquire a richer meaning far beyond intrinsically meaningless resources awaiting optimization (Thomson, 2004, p. 400).

The phenomenological view of the environment is based on a number of principles, most notably and recently articulated in the book *Eco-phenomenology: Back to the Earth Itself* (Brown and Toadvine, 2003) and discussed at length in Thomson (2004). Postulating a close link between our modern world view and the environmental crisis, eco-phenomenologists challenge the principles that underlie this world view. Thomson singles out two such principles as metaphysical and ethical. These principles have to do with, respectively, the mind/world dualism and the fact–value divide (Thomson, 2004, p. 382). Metaphysically, eco-phenomenology seeks to reunite the mind with the world by returning us to the experience of a pre-differentiated mind–world unity. In so doing, it also proffers an ethical principle that makes it possible for us 'to recognize the reality of environmental "values," the alleged "fact" that certain environmental values are "always already in the world"', and so on (Thomson 2004, p. 383). In short, eco-phenomenology challenges two major rationalist principles of the modern world view that are built into the scientific aspirations of ecosystem management. We believe that the formulation of the ecosystem management problem as wicked (see the earlier section on this topic) derives from these rationalist principles, which, as we discussed, also lie at the foundation of most DSS. By assigning inherent values to certain environmental entities, the phenomenological world view would characterize the issues of the ecosystem as things to be accommodated and organically dealt with, rather than as wicked problems to be avoided and overcome. This shift of perspective would have a deep impact not only on ecosystem management and forestry, but on the design of computer-based systems (DSS or otherwise) built for this purpose.

Note

1. The use of trade or firm names in this publication is for reader information and does not imply endorsement by the US Department of Agriculture of any product or service.

References

ACE (2005) Agent-based Computational Economics website. Retrieved on 8 June 2005 from http://www.econ.iastate.edu/tesfatsi/ace.htm

Beckman, T. (2000) *Martin Heidegger and Environmental Ethics*. Retrieved on 1 January 2005 from: http://www2.hmc.edu/~tbeckman/personal/Heidart.html

Becu, N., Bousquet, F., Barreteau, O., Perez, P. and Walker, A. (2003a) A methodology for eliciting and modelling stakeholders' representations with agent-based modelling. *Lecture Notes in Artificial Intelligence* 2927, 131–148.

Becu, N., Perez, P., Walker, B., Barreteau, O. and Le Page, C. (2003b) Agent-based simulation of a small catchment water management in northern Thailand: description of the CATCHSCAPE model. *Ecological Modelling* 170, 319–331.

Bousquet, F. and Le Page, C. (2004) Multi-agent simulations and ecosystem management: a review. *Ecological Modelling* 176, 313–332.

Bousquet, F., Bakam, I., Proton, H. and Le Page, C. (1998) Cormas: Common-pool Resources and Multi-agent Systems. *Lecture Notes in Artificial Intelligence* 1416, 826–838.

Bousquet, F., Barreteau, O., Le Page, C., Mullon, C. and Weber, J. (1999) An environmental modelling approach: the use of multi-agent simulations. In: Weill, F.B.A. (ed.) *Advances in Environmental Modelling*. Elsevier, London, pp. 113–122.

Brown, C. and Toadvine, T. (2003) *Eco-phenomenology: Back to the Earth Itself*. Syracuse University, Albany, New York.

Carver, S. (1991) Integrating multicriteria evaluation with GIS. *International Journal of Geographical Information Systems* 5, 321–339.

Courtney, J.F. (2001) Decision making and knowledge management in inquiring organizations: toward a new decision-making paradigm for DSS. *Decision Support Systems* 31, 17–38.

Daniels, S.E. and Walker, G.B. (1996) Collaborative learning: improving public deliberation in ecosystem-based management. *Environmental Impact Assessment Review* 16, 71–102.

Dörner, D. (1996) The *Logic of Failure: Recognizing and Avoiding Error in Complex Situations*. Perseus Books, Cambridge, Massachussetts.

Elgarah, W., Courtney, J.F. and Haynes, J.D. (2002) A dialectical methodology for decision support systems design. Paper presented at the 35th Annual Hawaii International Conference on System Sciences, Big Island, Hawaii.

Feick, R.D. and Hall, G.B. (1999) Consensus-building in a multi-participant spatial decision support system. *URISA Journal* 11, 17–23.

Garfinkel, H. (1967) *Studies in Ethnomethodology*. Prentice-Hall, Englewood Cliffs, New Jersey.
Grumbine, E. (1994) What is ecosystem management? *Conservation Biology* 8, 27–38.
Heidegger, M. (1975) *Poetry, Language, Thought*. Trans by Albert Hofstadter. Harper & Row, New York.
Kamenetzky, R. (1982) The relationship between the analytical hierarchy process and the additive value function. *Decision Sciences* 13, 702–716.
Keen, P.G.W. and Scott Morton, M.S. (1978) *Decision Support Systems: an Organizational Perspective*. Addison-Wesley Publishing, Reading, Massachussets.
Kling, R. (1980) Social analyses of computing: theoretical perspectives in recent empirical research. *Computing Surveys* 12, 61–110.
Li, Z., Cheng, K., Wang, Y., Hiltz, S. and Turoff, M. (2001) Thurstone's law of comparative judgment for group support. In: *Proceedings of the Seventh Americas Conference on Information Systems*. Boston, Massachusetts, pp. 241–244.
Malczewski, J. (1999) *GIS and Multicriteria Decision Analysis*. John Wiley and Sons, New York.
Maser, C. (1994) *Sustainable Forestry: Philosophy, Science, and Economics*. St Lucie Press, Delray Beach, Florida.
Newell, A. and Simon, H.A. (1972) *Human Problem Solving*. Prentice-Hall, Englewood Cliffs, New Jersey.
Pereira, G.A. and Corral Quintana, S. (2002) From technocratic to participatory decision support systems: responding to new governance initiatives. *Journal of Geographic Information and Decision Analysis* 6, 95–107.
Peters, M. and Irwin, R. (2001) Earthsongs: ecopoetics, Heidegger and dwelling. *Trumpeter: Journal of Ecosophy* 18(1), http://trumpeter.athabascau.ca/content/v18.1/
Rauscher, M.H. (1999) Ecosystem management decision support for federal forests in the United States: a review. *Forest Ecology and Management* 114, 173–197.
Reynolds, K.M. (2001) Using a logic framework to assess forest ecosystem sustainability. *Journal of Forestry* 99, 26–30.
Reynolds, K.M. (2005) Decision support for evaluating the US national criteria and indicators for forest ecosystem sustainability. In: Aguirre-Bravo, Celedonio, *et al.* (eds) *Monitoring Science and Technology Symposium: Unifying Knowledge for Sustainability in the Western Hemisphere, 20–24 September 2004, Denver, Colorado.* Proceedings RMRS-P-37CD, US Department of Agriculture, Forest Service, Rocky Mountain Research Station, Ogden, Utah.
Reynolds, K.M., Rodriguez, S. and Bevans, K. (2003) *EMDS 3.0 User Guide*. Environmental Systems Research Institute, Redlands, California.
Saaty, T.L. (1992) *Multicriteria Decision Making: The Analytical Hierarchy Process*. RWS Publications, Pittsburgh, Pennsylvania.
Shim, J.P., Warkentin, M., Courtney, J.F., Power, D.J., Sharda, R. and Carlsson, C. (2002) Past, present and future of decision support technology. *Decision Support Systems* 33, 111–126.
Simon, H.A. (1960) *The New Science of Management Decision*, Vol. 3. Harper, New York.
Thomson, I. (2004) Ontology and ethics at the intersection of phenomenology and environmental philosophy. *Inquiry* 47, 380–412.
Turoff, M., Hiltz, S.R., Cho, H.-K., Li, Z. and Wang, Y. (2002) Social decision support systems. Paper presented at the 35th Annual Hawaii International Conference on System Sciences, Big Island, Hawaii.
Vincke, P. (1992) *Multicriteria Decision Aid*. John Wiley and Sons, Rexdale, Ontario.
Walters, C.J. 1986. *Adaptive Management of Renewable Resources*. Macmillan, New York.
Weick, K.E. (1995) *Sensemaking in Organizations*. Sage Publications, Thousand Oaks, California.
Weick, K.E. (2001) *Making Sense of the Organization*. Blackwell Publishers, Oxford, UK.
Zimmerman, M.E. (1983) Toward a Heideggerean ethos for radical environmentalism. *Environmental Ethics* 5, 99–131.

Index